Ecology Revisited

Astrid Schwarz • Kurt Jax
Editors

Ecology Revisited

Reflecting on Concepts, Advancing Science

Editors
Astrid Schwarz
Institute of Philosophy
Technische Universität Darmstadt
Schloss
64283 Darmstadt
Germany
schwarz@phil.tu-darmstadt.de

Kurt Jax
Department of Conservation Biology
Helmholtz Centre for Environmental
 Research (UFZ)
Permoserstr. 15, 04318 Leipzig
Germany
kurt.jax@ufz.de

ISBN 978-90-481-9743-9   e-ISBN 978-90-481-9744-6
DOI 10.1007/978-90-481-9744-6
Springer Dordrecht Heidelberg London New York

Library of Congress Control Number: 2011920689

© Springer Science+Business Media B.V. 2011
No part of this work may be reproduced, stored in a retrieval system, or transmitted in any form or by any means, electronic, mechanical, photocopying, microfilming, recording or otherwise, without written permission from the Publisher, with the exception of any material supplied specifically for the purpose of being entered and executed on a computer system, for exclusive use by the purchaser of the work.

Printed on acid-free paper

Springer is part of Springer Science+Business Media (www.springer.com)

Acknowledgements

This book is the outcome of a plurality of activities and all of them informed the original idea of an edition and research project "Handbook of Ecological Concepts (HOEK)". From the very beginning, we envisioned the HOEK as a collective project that would provide a forum for exchange and debate of ideas in the rather scattered field of the history and philosophy of ecology, including theoretical ecology. Alongside concerns about woolliness, lack of transparency and contradictions in the use of ecological concepts, another important motivation for creating the HOEK, and constructing it as a genuine collective project, was our feeling that there was a pressing need to investigate the diversity in the history and theory of ecology in Europe, with its different national traditions, histories and scientific styles. This is true in particular for the early history of ecology from the late nineteenth century through to World War II, but pertains also to more recent developments. Such later developments include, for instance, the different national legislations in biological and environmental conservation, and the changes initiated by national governmental policies, e.g. measures for the increased use of renewable energies. Political developments such as these influence ecological science in the different countries, including the selection and development of concepts and the specific practices of ecology. Equally, however, the dynamics of concepts and theories also influence politics. This two-way interaction occurs, for example, in the context of specific implementation strategies for pan-European laws and directives, such as the European Water Framework Directive, as well as in the far-reaching and rapid changes in land use patterns and large-scale ecological restoration projects which started after the political upheavals of the 1990s, especially in East Germany and Eastern European countries.

In placing such specific emphasis on the European dimension of ecology and the framing of environmental issues in general – without, of course, excluding other geographical regions – we were immediately confronted with a rich diversity of traditions, scientific habits, science policy practices, and languages. Given the additional fact that the editors' own native language is German, we had to deal from the very start with the difficulties – if not impossibilities – and the challenges of translation, not to mention the point of transition and the semantic hiatus between natural and technical languages. All this served to raise our awareness even more of the issues relating to the history of concepts (Begriffsgeschichte) and its importance for the entire field of ecology and its context of application.

Bearing all these complex problems in mind, the project soon took on the dimensions of a rather gargantuan and unmanageable enterprise. So the first and most pressing question was how to make this project workable at all. We decided to start with a mailing campaign, asking our colleagues and friends from ecology, philosophy, history, linguistics, geography and sociology to give their comments and opinions about the project. The feedback was overwhelmingly positive. Not only did they all welcome the project, but by and large most of them got involved with it, be it as participants in workshops, as members of the editorial board, as authors, or as reviewers. Our most sincere thanks go out to them. Without their continuing support the project would never have become a reality. We greatly hope that this first volume justifies and merits the confidence they had in us.

The project was strongly encouraged and supported from the start by Ludwig Trepl and Wolfgang Haber from the Lehrstuhl für Landschaftsökologie, Technische Universität München in Freising-Weihenstephan. Our discussions with them and several other people from the department helped to sharpen our ideas.

In the project's very early stages – at the time of the "proto-HOEK", as it were – we studied some of the established and excellent working encyclopaedic projects in Germany very closely. We are grateful to Albrecht von Massow and also to Markus Bandur who made us familiar with the work of the HmT, the Handwörterbuch der musikalischen Terminologie.

The post-doctoral project carried out by one of us (Astrid Schwarz) in Paris at the Maison des Sciences de l'Homme (MSH) and the Institut d'Histoire et de Philosophie des Sciences et des Techniques (IHPST) enabled the project to take a great leap forward. My special thanks go to Pascal Acot and also Jean-Pierre Drouin for spending their time with me in libraries or nice Parisian coffeehouses and repeatedly asking important, sometimes inconvenient, but always helpful questions. Many thanks go also to Patrick Blandin, Donato Bergandi, Serge Frontier, Catherine Larrère, Patrick Matagne, Denise Pichod-Viale, and a few other colleagues from the Muséum Nationale d'Histoire Naturelle who engaged with me in many and inspiring discussion. As an outcome of this warm reception, the first workshop of the project took place in Paris at the MSH, where it was supported especially by Hinnerk Bruhns and also Caroline zum Kolk.

Two other workshops followed, one in Leipzig (2004) and one in Darmstadt (2006) both of them generously supported by the Volkswagen Foundation. They helped to consolidate and extend the "HOEK community" and provided both further encouragement and constructive criticism for our joint endeavour.[1]

Thanks go also to the colleagues at the institutes where we, the editors, are based. Colleagues from the Institute for Philosophy in Darmstadt were ready to accept my (Astrid Schwarz) occasionally intense absorption in the project. Petra Gehring made available her experience as a local editor of the huge editing project Historisches Wörterbuch der Philosophie, which lasted about 25 years and was

---

[1] See www.hoekweb.net for a description of the workshops and a list of participants.

completed in 2005. My many conversations with Alfred Nordmann were not only inspiring but are also reflected directly in my contributions to this volume.

Many colleagues at the Helmholtz Centre for Environmental Research also contributed their ideas to the project in the course of discussion. They helped me (Kurt Jax) in particular to anchor our theoretical ideas in the practice of ecology and its fields of application. A special word of thanks here goes to Klaus Henle, who accompanied the project with great sympathy and support, not least by providing the atmosphere and freedom for me to devote time to what is still a rather unusual topic for an institution dedicated to ecology and environmental conservation, at least within the German context. Special thanks go posthumously to the philosopher Heidrun Hesse. She was never afraid to go beyond her disciplinary boundaries and inspired many scientists to reflect on concepts and ideas in ecology and its applications. Her critical voice will be missed.

We are grateful to the following members of the Editorial Board for their assistance in preparing the book: Pascal Acot, Paris; Sandra Bell, Durham; Patrick Blandin, Paris; Alexej Ghilarov, Moscow; John Gowdy, Troy, NY; Volker Grimm, Leipzig; Wolfgang Haber, Freising; Yrjö Haila, Tampere; Getrude Hirsch-Hadorn, Zürich; Andrew Jamison, Aalborg; Alan Holland, Lancaster; Chunglin Kwa, Amsterdam; Thomas Potthast, Tübingen; Peter J. Taylor, Boston; Ludwig Trepl, Freising; Gerhard Wiegleb, Cottbus.

Many authors contributed to this volume whose first language is not English, but French, Russian, Finnish, Spanish, Norwegian, Italian or German. These manuscripts have been revised and in some cases translated in their entirety by Kathleen Cross, Susan Haak, Patrick Hamm, and Paul Ronning. All four worked with unbelievable care and patience – many thanks to all of them. Special thanks go to Kathleen Cross, who not only did most of the translation work, but contributed greatly in many conversations and e-mail exchanges towards strengthening and clarifying a number of the articles.

Last but not least we want to thank our former collaborator Christian Haak for his invaluable work in commenting on and editing so many pages of this volume. He did a wonderful housekeeping job with the information and data management.

The project website can be visited at http://www.hoekweb.net.

Darmstadt	Astrid Schwarz
Leipzig	Kurt Jax

Forwarded Foreword[2]

The initiators of the Handbook of Ecological Concepts (HOEK) argue that the HOEK should help in clarifying the relation between concepts in scientific ecology and the objects that are defined by them.[3] They do not attempt to tackle the old metaphysical problem of the world order, nor do they suppose that it is resolved. In fact, the question of whether a scientist *discovers* discontinuities, identities or regularities existing independently from himself that are rooted in reality, and from which he then extracts laws, or whether a scientist *explains* the world using a conceptual framework developed by himself or others is not a matter of epistemology: in practice, scientists do observe regularities in nature. Therefore, at the scale of a human being, oak trees are always, and in the same manner, different from beech trees and birch trees, etc. At the same time these regularities are prescribed by using the complex concept "species", the definition of which relates to meaning ascribed to the perceived discontinuities.

One can see that this approach to reality is determined by history; the nature of observations changes over time, as do the meanings of a concept that give sense to the observations. Consequently, the decision by those responsible for the HOEK to focus on the meaning of concepts and of what precisely they relate to, by retracing the process of their construction and their subsequent uses over time, is an innovation in the lexicon of scientific ecology.[4] To my knowledge, most of the scientific

---

[2] This is a slightly changed version of a paper given in French on the occasion of the first workshop of the project HOEK in Paris in 2002. The workshop took place at the Maison des Sciences de l'Homme Paris and was entitled "HOEK is going to come true". The editors are very grateful to Pascal Acot for his generous encouragement on the ground. He accompanied attentively the first steps of the HOEK, when Astrid Schwarz was a postdoctoral fellow of the DAAD (German Academic Exchange Service) in Paris.

[3] Astrid Schwarz and Kurt Jax, "Outline of the project", see www.hoek.tu-darmstadt.de

[4] Astrid Schwarz and Kurt Jax propose to present general ecological terms (as e.g. ecosystem or niche) that have developed in this way. Two decades ago a similar project directed by Jacques Roger (1920–1990) appeared in France. Its failure was its far too great generality and the early loss of the scientific person in charge (cf. Cahiers pour l'Histoire du Vocabulaire Scientifique, CNRS-INALF, 1981–1990). Recently a Dictionnaire d'Histoire et Philosophie des Sciences, edited by Dominique Lecourt (1999), also "generalist" in nature, has been published by Presses Universitaires de France.

thesauri dealing with this domain *juxtapose* past definitions and different historical uses of concepts, without really discussing the significance and importance of changes in signification. Inversely, the historical approach protects against many ruinous – and let us concede very French – temptations of normalisation and fixation of the vocabulary.

The project HOEK will certainly confront obstacles. After all, if it were not a difficult project, others would already have undertaken the risk to carry it out. Let us begin with the best known but wrong obstacle: the choice of concepts. Most past attempts have met with criticism on this front, and sometimes justly: the dictionary of J. Richard Carpenter,[5] for example, first published in 1938, does not contain the word "ecosystem"; the "Vocabulaire d'écologie" of Daget and Godron[6] contains neither the entry "struggle for existence" nor the entry "eutrophication", nor the entry "homoeostasis". Even the current list[7] of the HOEK, one might say, shows some gaps. It does not include, for example, the word "biosphere" but has the entry "cosmos". Despite future revisions the project can expect to face criticism of this kind of fault, be it justified or not. The first thing for our colleagues to do is determine which words should be in the HOEK that are not yet included, and which terms are included but should not be …

Some real difficulties will, nevertheless, remain. First, ecologists work within a historically as well as geographically specific cultural and environmental framework. Their thinking, and therefore the contents of the concepts they are using is inevitably marked by factors that are not directly scientific. Secondly, ecology is a discipline in which, for a century, specialists have tended to think in *processes* rather than in terms of *objects*: they speak, for example, of plant *successions* rather than of static *vegetation*. It thus becomes a complex problem to grasp the meaning of dynamic concepts. In the following I briefly develop these two points.

The Influence of Non-Scientific Factors

Ramon Margalef, the famous Catalan ecologist, remarked that

> All schools of ecology are strongly influenced by a *genius loci* that goes back to the local landscape [...] The mosaic-like vegetation of the Mediterranean and Alpine countries, subjected to millennia of human interference, has assisted at the birth of the plant-sociology school of Zürich-Montpellier [...] Scandinavia with a poor flora, has produced ecologists who count every shoot and sprout [...] And it is only natural that the vast spaces and smooth transitions of North America and Russia have suggested a dynamic approach in ecology and the theory of climax.[8]

---

[5] Carpenter (1938).
[6] Daget P, Michel G, with collaboration from David P, Riso J (1974) Vocabulaire d'ecologie. Hachette, Paris.
[7] I refer here to the list of 26.10.2001.
[8] Margalef (1968), p. 26 (emphasis P.A.).

This is not just a joke about the plant sociology of Uppsala (its promoters have been accused of artificially multiplying the number of associations by using a method of multistrata analysis of vegetation) or that of Zürich-Montpellier. It is rather a way of saying that the conceptual systems in ecology are not universal or at least difficult to generalize: long ago, when I stayed in French Guyana, one of the problems occupying some botanists there was whether it was possible to apply the method of *minimum area* of Josias Braun-Blanquet (1883–1980) to the rich diversity of the humid tropical primary forest. The answer was it would have been possible theoretically, but would have required a sampling area too large to make the method applicable in practice.

One of the questions for the collaborators of the HOEK might therefore be whether particular ecological conditions necessarily shape the development of the system of concepts from which meaning is being defined, or whether it is possible that universal significance actually exists. This is a crucial question, because it can address the issue of scientific styles – German, French, etc. – as well as that of institutional influences.

During the first decade of the twentieth century Charles Flahault (1852–1935), a French botanist ranger, failed to unify the vocabulary of plant geography because of the above mentioned problem. At this time, great disorder dominated the nomenclature of ecological units. The terms "association" or "formation" were not clearly defined and were used very differently depending on the respective researcher. In 1899 the "VII. Internationale Geographenkongress" in Berlin called for a dialogue on the vocabulary of vegetation geography and installed a commission to work on the question. Charles Flahault was invited by Otto Warburg (1859–1938), Adolf Engler (1844–1930) and the geographer Oscar Drude (1852–1933), but Flahault claimed to be "unable to come", probably because of the French-German conflict during this time concerning the Alsace and Lorraine.

One year later, in Paris this time, Flahault proposed a *project for phytogeographical nomenclature* at the first International Botanical Congress (1900). He had established a "Nomenclature of geographical and topographical units" (that went from "le groupe de regions" (group of regions) to "le station" (station) by passing through all possible intermediates such as "le domaine" (domain), le district (district), le sous-district (sub-district), etc.). He then attempted to assign to them a corresponding "série des termes phytogéographiques d'ordre biologique" (series of phytogeographical terms of biological order). Those terms went from the "type de vegetation" (vegetation type) to the "forme biologique" (biological form), through "groupes d'associations" (groups of associations), "associations" (associations), etc.

The congress concluded by proposing a huge consultation to be undertaken through the press and through correspondence. But nothing arose from the consultation and in 1905, at the "II. Internationaler botanischer Kongress" in Vienna, a commission, headed by Charles Flahault and the Swiss Carl Schroeter[9] (1855–1939), was named to put forth proposals at the Congress in Brussels in 1910. A comment

---

[9]Carl Schroeter is the inventor of the word autecology (or autoecology), to denote the ecology of the single isolated plant, and synecology to denote the ecology of an assemblage of plants.

of the botanist Jules Pavillard (1868–1961) from Montpellier sheds some light on the appreciation granted to the results of the phytogeographical commission within the botanist's community:

> Une grosse déception nous attendait à l'issue du IIIe Congrès International de Botanique tenu à Bruxelles du 14 au 20 mai 1910. La discussion ouverte devant la section de phytogéographie n' a pu conduire à aucune solution définitive des problèmes essentiels de la nomenclature.[10]

Beyond the fact previously mentioned, that it was difficult to come to an agreement between a phytosociologist from Montpellier and one from Scandinavia, important cultural influences sterilised the final result. The "English Committee" for example – speaking with one voice – held the position of the botanist Charles Edward Moss (1870–1930), who was very influential in Great Britain. Moss was influenced by the work on succession of the North American Frederic Edward Clements (1874–1945). Moss presumed the dynamics of vegetation to be the essential reality of groups of vegetation. He proposed to define an association as a "stage in a successive series" and a formation as "the totality of all stages of a successive series"!

Clements himself had already succumbed to the obsession of neologism in 1902 when he had reacted to the first propositions from Flahault – those of 1900 – in publishing an article in the journal edited by Heinrich Gustav Adolf Engler (1844–1930).[11] Flahault had recommended using terms limited to particular regions, according to him not translatable, such as *maquis, garrigue, toundra, llanos, etc.* Clements, having a profound systematic – not to mention a little dogmatic – view, did not miss this nice opportunity; he accused Flahault of forgetting that these names denote particular types of principal formations one could also find elsewhere, and pointed out that some of the terms proposed, such as "ecological series of groups of associations" or "type of vegetation", were much too long. Grasping the opportunity with both hands, he offered his own system, strictly based on Greek and Latin, as well as a rule for possible neologisms. As one could expect, this resulted in linguistically monstrous concepts, sometimes pedantic and certainly, for the most part, never used to this day! He proposed, for example, *ochtophilus* instead of "ripicole" or *conophorophilus,* to denote a plant occuring in coniferous forests. His project was not carried out: the phytogeographers already had their habits – but Clements' criticisms had been noted. And in 1910, because unanimity would have been necessary to pass a new nomenclature, Flahault's initiative failed: there were 14% abstentions or "no" on the final vote.

Certainly the HOEK does not intend any normativity in the sense of the project just reviewed. But the same or similar reasons that made Flahault's project fail could influence the development of the project initiated by Astrid Schwarz and Kurt Jax; or perhaps just the selection of entries that will be finally retained. I think that we should keep watch over this together.

---

[10] "A big disappointment was awaiting us at the end of the third International Botanical Congress, held in Brussels between Mai 14th and 20th in 1910. The discussion in the geobotanical section did not result in any definite solution of the essential problems in nomenclature."

[11] Clements (1902).

The Dynamic Point of View in Ecology: Difficulties and Fertility

The second topic I would like to address is the difficulty in developing a terminology of movement. A dynamic vocabulary does not denote a being, but, to take up the old Aristotelian terminology, "a being and a non-being at once". When I have a moving object in mind, I am thinking of something that is no longer and at the same time of something that I believe it will become. To go from a static point of view to a dynamic perspective in biology means to think in processes rather than in objects; this can pose serious problems, which I will attempt to demonstrate by looking closely at the ecology of plant succession.

While at the beginning of the twentieth century most European ecologists adopted a static perspective, a "photographic" view of a situation, in the United States a "cinematic" ecology had developed at the start of 1900s. In 1897, in a pioneering article, the botanist Conway McMillan showed how the physiognomy of even plant formation could suggest a progressive dynamic: "[Sphagnum moors or ponds] may be regarded as such glacial ponds or lakes in process of conversion to forest [...] and almost every imaginable transition may be found from open lakes with sandy beach-lines continuous on all sides [...] to solid masses of spruce and tamarack timber."[12] One can see how the *physiognomy of succession* eventually suggests the *evolution* of the vegetation group: the juxtaposition of states caught in a certain moment in time reflects the process of transformation of the vegetation.

This connection between the physiognomy of a landscape and its successive development is also expressed in an earlier text of Henry Chandler Cowles (1869–1939), father of the theory of plant succession: "In the dune region of lake Michigan, the normal primary formation is the beach; then, in order, the stationary beach dunes, the active or wandering dunes, the arrested or transitional dunes, and the passive or established dunes. The established dunes pass through several stages, finally culminating in a deciduous mesophytic forest, the normal climax type in the lake region [...]".[13]

Each time, the same type of explanation of movement is proposed: The conditions under which a certain process proceeds is modified by the development of the process itself: The pioneering vegetation is necessary to reinforce the dunes in order to assemble the conditions of development of a higher and denser pioneering vegetation that will then protect the more important shoots of trees and so on. This type of reflection was very fruitful in the history of ecology. It could explain, for example, why the vegetation of North America is three times richer in species than the vegetation of Occidental Europe. During the last glacial epoch, North American vegetation could recede slowly to Central America and then come back with the beginning of the current interglacial epoch, because of the North-South orientation of the Cordilleras. This was not the case in Europe where the chain of the Pyrenees, the Mediterranean Sea and the Alps formed a barrier difficult to surmount. It was

---

[12]McMillan (1896).
[13]Cowles (1899), p. 20.

the introduction of thinking in terms of successional movements that allowed for this type of analysis (and many others).

But the adoption of a dynamic perspective is very delicate. Certainly, we know how to represent, or *imagine* continuous processes. But scientific work is limited to decomposing the movement into a succession of distinctive, differing states. We have seen this with plant succession, but the same would apply to, for example, the morphogenesis of trees. It is necessary to establish discontinuities – but with which criteria, acceptable for the greatest possible number of scientists, should one establish the divisions? Once again one is faced with the question concerning thought styles and scientific schools.

Moreover, the thinking of Clements, who was one of the most important succession ecologists in the history of ecology, was criticised because of its rigidity. His organicist conception of communities and their transformations as well as the reality of the climax as an ultimate state – without any turning back – of the successional evolution, were also criticised vividly in the epoch between the two World Wars. This is precisely the reason why the nature of ecological systems is still discussed, and why the word "ecosystem" is so difficult to define. How, in this case, should one come to an understanding in conceptual questions?

Under the premise that no unifying effort is envisaged, I see the engagement of the participants of the HOEK as a testimony, undertaking nothing more than an attempt to understand the reasons for existing controversies. Some of the difficulties we will meet are not new. I think, without sinking into paranoia, that we will be accused, as I have mentioned, of adopting a "normative", "voluntarist" approach, with selections that are "arbitrary", "artificial" or even "chauvinist". But history also teaches us that difficulties and failures are constitutive for scientific progress. So it was that the failure of 1910 (the attempt of Flahault) resulted immediately in the nearly simultaneous foundation of the schools of plant sociology of Uppsala and of Zürich-Montpellier (and much could be said about them). These schools continued to dominate the European scientific landscape in the field of nomenclature of vegetation groups till the 1950s. And the justified criticism of them (originating principally from the Anglo-Saxon world) played an equally important role in the construction of system and dynamics oriented ecology, which is practiced today in Europe and all over the world.

The HOEK is obviously much more modest than the great attempts at the beginning of the twentieth century to reflect on vocabularies. Nonetheless, in this field history teaches us at least one thing: it is better to be criticised than to be unproductive. It follows that the idea of the HOEK initiators and the HOEK participants, in spite of the difficulties they will have to surmount, is unlike the practice of Charles Flahault at the beginning of the last century, declaring unashamedly about the concept of (plant) formation: "I never used this word, because I could not decide which opinion I should side with and which meaning I should give to it; I simply managed to get along without it!"[14]

Pascal Acot

---

[14]Flahault (1900), p 443.

References

Carpenter JR (1938) An ecological glossary. The University of Oklahoma Press, Norman (Reprinted in 1962. Hafner Publishing Company, New-York/London)

Clements FE (1902) A system of nomenclature for phytogeography. Englers Botanische Jahrbücher 31:1–20

Cowles HC (1899) The ecological relations of the vegetation on the sand dunes of lake Michigan. The University Press, Chicago, IL

Daget P, Michel G (ed) (1974) Vocabulaire d'ecologie. Hachette, Paris

Flahault C (1900) Projet de nomenclature phytogéographique. Actes du Congrès International de Botanique, Paris

Lecourt D (ed) (1999) Dictionnaire d'histoire et philosophie des sciences. Presses Universitaires de France, Paris

Margalef R (1968) Perspectives in ecological theory. University of Chicago Press, Chicago, IL

McMillan C (1896) On the formation of circular muskeags in Tamarack swamps. Bull Torrey Botanical Club 23:502–503

Contents

Part I Design of the Handbook of Ecological Concepts

1 Why Write a Handbook of Ecological Concepts? 3
Astrid Schwarz and Kurt Jax

2 Structure of the Handbook 11
Kurt Jax and Astrid Schwarz

3 History of Concepts for Ecology 19
Astrid Schwarz

Part II The Foundations of Ecology: Philosophical and Historical Perspectives

4 Multifaceted Ecology Between Organicism, Emergentism and Reductionism 31
Donato Bergandi

5 The Classical Holism-Reductionism Debate in Ecology 45
Ludwig Trepl and Annette Voigt

Part III About the Inner Structure of Ecology – Some Theses

6 Conceptualizing the Heterogeneity, Embeddedness, and Ongoing Restructuring That Make Ecological Complexity 'Unruly' 87
Peter Taylor

7 A Few Theses Regarding the Inner Structure of Ecology 97
Gerhard Wiegleb

8 Dynamics in the Formation of Ecological Knowledge 117
Astrid Schwarz

Part IV Main Phases of the History of the Concept "Ecology"

9 Etymology and Original Sources of the Term "Ecology" 145
Astrid Schwarz and Kurt Jax

10 The Early Period of Word and Concept Formation 149
Kurt Jax and Astrid Schwarz

11 Competing Terms 155
Kurt Jax and Astrid Schwarz

12 Stabilizing a Concept 161
Kurt Jax

13 Formation of Scientific Societies 171
Kurt Jax

14 The Fundamental Subdivisions of Ecology 175
Kurt Jax and Astrid Schwarz

Part V "Ecology", Society and the Systems View in the Twentieth and Twenty-first Century

15 The Rise of Systems Theory in Ecology 183
Annette Voigt

16 Ecology and the Environmental Movement 195
Andrew Jamison

17 Ecology and Biodiversity at the Beginning of the Twenty-first Century: Towards a New Paradigm? 205
Patrick Blandin

18 An Ecosystem View into the Twenty-first Century 215
Wolfgang Haber

Part VI Local Conditions of Early Ecology

19 Early Ecology in the German-Speaking World Through WWII 231
Astrid Schwarz and Kurt Jax

20 The History of Early British and US-American Ecology to 1950 277
Robert McIntosh

21	**The French Tradition in Ecology: 1820–1950** .... Patrick Matagne	287
22	**Early History of Ecology in Spain, 1868–1936** .... Santos Casado	307
23	**Plant Community, Plantesamfund** .... Peder Anker	325
24	**Looking at Russian Ecology Through the Biosphere Theory** .... Georgy S. Levit	333

Part VII   Border Zones of Scientific Ecology and Other Fields

25	**Geography as Ecology** .... Gerhard Hard	351
26	**Border Zones of Ecology and the Applied Sciences**............................ Yrjö Haila	369
27	**Border Zones of Ecology and Systems Theory**.................................... Egon Becker and Broder Breckling	385
28	**Economy, Ecology and Sustainability** .... John M. Gowdy	405

Picture Credits... 413

Glossary 415

Author Biography 419

Author Index.. 427

Subject Index 437

Part I
Design of the Handbook of Ecological Concepts

Chapter 1
Why Write a Handbook of Ecological Concepts?

Astrid Schwarz and Kurt Jax

Ecology has made considerable progress over the last few decades. Huge amounts of data have been collected and theories, concepts and practices elaborated, greatly increasing our understanding of living nature and our own influence on it. At several points during the twentieth century ecology became the focus of high expectations that it should help to solve the pressing – and now global – environmental problems that we face, and indeed these expectations appear still to be growing even today. So is now the right time to write a *Handbook of Ecological Concepts* that embraces a philosophical and historical perspective on ecology and its concepts? Is there not rather a need to produce more ecological data and models pertinent to the environmental crisis we are experiencing today, with climate change and other global processes of change? Can we really learn from previous environmental crises, such as Germany's "Waldsterben" in the 1980s,[1] or the idea of limits to growth, revived in the US of the 1960s,[2] or the discourse on water pollution in the nineteenth century,[3] or again the fear of wood scarcity ("Holznot") in the eighteenth century?[4] Is it really theories and concepts that play the main organising and disciplining role in science? And, finally, even if we agree on this, is the seemingly "old-fashioned" form of a handbook the right

---

[1] With regard to the "Sterben of the Waldsterben" see, for instance, the conference in July 2007 at the University of Freiburg (Germany)"Und ewig sterben die Wälder. Das deutsche Waldsterben in multidisziplinärer Perspektive". Organised by the Lehrstuhl für Wirtschafts- und Sozialgeschichte des Historischen Seminars (Franz-Josef Brüggemeier, Jens Ivo Engels) and the Institut für Forstökonomie (Gerhard Oesten, Roderich von Detten), both University of Freiburg.
[2] Höhler (2005); Schwarz (2004); Anker (2005).
[3] Kluge (1986); Luckin (1986).
[4] See in particular the work of Sieferle on Austria, but also on Switzerland and the UK (Sieferle et al. 2008).

A. Schwarz (✉)
Institute of Philosophy, Technische Universität Darmstadt, Schloss, 64283 Darmstadt, Germany
e-mail: schwarz@phil.tu-darmstadt.de

K. Jax
Department of Conservation Biology, Helmholtz Centre for Environmental Research (UFZ), Permoserstr. 15, 04318 Leipzig, Germany
e-mail: kurt.jax@ufz.de

way to go about it – in times of collective Wikipedian knowledge production and a widely acknowledged unravelling of traditional scientific categories and institutions?

We think the time is exactly right for a *Handbook of Ecological Concepts*.

There are a number of crucial and indeed pressing reasons for acquiring a deeper understanding not only of ecological concepts but also of the epistemology and history of ecology as a whole. Research aimed at solving environmental problems as well as the communication of this research, be it in interdisciplinary dialogue or between scientists and users in non-academic fields, requires not only evaluating more and more new data but also establishing clarity and transparency in relation to the conceptual foundations of ecology. Ecologists themselves have frequently and repeatedly deplored the ambiguity and vagueness of concepts.[5] A large number of dictionaries of ecology have been written and there have even been commissions created to debate ecological terminology.[6] The fact that philosophers and historians of ecology have been quick to join in this chorus of terminological critics is less surprising.[7]

However, the issue at hand, including the problem of communication, is not just a matter of terminology, i.e. of finding the right terms or simply defining concepts. It goes much deeper than this. In order to disentangle the conceptual knots and strands in ecology and to infuse ecological concepts with greater power, it is necessary to trace the fluctuations and transformations of concepts and the epistemological questions related to them. This is especially true of traditionally heterogeneous fields such as the environmental sciences or ecology.

Thus, our handbook is designed to serve a variety of purposes and interests. Overall, however, the aim of the HOEK is to contribute towards a better understanding of the plurality of meanings and epistemological changes in the field of ecological and environmental knowledge. We argue that ecology and the environmental sciences are not only driven by instruments or experimental techniques but are primarily organised around concepts and metaphors. It is precisely the concepts that embody the different epistemic, normative and cultural regimes or styles of thinking that are important in the field.

It bears noting, though, that our aim is not to provide "correct" definitions of ecological concepts or to debunk misunderstandings as misconceptions. On the contrary, we are interested in tracing back and analysing particularly those misunderstandings and displacements that are often productive in the transfer of concepts from one discipline to another or from everyday language to technical language. Consequently, we are equally concerned to clarify and sharpen concepts by offering a systematic and historical survey of their different uses and by highlighting the ways in which concepts are blurred when they transgress disciplinary borders or even just community borders in ecology itself. The phenomenon of concepts that

---

[5] Looijen (1998); Mayr (1984); Frazier (1994); Peters (1991); Grimm and Wissel (1997); Jax (2006).

[6] Such as the working group for phytogeographic nomenclature established by the III. International Botanical Congress (Flahault and Schröter 1910) or the Committee on Nomenclature created by the Ecological Society of America in 1931 (Eggleton 1942).

[7] Shrader-Frechette and McCoy (1993), or Greg Cooper in a presentation at ISHPSSB in Exeter 2007; see also Sagoff (2003).

1 Why Write a Handbook of Ecological Concepts? 5

straddle such borders or even transgress them has recently prompted particular interest and has been addressed from different theoretical positions. Thus, concepts are variously described as boundary concepts,[8] as being nomadic[9] or commuting[10] between several fields of knowledge; and concepts can function as destabilising stabilisers[11] or immutable mobiles.[12]

Ecology and the environmental sciences were confronted from the very beginning with this kind of co-production of the social and the epistemic and with the hybridisation of concepts, objects and institutions. They often failed adequately to fulfil the work usually required of disciplinary and epistemological purification. One might say that ecological knowledge is and was produced in a border zone of institutional, epistemic and metaphysical divisions.[13] Consequently ecology has to deal with openness and indeterminacy in theory building, practices and knowledge production in general.

Occasionally, ecologists themselves refer more or less explicitly to this situation and raise concerns about the quality of their own concepts, calling for greater self-reflexivity with respect to the conceptual framework of the discipline.[14] The lack of clarity regarding the basic conceptual foundations of ecology actually impedes the construction of a strong theoretical framework and – to an even greater extent – communication in inter- and transdisciplinary discourse. Above all, however, there is considerable pressure from society to clarify concepts, e.g. in the context of political environmental decisions or legal frameworks that make use of ecological concepts and knowledge.[15] A clearer and more conscious use of concepts will obviously contribute towards solving environmental problems and improving ecological research. Thus, as Steward Pimm was able to demonstrate,[16] one reason why the exhaustive debate on the relation between diversity (or complexity) and stability during the 1960s and 1970s,[17] failed to produce any satisfactory outcome was because it was dealing with several different concepts of "complexity" and "stability" and thus with several questions instead of just one. These considerations illustrate

---

[8] Star and Griesemer (1989).

[9] Stengers (1997).

[10] Schwarz in a paper entitled "Commuting concepts and objects in scientific ecology", paper given at the first conference of the European Philosophy of Science Conference in Madrid 2007 (http://www.ucm.es/info/epsa07/misc/EPSA07_BookOfAbstracts.pdf).

[11] Kaiser and Mayerhauser (2005): "Destabilising Stabilisers", paper presented at the conference *Imaging Nanospace*, Bielefeld.

[12] Latour (1993).

[13] Ecological knowledge has been and still is produced in physiology or forest science laboratories, in applied and theoretical contexts, and with a philosophical background rooted in systems theory or in complexity theories, in reductionism, holism, emergentism, as well as in other -isms.

[14] E.g. Haila and Järvinen (1982); Peters (1991); Pickett et al. (1994/2007).

[15] For example, the various Ecosystem Management approaches or the Convention on Biological Diversity and its "Ecosystem Approach".

[16] Pimm (1984).

[17] Goodman (1975); Trepl (1995).

the point that both internal and external reasons exist that not only justify but actually necessitate a project such as the HOEK.

What Is Innovative About the HOEK?

The idea of the HOEK has taken a long time to develop. In our own earlier work as ecologists we often sought what we are now attempting to present, namely a guide to the basic concepts of ecology that helps us to understand a whole range of issues, including how they originated, what different expressions they take, why there is so much confusion around some of them, how they influenced the development of ecological theory and practice, what the problems are in their application, and how we can best make use of the underlying ideas to create a theoretical ecological framework and to help solve environmental problems.

The innovative character of the work presented here is that it facilitates rapid access to the (sometimes multiple) conceptual content of the terms as well as providing in-depth information about their philosophical and historical context. The structure and approach of the Handbook (as explained in more detail in Chaps. 2 and 3) differ substantially from those of common dictionaries of ecology. The most important difference in comparison with existing reference works is that the HOEK does not seek to offer short "technical" definitions for alphabetically ordered concepts. Indeed, there are no standard definitions such as niche being "the functional position of an organism in its environment, comprising the habitat in which the organism lives, the periods of time during which it occurs and is active there, and the resources it obtains there".[18] Instead, every meaning of the concept "niche" that has ever occurred is discussed in its historical context; changes, trends and fashions are elaborated and linked with persons, institutions, instruments and theories – in short: the concepts are discussed in an epistemologically explicit space, elucidating the historical, logical and semantic processes that link a given concept with its object. The Handbook does not cover each and every term in ecology but only a limited number that are of major theoretical and practical relevance (see below). In the glossary at the end of the book those concepts are listed that are relevant in the volume. Being a trilingual catalogue it gives in the same time an impression of the richness of terms in different languages.

Who Are the Authors?

The overarching goal of this project is to write a philosophically and historically informed encyclopedic reference work. In calling it a project, we also want to emphasise that the HOEK is more than a book; it is also an enterprise – indeed we

---

[18] This is the style of definitions given, for instance, in the Concise Oxford Dictionary of Ecology (Allaby 1994, p. 269).

1 Why Write a Handbook of Ecological Concepts?

might call it an adventure – aimed at bringing together scholars from fields as varied as ecology, philosophy, history of science, conservation biology, anthropology, linguistics and other disciplines to chart the field of ecological knowledge. This they do both by contributing to this book series in various ways – as authors, referees or advisors – and also by coming together in workshops and projects related to the topics of the HOEK whose outcomes feed back into it again. The HOEK thus seeks to serve as a platform to further the self-reflexivity of ecology and the environmental sciences and to foster the development of a strong theoretical core for these disciplines, a core with a sound philosophical basis.

A further important particularity of the HOEK should be mentioned. The treatment of the different concepts of ecology draws strongly on the European history of ecology. Although this geographical region was decisive for the creation and early development of ecology as a science, it has frequently been neglected in the past,[19] not the least on account of language barriers. Many fascinating ideas from early ecology await rediscovery – ideas which, if reflected on and made more widely known, can help us to avoid reinventing the ecological wheel time and again and may even spawn new conceptual developments and aid current debates on ecological theory and practice.

Who Are the Intended Readers?

The HOEK is aimed at people interested in ecology and in the wider realm of environmental research and management, as well as people in the area of environmental policy making. Thus, the researcher from the discipline of ecology may use it to look up a classic reference or an ill-remembered meaning, or use the information to help in a clearer structuring of complex research questions (or even research projects). Students can use the Handbook to gain clarification of an unfamiliar term or to improve their understanding of the conceptual foundations of ecology. Specialists looking from a different perspective (such as policy making, management or law) will also find information for their purposes, e.g. the lawyer's clerk fishing for arguments with which to impugn some miscreant in court. Equally, though, the HOEK seeks to exert a powerful theoretical impact and is thus also written for all those interested in the philosophy and history of ecology and the environmental sciences. It can contribute to debates aimed at adequately describing engineering or the applied and fundamental sciences; the HOEK might also turn out to make an important contribution to the debate on models and simulation or else offer new insights in relation to the field sciences, which have as yet received relatively little epistemological and cultural attention in contrast to the laboratory sciences.

---

[19] For instance the collection of seminal papers "Foundations of Ecology" compiled by Real and Brown in 1991.

These two goals – reaching a relatively wide and non-homogeneous public and using a methodology not very common in the field under investigation – require a degree of systematic reflection on how best to proceed with such a project. The following sections will therefore introduce the general structure of the HOEK and the ideas behind it (Chap. 2), and then move on in a more theoretical part to present some thoughts about the history of concepts (Begriffsgeschichte), its methodology and its possibilities and limits in a hybrid scientific field such as ecology (Chap. 3).

As we have shown above, then, it is both timely and necessary to write a *Handbook of Ecological Concepts*. The field requires systematic and coordinated treatment. While "wikis" offer an interesting and important new form of common knowledge production, they do not guarantee the systematic character provided by a thoroughly edited handbook. This is exactly what the HOEK seeks to offer and what is needed to better understand and hopefully also improve the theory and practise of ecology and the environmental sciences. By the same token, we are persuaded that this necessary and pressing endeavour can be accomplished methodologically and epistemologically in a proper and productive way by not just referring to but by building on the history and philosophy of science.

References

Allaby M (ed) (1994) The concise Oxford dictionary of ecology. Oxford University Press, Oxford
Anker P (2005) The ecological colonization of space. Environmental history (http://www.historycooperative.org/cgi-bin/justtop.cgi?act=justtop&url=http://www.historycooperative.org/journals/eh/10.2/ankcr.html) (last accessed 10/11/2010)
Eggleton F (1942) Report of committee on nomenclature. Ecology 23:255–257
Flahault Ch, Schröter C (eds) (1910) Phytogeographische Nomenklatur. III. Int. Bot. Kongress, Brüssel 1910. Zürcher & Furrer, Zürich
Frazier JG (1994) The pressure of terminological stresses – urgency of robust definitions in ecology. Bull Br Ecol Soc 25:206–209
Goodman D (1975) The theory of diversity-stability relationships in ecology. Q Rev Biol 50: 237–266
Grimm V, Wissel C (1997) Babel, or the ecological stability discussions: an inventory and analysis of terminology and a guide for avoiding confusion. Oecologia 109:323–334
Haila Y, Järvinen O (1982) The role of theoretical concepts in understanding the ecological theatre: a case study on island biogeography. In: Saarinen E (ed) Conceptual issues in ecology. D. Reidel, Dordrecht, pp 261–278
Höhler S (2005) Raumschiff 'Erde': Lebensraumphantasien im Umweltzeitalter. In: Schröder I, Höhler S (eds) Welt-Räume. Geschichte, Geographie und Globalisierung seit 1900. Campus, Frankfurt a.M, pp 258–281
Jax K (2006) The units of ecology: definitions and application. Q Rev Biol 81:237–258
Kaiser M, Mayerhauser T (2005) Nano-Images as Destabilizing Stabilizers. Paper given at the conference "Imaging NanoSpace – Bildwelten der Nanoforschung", Center for interdisciplinary Research Bielefeld
Kluge T (1986) Wassernöte. Alano, Aachen
Latour B (1993) We have never been modern. Harvard University Press, Cambridge
Looijen RC (1998) Holism and reductionism in biology and ecology: the mutual dependence of higher and lower level research programme. Kluwer, Dordrecht

Luckin B (1986) Pollution and control: a social history of the Thames in the ninetheenth century. Hilger, Bristol
Mayr E (1984) Die Entwicklung der biologischen Gedankenwelt. Vielfalt, Evolution und Vererbung. Springer, Berlin
Peters RH (1991) A critique for ecology. Cambridge University Press, Cambridge
Pickett STA, Kolasa J, Jones CG (1994/2007) Ecological understanding, 2nd edn. Academic, San Diego, 2007
Pimm SL (1984) The complexity and stability of ecosystems. Nature 307:321–326
Real LA, Brown JH (eds) (1991) Book foundations of ecology: classic papers with commentaries. University of Chicago Press, Chicago
Sagoff M (2003) The plaza and the pendulum: two concepts of ecological science. Biol Philos 18:529–552
Schwarz AE (2004) Shrinking the ecological footprint with nanotechnoscience? In: Baird D, Nordmann A, Schummer J (eds) Discovering the nanoscale. IOS Press, Amsterdam, pp 203–208
Shrader-Frechette KS, McCoy ED (1993) Method in ecology: strategies for conservation. Cambridge University Press, Cambridge
Sieferle R-P, Krausmann F, Schandl H (2008) Socio-ecological regime transitions in Austria and the United Kingdom. Ecol Econ 1:187–201
Star SL, Griesemer JR (1989) Institutional ecology, translations and boundary objects: amateurs and professionals in Berkeley's museum of-vertebrate-zoology, 1907–39. Soc Stud Sci 19:387–420
Stengers I (1997) Power and invention: situating science. University of Minnesota Press, Minnesota
Trepl L (1995) Die Diversitäts-Stabilitäts-Diskussion in der Ökologie. Berichte der Akademie für Naturschutz und Landschaftspflege. Beiheift 12:35–49

Chapter 2
Structure of the Handbook

Kurt Jax and Astrid Schwarz

The *Handbook of Ecological Concepts* deals with fundamental terms that are or have been of theoretical relevance in scientific ecology. They are discussed using an approach that to some extent builds on the methodology of history of concepts. Approaches using such a methodology were developed during the second half of the twentieth century in various encyclopaedic projects in the fields of history, politics, musicology and philosophy, among others (for a more detailed account, see Schwarz, Chap. 3 this volume). Rather than providing simple definitions and explanations, these approaches seek to trace and reconstruct the dynamics of concept building and conceptual transformation. This is exactly what this Handbook aims to do and is also reflected in the structure of the first volume. The following thoughts are rather provisional but confidently assume that this first volume will be followed by other volumes that allow to unfold the already existing blueprint entirely.

In general terms the articles follow a common scheme. This allows both for quick and easy reference as well as for in-depth analysis that includes both historical and philosophical analysis of the concepts concerned. Generally, the concepts will not to be arranged in alphabetical order but in so-called conceptual clusters, such as "ecological units" or "ecological interactions", enabling entries to be structured in terms of both form and content (see below). These projected volumes are to trace scientific discourses by strictly tracking a particular term, such as "niche" or "organism". The articles follow the scheme described below and address one key concept each.[1] Other volumes which, like this first one, deal with particular

---

[1] The second volume is planned to deal with "ecological units" and would address in four entries the concepts "organism", "population", "community" and "ecosystem" as a common conceptual cluster. The concepts that appear in this volume are given in a multilingual glossary at the end of the book. About two thirds of these concepts are so-called main concepts. A more complete list of entries is given at http://www.hoekweb.tu-darmstadt.de. It is so far limited to approximately 200.

K. Jax
Department of Conservation Biology, Helmholtz Centre for Environmental Research (UFZ),
Permoserstr. 15, 04318 Leipzig, Germany
e-mail: kurt.jax@ufz.de

A. Schwarz (✉)
Institute of Philosophy, Technische Universität Darmstadt, Schloss, 64283 Darmstadt, Germany
e-mail: schwarz@phil.tu-darmstadt.de

epistemic, ontological, and socio-political issues of ecology, and require a more flexible approach in order to describe their subjects adequately. Notwithstanding this flexibility, all entries are characterised by the fundamental (and, for the Handbook, essential) focus on both historical and philosophical (epistemological) perspectives on ecology and its concepts.

What Is the Structure of Each Entry?

Each entry in the Handbook comprises five sections. The first offers some brief philological information and a summary of the full entry. The second section constitutes the main bulk of the entry, providing an overview of the historical and epistemological patterns and features of the term, which are then explained in greater detail in the third section. Up to this point, the sections for the entries follow a structure that is more or less familiar from encyclopaedical handbooks, such as the *Handwörterbuch der musikalischen Terminologie* (HmT 1972–2006) or the *Historisches Wörterbuch der Philosophie* (1971–2007). However, the fourth section is rather unusual: it brings an openness and flexibility to the project by providing the possibility for commenting on and thereby supplementing articles by other authors. For instance, it allows experts from different fields to add comments relating to previous sections of the entry, to insert important cross-references that need to be discussed in a more extended form and, finally, it can be also used to invite authors to present counter-arguments that may have emerged in the process of designing the entries or during the peer review process. The fifth and final section lists the literature references for the entire entry.

In the following, the common HOEK scheme for each article, or entry, is given in more detail:

1. The top of the article always consists of a literal translation of the concept and of the different forms existing in other languages (at least in English, French and German), a description of the etymology including the pre- and extra-ecological uses of the term, an account of the sources containing the first use of the concept in ecology; and finally a summary of the whole article.
2. The main section of the discussion of each key concept is divided into several sub-sections:
    (a) Main phases of the history of the concept.
    (b) Brief account of epistemological changes and influences.
3. This section provides detailed explanations of the main elements as described in 2. Related aspects are addressed while remaining close to the core elements of the concept in question. Cross–references are established to problems that occur when the concept is used outside ecology, e.g. in environmental protection or biological conservation.
4. Comments by other authors on particular aspects of an entry.
5. Sources/literature.

The form and content thus meet the demands of primary research and offer a quick guide to specific concepts. Readers who are interested mainly in acquiring a brief overview of a concept will find this in section 1 of the articles, while sections 2 to 4 offer greater depth by way of more detail and theoretical complexity. Perspectives that offer a close-up view of contrasting opinions and issues around each concept alternate with "long shots" that present a broader outlook on how and when the term entered ecological discourse and the way in which ecological concepts might be related to other historical entities or even epochs.[2] Our intention is that switching the focus back and forth between perspectives rooted in different times and spaces will stimulate new insights into the general polyvalent character of ecology.

What Are Conceptual Clusters?

The classical way of structuring the Handbook would be to start with the letter "A" for "abundance" and proceed through the alphabet towards "W" for "water cycle". One thing this would lead to, however, is a great deal of repetition in the description of concepts that are similar to one another; what it would also do, though, is leave largely untouched a discussion of the links between related concepts. These are the first and most straightforward reasons why the Handbook is structured by what we have called "conceptual clusters". A conceptual cluster[3] assembles concepts with common properties in terms of their epistemology, their meaning and function in ecology, and the phenomena they describe. Examples of such clusters are "ecological units", which brings together concepts such as population, community, formation, biogeocenosis and ecosystem, and "ecological interactions", which includes concepts such as predation, competition and mutualism.

Conceptual clusters make it possible to describe and compare different but related terms and concepts more efficiently and conveniently. They avoid repeated discussions of the same conceptual problems for each concept and reduce the complexity created by the existence of a multitude of similar concepts. They may even help to structure ecological theories, of which concepts are the basic building blocks. Finally, the process of comparing and contrasting different concepts that make up a conceptual cluster also contributes towards a better understanding of each particular concept. Biocoenosis and population, for example, can be distinguished very clearly in almost every definition; the same goes for ecosystem and plant association. Nevertheless, common to all these concepts is that they describe units relevant to ecological research containing (usually) more than a single individual organism[4] and that they have all been subject to the same questions during their histories. Such questions include, for example, whether they are delimited by

---

[2] See for more details in Pomata (1998).
[3] The term for this idea, which we previously had called "conceptual fields", arose during discussions at the first HOEK workshop in Paris in 2002.
[4] Very rarely, definitions of ecosystems require at minimum only one organism (e.g. Stöcker 1979).

topographical or process-based ("functional") boundaries, whether interactions between the elements are necessary – and, if so, to what extent – to call them ecological units.[5] In addition, some of the concepts are considered – rightly or wrongly – to be synonymous or have been developed conceptually from each other (e.g. on the basis of analogies or oppositions). Finally, different ecological units are often considered to be connected hierarchically with each other, for example in a nested hierarchy from the population to the biosphere. All these reasons suggest that it is useful first to address the different ecological units together before any new differentiation can be made.

Conceptual Clusters and Semantic Fields

The idea of conceptual clusters, as introduced here, has some affinity with the concept of Wortfelder (semantic fields), which has been developed in semantics. A brief look at the similarities and differences between the concept of semantic fields and that of conceptual clusters will help to sharpen our understanding of the latter and the purposes it can serve.

The concept of semantic fields was first developed by German linguist Jost Trier.[6] It was intended to describe areas, or clusters, within natural languages in which words belong and are related to a common conceptual domain (Sinnbezirk). The meanings of the different words in a semantic field were considered to be determined by their mutual relations rather than through isolated analysis of each word. These basic assumptions of Trier's theory of semantic fields continue to hold even today in the different forms in which the approach is used.[7] Conceptual clusters likewise delimit a conceptual sphere but, unlike semantic fields, are focused less on the semantic aspects of the terms described and – as the name emphasizes – more on the conceptual aspects and on the phenomena to which the concepts pertain. In addition, conceptual clusters do not deal with natural language but with a technical language. We are dealing here with terms and concepts that were either newly created, such as the neologism "ecology", or are used in a technical way, such as the word "niche". Also, our conceptual clusters include not only terms from one language, English, but where appropriate also from other languages, in particular from German and French. The "same" term sometimes has distinctively different meanings when used in different ("natural") languages.

Conceptual clusters also share the assumption of the Wortfeld theory that (ecological) terms and concepts assembled within a cluster can be understood better

---

[5] See Jax et al. (1998); Jax (2006).
[6] Trier (1931).
[7] See especially Gloning (2002). Other components of Trier's original theory, however, have since been largely rejected, in particular the assumption that the words in a semantic field should cover the whole conceptual sphere *exhaustively* in a mosaic-like manner, and that it is possible to structure clearly the entire vocabulary of a language into semantic fields.

when they are viewed in the light of their interrelations with each other. In other words, the purpose of conceptual clusters is to provide a fertile context for individual concepts, in which the meanings of the concepts are able to emerge by means of contrast, juxtaposition and interconnection, that is, by highlighting the relations between different concepts.

How to Construct and Use Conceptual Clusters

The difficult question regarding the criteria according to which conceptual clusters should be defined, along with what qualifies words to be included in a particular cluster, is one that our concept also shares with the concept of semantic fields. There are, of course, many different ways to divide up the multitude of ecological concepts based on content. Conceptual clusters as we conceive of them here bring together different terms whose meanings are closely related. These may be different words with an identical meaning (synonyms), identical terms with partially overlapping meanings, overarching concepts and specific instances of concepts (such as "association" or "formation" as a special expression of the overarching concept of "community"). In some cases, we can even observe the same word having developed significantly different meanings (such as the word "function" in ecology, which can denote both "process" and "role"). Such semantic processes may occur within ecology or in the course of a shift between scientific and extra-scientific uses and vice versa, for instance through metaphorical usage.[8]

In some cases – indeed, ideally – these concepts can also be ordered in a hierarchical manner. Thus a cluster may have the heading "ecological units",[9] this being the overarching concept within which – according to the structure of Volume 2 – four different key concepts are subsumed, namely "individual organism", "population", "community" and "ecosystem".[10] Below this level are further specialised concepts (including "biocoenosis", "association" and "formation" in the case of "community") which share the generic meaning of one of the key concepts but have

---

[8] Sometimes it is the scientific word that is derived through a change of meaning from ordinary language (e.g. the terms "niche" or "guild"), and sometimes ordinary language uses ecological terms in a metaphorical or broader sense (e.g. the "political ecosystem").

[9] "Ecological units" are defined here as all those units that are objects of ecological research. *Unit* is to be understood here as an aggregation of objects (particulars), which are chosen and arranged according to such criteria that they can be characterised as new and interesting objects in their own right.

[10] In the case of ecological units, the four key concepts in fact closely approximate to "basic level concepts" as described by cognitive psychology (see, e.g. Medin and Smith 1984, p. 124): a middle level of categorisation around which most knowledge is organised and which is the preferred level of usage in communication (Löbner 2003, pp. 274 ff.). Like basic level concepts in natural languages, "community" or "ecosystem" have different and diverse uses, so it comes as no surprise that different meanings and ambiguities arise when the basic concept is mistaken for the more specialised one derived from it.

more specific characteristics not shared by all the definitions of the key concept. Sometimes such specialised concepts have specific designations and sometimes not (simply being called "community", for example, but having a more specialised meaning).

The concepts dealt with in the HOEK will identify and discuss such differences below the level of the basic terms, but will generally not address terms such as "association" separately.[11]

Our approach to conceptual clusters covers both the synchronic and the diachronic dimension of concepts, i.e. we also include previous meanings of the terms we discuss. The dynamics of concepts and their most recent manifestations are always linked in a network of previous and current concepts and can best be understood within such networks.

Our aim in using this analytical and conceptual framework is

1. To understand the dynamics of concepts and the structure of ecological theory, and
2. To find frameworks that are appropriate for rendering concepts operational, in the knowledge that different expressions of the "same" concept may be adequate for the many different contexts in which they are used

The internal structure of conceptual clusters is not fixed in any general way. Initially the cluster is constituted by a selection of concepts gathered together under a general heading. The structure of the cluster can in principle be analysed in a variety of ways. Although arranging ecological concepts into conceptual clusters always implies certain theoretical assumptions about the overall structure of ecological knowledge, the selection of concepts included within a particular cluster is not a statement about the usefulness of this or that concept or theory. On the contrary, a conceptual cluster will almost always bring together concepts derived from completely opposing theories. These concepts cannot be applied to the same questions with unambiguous results.[12] As a consequence, our analysis often may not even lead to a completely consistent (re-)ordering of concepts within a conceptual cluster. The development of conceptual clusters for ecological concepts is thus to some extent itself a research activity. Conceptual clusters thus also provide fresh impetus for conceptual research within ecology and for the improvement of a general theoretical framework for ecology.

---

[11] The method described here creates a kind of *hierarchical structure of concepts*. This hierarchy exists only at the *conceptual* level, of course, and does not imply a physical hierarchy, as, for example, in commonly postulated "nested hierarchies" of ecological units (population as part of communities as part of ecosystems). Note that "population" and "ecosystem" are on the *same* level of the conceptual hierarchy of ecological units.

[12] For example, "ecological units" includes both the "holistic" community concepts of Clements and Thienemann as well as the "reductionistic" community concepts of Gleason and Ramensky.

References

Gloning T (2002) Ausprägungen der Wortfeldtheorie. In: Cruse A, Franz H, Michael J, Peter Rolf L (eds) Lexicology. An international handbook on the nature and structure of words and vocabularies, vol 2. Walter de Gruyter, Berlin, pp 728–737

Jax K (2006) The units of ecology. Definitions and application. Q Rev Biol 81:237–258

Jax K, Jones CG, Pickett STA (1998) The self-identity of ecological units. Oikos 82:253–264

Löbner S (2003) Semantik. Eine Einführung. De Gruyter, Berlin

Medin DL, Smith E (1984) Concepts and concept formation. Annu Rev Psychol 35:113–138

Pomata G (1998) Close-ups and long shots. In: Medick H, Anne-Charlott T (eds) Geschlechtergeschichte und allgemeine Geschichte. Herausforderungen und Perspektiven. Göttingen, Wallstein, pp 57–98

Stöcker G (1979) Ökosystem - Begriff und Konzeption. Archiv für Naturschutz und Landschaftsforschung 19:157–176

Trier J (1931) Der deutsche Wortschatz im Sinnbezirk des Verstandes: Die Geschichte eines sprachlichen Feldes. Winter, Heidelberg

Chapter 3
History of Concepts for Ecology

Astrid Schwarz

The *Handbook of Ecological Concepts* is particularly interested in the ways in which ecological concepts are used. Its main concern is to trace the dynamics and continuity of concepts, that is, to analyse processes of conceptual transformation as well as strategies for rendering concepts robust in both current and historical ecological knowledge. Concepts are not discussed in terms of being either false or true but rather as being more or less appropriate to their intended task. One important criterion for the adequacy and usefulness of a concept is its functional efficiency and operational reliability,[1] which hinges on the power of the concept to classify, characterise and differentiate processes or phenomena. The greater this power of differentiation, the more robust the concept.

In this book we seek to "follow the concepts", that is, looking at them as they go about their work as part of a language game – and not when they are on holiday, as Ludwig Wittgenstein once noted by way of critiquing philosophical interest in the various definitions and historical meanings of a concept rather than in its actual use. The meaning of a word lies in its use in practice; meaning lies in the act of expression – and not behind it.[2] After all, it is the context and use of a technical concept which determine its meaning, be it one in current use or one used in the past.

This emphasis on use in the genealogy of scientific concepts has been described elsewhere in terms of a series of discontinuities. Such discontinuities – some of them highly significant – are encountered from time to time when tracing the history of certain concepts. This has led philosophers such as Gaston Bachelard and Georges Canguilhem to the insight that "scientific thinking is constantly reshaping

---

[1] Lübbe 2000: 36: "…die Funktionstüchtigkeit eines Begriffs für einigermaßen randscharfe Unterscheidungs- und Zuordnungsleistungen ist ein besonders wichtiges Kriterium für die Zweckmäßigkeit eines Begriffs."

[2] As Gordon Baker points out, Wittgenstein seeks to establish a different form of representation in the thinking of his reader: "speaking and thinking are *operating* with signs, and it is use that gives life to 'dead' signs" (Baker 2001, p. 16).

A. Schwarz (✉)
Institute of Philosophy, Technische Universität Darmstadt, Schloss, 64283 Darmstadt, Germany
e-mail: schwarz@phil.tu-darmstadt.de

its past, because its character is one of constant revolution" (Canguilhem 1979, p. 18).[3] Thus to achieve a better and more accurate understanding of science, it is essential that philosophers and historians look at conceptual discontinuities rather than asserting false continuities by producing collections of biographies or "tableaus of doctrines" in the style of a natural history (ibid., p. 17). It is crucial to understand – and to render comprehensible – "the extent to which concepts, attitudes and methods that are outdated nowadays were outdated even in their own day and, consequently, how far the outdated past remains the past of an activity which continues to require scientific naming" (ibid., p. 27).

Thus it is not primarily the genesis (introduction and definition) of a concept that is the key indicator of its usefulness in a scientific context; instead, it is its continual reprocessing – its adaptive malleability – that makes a concept useful and thus robust. By pointing out discontinuities rather than continuities, philosophical attention is focused on the need for constant regeneration and the accommodation of scientific concepts in different conceptual and theoretical environments. Since the retrospective study of historical meaning cannot contribute substantially towards establishing solid evidence for the usefulness and appropriateness of a concept in the present day, it is the current use of concepts that ought to be observed and analysed.

In light of these considerations one may want to ask more probingly what, if anything, a historiography of concepts can contribute to the study of scientific concepts. Can it make a contribution even despite its limited focus on the genesis of concepts and even though it generates historical knowledge about the meaning of these concepts at a given time rather than knowledge about their use? Or does the historiography of concepts surpass these limits and offer a blueprint for a historiography of concepts in science?

To begin, we might say that the historiography of concepts undermines any bias we hold towards our own present, our "*Gegenwartsbefangenheit*" (Lübbe 2000, p. 41). The historiography of concepts can sensitize us to the fact that the proliferation of publications, contexts and techniques for the use of scientific concepts affords only a very narrow and restrictive "present" in which a concept has a stable meaning. It also establishes a critical distance to the dynamics of current scientific development, along the lines of the dictum, "being aware of the lessons of history … makes one wary of the effects of fashion and error" (Horder 1998, p. 186). In a similar vein philosopher and historian of biology Jane Maienschein argues that "… good science requires an historical perspective […] we make progress in science […] by looking back as well as staring immediately forward at the cutting edge" (2000, p. 341).

Our chief task could be to use knowledge about the methods and concepts of the "outdated past" in order to properly acknowledge an adequate vantage point from which to identify our own potentially "outdated present" – outdated in the sense of an expired fashion, nothing more dramatic than an error, a regressive theory, a dead concept.

Another important impact of a historiography of concepts is that it elevates attention not only for the distinction but for the character of relation between the first

---

[3] If not otherwise marked, translation of citations was done by Kathleen Cross.

world of reality and the third world of concepts and ideas, to paraphrase Popper's three world model. The historiography of concepts allows to escape "naïve circular reasoning from word towards object and back" (Koselleck 1998, p. 121).

All this can be considered as an initial illustration of the thesis put forward in the subtitle of this book, namely that to reflect on concepts serves to advance science.

Building Blocks of a Historical-Systematic Handbook

Accordingly, one critical feature of *a historical-systematic handbook of scientific concepts* needs to be that it adopts an attentive and thoughtful attitude to the particular, contingent present that is reflected in the interrelated concepts and theories of a research programme. At the same time, a previous historical or colloquial meaning may still be present in a concept's use and may therefore prove relevant in the formation of scientific concepts, even if these meanings are not always expressed.[4] They might inform the special features of character of nature (Gernot Böhme) or the hard core of a research programme (Imre Lakatos), working behind the scenes of the epistemic operations and ascribing meaning to nature implicitly (see Schwarz, Chap. 8, this volume). "The great tradition of a balance of nature, going back to antiquity, imputed to nature homogeneity, constancy, or equilibrium and abhorred thoughts of extinction and randomness. Order and coherence were commonly believed in Christian tradition to be characteristic of Divine providence. Such ideas die hard" - notes historian of ecology Robert McIntosh (1991, p. 26). Previous meanings from either a scientific or a non-scientific context may also influence the formation of scientific concepts. As Koselleck noted, "a new term may be coined which expresses in language previously non-existent experiences or expectations. It cannot be so new, however, that it was not already virtually contained in the respective existing language and that it does not draw its meaning from the linguistic context" (1998, p. 30).

The neologism "ecosystem" is a good example, in that it draws together meanings that are implicit in other chosen words (Begriffswörter) already in use, such as holocoen or biosystem (for a more detailed discussion, see Jax and Schwarz, Chap. 11, this volume). The concept "niche" is an example of the power of the pre-conceptual meanings present in debates about the environmental requirements of an organism ("place" niche), the "role" of the organism in the community, and the functional notion of niche (Haefner 1980, p. 125) as a hypervolume in N-dimensional space.

The special task of *a historical-systematic handbook of ecological concepts* is to identify the basic concepts in the field of ecological knowledge and to do justice to the epistemic and institutional pluralism in ecology. In order to do so, one needs

---

[4] The differences between spoken and written language and between different public spaces become relevant here. A scientist might talk about his or her background assumptions (and even get them published, for example, in an interview), but they will not publish them in a scientific journal. They might write down their notes in a lab diary or even publish their seminar or lecture notes on the internet; but they will not make them available without these "brackets of place".

to acknowledge fully that ecological knowledge is also produced outside the scientific discipline of ecology. This is knowledge which may, conversely, acquire relevance for institutionalised ecology. In order to do justice to this institutional openness, ecology has more recently been described as a border zone and thus as a discipline located constitutively at the border between institutional and epistemic fields, as both laboratory and field science. Ecology – or "Border Biology" as Robert Kohler (2002) proposes – is a culture of layers and mosaics. According to Kohler, the tension between laboratory and field sciences doesn't simply disappear by the border practices effectively becoming a new disciplinary core. Instead, object constitution and theory building "on the border" is retained in ecology to the same extent as its institutional character *qua* border discipline is upheld, with all the attendant difficulties regarding the status of field sciences in the topology of the sciences. The fact that ecology was conceived of as a "bridging science" from the start and has retained this motif even today (see Schwarz and Jax, Chap. 19, this volume) might be regarded as an ongoing attempt to give border constellations a positive turn. These constellations are rooted partially in 19th century natural philosophy and are characterised by epistemic and ontological antagonisms. It is only since around the 1980s that there has been a broader, reflexive border discourse that addresses the innovative and creative character – and above all the ubiquity – of borders and mixes in science and technology.

A second feature of an ecological handbook with a historical-systematic perspective is that it can show that, throughout the history of the discipline, ecologists have developed a high degree of sensitivity and attention to what they feel to be conceptual flaws and fuzzy elements in their technical language (see Acot, *Forwarded Foreword*, this volume). This is certainly one consequence of the epistemic and institutional difficulties scientists inevitably come across in ecological knowledge production. But it also signals an awareness that in ecology, as in biology more generally, it is principally concepts and not theories that are at the centre of epistemic strategies.

The special feature of *this historical-systematic handbook of ecological concepts* is that the concepts are discussed not in alphabetical order but rather according to so-called conceptual clusters (see Jax and Schwarz, Chap. 2, this volume). The particular advantage of this approach is not only that the relationship between the chosen term (Begriffswort) and the concept (Begriff) is rendered more flexible but also that the history of the concept and the history of the terminology can be brought together to forge a systematic reconstruction of a concept's use in the ecological field. A further advantage is that the aggregation in conceptual clusters facilitates a productive crossover between conceptual history and other methods, such as the history of ideas (Lovejoy 1948), historical semantics (Busse 1987, Busse et al. 2005), the history of metaphors, and the history of discourse (Bödeker 2002).

Our handbook pursues a programme that is at once hedonistic and ascetic. It is hedonistic in its use of methods and in its appropriation of various systematic schemes: while the conceptual history perspective is central, it functions above all as a method that is open to critical negotiation in relation to the history of discourse and the history of ideas, with consideration also to historical semantics in linguistics and the history of metaphor. Insights and descriptive tools from the history of science and philosophy are made use of, as are systematic representations from philosophical anthropology and the philosophy of history. The handbook is ascetic in its selection of concepts and, in particular, in its mode of presentation of those concepts: just a few concepts – those that structure the discipline – are discussed in this way (see Annexe), and they are presented in accordance with a particular schema (see Jax and Schwarz, Chap. 2, this volume).

Closing and Opening the Debate

Since the foretelling of its demise a few years ago (see, for example, Gumbrecht 2006), conceptual history is now said to be in a "state of transition" (Müller 2005), as heralded by the title of a recent edition of the German-language journal *Archiv für Begriffsgeschichte* dedicated to the discipline. It is placed in a relationship of productive tension to metaphorology and epistemology, to the history of objects (things) and the history of discourse; the issue of things *in* language and things *for* language is raised with renewed emphasis, the pros and cons of its methodological exigencies explored and examined. Conceptual history is effectively undergoing a revision driven by cultural studies. The idea is that it should thus "serve to overcome the ubiquitous barriers to communication that exist between the representatives of different disciplines and cultures".[5] The interdisciplinary configuration of the objects of conceptual history (Müller and Schmieder 2008, p. XII f.) is discovered and trialled as "dispositif history" (Berg 2008, p. 329) and as a medium of reflexivity per se (Mayer 2007).

Regardless of the extent to which the specifics of these methodological corrections and readjustments in perspective may or may not prove convincing, there is unanimous agreement about the fact that a conceptual history of scientific concepts needs to be written differently than a history of "basic historical concepts", or of "musicological terminology",[6] and even than the history of a "philosophia perennis" as conceived by Joachim Ritter, the originator of the *Historisches Wörterbuch der Philosophie* (Historical Dictionary of Philosophy HWPh). A conceptual history of the sciences raises questions, above all, as to the transformability and robustness of concepts in the formation of scientific objects and scientific theory building. It calls for consideration of the possibilities and opportunities afforded by conceptual history to reflect on discontinuities, as these are associated, for example, with the terms "epistemic break" or "paradigm change". And, not least, it raises the question

---

[5] From a report by Ernst Müller and Falko Schmieder following the workshop "Begriffsgeschichte in den Naturwissenschaften" (Conceptual history in the sciences), held 9–10 February 2007 at the Centre for Literary and Cultural Research, Berlin (Archiv für Begriffsgeschichte 49 (2007), p. 210).

[6] The *Handbook of Musicological Terminology* (Handwörterbuch der musikalischen Terminologie HmT) was one of the big German conceptual history projects, founded by Hans-Heinrich Eggebrecht (1970). For our handbook we mainly adopted the structure of the articles that offer different levels of information.

of whether conceptual history can do justice to the intermeshing of scientific practice, the objects of science and their concepts (Müller und Schmieder 2007, p. 210) – and, if so, how.

A History for Scientific Concepts Versus a History for Biological or Ecological Concepts?

Emphasis is often placed on the importance of a conceptual history for biology in contrast to, say, physics. This is because the former has no language amenable to formalisation and therefore no purely relational concepts at its disposal. Georg Toepfer, author of the three-volume series *Historisches Wörterbuch der Biologie* (2009 preprint), claims (following Georges Canguilhem) that biology is a "concept-centered science" (begriffszentrierte Naturwissenschaft) (see below, p. xii). Indeed, Georges Canguilhem gives a positive turn to the fact that it is not possible to transfer biological terminology wholesale into a formalised mathematical language; instead, he elevates this incompleteness to a privileged feature of biology.

This characterisation of indeterminate concepts, often valid only locally and limited in their scope and systematisation, applies to ecology as well. It appears to be of little import whether these are everyday words, such as "niche" and "energy", or neologisms, which are apparently more easy to control. The property of generality becomes relevant in a different way here. The language game "ecosystem" provides an impressive demonstration of the kind of semantic fecundity a concept is capable of generating which at first appears sterile by definition. The ecosystem has since carved a path through a large number of different discourses and disciplines, ranging from linguistics, economics and medicine to descriptions of the social role of innovative technologies. Given this, it is almost self-evident that concepts originating in other spheres of science and technology should also be appropriated by the ecological community, "landscape" and "carrying capacity" being just two examples. Ecological concepts can also take the reverse route: specialist terminology becomes a part of everyday language once more. "Biotope" and "biodiversity" are nowadays in evidence everywhere, and in an environmentally aware society everybody knows what an ecological niche or an ecosystem is, and what dying forests and climate change mean. "Useful terms taken from a major scientific discipline (can) become the master keys to an epoch" (Pörksen 2002, 15): the scientific concept becomes a metaphor in non-scientific usage.

Concept or Metaphor?

This popularisation could be described as a powerful extension of potential relations of similarity, which are now no longer determined by the conceptual cluster of a scientific discipline, but are rather opened up in an uncontrolled way. Concepts lose their contextualisation as nodal points in conceptual networks in which they "are related to one another as superordinate or subordinate, contrasting or correlative concepts" (Dutt 2008, 244). There is now a consensus regarding the fact that metaphors are indispensable in scientific speech and writing and that there is a close connection between concept building and metaphorical usage. What remains in dispute, though, is how the distinction between concept and metaphor can be explained. There is a tendency in cultural studies circles to abandon the hotly contested difference between concept and metaphor in favour of metaphor. However, the reason frequently given – that, after all, many concept words are polysemic and therefore vague – seems neither necessary nor sufficient to abandon determinate conceptual content in favor of metaphorical transfer and hence the concept as a conception or even as a concept – the more so as one might argue that "vagueness as well as non-ambiguity are determined by context" (Teichert 2008, 100). It turns out that the conceptual demarcation of metaphor and concept itself depends to a large extent on background theoretical assumptions and, as such, is also context-driven or, one might argue, is also part of a politics of words. However, it is one of the doyens of the historiography of concepts, Reinhart Koselleck, who has pointed out that a concept "must be ambiguous", thereby highlighting this feature as one necessary and applicable to concepts, whereas elsewhere it would be applied only to metaphors.

> (A) concept must remain ambiguous to be capable of being a concept. […] Thus a concept may be clear-cut, but it must be ambiguous. *All concepts in which an entire process is semiotically concentrated defy definition; only something which has no history can be defined* (Nietzsche). A concept brings together the diverse array of historical experience and a summation of theoretical and practical factual references in a context which is only given and can only be truly experienced as such by means of the concept . […] Each concept sets certain horizons as well as certain limits to possible experience and conceivable theory.[7]

Clearly, this is more of a hypothesis about the concept as it qualifies to be *used* in Koselleck's dictionary of basic historical concepts (Geschichtliche Grundbegriffe),[8] but it is anything but a theoretically satisfying definition of a concept (Knobloch 1992, p. 8) insofar as there is no theory of a historiography of concepts (Teichert 2008, p. 111). This is also the case for the *Historical Dictionary of Philosophy* (HWPh) that explicitly abandoned the idea of an integrating theory in favour of a pragmatic procedure; Müller points out that the "HWPh offers not so much a history of concepts in the strict sense, but instead documents a history of applications of philosophical terms" (2004, p. 9). The result is that the invariance of the concepts and the continuity of philosophical meanings is accentuated.

---

[7](Ein Begriff […] muss vieldeutig bleiben, um ein Begriff sein zu können. […] Ein Begriff kann also klar, muß aber vieldeutig sein. *Alle Begriffe, in denen sich ein ganzer Prozeß semiotisch zusammenfaßt, entziehen sich der Definition; definierbar ist nur, das, was keine Geschichte hat* (Nietzsche). Ein Begriff bündelt die Vielfalt geschichtlicher Erfahrung und eine Summe von theoretischen und praktischen Sachbezügen in einem Zusammenhang, der als solcher nur durch den Begriff gegeben ist und wirklich erfahrbar wird. […] Mit jedem Begriff werden bestimmte Horizonte, aber auch Grenzen möglicher Erfahrung und denkbarer Theorie gesetzt (Koselleck 1995, p. 119 f.).

[8] Obviously, Koselleck is not describing the use of concepts in a research field. What he gives here instead is, as Bernhard F. Scholz points out, "a very adequate description of the manner in which concepts (and words) circulate in the strand of conversation which serves in the construction and maintenance the social reality of the lifeword" (1998, p. 89).

The role and impact of a concept is viewed completely differently when it is looked at from the perspective of logic or philosophy of science (Busse 1987, p. 49 ff.). The focus is obviously not so much on delineating the scope and intention of a concept (something that may indeed turn out to be difficult, if not impossible, given concepts such as democracy or liberty). Instead, attention is drawn towards the logic of a socio-political discourse that gives these concepts their attractive mode of formation: "Those (basic) concepts are ennobled which point to, crystallise and attract contradictions" (Knobloch 1992, p. 12). Knobloch even presumes that it might be worthwhile to talk instead about "functional elements of a historically distinct speech act" (ibid.) and thus to abandon the language game "concept" completely in favour of a more pronounced conception of the objects of a historiography of concepts. This orientation towards the speech act and the discursive per se might be one way of coping with the complexity of historical (and current) constellations and of removing the dilemma of designation.

Another means of mitigating the disputed difference between metaphor and concept might be to consider both of them as explanatory models. In the sentence cited above, "concept" might be regarded as being construed precisely in the sense of an explanatory model: "Each concept sets certain horizons as well as certain limits to possible experience and conceivable theory." The conceptual model determines how empirical data, associated concepts and hypotheses are linked and how they might eventually develop explanatory power. From the metaphor side, the conception of the so-called interactive metaphor, as discussed by Mary Hesse (1980), might fulfil a similar function. The crucial point is first of all that metaphors are not looked at initially as rendering similarities visible but rather in the sense of creating them. The metaphor here is not the result but rather the *cause* of a relation of similarity. One frequent objection to this, however, is that such a metaphor would preclude anything further being said about the scope of the metaphorical predication and would become arbitrary: once having been launched as a scientific model, it can no longer be controlled, meaning that it can no longer be foreseen which of the associated concepts and ideas will ultimately be relevant and perhaps become "conceivable theories". But because a metaphor must be successful as a scientific model, it must also somehow be related to scientific action. Therefore, the relation of similarity cannot be completely arbitrary but is merely unpredictable. It is ultimately this property of the unforeseeable extension and modification of associated concepts and ideas which gives rise to a positive heuristics. It is possible to sum up in essence what is meant by the distinction between unpredictable and arbitrary extension in three points, which in turn come close to being a definition of a scientific concept. A successful metaphor is not bold but rather reserved; it is interactive and works in different contexts, and it can be proven to be inconsistent or even wrong.[9] The usefulness of scientific metaphor as a model can be measured by the degree of "interpretative resonance along with a simultaneous internal suitability to the object" (Debatin 1990, p. 805).

---

[9] For a more extended discussion of the difference between a scientific and a literary metaphor and of the concept of explanation as a metaphorical re-description, see Schwarz 2003, pp. 265 ff.).

In the end we might want to conclude that the controversy concerning the respective role of metaphors and concepts draws attention to their common features which are most relevant for the method of tracking their use: (1) the close interrelation between concept building and scientific practice, (2) the rule of reflexion given either by a concept or by a metaphor, and finally, (3) the role of conceptual clusters in integrating different types of concepts, theories *and* metaphors.

References

Baker G (2001) Wittgenstein: concepts or conceptions? Harv Rev Philos IX:7–23
Berg G (2008) Die Geschichte der Begriffe als Geschichte des Wissens. Methodische Überlegungen zum 'practical turn' in der historischen Semantik. In: Müller E, Schmieder F (eds) Begriffsgeschichte der Naturwissenschaften. Zur historischen und kulturellen Dimension naturwissenschaftlicher Konzepte. Walter de Gruyter, Berlin, pp 327–343
Bödeker HE (ed) (2002) Begriffsgeschichte, Diskursgeschichte, Metapherngeschichte. Wallstein Verlag, Göttingen
Busse D (1987) Historische Semantik. Analyse eines Programms. Klett-Cotta, Stuttgart
Busse D, Niehr T, Wengeler M (eds) (2005) Brisante Semantik. Neue Konzepte und Forschungsergebnisse einer kulturwissenschaftlichen Linguistik. Max Niemeyer Verlag, Tübingen
Canguilhem G (1979) Wissenschaftsgeschichte und Epistemologie. Gesammelte Aufsätze (trans by Bischof M, Seutter W), Wolf Lepenies (ed). Suhrkamp, Frankfurt/M
Debatin B (1990) Der metaphorische Code der Wissenschaft. Zur Bedeutung der Metapher in der Erkenntnis- und Theoriebildung. S Eur J Semiotic Stud 2:793–820
Dutt C (2008) Funktionen der Begriffsgeschichte. In: Müller E, Schmieder F (eds) Begriffsgeschichte der Naturwissenschaften. Zur historischen und kulturellen Dimension naturwissenschaftlicher Konzepte. Walter de Gruyter, Berlin, pp 241–252
Eggebrecht H-H (1970) Das Handwörterbuch der musikalischen Terminologie. Archiv für Begriffsgeschichte 14:114–125
Gumbrecht HU (2006) Dimensionen und Grenzen der Begriffsgeschichte. Wilhelm Fink Velag, München
Haefner JW (1980) Two metaphors of the niche. Synthese 43:123–153
Hesse M (1980) Revolutions and reconstructions in the philosophy of science. Harvester Press, Brighton
Horder TJ (1998) Why do scientists need to be historians? Q Rev Biol 73:175–187
Knobloch C (1992) Überlegungen zur Theorie der Begriffsgeschichte aus sprach- und kommunikationswissenschaftlicher Sicht. Archiv für Begriffsgeschichte 35:7–24
Kohler R (2002) Labscape and landscape. The University of Chicago Press, Chicago
Koselleck R (1995) Vergangene Zukunft: Zur Semantik geschichtlicher Zeiten (1st edn 1979). Suhrkamp, Frankfurt/M
Koselleck R (1998) Social history and *Begriffsgeschichte*. In: Hampsher-Monk I, Tilmanns K, van Vree F (eds) History of concepts: comparative perspectives. Amsterdam University Press, Amsterdam, pp 23–36
Lovejoy AO (1948) Essays in the history of ideas. The John Hopkins Press, Baltimore
Lübbe H (2000) Begriffsgeschichte und Begriffsnormierung. In: Scholtz G (ed) Die Interdisziplinarität der Begriffsgeschichte. Meiner, Hamburg, pp 31–41
Maienschein J (2000) Why study history for science? Biol Philos 15:339–348
Mayer H (2007) Nomadisch unscharf. Vorschläge zur Begriffsgeschichte der Naturwissenschaften. Frankfurter Allgemeine Zeitung 14.02.2007

McIntosh RP (1991) Concept and terminology of homogeneity and heterogeneity in ecology. In: Kolasa J, Pickett STA (eds) Ecological heterogeneity. Springer, New York, pp 24–46

Müller E (2005) Einleitung. Bemerkungen zu einer Begriffsgeschichte aus kulturwissenschaftlicher Perspektive. In: Ernst Müller (ed) Begriffsgeschichte im Umbruch. Archiv für Begriffsgeschichte, Sonderheft Jg. 2004, pp 9–20

Müller E, Schmieder F (2008) Einleitung. In: Müller E, Schmieder F (eds) Begriffsgeschichte der Naturwissenschaften. Zur historischen und kulturellen Dimension naturwissenschaftlicher Konzepte. Walter de Gruyter, Berlin, pp 11–23

Müller E, Schmieder F (2007) Begriffsgeschichte in den Naturwissenschaften – die historische Dimension naturwissenschaftlicher Konzepte. Archiv für Begriffsgeschichte 49:210–214

Pörksen U (2002) Die Umdeutung der Geschichte in Natur. Gegenworte 9:12–17

Scholz BF (1998) Conceptual history in context: reconstructing the terminology of an academic discipline. In: Hampsher-Monk I, Tilmanns K, van Vree F (eds) History of concepts: comparative perspectives. Amsterdam University Press, Amsterdam, pp 87–102

Schwarz AE (2003) Wasserwüste – Mikrokosmos – Ökosystem. Eine Geschichte der Eroberung des Wasserraumes. Rombach, Freiburg/Br, pp 273–281

Teichert D (2008) Haben naturwissenschaftliche Begriffe eine Geschichte? Anmerkungen zum Zusammenhang von Metaphorologie und Begriffsgeschichte bei Hans Blumenberg. In: Müller E, Schmieder F (eds) Begriffsgeschichte der Naturwissenschaften. Zur historischen und kulturellen Dimension naturwissenschaftlicher Konzepte. Walter de Gruyter, Berlin, pp 97–116

Toepfer G (forthcoming). Historisches Wörterbuch der Biologie. Geschichte und Theorie der biologischen Grundbegriffe, Vol 1 (preprint version June 1, 2009). Verlag J.B. Metzler, Stuttgart

Part II
The Foundations of Ecology: Philosophical and Historical Perspectives

Chapter 4
Multifaceted Ecology Between Organicism, Emergentism and Reductionism

Donato Bergandi

The classical holism-reductionism debate, which has been of major importance to the development of ecological theory and methodology, is an epistemological patchwork. At any moment, there is a risk of it slipping into an incoherent, chaotic Tower of Babel. Yet philosophy, like the sciences, requires that words and their correlative concepts be used rigorously and univocally. The prevalent use of everyday language in the holism-reductionism issue may give a false impression regarding its underlying clarity and coherence. In reality, the conceptual categories underlying the debate have yet to be accurately defined and consistently used. There is a need to map out a clear conceptual, logical and epistemological framework.

To this end, we propose a minimalist epistemological foundation. The issue is easier to grasp if we keep in mind that holism generally represents the ontological background of emergentism, but does not necessarily coincide with it. We therefore speak in very loose terms of the "holism-reductionism" debate, although it would really be better characterised by the terms emergentism and reductionism. The confrontation between these antagonistic paradigms unfolds at various semantic and operational levels. In definitional terms, there is not just emergentism and reductionism, but various kinds of emergentisms and reductionisms. In fact, Ayala (1974; see also Ruse 1988; Mayr 1988; Beckermann et al. 1992; Jones 2000) have proposed a now classic trilogy among various semantic domains – ontology, methodology and epistemology. This trilogy has been used as a kind of epistemological screen to interpret the reductionist field. It is just as meaningful and useful, however, to apply the same trilogy to the emergentist field. By revealing the basic assumptions of each, we should be better able to understand the points that are similar and shared, as well as the incommensurable ones.

The first question regarding the emergentism and reductionism debate concerns the type of explanation the sciences are seeking. At present in the sciences – from physics to the human sciences – the ontological and epistemological foundation is essentially naturalistic and materialistic, meaning that all natural (or social) objects, events and processes can be understood without reference to extra- or supernatural

---

D. Bergandi (✉)
Muséum National d'Histoire Naturelle, Paris, France
e-mail: bergandi@mnhn.fr

(vitalistic or theological) entities, causes, aims or explanations. The order and laws structuring natural reality are intelligible and, in principle, there is no limit to naturalistic explanations. The existence of this philosophical substrate – the existence of a scientific and naturalistic epistemology – should be taken into account every time the key words 'emergentism' and 'reductionism' appear.

In ecology – and, without exception, in the other natural and human sciences – the classical confrontation between emergentism and reductionism plays a very important and structuring role. It is necessary to be aware that their basic assumptions involve different and generally antinomian ontologies (worldviews, the "true" structures of reality, or in other words, our "bets" on the structure of reality), methodologies (research strategies) and epistemologies.[1] The existence of these specific semantic domains should be kept in mind every time we approach this issue.

Holism and Reductionism: An Epistemological Confrontation?

Today's perspective of reductionist cosmological ontology has its antecedents in the mechanistic worldview of previous centuries. Gradually, from Leucippus and Democritus to Dalton and, among others, Bohr, reality has been defined from an atomistic perspective: reality consists of distinct, discrete, indivisible atoms with a fixed spatio-temporal amplitude. Unlike reductionism, the holistic ontological perspective of emergentism is continuistic and relational: reality consists of a *continuum* of events and processes that are intrinsically interconnected and interdependent. At first sight both reductionism and emergentism currently share a common scientific philosophy, namely that all biological phenomena are fundamentally physico-chemical and that the laws of physics and chemistry are applicable to biological phenomena. Nevertheless, emergentism holds that the various levels of organisation (physical, biological and psycho-sociological) are characterised by the acquisition of new and specific properties (emergent properties). These properties increase the degree of complexity of a given level compared with the various levels of which it is composed (hierarchical organisation). For this reason, even if physics and chemistry are normally applicable to, say, ecological phenomena, each level of organisation requires appropriate laws and theories that allow for an understanding of the specific properties of that particular level. By contrast, reductionism denies the existence of emergent properties or else considers them an epiphenomenon strictly dependent on the state of our knowledge – what is emergent today will lose its emergent character tomorrow (Hempel and Oppenheim 1948, pp. 149–151).

These ontological assumptions have, of course, significant consequences in the methodological and epistemological domains. In the methodological domain, the two perspectives view the analytical method in a very different way.

---

[1] In this context the word 'epistemology' connotes the more limited and specific meaning of the research domain concerning the relationships among theories and laws that belong to different organisational levels. In other words, it is characterised by the epistemic challenge of "heterogeneous reduction", or "theoretical reductionism" (Ruse 1988).

Reductionism considers that at a given level of organisation, analytical study of constituent parts and their relationships is necessary and sufficient to predict, or at least explain, all the properties of that level. Fundamentally, reductionism is a "bottom-up" strategy. It takes into account the level at which the events to be explained occur (ecological phenomena, for instance) as well as the lower levels that contribute to that explanation (for example, genetics, chemistry or physics). An analytical and additive method, therefore, dissects the entity, or decompose the process, under examination into its component parts, or phases, and attempts to take into consideration the relationships among them. A successive summation of the individual component properties or interactional properties should allow extrapolation of the global properties of the entity as a whole. In some cases, this dissective and synthetic process should allow us to formulate some more general theories or laws.

Methodologically, the emergentist approach, while recognising the need for analysis, considers its explanatory power limited. In fact, according to an emergentist and hierarchical perspective, the feedback loops that link different levels of organisation play a role of utmost importance in the determination and causation of the emergent properties. From a methodological point of view, the higher and lower levels adjacent to the primary object of study are considered differently than in methodological reductionism. This approach does not limit the analysis to the constitutive parts of – or their relationships in – a specific level of organisation. In other words, for this "top-down" approach, both the higher levels (downward causation) and the lower ones participate in determining the properties of specific levels. Thus, a multi-level triadic approach – where at least three levels of organisation are considered simultaneously – is held to be a methodological necessity and is the main characteristic of the emergentist methodology (Feibleman 1954; Campbell 1974; Salthe 1985; Bergandi 1995; El-Hani and Pereira 2000).

Epistemologically, reductionism is a mono-directional bottom-up explanatory strategy. This approach is directly descended from nineteenth century positivism and from neo-positivism (1920s and 1930s). In its struggle against the intrusiveness of metaphysics in science, neo-positivism sought a unification of science based on the language, laws and theories of physics. Epistemological reductionism maintains that the theories and laws of a specific organisational level can be – and sometimes must be – "reduced" to the theories and laws of a more "fundamental" field of science (Woodger 1952; Nagel 1961; Levins and Lewontin 1980; Bunge 1991; Jones 2000). According to this epistemological perspective, an ideal scientific development will involve, in the long run, the "de-substantialisation" of non-fundamental sciences. For instance, taking into account the relationships between ecology (secondary science) and physics (primary science), ecological laws and theories could be reduced to physical laws and theories (heterogeneous reduction). Were this to occur, the process of integration, incorporation and absorption of ecological phenomena in the physical domain would provide a larger and clearer understanding of all the phenomena that previously constituted the objects of ecological research. Such a hypothetical reduction would determine the birth of a new and more meaningful physical science, emerging from the "dilution" of biology into physics. And, as Popper pointed out, such a successful reduction is substantially unattainable because it would imply a "complete" theoretical understanding of life in physical terms (1972).

Epistemological holism, on the other hand, posits a more dialectical relationship between laws and theories belonging to different organisational levels. On the one hand, this perspective holds that there is no scientific domain to which the other sciences should be reduced. According to emergentist ontology, every organisational level has one or more emergent properties that are correlated to specific laws and theories which, in turn, are assumed to be intrinsically non-reducible. On the other hand, according to Quine (1961, p. 42) "the unit of empirical significance is the whole of science".[2] This means that the existence of anomalies that cannot be explained in terms of existing knowledge requires us to make adjustments to science as a whole. In other words, a transformation in any scientific domain, and not only in the "fundamental" sciences, can determine changes in any other domain of science. This perspective entails rejecting the physical explanation as the fundamental and preferred form of explanation to which the other sciences have to be reduced.

In sum, it is possible to identify the foundational, philosophical core of all materialistic emergentist views of reality using the following criteria, which correspond to different semantic domains:

Ontology

1. *Holism*: Not all holistic positions are emergentist, but all emergentist views are holistic. *Holism fundamentally means the intrinsic, structural, spatio-temporal interdependence of phenomena*[3] and constitutes the major and inescapable ontological presupposition of emergence.
2. *Levels of organisation*: *Reality is a hierarchical, multi-layered, multi-level process*. According to this interpretation of reality every level of organisation (or integration) is characterised by specific emergent properties, qualities or behaviours. This ontological perspective can be interpreted according to a realistic view – the levels with their emergent properties definitely represent reality – or a constructivist one – the levels of organisation are *"levels of description"* of reality: we identify levels, and attribute specific properties to them, according to the purpose of our research.
3. *Novelty*: *The emergent properties of every level of organisation express new qualities and a new order of phenomena compared with the level of organisation on which they depend and from which they emerge.*

---

[2]It is worth pointing out that, unlike the Quine thesis, the holistic reference of the Duhem thesis is the whole of physics. Its working has been described according to an organicist perspective: in physics, as in an organism, all the theories work together, even if they are not all called into play at the same level of intervention (1977, pp. 187–188).

[3]To avoid any risk of misunderstanding, it would be more appropriate to use the term 'holism' to indicate specifically the relational view of reality according to which natural (or social) reality is constituted by spatio-temporal interdependent entities. Its logical opposite is the ontological atomistic view.

Methodology

4. *Avoiding the fallacy of "misplaced concreteness"* (Whitehead 1925; Dewey and Bentley 1949). This is a basic prerequisite for any emergentist constructivist methodology. There is a preliminary heuristic assumption that all analytical distinctions concerning "wholes", "parts", and "relations" are pure theoretical "mind constructions" which have meaning only in relationship to the specific aims of the inquiry. Consequently, wholes, parts and relations must not necessarily be considered to have an intrinsic ontological reality, merely an epistemic one (see Bergandi 2007).[4]
5. *Multi-level approach*: To explain the emergent properties of a specific level of organisation or system, the adjacent lower and higher levels must be considered as significant as, and simultaneously with the level of the primary object of research. This triadic approach is not a luxury but a necessity for any research claiming an emergentist approach. In fact, to restrict to take into consideration the lower level relationships among elements is equivalent to enacting a reductionist methodology.
6. *Fallacious attribution of emergent properties*: The constructivist background (see (4) above) should always be borne in mind in the attribution of emergent properties to a level of organisation. The hypothesis that these properties cannot, in reality, be effectively attributed to the constituent parts, sub-systems or higher inclusive levels must be carefully refuted. In fact, any potentially erroneous attribution of an emergent property could result from an incomplete or wrongheaded analysis of the whole hierarchical structure.

Epistemology

7. *Unpredictability*: The emergent properties of a level of organisation cannot be predicted, even in principle, by even the most complete knowledge of the parts, properties and relationships among the parts.[5] In other words, a specific organisation of matter is correlated to exclusive properties. To be able to explain them would require the constitution of a new or reorganised scientific discipline which would use new postulates, theories and laws that introduce new terms and patterns suited to the emergent phenomena and properties.[6]

---

[4] It is worth recalling that constructivism does not deny the existence of a reality (natural, social, and so on). Rather, this perspective foregrounds the idea that *within* this reality, thanks to our epistemic constructs, we identify or recognise certain characteristics, aspects and processes that are functional to our aims and objectives (scientific, social, and so forth).

[5] This is an elliptic formulation; the correct one is the following: the laws concerning the emergent properties of a level of organisation cannot be predicted, even in principle, by the laws concerning the lower level relations between the constituent parts.

[6] For instance, even the most radical reductionist could not explain biological evolution by referring only to the overall theoretical package of physics and chemistry; according to Williams: "*at least the one additional postulate of natural selection and its consequence, adaptation, are needed*" (Williams 1966, p. 5; see also 1985, p. 1).

From Organicism to the Oxymoronic "Reductionist Holism" of Ecosystem Ecology

From its beginning, ecology has been structured within a holistic ontological framework. Ecology is most widely known as the science that concerns the relationships between organisms and their environment, that is, a science interested in all the conditions that permit organisms to live (Haeckel 1866). Early on, this holistic framework mainly took the form of an organicist worldview. Representative in this regard are the works of Stephan A. Forbes, Frederic E. Clements and John Phillips.

Some years after the far-sighted definition of ecology by Ernst Haeckel, Stephan Alfred Forbes wrote two papers that vividly portrayed the complex, intricate relationships between organisms and their environments. In *On some interactions of organisms* (1880) and *The lake as a microcosm* (1887; for the concept of "microcosm" as a central metaphor in ecology, see Schwarz 2003), Forbes was among the first to put forward the idea that natural systems exist in a state of equilibrium and must be studied "as a whole". He also delineated a strict connection between natural selection and the laws of oscillation of plant and animal species. He suggested that the functional relations among organisms were comparable to the relations between organs within an animal's body. Any change (in numbers, habits or distribution) within a specific plant or animal group will impact various other groups "in a far extending circle" (1880, p. 3). In the struggle for existence under the influence of natural selection, predator and prey species ordinarily find a balance and, to a certain extent, adjust their rates of reproduction accordingly. They have common interests: an excessive increase in a predator species will inevitably determine a decrease in the very species that constitute its food supply and consequently a decrease in its own species. However, Forbes also thought that in the intricate network of relationships between organisms on the one hand and between organisms and their environments on the other, the real limits to excessive multiplication of a species are to be found in the inorganic features of its environment (*Ivi*, 11, p. 16).

The "lake" was presented by Forbes as the paradigmatic case of a relatively isolated system in which the "organic complex", the species assemblage, could not be studied without taking into account all the forms of relationship between different species (predator/prey, competition, mutualism, and so forth) belonging to the lake and the surrounding terrestrial system (1887, p. 537). In other words, prior to the trophic ecology of Elton (1927) and Lindeman (1942), Forbes considered that when studying a carnivorous lake fish, one must also take into account the species upon which it depends for its existence, the organic and inorganic conditions upon which these species depend, the other competitor species, as well as the entire system of conditions affecting the existence of the plant and animal species that contribute to the existence of a specific group of related species (see also Forbes 1914).

The work of other ecologists, including Frederic E. Clements, John Phillips, Henry A. Gleason and Arthur G. Tansley, shows traces of the influence of a specific version of what we nowadays call the holism-reductionism debate. In the competition over the epistemological determination of ecology, individualistic (Gleason), anti-organicist and anti-emergentist (Tansley) supporters were ranged against the upholders of organicist holism (Clements, Phillips; Bergandi 1999; see also Chap. 5).

In the search for the fundamental units of nature, plant ecology played a role of utmost importance. Various units succeed each other: the biome, the climax, plant associations and the biotic community. According to Clements (1916) the climax formation is an organic entity. The formation grows, develops and dies as an organism. Later, Phillips, following his committed organicist position (1931, 1934, 1935a, 1935b), was to consider the same biotic community as an organism. The analogy between the organism and the unit of vegetation, the formation or the biotic community enabled Clements and Phillips to extrapolate certain characteristics from the first element to the second, albeit with the risk of transforming a relative similarity into an identity relationship for all aspects – in doing so, there is always the danger of running into an intellectual dead end. While we have never seen an organism grow younger, an environmental modification (soil desegregation, for example) can determine an ecological regression, in other words, species impoverishment. However, the phenomenological reading of ecological organicism hides a more fundamental level of interpretation. These authors, in reality, wanted to point out the holistic ontological dimension of ecological entities. In other words, they sought to underscore the "organisational" idea that is inherent in biotic entities. From this point of view, the influence of philosophical organicism is not to be completely ruled out. It is interesting to note that the organicist and emergentist philosophical works of Herbert Spencer, Alfred N. Whitehead, Samuel Alexander, Conwy L. Morgan and Jan Smuts are all quoted by Phillips and Clements, even if in relatively late papers (Phillips 1931, 1935b; Clements 1935: see Bergandi 1999).

Moreover, it is noteworthy that Forbes on the one hand and Clements and Phillips on the other support different forms of organicism. Forbes supports a conception of a biotic community that, while certainly holistic and expressed in organicist terms, is substantially pre-emergentist. His analysis stresses the interactional dimension between organisms and between organism and environment, whereas Clements and Phillips are proponents of an organicist perspective which clearly involves the idea of emergence. For instance, Clements emphasises not only that: (1) a plant formation is of itself an organism; (2) the climax is the mature stage of the formation; but also that (3) "the reaction of a community is usually more than the sum of the reactions of the component species and individuals", in the sense that the community naturally produces a cumulative amelioration of the habitat that would not be possible without the combined action of the individual plants belonging to the group (1916, pp. 3, 79, 106).

By contrast, Tansley, in a highly paradoxical way, departed from Clements and Phillips' organicist perspective by proposing the "ecosystem" concept, which revealed itself to be a more integrative, holistic entity – the physical system constituted by the organisms and physical factors. But this proposition neither involved the disappearance of the other proposed units nor, in its refusal of organicism, was it able definitively to overcome this epistemological framework (on the definitions of ecological units, see Jax et al. 1998; Jax 2006). In fact, Tansley identified the ecosystem as a "*quasi-organism*" (1935). In reality, what was at stake was not only the potentially misleading use of the word "organism", but above all the principle's unpredictability as implied in the organicist community worldview of Clements and Phillips (Tansley 1935, pp. 297–298). According to Tansley, even if the community is composed of organisms in mutual association, examination of this entity must be conducted using an analytic and anti-emergentist perspective. The Tansley refusal of the Clementsian worldview followed in the footsteps of Gleason's refusal. Gleason (1917, 1926) maintained an atomistic and individualistic point of view on plant association. The lack of limits and structure of the associations was the fundamental reason that pushed him to see in these ecological entities the result of random immigration and environmental variations. This unoriented, random juxtaposition of plants determined structurally different forms of associations, and that required the total acceptance of analysis as a direct methodological consequence. Organisms and populations were studied separately, and their association was reducible to the various isolated plant functions.

The Tansleyan ecosystem concept has had a decisive influence upon successive phases of the development of ecology. His categorisation of the "basic unit of nature" was later to be rendered dynamic thanks to Lindeman's energetic thermodynamics approach (1942), an analytical and additive method that explained the ecosystem in terms of energy exchanges among the different compartments in the biotic community and between the community and the physical environment. Between the 1950s and 1960s, the Odum brothers developed an ecological paradigm that combines this energetic ecosystem framework with a holistic and emergentist ontology (1953, 1959, 1971: see also 1983, 1993).

The following phrase clearly sums up Eugene Pleasants Odum's ontological, methodological and epistemological assumptions.

> Just as the properties of water are not predictable if we know only the properties of hydrogen and oxygen, so the characteristics of ecosystems cannot be predicted from knowledge of isolated populations; one must study the forest (i.e., the whole) as well as the trees (i.e., the parts). Feibleman (1954) has called this important generalization the 'theory of integrative levels'. (1971, pp. 5–6)

In other words, ecosystems are complex entities characterised by emergent properties that cannot be predicted by strictly applying the analytical method. At the same time, Odum considers his ecology to be the true expression of a holistic approach: "Practice has caught up with theory in ecology. The holistic approach and ecosystem theory, as emphasized in the first two editions of this book, are now matters of world-wide concern." (Odum 1971, p. VII). The issue here is the following: the Odumian holistic approach takes into account "the ecosystem as a whole"; but what, precisely, is this "whole"? Is it a matter of ecology, physics or some other scientific discipline?

In addition, it is interesting to note that Odum considers that "the findings at any one level aid in the study of another level, but never completely explain the phenomena occurring at that level" (1959, p. 7; 1971, p. 5). Having said this, Odum seems to deny any value of epistemological reductionism, considering that ecosystem ecology is not reducible to physics. At the same time, it is a matter of fact that

Eugene Paul Odum, collaborating with his brother Howard Thomas Odum, locates the theoretical core of systems ecology in energetic analysis:

> In ecology, we are fundamentally concerned with the manner in which light is related to ecological systems, and with the manner in which energy is transformed within the system. Thus, the relationships between producer plants and consumer animals, between predator and prey, not to mention the numbers and kinds of organisms in a given environment, are all limited and controlled by the same basic laws which govern nonliving systems, such as electric motors or automobiles. (1971, p. 37; see also Chap. 18)

The Odums' epistemological manifesto has been so effective that from then onwards ecology has been perceived and presented as the holistic science *par excellence*.[7] In referring to a philosopher of science, Jerome K. Feibleman, they outline a hierarchical worldview where every level of organisation is characterised by a specific degree of complexity and properties that are not predictable or explicable from the study of the lower levels alone (epistemological holism). Implicitly in their early works and explicitly in the later ones (Odum 1993), the emergence concept and an emergentist ontology are the cornerstones of the Odumian ecosystem paradigm. Methodologically, however, they ran into an incoherence that unbalances their whole theoretical edifice. There are three reasons for this. First, the difference between collective and emergent properties escapes the Odum brothers, at least in their early work. Some population and community properties (density, age distribution, natality, mortality, species diversity, etc.) – even if expressed as statistical functions – are considered as unique characteristics of the group. In all these cases the properties, even if they must be considered as group statistical functions, are determined through the examination of the components using classical analytical and additive methodology (Salt 1979). Second, the physicalist background of the Odums' systems ecology stands in contradiction to emergentist ontological assumptions. For instance, they consider the outcome of the Eniwetok Atoll energy evaluation (Odum and Odum 1955; Odum 1977) to be an emergent property. Thus, they are considering ecological systems as structured physical entities, forgetting that their specificities are not reducible to the physical domain. Finally, a true emergentist approach will be necessarily a multi-level triadic approach that considers simultaneously at least the lower and higher adjacent levels in addition to the level at which the main object of research is to be studied phenomenologically. Instead, in the Odums' work the affirmed importance of the emergent properties of ecological systems is not coupled with a corresponding emergentist methodology — at least not until Odum and Barrett (2005, p. 8), where the necessity of a genuinely triadic emergentist methodology can be clearly recognised. However, the previous Odumian approach is fully legitimate (see the article by Chap. 15). It is nevertheless a kind of crypto-reductionist systemism or, to put it in oxymoronic terms, a kind of reductionist holism, that can at best be considered as "*holological*" (Hutchinson 1943), and not as the true expression of a holistic, emergentist methodology and epistemology. Hutchinson proposes making the distinction between holological and

---

[7]The term "holism" was to appear from the third edition (1971) onwards, even if the corresponding worldview had already been outlined in previous works.

merological approaches. In a system investigated with a holological approach "(...) matter and energy changes across its boundaries are studied", whereas with a merological approach "(...) the behavior of individual systems of lower order composing (the system) S are studied" (1943, p. 152). However, it is worth noting that the holological approach is an expression of a systemic and yet physicalist perspective, while the mereological one, methodologically speaking, is a strict expression of an analytical-additive reductionist perspective. McIntosh (1985, pp. 199–213), Taylor and Blum (1991, p. 284; see also Taylor 2005) were among the first to analyse the Janus-like character of the ecosystem ecology represented by E.P. Odum: they saw it as a *"functionally holistic"* new ecology, which was essentially, however, expressed through system modelling involving the physical attributes of ecosystems.

Conclusion

The holism-reductionism debate in ecology is, without a doubt, a protean issue. In ecological studies, first, the debate took a number of forms: an organicist worldview that expressed the holistic, systemic relations existing between organisms, and between organisms and their environment (Forbes); an organicist and emergentist view of plant communities (Clements, Phillips); a view of plant associations as individualistic, atomistic, randomly generated entities (Gleason); and, finally, Tansley's integrative "ecosystem" concept that expressed the epistemological refusal of Clementsian organicism and emergentism.

Second, it showed itself in the form of the acceptance or refusal of physicalism. If we broach the epistemological nature of ecology and come to the conclusion that ecology is fundamentally a holistic science, we would be mistaken in thinking that, methodologically, ecology necessarily embodies an emergentist approach. In fact, an "emergentist holistic" approach need be understood neither as a reiteration nor as a tautology. This distinction is of utmost importance. It enables us to avoid all the inconsistencies inherent in the Odumian paradigm and all paradigms that propose a holistic ontology but that, in practical methodological research, deploy the full panoply of reductionism. In fact, in the history of science, a holistic and emergentist ontology is not always applied consistently in emergentist methodology and epistemology. For instance, once we cease to consider ecosystem ecology as the expression of a "holistic" attitude and recognise in it instead a "holological" framework, a kind of oxymoronic "reductionist holism", then we will avoid misunderstandings and be back on track. We will then be free to construct a truly consistent holistic and emergentist ontological, methodological and epistemological framework.

Finally, to sum up from a strictly epistemological point of view, one of the major implications of the holism-reductionism debate is the confrontation between two philosophies that at first sight support a shared hierarchical worldview of natural reality. However, there is one very important difference. From the reductionist point of view, the ideal point to reach is that all the scientific disciplines will, sooner or later, be formulated, interpreted and reduced to the more fundamental sciences, particularly physics. From the holistic or, more correctly, emergentist point of view, the supposed ontological natural hierarchy does not involve a hierarchical relationship between the scientific disciplines but rather a systemic one. The sciences with their specificities and particularities allow us to grasp different aspects of reality which we cannot reduce to one another, but which we can combine in order to arrive at a non-definitive, ever-changing picture of reality.

For emergentists, the universe is a growing entity that generates ever new phenomena, events and qualities which can be neither predicted nor deduced from those that preceded them. We must remember, however, that an a priori unpredictability does not necessarily involve the refusal of an (ideal) a posteriori explanation. On the contrary, according to anti-emergentists "nothing is new under the sun" and, above all, any so-called novelty is predictable and explicable: the same universe, yet two antinomic and incommensurable worldviews. Which paradigm is closer to reality? Are the reductionists correct when they claim emergents are epiphenomena? Are the emergentists wrong when they attribute an ontological status to emergence and, above all, axiomatically assert its a priori unpredictability? These are open questions to which the answers will probably never be given once and for all but always case by case.

Finally, to grasp the logical structure of alleged emergence, we must ask ourselves: what are the emergents – properties, relations, entities or laws? What is the level of organisation that bears the property which is supposed to be emergent? In addition, the levels of organisation – and among them, significantly, certain ecological levels such as the ecosystem or landscape – must be understood as levels of description, or "methodological abstractions". These epistemological fictions sometimes make it possible to develop models that allow us to approach natural reality "asymptotically". Otherwise we risk an insidious epistemological fallacy: a hypostatisation of our abstractions that brings us to project our hypotheses and theories onto reality, forgetting that they are merely notional tools by which to approach it. A metaphor may help to clarify this idea: it is like a dog that starts to play with you but forgets, in the excitement of the game, that it is playing and begins to bite in earnest. In other words, a constructivist epistemology prevents us from being bitten by the rock-hard certitudes of naive realism. Our scientific constructs make it possible to approach natural reality without ever fully grasping it. These constructs allow us to understand certain aspects of reality in a non-definitive way. They remain valuable until such time as new constructs allow us to get even closer. This is a genuine process of scientific knowledge where the syntagm "The End" will never be written.

References

Ayala FJ (1974) Introduction. In: Ayala FJ, Dobzhansky T (eds) Studies in the philosophy of biology. Reduction and related problems. MacMillan, London, pp vii–xvi
Beckermann A, Flor H, Kim J (1992) Emergence or reduction? De Gruyter, Berlin
Bergandi D (1995) 'Reductionist holism': an oxymoron or a philosophical chimaera of E.P. Odum's systems ecology. Ludus Vitalis, 3, 5, pp 145–180; reprinted in Keller DR, Golley

FB (eds) (2000) The philosophy of ecology: from science to synthesis. University of Georgia, Athens (abridged version), pp 204–217

Bergandi D (1999) Les métamorphoses de l'organicisme en écologie: de la communauté végétale aux ecosystems. Revue d'histoire des sciences 52(1):5–31

Bergandi D (2007) Niveaux d'organisation: évolution, écologie et transaction. In: Martin T (ed) Le tout et les parties dans les systèmes naturels. Vuibert, Paris

Bunge M (1991) The power and limits of reduction. In: Agazzi E (ed) The problem of reductionism in science. Kluwer, Dordrecht, pp 31–49

Campbell DT (1974) 'Downward causation' in hierarchically organized biological systems. In: Ayala FJ, Dobzhansky T (eds) Studies in the philosophy of biology. MacMillan, London, pp 179–186

Clements FE (1916) Plant succession: an analysis of the development of vegetation. Carnegie Institution, Washington, DC, p 242

Clements FE (1935) Experimental ecology in the public service. Ecology 16:342–363

Dewey J, Bentley AF (1949) Knowing and the known. Beacon, Boston

Duhem P (1977) The aim and structure of physical theory. Atheneum, New York

Elton CS (1927) Animal ecology. Sidgwick & Jackson, London

El-Hani CN, Pereira AM (2000) Higher-level descriptions: why should we preserve them? In: Andersen PB, Emmeche C, Finnemann NO, Christiansen PV (eds) Downward causation: minds, bodies and matter. Aarhus University Press, Aarhus, pp 118–142

Feibleman JK (1954) Theory of integrative levels. Br J Philos Sci 5:59–66

Forbes SA (1880) On some interactions of organisms. Ill Nat Hist Surv Bull 1(3):3–17

Forbes SA (1887) The lake as a microcosm. Ill Nat Hist Surv Bull 15(9):537–550

Forbes SA (1914) Fresh water fishes and their ecology. Illinois State Laboratory of Natural History, Urbana (read at the University of Chicago, August 20, 1913)

Gleason HA (1917) The structure and development of the plant association. Bull Torrey Bot Club 44:411–462

Gleason HA (1926) The individualistic concept of the plant association. Bull Torrey Bot Club 53:7–26

Haeckel E (1866) Generelle Morphologie der Organismen. Allgemeine Grundzüge der organischen Formen-Wissenschaft, mechanisch begründet durch die von Charles Darwin reformirte Descendenz-Theorie. Reimer, Berlin

Hempel CG, Oppenheim P (1948) Studies in the logic of explanation. Philos Sci 15:135–157

Hutchinson GE (1943) Food, time, and culture. N Y Acad Sci 15:152–154

Jax K (2006) Ecological units: definitions and application. Q Rev Biol 81(3):237–258

Jax K, Jones CG, Pickett STA (1998) The self-identity of ecological units. Oikos 82(2):253–264

Jones R (2000) Reductionism: analysis and the fullness of reality. Bucknell University, Lewisburg

Levins R, Lewontin R (1980) Dialetics and reductionism in ecology. In: Saarinen E (ed) Conceptual issues in ecology. D. Reidel, Dordrecht, pp 107–138

Lindeman RL (1942) The trophic-dynamic aspect of ecology. Ecology 23:399–418

McIntosh RP (1985) The background of ecology. Concept and theory. Cambridge University Press, Cambridge

Mayr E (1988) Toward a new philosophy of biology. Belknap/Harvard University Press, Cambridge

Nagel E (1961) The structure of science: problems in the logic of scientific explanation. Brace and World, New York, Harcourt

Odum EP (1953) Fundamentals of ecology. W.B. Saunders, Philadelphia

Odum EP (1959) Fundamentals of ecology. W.B. Saunders, Philadelphia

Odum EP (1971) Fundamentals of ecology. W.B. Saunders, Philadelphia

Odum EP (1977) The emergence of ecology as a new integrative discipline. Science 195:1289–1293

Odum EP (1983) Basic ecology. W.B. Saunders, Philadelphia

Odum EP (1993) Ecology and our endangered life-support systems. Sinauer, Sunderland

Odum EP, Barrett GW (2005) Fundamentals of ecology, 5th edn. Thomson brooks, Belmont
Odum HT, Odum EP (1955) Trophic structure and productivity of a windward coral reef community on Eniwetok Atoll. Ecol Monogr 25:291–320
Phillips J (1931) The biotic community. J Ecol 19:1–24
Phillips J (1934) Succession, development, the climax and the complex organism: an analysis of concept. J Ecol 22(1):554–571
Phillips J (1935a) Succession, development, the climax and the complex organism: an analysis of concept. J Ecol 23(2):210–246
Phillips J (1935b) Succession, development, the climax and the complex organism: an analysis of concept. J Ecol 23(3):488–508
Popper KR (1972) Objective knowledge: an evolutionary approach. Clarendon, Oxford
Quine VOW (1961) From a logical point of view. Harvard University Press, Cambridge
Ruse M (1988) Philosophy of biology today. State University of New York, Albany
Salt GW (1979) A comment on the use of the term emergent properties. Am Nat 113:145–149
Salthe SN (1985) Evolving hierarchical systems: Their structure and representation. Columbia University Press, New York
Schwarz AE (2003) Wasserwüste – Mikrokosmos – Ökosystem. Eine Geschichte der Eroberung des Wasserraumes. Rombach-Verlag, Freiburg
Tansley AG (1935) The use and abuse of vegetational concepts and termes. Ecology 16(3):284–307
Taylor PJ (2005) Unruly complexity: ecology, interpretation, engagement. The University of Chicago, Chicago
Taylor PJ, Blum AS (1991) Ecosystem as circuits: diagrams and the limits of physical analogies. Biol Philos 6:275–294
Whitehead AN (1925) Science in the modern world. MacMillan, New York
Williams GC (1966) Adaptation and natural selection: a critique of some current evolutionary thought. Princeton University Press, Princeton, New Jersey
Williams GC (1985) A defense of reductionism in evolutionary biology. In: Dawkins R, Ridley M (eds) Oxford surveys in evolutionary biology, vol 2. Oxford University Press, Oxford, pp 1–27
Woodger JH (1952) Biology and language. Cambridge University Press, Cambridge

Chapter 5
The Classical Holism-Reductionism Debate in Ecology

Ludwig Trepl and Annette Voigt

Introduction

Controversies between holism and reductionism are a familiar feature in many fields of inquiry besides ecology. Although seldom described in these terms,[1] the issue has long played – and continues to play – a significant role in science, philosophy and political ideology; indeed, one could almost say that it has always existed. It played a major part, for example, in shaping the ideological conflicts between conservatism and liberalism in the nineteenth and twentieth centuries. Holistic ideas were present especially in "Lebensphilosophie" (philosophy of life) and in other philosophies critical of science, such as historicism, that shaped the zeitgeist at the time of ecology's emergence. Towards the end of the twentieth century, holism became a significant force in "political ecology". This is true not only of those strands of political ecology and related fields that are explicitly committed to the task of "renewing" the world view of their time, such as "Deep Ecology" (Naess 1973; Drengson and Inoue 1995) and the "New Age Movement" (Capra 1982); rather, the view of nature within all political ecology is essentially a holistic one. However, a range of conceptual figures associated with holism can be found in many older philosophies as well. The macro-microcosm figure, for example, appears as far back as Plato (cf. Schwarz 2003) and came to exert influence within modernity mainly through the world view that underlies Leibnizian rationalism (cf. Eisel 1991; Langthaler 1992). Nowadays research programmes in most of the natural sciences are shaped by reductionist ideas; they are linked primarily to neo-positivist philosophies and can be traced back to older empiricist philosophies, as well as to Cartesian rationalism.

The holism-reductionism debate in ecology can be properly understood only against this non-scientific backdrop, because what is at issue in ecology is more than

---

[1] The term "reductionism", for example, became common only in the middle of the twentieth century, even though it covers older philosophical problems, which previously came under the rubric of "materialism" or "mechanism", and the methodology of specific sciences (cf. Stöckler 1992).

A. Voigt (✉)
Urban and Landscape Ecology Group, University of Salzburg, Hellbrunnerstraße 34,
5020 Salzburg, Austria
e-mail: annette.voigt@sbg.ac.at

just whether scientific theories of a certain type provide a correct description of certain natural phenomena. What these philosophical and political-ideological controversies demonstrate above all is that the issue is of relevance in areas that lie well beyond the confines of the discipline itself. It therefore appears justifiable to us to analyse a wide range of debates on the basis of the holism-reductionism complex, including those that are not explicitly about it at all.[2] The significance of the complex is by no means exhausted in the impact it has on ecology or on understandings of ecology.

The holism-reductionism controversy in ecology is all about the relationship between wholes and their parts. This is a major problem in many sciences – in physiology (the organism), geography (the landscape), psychology (the soul) and sociology (society), as well as in physics, linguistics and epistemology. However, the debate in ecology is of particular interest when it comes to understanding the way holistic and reductionist ideas about nature and society work in the context of political-ideological controversies. The ecological debate has one thing in particular in common with a very few sciences (with sociology above all, though not with other natural sciences), something that links it very closely to those ideological struggles, and that is this: ecology is one of those sciences whose objects of inquiry are constituted in such a way that the parts are usually individuals and the whole a community. The contrast between reductionism and holism in ecology tends to take the form of an opposition between individualism and organicism. In turn, individualism, as a form of reductionism, is holistic in the sense that while individuals are thought of as wholes (as organisms), a community is not. "Community", according to individualism, is merely a name for a certain number of individuals, gathered together more or less at random by the scientist, who are considered to be "autonomous" and who alone are seen as being real. In organicism, by contrast, "community" is conceived of as an organic community or as a superorganism, in other words, the relationship between the part and the whole is conceptualised in analogy to the relationship between organ and organism. – Whenever organicism or the "organismic concept" is mentioned in ecology, the reference is to something different from what, in biological terms, is commonly called the "organismic approach" or "organismic biology". The idea entailed by the former is that a "Lebensgemeinschaft" (biotic community) is of the same character as an individual organism; in the latter, importance is generally attached instead to the level of the *individual organism*, the point here being to counter the tendency to focus solely on the molecular level.

In the following we shall refer to only a few of the many variants of the holism-reductionism debate in ecology and to certain stages of their transformation. Our main concern is to reconstruct the *logic* of the debate and to determine the *conceptual structures* on which it is based. Given that the positions adopted address a problem that not only exists within science but is above all of a fundamental philosophical nature, we can interpret the actual emergence of such positions and their transformation in ecology in the following way. There are certain ways of conceptualising the relationship between parts and wholes, or between individuals and community – there is not an indefinite number of variants of these conceptual

---

[2] Cf. Trepl (1994) and the critical response by Levins and Lewontin (1994).

figures. What we are interested in are the conditions in which certain combinations of their elements and certain transformations in their structure are possible; above all, we are interested in the practical and ideological implications of these conceptual figures. We do not intend to present every single "important" theory that has ever made an appearance in the history of ecology; instead, our aim is to construct ideal types. These enable us to present the positions that were actually adopted throughout history and to compare them in a systematic way. Our main criterion for selecting the examples is less that they are considered, for whatever reasons, to be important in ecology – the fact that they have been influential, for example – but rather that they are suitable for explicating the ideal type constructions.

Numerous potential variants of holism and reductionism exist and can be found in biology. We mention this briefly at the start. The main point here is that the spectrum of potential variants is by no means exhausted by organicism and individualism. Than, we present reconstructions of ideal types of each in its classical form as they emerged during the first few decades of the twentieth century, using examples by way of illustration. Since the dispute between them cannot be resolved empirically, we inquire as to whether this might be possible at the methodological level. It proves to be difficult at this level as well. We choose an approach based on a theory of constitution,[3] which enables us to regard both holism and reductionism as being "inspired" by certain world views. On this basis it becomes easier to understand the utterly different practical consequences entailed by holistic ecological theories on the one hand and reductionist theories on the other. We conclude with an example that describes the dynamic through which both approaches (usually in response to one another) change.

Variants of Holism and Reductionism

The literature refers to holism and reductionism in many very different ways.[4] In the following we take a few examples and explain briefly how they work conceptually on the basis of what is common to all forms of holism and reductionism. Our examples are restricted to those variants that play a role in biology, so they are about explaining "life".

---

[3] Constitution here is understood differently from the sense common in philosophy, that is, the way Kant, for example, used it. Instead, it refers to the idea that scientific theories are not simply generalised depictions of specific empirical observations but that they owe their existence to non-scientific conceptual structures already in existence. Thus, we can speak here of a "l´a priorí historique" (historical apriori) (Foucault 1969), of "bereitliegenden kulturellen Deutungsmustern" (cultural patterns of interpretation that are already available) or of "Konstitutionsideen" (ideas of constitution) (Eisel 2002, p. 130), which are the conditions of possibility – realized through culture – for scientific concepts and for their corresponding objective experiences. Indeed it is these conditions which ensure that new facts do not destroy the old conceptions in general but consistently confirm the theory (or the paradigm) (Eisel 2002; cf. Kuhn 1962).

[4] Some examples of literature that takes this issue further include: (science in general) Nagel 1949, 1961; Bueno 1990; Agazzi 1991; (biology) Ayala 1974; Ayala and Dobzhansky 1974; Ruse 1973; Hull and Ruse 1998; Bock and Goode 1998; Looijen 2000; (ecology specifically) Saarinen 1982; Bergandi 1995; Bergandi and Blandin 1998; Keller and Golley 2000; Kirchhoff 2007; Voigt 2009.

It is probably true to say that what links all those things together that are referred to as holism is not much more than the principle that the "whole" has "priority" over the "parts" – whatever "priority" might mean exactly – and a set of reservations about any form of "simplification". On the reductionist side, the commonality between different positions probably consists above all in their emphasising that statements about phenomena of a complex nature should be derived from statements about phenomena of a simpler nature, and that science essentially consists in this kind of "reduction".

Different forms of holism and reductionism also come about depending on the aspects of the research object (or of the scientist's relationship to that object) to which these principles are applied. In other words, it is not only certain methods or research programmes that can be called holistic or reductionist, but also certain views about the "nature" of their objects. This means that whether a certain position appears to be holistic or not depends on the perspective taken. We shall discuss here just a few of the numerous permutations possible: "wholeness" can be taken to mean a variety of very different things (2.1); "simplification" can mean very different things (2.2); the assertion that something is reductionistic or holistic may be a reference – among other things – to the nature of reality or to how we should proceed in order to find something out about it (2.3).[5]

Aspects of the Concept of "Wholeness"

Those methods and theories that are termed holistic differ greatly according to which aspect of the concept of wholeness they highlight. In biology, such aspects include totality, gestalt, uniqueness, system character and "Lebendigkeit" (aliveness) – although often many of these aspects cannot be separated from one another.

In the case of "aliveness", the choice to focus on one or the other aspect has far-reaching methodological consequences. For example, the whole can be identified as an "inner essence" (e.g. "soul"); one point of access to this wholeness can be seen as being that the relationship between the inside, which remains hidden, and the outside, which is perceived, is one of the latter giving expression to the former. It then becomes possible to draw on methodologies from the human sciences whose aim is to "understand" this inner essence through its representation in the external world (especially Dilthey's "Ausdrucksverstehen" around the turn of the twentieth

---

[5] Other levels in addition to the ontological and methodological would be the epistemological level ("Erkenntnistheorie"), which is about the validity of knowledge, and the level of theory of science in the narrow sense ("Wissenschaftstheorie im engeren Sinne"), which is about the character of the empirical phenomenon of science. One might also add a level of theory of constitution, at which the independence of the ontological and epistemological level disappears. One could, for example, describe the theory of Thomas Kuhn as on the level of theory of science holism in the narrower sense mentioned. Coherence theories of truth (e.g. the Duhem-Quine theory) might be called epistemological holism (cf. e.g. Oppenheim and Putnam 1958; Ayala 1974; Putnam 1987).

century, cf. Dilthey 1883). External forms are largely understood in terms of a gestalt. Even if every view of the whole as a gestalt does not imply such a relationship of representation, this is nonetheless very often the case. Examples include work by Portmann (e.g. 1948) and Troll (e.g. Wolf and Troll 1940; Troll 1941), who refer to the morphology of individual organisms. But even in relation to objects such as vegetation there are "physiognomic" approaches that correspond to this model. Of particular significance, historically speaking, is Grisebach's (1838) concept of formation, whose relationship to Humboldt's "Physiognomics of Plant Life" (Humboldt 1806) clearly places its origins in an approach based firmly on "Ausdrucksverstehen" (cf. Trepl 1987, pp. 103 ff.). The *gestalt* aspects of this are usually linked to other aspects of wholeness, such as the organic interaction of parts. Holistic positions of this kind refuse to conform – sometimes avowedly so – to scientific demands, insofar as they counter the latter with a "a vivid and clear idea",[6] a holistic "Naturschau" (contemplation of nature, Thienemann 1954, p. 322) or a "contemplative look at nature of a morphological kind"[7] and declare these to be the goal of biology.[8]

Despite having what is, in principle, a similar conception of inner essence as a wholeness, others claim not to depart at all from the scientific methodological ideal. In neovitalism, for example, the specificity of life was seen in it being characterized by a "life force" (entelechy). This is not regarded as a physical force and certainly not one that can be measured scientifically; instead, it is seen as being a living, soul-like force, which is gestalt-forming and therefore holistic. Despite this, entelechy (known as "Factor E") is claimed to be "empirically real".[9] Vitalism has been accused of being dualistic by authors who have been described as holists in the history of biology (e.g. Bertalanffy, Haldane). Vitalism, they say, sees in the living organism only a sum of so many parts that are complemented and monitored by a kind of soul in the role of engineer, rather than seeing the essence of life in the interactive structure of the whole (Bertalanffy 1949, p. 30). It is this, namely, the organic interaction of the parts, that constitutes the holistic element in life. In this view, *biological* holism – that is, what was explicitly known as holism in the history of biology and in philosophies related to biology (the views of J. S. Haldane, Smuts and A. Meyer-Abich, for example), as well as in systems theories with a holistic orientation in the tradition of Bertalanffy (see also Chap. 15) – is given when the key element of life is *not* seen to lie in an inner force inaccessible to scientific methods. Biological holism consists, instead, in the view that the characteristic of being alive can only be attributed to objects that are a whole, and that this whole exists in a special relationship to its parts that is not found in non-living objects. These wholes, so the theory goes, require an approach of their own that is different from that of physics. To the extent that holistic theories divide reality into different

---

[6] "bildhafte, anschauliche Vorstellung" (Friederichs 1957, p. 120, cf. also 1937).

[7] "anschauende Naturbetrachtung morphologischer Art" (Thienemann 1954, p. 317).

[8] Cf. on biology as a whole: Köchy 1997, 2003; cf. on ecology: Trepl 1987; Jax 1998, 2002.

[9] "empirisch wirklich" (Driesch 1935, p. 75, cf. also Mocek 1974, 1998).

levels or autonomous wholes to which different scientific methods have to be applied respectively, they can be called pluralistic.

Different Kinds of Reduction

Reductionism is used to refer to a situation in which, during the course of development of a science, a requirement is made of all its theories that they should be based on the theories of a basic science. In biology, reduction is understood principally in terms of tracing back something that is living to something that is not living by means of a physical-chemical explanation of specific metabolic processes, for example. Reductionism essentially coincides with what is often called "mechanicism" (or "mechanism") or else "physicalism".[10]

Two forms need to be distinguished here in particular:

1. Some hold the view that the whole needs to be explained by acquiring knowledge about its *parts* (a "bottom up" approach). However, reduction is deemed to have been successful only when these parts have been reduced to certain "things", namely "fundamental units" ("atomistic" reductionism). These fundamental units are *not alive*; in other words, this form of reductionism assumes that even if the whole is an organism, there is always a subordinate level in the hierarchy (in the sense of a "nested hierarchy") at which the parts can no longer be considered to be living (subcellular level, molecular level). However, by conducting research on these, it is possible to obtain full knowledge of the whole, the living organism. Examples are superfluous here, as this approach constitutes the mainstream in biology. Indeed it was this form of reductionism, known as "mechanism" (cf. e.g. Roux 1895; Loeb 1916), which was discussed most among biologists during the period when ecology was emerging: the phenomena that could be observed in organisms could ultimately be explained causally at the molecular level. Both Darwinism and neo-Darwinism were also described in terms of mechanistic reductionism. In this case, however, the individual organism is not reduced to the molecular level; instead, in its mechanistic explanations of evolution, Darwinism always presupposes the organism as a whole. Events occurring at the molecular level only become relevant in terms of evolutionary biology when they are viewed in relation to the organism (e.g. as a contribution to its fitness). Evolution is explained mechanically on the basis of interactions between organisms[11] and of those between organisms and their abiotic environment.

---

[10] The *accusation* of reductionism means that the simplification is carried out in such a way that it leads precisely to *not* explaining, or explaining wrongly, the matter to be explained, such as a living organism.

[11] The fact that in the context of Darwinism other levels – the individual gene or the population – are shifted to centre stage changes nothing of this fundamentally organism-centred structure of Darwinism. Even if individual genes are taken as a starting point, they nonetheless "want" something, as "selfish" genes (Dawkins 1976) and are not simply chemical-physical phenomena.

2. It is also possible to undertake a physical-chemical reduction quite independently of the issue of levels in a "nested hierarchy". One example of such a reduction is the "physicalisation" of the organism, for example by looking at blood circulation as a hydraulic system (Harvey in the seventeenth century). Unlike reduction to the molecular level, the whole here is reduced primarily not to its parts but to processes or characteristics of *all* its parts (e.g. flow speed). Measuring these is seen as a way of comprehending the whole. Parts of the system are addressed from a common *functional* perspective, where the function often lies in contributing towards maintaining particular processes. It is thus possible for an extreme form of reductionism to appear, from a different perspective, to be holism[12] insofar, for example, as everything about an object is expressed in energetic terms.

Methodological and Ontological Holism/Reductionism

According to methodological reductionism, traditional (or typically) biological explanations should be replaced by physical-chemical ones, i.e. functional explanations should, in general, be replaced by causal ones. This generates the possibility of explaining biological phenomena in a strictly scientific way.[13] However, this says nothing about the way of being ("Seinsweise") of the object in question, because all that is being argued is that it seems advisable to undertake such a reduction on methodological grounds (e.g. the call for the greatest possible simplicity of the entire body of theory). Methodological reductionism here is diametrically opposed to methodological holism,[14] which insists that the specific biological mode of explanation should be retained. Assuming we are dealing with *methodological holism only*, the claim being made is not that specifically biological terms (e.g. maturity, stimulus, mating instinct) describe objective facts which cannot be captured in physical-chemical terms, but rather that their use is merely understood to be methodologically useful or necessary. Even explicitly teleological explanations, which ultimately always refer to the phenomenon to be explained in terms of its significance in the context of a whole, are permissible in this sense, because without them we would have difficulty – indeed, it would be impossible – to ask questions that are relevant in view of the specific phenomena with which biology is concerned.[15] Thus, teleological explanations are of heuristic value – in fact,

---

[12] "The thermodynamic approach is particularly well fitted as a tool to describe ecosystems from an holistic point of view because it is based on the macroscopic flows of energy and mass" (Jørgensen 2000, p. 113).

[13] On whether and to what extent functional explanations are also teleological, see e.g. Nagel (1979); Rosenberg (1985); Mayr (1988); McLaughlin (2001).

[14] "Methodological holism" is often also used to describe a social scientific view according to which social relations can only be interpreted and explained in terms of social wholes (e.g. classes, but especially the society as a whole) (Mittelstraß 1995, p. 123).

[15] Cf. especially Cassirer (1921), as well as the whole tradition of meta-theory in biology based on Kant; it can generally be seen as a form of methodological holism.

in this respect, they are indispensable. This kind of methodological holism is compatible with ontological reductionism, just as methodological reductionism is compatible with ontological holism (cf. also Mayr 1982).

According to ontological reductionism, everything that exists consists of "fundamental" elements: "an organism is essentially nothing *but a collection* of atoms and molecules" (Crick 1966).[16] This assumption about the character of being implies that the characteristics of higher forms of organisation can generally be explained (completely) causally as being due to the mutual influence of fundamental elements ("micro-determination"). This also provides the grounds for limiting the types of terminology allowed in science. Biology can be expressed in terms of physics and chemistry. Contrary to this position we have a form of ontological holism whose core assertion (as far as biology is concerned) is this: organisms *are* not the way they appear in their reductionist physical-chemical guise. It is generally assumed that units found at higher levels of organisation are comprised of nothing other than units at lower levels (that is, for example, organisms are made up of organs, the latter of cells, and cells of molecules); there is no suggestion here that there is *also* a vital force at work. However, the complexity and organisational structure of such higher-level units make it impossible to say that they are "nothing but" a collection of "fundamental" units. A hierarchical order of increasing complexity can, it is said, be set up (cf. the familiar representations of life as a "layered structure": atom, molecule, organelle, cell, tissue, organ, organism, community etc.). At every higher level of organisation, one finds *emergent,* or *irreducible,* characteristics – hence the insistence on the autonomy of biology. But as far as methodology is concerned, it is perfectly possible in this context to hold the view that a physical-chemical reduction is useful or even necessary, namely as a connecting theme for research (methodological reductionism).[17] It is just that this will not be linked to the view that higher-level units can be explained in their entirety and that biology will *ultimately* prove to be explicable in terms of physics and chemistry: life is essentially something other than the objects that are accessible to these sciences. This ontological holism, then, could be compatible with methodological reductionism (see Putnam 1987). In the context of ontological holism, though, it is also possible to hold the methodological-holistic view, mentioned above, that reduction is *not an option at all* if one wishes to acquire insight into a *living* object of nature.

As in biology as a whole, the holism-reductionism debate in ecology is all about the "special conditions for interpretation and explanation of *partial* groups of objects"[18] and, in particular, about the question of whether irreducible wholes exist or not. This controversy acquires a special twist in ecology, however: If reducing a

---

[16] What was described above as "atomistic reductionism" may be such an ontological reductionism, but it may also be meant merely in the sense of a methodological rule.

[17] If we emphasise *this* and not the heuristic indispensability of teleology, then the Kantian tradition mentioned above should be described as methodological *reductionism*.

[18] "besonderen Deutungs- und Erklärungsbedingungen *partieller* Gegenstandsbereiche" (Mittelstraß 1995, p. 123).

unit consisting of interacting organisms to its parts is not continued to a level at which the parts are no longer considered to be living, then the relationship between whole and parts becomes a relationship between community and individuals (both community and individual being understood in the broadest sense). In its relationship to individuals, community can be conceptualised according to several different models. Let us look at two of these first of all. Community can be conceptualised:

1. as "organische Gemeinschaft" (organic community), or rather as an organism of a higher order, so that the parts (individuals) become (components of) organs of the whole;
2. as a community in the narrower sense ("Gesellschaft" as opposed to "Gemeinschaft"), i.e. as an interactive network that comes about when individuals that are independent in principle enter into relations of cause and effect with other individuals (or, in borderline cases, when no relations of cause and effect exist anymore and the community becomes an aggregation of unconnected individuals).

The form of holism mentioned in (1) above is the one that predominates in ecology. Whenever ecologists speak of organicism, the view being referred to is the one that communities of organisms are themselves organised like organisms. In the context of biology, this is a form of holism specific to ecology, because within biology communities of organisms are the object of ecological inquiry by definition. (2) is the form of reductionism typical in ecology, namely individualism. The model of the machine takes up a mediating position between the two models of organism and community. The machine is a whole, only it is one that functions not organically but mechanically, one whose parts (are supposed to) fulfil a function together. That function does not lie, however, in generating or maintaining itself, and the whole is not grown but rather constructed. This model is relevant to the dynamic that arises when the individualistic and organicist approaches encounter one another and are exposed to different kinds of external conditions. First of all, however, we are concerned only with the first two models mentioned.

Holism and Reductionism in Classical Ecology

The development of holism in biology – and therefore in ecology as well – is attributed mainly to the biologists J. S. Haldane, Bertalanffy and Needham,[19] as well as to certain non-biological influences.[20] Nonetheless, the role of these authors ought not to be overestimated, despite the fact that they were embraced by ecologists themselves and, probably on account of this, were considered in history of science accounts to be particularly influential. Certain less direct influences were almost

---

[19] E.g. Haldane (1931); Needham (1932); Bertalanffy (1932, 1949).
[20] E.g. Smuts (1926); cf. his influences on ecology in Phillips (1934, 1935), Bews (1935); on this, see e.g. McIntosh (1985); Trepl (1987); Hagen (1992); Anker (2001).

certainly of much greater significance. When ecology emerged towards the end of the nineteenth century, and even in the first few decades of the twentieth, the zeitgeist was strongly influenced by holistic views,[21] which were an essential aspect of conservative philosophies and ideologies of the time that were critical of civilisation and were anti-mechanistic (cf. Harrington 1996, Müller 1996). Older philosophies also played a not inconsiderable role (e.g. those of Herder and, by way of the latter, Leibniz; cf. e.g. Eisel 1980, see also Chap. 25), as they were of paradigmatic significance for the new geography – ecology initially developed in very close association with physical geography. The paradigmatic core of geography was the unit of culture and concrete nature, conceived of as an organism and known as "Land" (land) or "Landschaft" (landscape) (cf. Eisel 1980).

Frequently named as the representatives of an early holistic strand among ecologists are Friederichs (1927, 1934, 1937), Thienemann (1941, 1944), Thienemann and Kieffer (1916), Clements (1916, 1936), Shelford (especially Clements and Shelford (1939)), Phillips (1934, 1935), Braun-Blanquet (1928) and Sukachev (1958).[22] While their views of the essence of "synecological" objects ("biocoenose", "community", "holocoen") differ in detail, they can still be characterised overall as organicist in the sense described above.

One important contrasting position to the "renewal" the sciences were undergoing on the basis of different concepts of wholeness was positivism and, later, in the twentieth century, neo-positivism (logical positivism). To the latter, statements about wholeness are considered to be metaphysical. This basic attitude is reflected in ecology – although it owes its existence by no means to the influence of (neo-) positivism alone, the latter being just one of several philosophies at the time that were oriented towards the exact sciences. Representatives of this strand include Forbes (1887), Gleason (1917, 1926, 1927), Ramensky (1926), Lenoble (1926), Gams (1918) and Peus (1954).[23]

In the following we reconstruct the classical positions of organicist holism and reductionist individualism as ideal types. An ideal type does not claim to be an entirely real position – even though our constructions draw on formulations put forward in particular by Clements (1916, 1936) on the one hand and Gleason

---

[21] These views did not constitute a single coherent position, but rather very different ones, of which only a few are (explicitly) conservative and critical of civilisation (certainly not gestalt theory – Ehrenfels 1890, 1916; Wertheimer 1912; Köhler 1920 – or holistic approaches in neurology, e.g. Goldstein 1934). The greatest influence was exerted by some authors who were adherents of the philosophy of life (such as Klages, Spengler, Bergson) and other philosophers (e.g. Husserl, Heidegger, Whitehead). Although only minimally influential in the public intellectual sphere, Smuts (1926) and Meyer-Abich (1934, 1948) were highly regarded in biology; the philosophical and social scientific works of biologists, such as Uexküll's "Staatsbiologie" (Uexküll 1920) (State biology) should also be mentioned here.

[22] Cf. also McIntosh (1985); Trepl (1987); Botkin (1990); Jax (2002); Kirchhoff 2007; Voigt 2009.

[23] On this, see also McIntosh (1975, 1985, 1995); Trepl (1987); Jax (2002); Schwarz (2003); Kirchhoff 2007; Voigt 2009.

(1926) and Peus (1954) on the other, which (at least in a certain mode of interpretation)[24] express very clearly the essence of organicist holism and reductionist individualism. An ideal type, after all, "*is* not a *description* of reality but it aims to give unambiguous means of expression to such a description".[25] It "is formed by the one-sided *accentuation of one* or *more* points of view and by the synthesis of a great many diffuse, discrete, more or less present and occasionally absent *concrete individual* phenomena, which are arranged according to those one-sidedly emphasized viewpoints into a unified *analytical* construct".[26] The idea is – and this can only be achieved through the method of constructing ideal types – to render classical organicist holism and reductionist individualism comprehensible on the one hand, and to formulate a foundation that makes it possible to describe and systematically compare different concrete theories on the other.[27]

The Classical Organicist-Holistic Position

According to organicist holism, supra-organismic units, when described as "biocoenoses", "communities" or "associations", are communities that are determined primarily by relationships of dependency between organisms.[28] This dependency refers to their development as well: Clements uses the term "reactions" to describe the dependency of subsequent stages of succession on those preceding them. The developing whole (see below) also depends heavily on the competitive activities of the parts (Clements 1936, p. 143). In a stable end state (climax), however, the relationship between parts and whole consists essentially in each individual part being brought forth by all the others, and therefore in a reciprocal bringing forth of the

---

[24] For a detailed discussion of this mode of interpretation, in relation to Clements, see: Wolf, Judith (1996): See also Eisel (1991).

[25] "ist nicht die Darstellung des Wirklichen, aber er will der Darstellung eindeutige Ausdrucksmittel verleihen" (Weber 1904, p. 234, emphasis in original).

[26] "wird gewonnen durch einseitige *Steigerung eines* oder *einiger* Gesichtspunkte und durch Zusammenschluß einer Fülle von diffus und diskret, hier mehr, dort weniger, stellenweise gar nicht, vorhandenen *Einzel*erscheinungen, die sich jenen einseitig herausgehobenen Gesichtspunkten fügen, zu einem in sich einheitlichen *Gedanken*bilde" (Weber 1904, p. 235, emphasis in original).

[27] For the holism-reductionism controversy at the start of the twentieth century in ecology, see, amongst others, McIntosh (1980); Tobey (1981); Worster (1985); Trepl (1987); Hagen (1992); Golley (1993); and Anker (2001).

[28] A biocoenosis and its habitat can also be seen together as comprising an organismic unit. "Jede Lebensgemeinschaft bildet mit dem Lebensraum, den sie erfüllt, eine Einheit, und zwar eine in sich oft so geschlossene Einheit, daß man sie gleichsam als einen Organismus höherer Ordnung bezeichnen kann." (Each community forms a unit with the habitat it fills; this unit is often so unified in itself that it can more or less be described as a higher-order organism.) (Thienemann and Kieffer 1916, p. 485).

parts and the whole. Thus, individual organisms or even units made up of several organisms, such as the functional groups of producers, consumers and decomposers,[29] exercise certain functions within the community as a whole, much like organs in an organism. These *functions* need to be carried out in order to preserve not just the whole but also the organs themselves which carry out the functions. This kind of community is a whole, which exists objectively, just like an individual organism, and is "naturally" isolated and individual in space and time – if somewhat less clearly than an individual organism.[30] Moreover it is accorded the essential characteristics of an organism – the "superorganism theory" is a prominent example of this. The consequence of this for methodology is that the characteristics and behaviour of individual organisms need to be examined according to the contribution they make towards the functioning (adaptation) of the community as a whole.

Development of the Community as a Goal-Oriented Process of Adaptation and Detachment

The organicist-holistic theory is a *theory of development*. This is usually not taken into account, which means that the image of nature in holistic ecology is erroneously seen as a static one whenever – as occurs very often – an (allegedly outmoded)

---

[29] "Nur die Pflanze kann, indem sie das Sonnenlicht als Energiequelle benutzt, aus anorganischen Stoffen organische aufbauen und da das Tier sich nur von Organischem ernähren kann, so ist die Tierwelt direkt oder indirekt fest mit der Vegetation verbunden. Wenn das Tier als Raubtier von anderen tierischen Wesen lebt, so müssen die Beutetiere zur Lebensgemeinschaft des Raubtieres gehören usw." (Plants alone are able to form organic material out of inorganic material, using sunlight as a source of energy, and since animals can only feed off organic material, the animal world is firmly tied, directly or indirectly, to the vegetation. If an animal lives off other creatures as a predator, then prey animals must be a part of the predator's community etc.) (Thienemann 1944, pp. 7.)

[30] "All diese Tiere und Pflanzen aber stehen nicht unvermittelt nebeneinander, sondern sind durch die mannigfachsten Beziehungen aneinander gebunden; jede Stätte im Lebensraum hat so ihre Lebensgemeinschaft oder Biocoenose." (Yet all these animals and plants do not exist alongside each other in an unmediated way; rather, they are tied to one another through many different relationships. Thus every site in the habitat has *its own* community or biocoenosis.) (Thienemann 1944, p. 7). "So ist die Natur aufgegliedert in eine ganze Stufenfolge, eine Hierarchie ineinander geschachtelter lebenserfüllter Räume, die nicht nur räumlich miteinander verbunden, sondern auch voneinander durch den Kreislauf der Stoffe [...] abhängig sind. Jede Lebensstätte ist wiederum Glied einer größeren, bis hinauf zur ganzen Erde." (Thus, nature is subdivided into a whole layered succession, a hierarchy of interlinked spaces filled with life, which are not only connected to each other spatially but are also dependent on one another due to the material cycle [...]. Every living site is in turn part of a larger site, right up to the Earth as a whole.) (Thienemann 1944, pp. 8). "Die Natur [...] ist, vom kleinsten Wiesenfleck angefangen bis zum ganzen Weltall, überall ein geschlossener lebender Organismus, in dem jedes einzelne kleinste Glied auf jedes andere abgestimmt ist; jede Veränderung eines Teils wirkt sich aus auf alle übrigen." (From the smallest spot of meadow to the whole universe, nature (...) is everywhere a unified living organism in which every single tiny element is linked in with every other; any change in one part has an impact on all the others.) (Thienemann 1944, pp. 35).

"ecological balance" is referred to (e.g. Pickett et al. 1992). This image of nature is considered, in turn, to be one reason behind the existence of curiously static conceptions in nature conservation (e.g. Reichholf 1993; Scherzinger 1996).

Just like an individual organism, the community develops according to intrinsic rules of development through different "stages" to reach a state of maturity (climax); it maintains itself in this state or else embarks anew on the process of development: "As an organism the formation arises, grows, matures, *and dies*." (Clements 1916, p. 3, our emphasis). The fact that this development is internally guided does not mean that the community is autonomous, for it entails the latter differentiating itself in accordance with external dictates while at the same time changing certain of them. In the early phases of succession, those species that are adapted to the site become established, that is, site conditions determine whether they will become established. The communities made up of these species change the site, however. They thus create conditions which are unfavourable for them but which will favour species at later stages of succession. Having lost out in the competition, they are replaced by the latter (cf. Clements 1916, 1936; Thienemann 1944). Unlike in Darwinism, competition is not regarded from the perspective of individuals who prove their worth in this "struggle for survival", so that "progress" occurs in the evolutionary line in terms of adaptation to particular environmental conditions; instead, it is regarded in terms of the function it performs for the developmental unit as a whole. The succession of different species at a certain site becomes an act of self-sacrifice[31] on the part of pioneer species for the benefit of "higher" stages and, ultimately, the climax community.

From the perspective of the community, this process of replacement means that it adapts as a whole to environmental conditions ("response"). This occurs above all through a change in the composition of its species, prompted by competition, and through internal differentiations, in particular an increase in the number of species and in the number of interactions. At the same time, the community adapts the locality *to itself* ("reactions"). "Each stage of succession plays some part in reducing the extreme condition in which the sere began" (Clements 1916, p. 98). Since the "stages" that manage to achieve this feat are superceded in this process, one can say that the community adapts the locality to its own *future* constitution. In this way the community becomes increasingly independent from specific site factors – from those that exist initially in the sense that they no longer exist, and from those that exist in later phases in the sense that they influence the organisms only indirectly, through the community. Thus, development is simultaneously a process of adaptation and detachment. The dependency of the whole on external factors decreases as the dependency of the parts on one another and as the way they adjust to one another increases. In contrast to the prior stages, the climax stage is such that it no longer changes the external and internal conditions for the individual species in the community. Instead, each individual organism "processes" the impacts of external

---

[31] Der "Erhaltung des Ganzen wird, wenn nötig, auch das größte Teilglied geopfert." (Even the largest part will, if necessary, be sacrificed for the preservation of the whole.) (Thienemann 1944, p. 9).

factors through their interactions in such a way that the state of the whole remains unchanged. Differentiation (that is, an increase in diversity), internal functional dependency and mutual adjustment, and independence from external factors (in line with the homeostasis of an individual organism) occur together in a necessary common context. The fact, that development can be characterised as *internal* means that the community realizes that which was "embedded" in it from the start.

In the process of the site factors adapting to the developing community, these factors are changed by the impact of the community in such a way that the original diversity present at the site in question disappears (through the emergence of a bioclimate, humus formation etc.). This then means that just one single climax community will exist in each area governed by a specific climate, regardless of the (original) small-scale differences at the site (what Clements called "monoclimax", and what Braun-Blanquet and Sukachev, for example, referred to in similar terms). At the same time, the community optimizes its own adaptation to the conditions presented by the regional climate in this development: in the climax state it exists in a state of balance with this climate.[32]

Thus, this theory of development implies that the concept of the environment divides into two categorically different concepts: (1) the many and varied small-scale, temporally different *site factors* where the community lives (especially edaphic and microclimatic factors), which in the course of succession relinquish their influence on the development of the community and which the community then adapts to *itself*; (2) the *regional climate*, over which it has no influence and to which the community *itself adapts* in the course of its development. Adaptation is complete in the climax state.

Thus, in organicist holism, succession is conceived of as being *goal-oriented*. Even if we were not to interpret it in explicitly teleological terms, it is still teleological in the sense at least that it proceeds towards a final state conceived of as being fixed from the start, because the regional climate – unlike the site factors – is not modified by the community. This final state becomes established in spite of different starting conditions (edaphic and micro-climatic differences, random migration patterns). This theory is happy to accept that there are synecological units which do not accord with the image of a highly integrated whole or with the "intended" climax state. But by virtue of being a *theory of development* it manages to conceptualise such units as partially developed phases or as deviations in the "ontogenesis" of a community that is generally holistic, that is, one conceived of as a superorganism.[33]

---

[32] "[T]here is but one kind of climax, namely that controlled by climate" (Clements 1936, p. 128). "Such a climax is permanent because of its entire harmony with a stable habitat. It will persist just as long as the climate remains unchanged." (Clements 1916, p. 99).

[33] Clements sees successions that do not lead to a monoclimax community as deviations or as a "subfinal stage of succession" (Clements 1936, p. 130). For example, there are very stable prior stages (proclimax), subclimax or disclimax stages in which external natural factors or human influences impede attainment of the climax stage (Clements 1916, 1936).

The Classical Individualist-Reductionist Position

For individualistic theories the individual organism is the fundamental unit which alone is granted the status of "real". The causal principle is regarded as being not only permissible but indeed adequate for explaining the links between individuals and between these and their abiotic environment. Thus, the notion of function in relation to a whole that encompasses so many individuals is not what grounds this explanation. In this respect, individualistic theories embody a form of biological mechanicism, even though the "fundamental units", as emphasized above, are living organisms. – Here again, we present an ideal typical reconstruction of this position.

We can take the following to be the starting point of the individualist position: "The existence and success of each species depends solely on the realization and quality of its own environment; it is left to fend for itself within that environment. There is nothing above or beyond this that influences the animal and its life, ecologically speaking."[34] A community does not exist for the organisms.[35] Other organisms are environmental factors, as are abiotic factors; another living creature does not "appear" to an animal to be such (Peus 1954, p. 300). From the perspective of the organism it is irrelevant, for example, whether water is available in the form of a puddle ("abiotic factor") or as a component of a prey organism, as long as it is equally useful to it.[36]

This radical individualist position sees those units which ecologists call "biocoenoses", "associations" etc. as "products of human imagination"[37] (Peus 1954, p. 300). Since they are fictitious they cannot be the object of scientific inquiry. According to Peus ecology should restrict itself to autecology: "Biocoenology has no grounding in reality as a science."[38] However, the logic of the individualistic approach grants a certain *heuristic usefulness* to the concept of community. On the one hand, communities must not be viewed as real entities, as they are in the holistic-organicist view: not only are they not superorganisms, they are also not "natural" units containing organisms that occur only in quite specific combinations and that may be discovered and described – as individual species of organisms are – by appropriate kinds of research. While it is true to say that individual organisms

---

[34] "Jede Spezies ist in ihrer Existenz und in ihrem Gedeihen allein von der Verwirklichung und Qualität ihrer eigenen Umwelt abhängig; sie ist darin auf sich allein gestellt. Darüber hinaus gibt es nichts, was das Tier und sein Leben ökologisch gesehen beeinflußt." (Peus 1954, p. 307).

[35] This is not to say that relationships with other organisms may sometimes be very close, indeed obligatory (obligatory predatory and mutualist relationships). What does not emerge, though, are communities, i.e. units of a higher order in relation to which it would make sense, for example, to say that the organisms were fulfilling certain functions for it. In the individualist view, it is only in relation to individual organisms that it makes sense (heuristically) to say that a function is being fulfilled for something.

[36] This is why the individualist position is often linked to the "organism-centred approach", which attempts rigorously to adopt the perspective of the organism concerned in descriptions of environmental factors (cf. Peus 1954; MacMahon et al. 1981; Jax 2002).

[37] "Gebilde des menschlichen Vorstellungsvermögens" (Peus 1954, p. 300).

[38] "Die Biozönologie als Wissenschaft hat keinen realen Grund" (Peus 1954, p. 300).

form groups with other organisms in a given area, the composition of these groups will change depending on environmental factors and on the random nature of migration patterns. On the other hand, however, it can be useful to give some of them (e.g. those that appear more frequently under current environmental circumstances and conditions of migration) names (for example, names taken from the system of plant sociology) in order to establish some point of reference amid the many different combinations that occur. One just has to be aware that quite different species combinations can also come about, and that the frequency of the species groupings identified as "associations" etc. is due only to the coincidental fact that the external circumstances required by these associations occur frequently – and not to any internal rule of development which, even given very different conditions at the start, gives rise time and again to specific combinations (such as that of the "mature" state). In this view, there is no "structural uniformity of vegetation", for the simple reason that any given area is subject to annual variations (Gleason 1926, p. 10). In addition, the issue of whether an area of vegetation is regarded as one single association or as a mixture of several depends on the perspective taken by the scientist concerned and on how the temporal and spatial boundaries are drawn (Gleason 1926, pp. 10). When communities are isolated out by the scientist from among the multitude of possible and real species combinations, other boundaries and therefore other units (known as associations, for example) emerge, depending on the focus of inquiry. A network of relationships will come to an end at different points, depending on whether mutualistic or predatory relationships are selected for study, so that the outcome is a different community. Viewed in this light, the scientist is *constructing* communities.[39]

Changes to the Community Brought about by Changed Environmental Conditions and the Randomness of Species Migration

Views concerning the way in which "communities" (which may be nothing other than groups of organisms occupying adjacent spaces) change through time differ fundamentally in several ways within the individualistic-reductionist approach

---

[39] The different individualist positions could be distinguished according to their starting assumptions. These may be (1) that communities can appear to be "naturally" demarcated – in other words, individuals arrange themselves into certain groups according to, say, the requirements they have of mutualistic partners. An *objective* boundary to the community is given at the point where the (necessary) interactions finish. Alternatively (2) the boundaries may appear to be drawn by the scientist *according to his or her research interests*. Within this, it is possible to distinguish between a realistic and a nominalist variant. The realistic variant entails the view that there is a (real) network of relationships that exists independently of an observer; this network has a specific constitution, which is so complex, however, that we cannot (yet) recognize it and are forced, for pragmatic reasons, to create "artificial" demarcations. The nominalist variant sees communities fundamentally as being constructed by an observer (see also Chap. 27).

compared with the organicist-holistic approach where, as described above, changes are seen as *developments* in the literal sense of the word:

1. "[Plants] can and do endure a considerable range in their environment" (Gleason 1926, p. 18). The occurrence of species in early phases of settlement is determined by the randomness of the arrival of diaspores and the existence of the environmental conditions necessary for them to develop. Changes in the plants' environment (site factors) are not primarily an outcome of the activity of organisms,[40] but occur above all by chance (e.g. climatic fluctuations, Gleason 1926, p. 18). Therefore, the succession of different species combinations at any one site is determined by the randomness of migration and by the randomness of changing environmental conditions for the individual species. "The next vegetation will depend entirely on the nature of the immigration which takes place in the particular period when environmental change reaches the critical stage" (Gleason 1926, p. 21). Succession does not proceed, therefore, towards a particular end state (meaning: it brings to fruition that which was "inherent" in it from the start); rather, it is random, it involves change but *not development*. If immigration and environmental selection remain unchanged, then the community attains stability; changes in one or both factors lead to a change in the composition of species. Climax – to the extent that this term makes sense in an individualistic context – can mean no more than a phase in which no change takes place for a certain period of time (Gleason 1926, p. 26). The further off in the future it lies, the less the state of a community can be predicted, because one cannot know enough about the environmental factors and migration events that determine it. This contrasts with the organicist-holistic view, in which the random and unpredictable nature of the *initial conditions* is emphasized, while the end state is always the same: if we look far into the future, we can predict which state will become established or, at any rate, which one ought "normally" to become established, namely the climax state.

2. While it is a *single* community that *develops* in the organicist view, in the individualist view it makes no sense to ask whether the phenomena observed constitute a change in *one* or the succession of *several* communities – or rather, it can be answered either one way or the other on account of the fundamentally arbitrary options for demarcation; it is simply that different combinations of species follow on from one another in a more or less continuous rotation.

3. The adaptiveness of the individual organism is the *starting point* for the creation of a community. Always and in every phase of succession, those organisms become established that are adapted to the conditions which exist in that place. It makes no sense to speak of the community adapting (with increasing success). The role accorded to adaptation here differs radically from the one it plays in the organicist approach. In the latter, adaptiveness (of the community to the regional

---

[40] "The plant individual [...] is limited to a particular complex of environmental conditions, which may be correlated with locality, or controlled, modified, or supplied by vegetation." (Gleason 1926, p. 17).

climate) is the *outcome* of a development that leads to the climax community, because the community increasingly gathers to itself those species which are useful to it in the homeostatic maintenance of its organic balance.
4. In the individualistic approach, the concept of "law" has the meaning of a natural law as commonly understood, in other words, a sentence containing the words "whenever..., then...", which always has to be known, along with the relevant boundary conditions, whenever one wishes to explain or predict something. Boundary conditions and (natural) law make it possible to explain why a particular change has occurred in the species combination at a site. In contrast to this, the term "law" is used in a completely different sense in the organicist approach. Here, changes in communities of organisms are described as "developments that follow a set pattern" ("gesetzmäßige Entwicklungen"), meaning the same as when the notion is applied to an organism, namely that in a "normal" or "typical" case (in the normative, not statistical sense) certain conditions follow on from one another. They are to be understood as a development of something which was already "inherent" in the "undeveloped" state.

Can the Controversy Be Resolved?

We shall begin by discussing the hypothesis that there is no way empirically to resolving the holism-reductionism debate in ecology in favour of one or the other position. We shall then discuss whether it might be possible to achieve a resolution of the issue at the methodological level – perhaps one of the approaches uses inadmissible methods or methods that are poorly suited to the specific object of inquiry. In the course of this discussion we will also deal with the problem of the admissibility of teleological explanations of natural phenomena. The outcome of this discussion also suggests that it is not possible to reach a clear decision on the issue. The controversy is therefore looked at from a third perspective: perhaps both approaches owe their existence to external influences – "world views" – or are at least "inspired" by them, which could mean that it is impossible to decide between them using scientific means.

The Controversy Cannot Be Resolved Empirically

All the evidence suggests that the dispute between organicist holism and reductionist individualism cannot be resolved empirically, that is, by presenting facts that would support the one and refute the other viewpoint. The assertions put forward by each position can, as a rule, be explained just as well from the other side's perspective. The holists maintain, for example, that a stable state is established at the end of succession, and indeed they have been able to supply a certain amount of empirical evidence to support this assertion. However, there are theories of the individualist persuasion, which are equally able to explain these results as one possible outcome of the course of successions (e.g. Horn 1976).[41]

To offer a second example, it has been argued from the individualist perspective that empirical findings, if anything, fail to support the holistic assertion that succession follows certain laws, as suggested by the analogy with the development of an organism (e.g. Drury and Ian Nisbet 1973). However, this argument basically goes nowhere, insofar as holism doesn't claim that the events formulated in the law – such as an increase in the parameters of a community and in diversity during the course of succession – can always or usually be observed in any given instance; rather, it claims that those events "accord with" the healthy development of a superorganism. Such *developmental laws* are thus of quite a different nature than natural laws, such as the law of falling bodies, which have no exceptions. Indeed, the claim that an increase in diversity follows a set pattern allows for any number of "exceptions" to the rule, where the term "rule" does not describe what always or normally happens (given specific parameters) but rather dictates what is *supposed* to happen.

Thus an empirical resolution to the debate does not appear to be possible. It is no surprise, therefore, that the dispute is often played out at a different level: the other side is accused not so much of failing to supply sufficient empirical evidence to support its claims as of using an *inadmissible method*.

Organicistic Holism, Individualist Reductionism and the Problem of Teleology

It is the organicist approach in particular that is affected by the accusation of having an inadequate methodology. While this approach is hard to refute at the empirical level, it is certainly rather vulnerable methodologically. Indeed this is the main objection that is levelled at every variant of holism. However, this objection is only valid because (or rather, insofar as) its adherents are committed to the methodological ideals of the modern natural sciences. Yet it is hard for them to avoid doing so[42]: It can be seen as constitutive of modern thinking that each of the three ways of contemplating nature – the aesthetic, the normative-evaluative and the scientific – are valid independently of one another. This means that the scientific perspective ought not to contain statements of the kind that characterise the normative-evaluative perspective. The constitutive role of the natural sciences is to provide causal explanations and to avoid teleological ones. Values – and thus goals as well – can only be

---

[41] On the reformulation of organicist-holistic arguments in the individualist context in general, cf. Gnädinger 2002.

[42] Explicit attempts of this kind have been made by Friederichs, for example (see above).

"attributed" to nature by us. Everything that is viewed from the scientific perspective therefore appears to be a value-free object of theoretical knowledge.[43]

However, in biology, the science that deals with *living nature*, nature is often spoken of in a way that seems to be diametrically opposed to the scientific method. The concept of the living organism seems to include the fact that terms such as benefit and harm, optimum, developmental goals etc. can be meaningfully applied in the scientific context, which is not the case with non-living natural objects. The organism (in modern societies, at least) is conceived of in such a way that each of its parts owes its presence to the agency of all the other parts and exists for the sake of the others and of the whole (Kant 1970, §§64–65). The parts are *organs* of the whole. When one speaks of the function of the organs being to maintain the organism, one is making a teleological judgement. The term "self-maintenance", when used in relation to the organism, assumes that the *state* of life is the purpose and desirable goal of the organism. Thus, it might appear that it is possible to attribute a value to natural phenomena that is objective and is not related to human value systems, but rather lies in the maintenance of the organism.

However, it is not possible in science to assert that things in nature happen according to some purpose. The orientation towards purposes presupposes the idea of a purpose which precedes a cause, i.e. action based on intention; yet we cannot insinuate that nature has intentions. If we do, we are not looking at it scientifically. Therefore, it is not only organicist holism in ecology but also the variant of holism that concerns itself solely with individual organisms as an organic wholes – which includes, in principle, the individualist position, as discussed above – that is vulnerable to criticism from the perspective of radical mechanicism. To claim that natural phenomena can be explained teleologically (and conceiving of them as the unity of an organism is nothing else) would mean leaving behind the foundations on which all the sciences rest.[44]

One possible objection to this radically mechanicist perspective is that its criticism is unjustified if teleological explanations are intended heuristically and not as objective explanations (cf. Kant 1970, §§69–78, see also above on methodological holism). It is necessary to regard the organism as a whole and its parts as elements that serve the function of maintaining it in order to have a point of reference for biological research, even if the latter then has to explain its object in causal terms. Indeed such explanations are indispensable in rendering the phenomenon of life

---

[43] However, value freedom by no means signifies that as a general rule modern natural science is not guided by interests, and hence by values. Its theories provide access to the "Wirklichkeit unter dem leitenden Interesse an der möglichen informativen Sicherung und Erweiterung erfolgskontrollierten Handelns." (reality under the guiding interest in assuring and extending instrumental action in potentially informative ways.) (Habermas 1968, p. 157).

[44] In the complex debate about this issue some positions claim that the problem of teleology is done away with by evolutionary theory providing a causal explanation for the emergence of functional characteristics (e.g. Mayr 1982), while others assert that teleological explanations can be reconstructed as deductive-nomological ones (in Nagel's sense), and that they should therefore definitely be accepted as scientific (cf. Looijen 2000).

"visible" at all (Cassirer 1921). Thus, if biology is to be a *natural* science, it must use causal explanations. In order to justify the need for seeing biology as a *special kind* of natural science, though, one is dependent on (heuristic) teleological judgements (and therefore on methodological holism). This, in turn, can be held against mechanicist reductionism: if the maintenance of the whole of an organism were *not* conceptualised as constituting the purpose of the processes going on within it – even if only in heuristic terms – then one would not be able to find a starting point for physical-chemical explanations. In other words: physical-chemical explanations would be irrelevant – they only acquire relevance when they contribute towards explanations of the phenomena we call "life".

The holistic approach in *ecology*, though, conceives not only of the individual organism as an individual whole but the *community* and indeed "nature" as well. This whole, for its part, consists of hierarchically interlocking individual wholes. It is the function[45] of the parts in the service of the whole (on which they in turn depend) to work in such a way that everything supports everything else – just like the relationship between organs and organism. Following Kant's argument further, however, it may be so that we are *forced* to use the notion of utility (even if only heuristically) in the case of the (individual) organism, because the organism appears to us to be "underdetermined" (that is, random) in its characteristics "according to mechanical laws alone" (i.e. causally); but this is not the case with *the community of organisms* (or with nature in general) (cf. Weil 2005). In order to explain why the heart, for example, is constituted in a particular way, one needs to know what function (what purpose) it fulfils in the organism. But in order to explain why certain plants grow in a community, one does not need to know their function in relation to other organisms, such as serving as food for animals – at most we need to know the function other organisms have in relation to *them*. Even harder to comprehend is the call for explanations to be sought in the function they have in relation to the *whole* of the community: while organs cannot live outside an organism, organisms can, as a rule at least, live outside their respective community, namely in other communities, or else in complete isolation. It is highly questionable, therefore, whether a teleological judgement relating to the whole of the community makes sense, even if it is intended heuristically only. Ecological holism thus has a bigger problem justifying its approach than does that of the physiologists, which relates to the individual organism. The reason why it was – and still is – so widespread in spite of this seems to call for explanation at a different level. This is what we shall discuss in the following. What we seek to shed light on, though, is not only the existence and even occasional dominance of organicist holism; rather, we shall argue that the difficulty (or impossibility) of deciding between the two approaches – the individualist approach is, as we have seen, just as vulnerable to critique at the methodological level – can be better understood when one takes into account the fact that both are constituted in part by external influences.

---

[45]Etiological function as referred to by Wright (1973), cf. McLaughlin (2001).

Ecological Paradigms as Partially Constituted by Ideology

One way of bringing such external influences to bear theoretically is the hypothesis that ecological holism is an effect of ideology – conservative ideology (cf. e.g. Eisel 1991). The major tenet of such ideologies is that "community" means an "organic" community or a higher-order organism (Greiffenhagen 1971). The structural parallels between conservative social theories and organicist ecological theories are striking and have been widely discussed (e.g. Eisel 1991, 2002; Trepl 1993, 1997; Körner 2000; Anker 2001; Schwarz 2003; Voigt 2009). The arguments fielded by ecologists against organicist holism often seem to be grounded in the desire to keep science pure from ideological contamination (e.g. Scherzinger 1995; cf. Körner 2000).

Looked at from the perspective of a theory of constitution, one important reason why the organicist-holistic and individualist-reductionist positions exist in ecology can be seen to lie in the fact that social relations exist which generate different cultural ideas and make nature *appear* to be like this or like that. In the empirical sciences, ideas that originated in culture are set up against one another and differentiated as scientific theories, a move that generally goes unnoticed.

When the first theories about ecological units were developed in ecology, there were two competing figures in philosophy, the social sciences and political theory which constituted the framework for conceptualising the relationship between individuals and society. These can be categorised according to the two types of opposing world views that are constitutive of modernity, the one progress-oriented (especially liberalism) and the other conservative. Even if these two basic figures have been greatly changed, modernised and their components recombined in different ways over time, both remain influential.

The liberal view (cf. e.g. Kühnl 1982; Arneson 1992) is that of a community of *autonomous individuals*. It is based on the idea of an autonomous subject that has liberated itself from the fetters of feudal society and of religion etc. so that now it is responsible only for itself, pursuing its own advantage by means of general rationality and general technical knowledge. Rather than according with tradition and nature, social development is related to individual emancipation and is therefore *open to progress*. Liberalism is grounded in the idea that the world is shaped by the struggle between individuals (Hobbes) who come together in communities because this benefits them as individuals in the long term. In liberal theory society is seen as the "sum" of individual people, who alone constitute reality. Society has no "inner unity"; it is a superficial mechanical agglomeration of individuals. Society is a system of interactions between individuals in which the interactions are geared towards their usefulness for those individuals; it is a system in which interests are reconciled in utilitarian fashion, premised on a "struggle for survival".

This liberal view can be equated with the individualist approach in biology, especially Darwinism and individualist ecological theories. Their structural connections and, to some extent, the causal links in their emergence have been described often.[46] One aspect of their structural connections is that they both view *individuals' needs* as being the factor that forces them (individuals) to come together in communities, that is, to establish relationships with others. In addition, they both hold that social change *has no ultimate goal*: the direction it takes depends upon chance factors, and if there is such a thing as development to a higher level, then it does not involve a community coming closer to a pre-given goal but rather an improvement from the perspective of individuals – of those who win through in the competition for survival. Their respective ontological and epistemological views also match one another: it is a fundamental tenet of both individualist ecology and liberalism and its related empiricist epistemologies that societies do not exist in reality but are merely the outcome of the ordering activities of human reason.

In contrast to this, the conservative view[47] emphasizes the fact that the individual is tied into a higher order – into an *organic* community ("Gemeinschaft") rather than just a community, or society, per se ("Gesellschaft").[48] The individual must accept his or her predetermined ties to religion, nation, tradition, family etc., as it is only in this constellation of relationships that the individual can develop his or her own "special nature" ("Eigenart"). This is how each individual makes his or her own contribution towards maintaining the God-given order of the whole (or that given by nature or history) and this is how individuals contribute towards the development of this whole, which is always organic and is always conceived of as "evolved", rather than as a construction. Each individual acts freely by recognizing and accepting the task allotted to him or her in the given order. Freedom comes about, then, through the recognition of ties. History is not an open process shaped by autonomous subjects and determined by the forces of chance. Rather, it consists in the perfection of the qualities inherent in "nations" ("Völker") – their "character" – and in the perfection, by these nations themselves, of the qualities inherent in the nature that exists in their "Lebensraum" living space. This is the task allotted to nations as well as to individuals. "Culture" develops in the process of the nations adapting to the dictates of nature in their "Lebensraum", which simultaneously represents a process of breaking free from the constraints of nature (cf. Kirchhoff 2005).

The structure of the counter-Enlightenment, conservative picture of community displays a remarkably precise correspondence to the structure of the organicist-holistic picture of ecological communities. Just as in political theories, the development

---

[46] For Darwinismus cf. e.g. Engels (1886); Nordenskiöld (1928); Desmond and Moore (1991); for the individualist approach in ecology e.g. Trepl (1994); Eisel (2002); Voigt (2009).

[47] This is a reference to the basic conceptual figure of conservatism that arose in the course of the counter-Enlightenment (cf. Eisel 1982). Later political movements grouped under the conservative heading (e.g. "technocratic conservatism") often deviate considerably from this (cf. Greiffenhagen 1971).

[48] Cf. the distinction between "Gesellschaft"and "Gemeinschaft" in Tönnies (1887).

of the community represents the realization of that which is dictated to them at a transcendental level (naturalised by Clements as the regional climate). This occurs by means of creative individual effort, in a process of differentiation and the development of particularity ("Eigenart"). In organicist ecological theory, too, the development of a community is a process in which breaking free from the dictates of nature (abiotic environmental conditions) occurs through adaptation to them. Here, too, individuals are conceptualised as being not only physically dependent on the whole per se (rather than on what they make for themselves out of the "resources" available), but as being dependent on it also insofar as their existence can only "make sense" in the context of the whole. This is because the meaning of each person's life lies in fulfilling the tasks given to them in relation to the whole (within the context of the latter's development towards the perfection of which it is capable); it is a whole to which they owe their existence. Equally, the "purpose" of individual organisms lies in exercising precisely those functions for the organic community that are appropriate to the task of the self-maintenance of the community as a whole and of its development towards the climax state.

Ecological paradigms can thus be read as political ideologies read back into the workings of nature. Conversely, though, it is also the case that a particular ecological paradigm entails particular political attitudes. Depending on how it is considered best to characterise ecological objects at a fundamental level, a certain set of policies will appear to be meaningful and necessary. This applies not only to political attitudes regarding the relationship between the individual and the society but also, as we shall see, to the relationship between the society and nature. Having said this, it would be rash to assume that a supporter of the organicist approach in ecology, for example, must necessarily hold politically conservative views.[49]

Implications of Holism and Reduction

Whichever of the two views is taken as the basis for action, both entail far-reaching practical consequences. This practical relevance may exist in relation to technology (e.g. measures in agriculture and forestry, fishery, nature conservation) or in relation to "knowledge for orientation", in which "nature" is viewed within a context of values, either as being subject to evaluation or as a value-setting authority.

The Organicistic-Holist Approach in Nature Conservation

The organicistic-holistic position is a teleological one. At least, it is an obvious move to interpret its empirical assertions in this way – that succession consists in a process

---

[49] This would presuppose at the very least that for such a person an ideal community is oriented towards an ideal image of nature. This is certainly usually the case, but is not necessarily so.

of differentiation leading to a climax state, and so on. In particular, it is hardly plausible that a theory concerning the essence and the development of synecological units could arise outside of a generally teleological conceptual structure, even if the desire for scientific rigour may prompt attempts at causal reformulation.[50] In a teleological interpretation of succession it is not simply that the community actually changes in such a way that it moves towards a state of balance; rather, succession is a development towards a goal.[51] This climax state is not only a normal state in the sense that it is generally reached; it is one that *ought* to be reached by the superorganism. Deviations from this are *maldevelopments* – similar developments in individual organisms are judged no differently. The attributes associated with development – i.e. internal differentiation and a consequent increase in diversity, functional integration and stability in the sense of homeostatic independence from disrupting environmental impacts – are not merely facts in this view but are necessary attributes and the *standard* for the development of the organic community to a *higher* development; the climax state is the state of *perfection* (in relative terms, dependent on what is possible for the community on the basis of its inherent "predispositions"). Thus, value judgements are being (implicitly) expressed here. Initially, these are "values" only from the perspective of the community (an end in itself). However, if "nature" is conceptualised as an organic whole, just like the community, and if "the human being" is seen as "a part of nature" – as is suggested by the world view generally associated with that of ecological holism – then these values take on an ethical quality, because they are dictating a norm to humans.[52]

If the varieties of organisms in an organic community are all mutually dependent on one another, then all the species are affected if one of them is removed, and the obvious conclusion would be that the community as a whole has suffered *harm*. However, every species (every organism) is judged on the basis of the function it fulfils for the maintenance of the community; conversely, the whole is important because it is indispensable for each individual species. Thus, the conclusion reached in a circular but rigorous way from the premise that communities are organisms is that "ecosystems" must be protected for the sake of the species and species for the sake of ecosystems (on this, cf. especially biocentric-holistic approaches to nature conservation of scholars such as Leopold 1949; Meyer-Abich 1984; Callicott 1989).

---

[50] States of equilibrium at the end of a succession can in principle be understood in exactly the same way as states of flow equilibrium found in the abiotic domain, that is, in purely causal terms. However, in organicist theories they are understood as an organic balance (for this term, see Weil 1999) as in an organism. The question of whether one regards the end-stage explanations given here as heuristic resources (see above) or whether one sees them as causes that are objectively at work in nature is the deciding factor in whether or not theories about superorganisms (or theories about organisms in general) can be regarded as scientifically admissible; whether they are correct is another matter.

[51] Cf. for early nature conservation e.g. Schoenichen (1942); Thienemann (1944).

[52] On the different meanings of what is and can be called "value" and "judgement" around ecology and nature protection, cf. e.g. Brenner (1996); Eser and Potthast (1997).

If change in a community constitutes a development much like that in an organism, then what emerges from this is an indication of the kind of technology appropriate to communities. The concept of the organism implies that the latter cannot be *constructed* but rather grows and develops. The organism can be *nurtured* – and it is advisable to nurture it, if one wants to make "sustainable" use of it. Its development can be *encouraged*, while deviations from the state it should be in require *healing*. Such holistic natural objects may also be wholes consisting of nature and humans (cultural landscapes).[53] (cf. also Eisel 1980)

Arguments of this kind are widespread in both nature conservation and ecology, as, for example, in current theories around the concept of "ecosystem health".[54] Indeed, it is hard to find any exceptions to what has become "common sense" within the nature conservation and environmental movements. But one caveat should be made at this point: the further away one goes from the sphere of commitments and calls for environmental sensitivity and the closer one gets to the sphere of administrative action, the more one finds elements that come from the opposite (reductionist) direction – often in "bizarre and paradoxical rhetorical combinations"[55] (Hard 1994, p. 126).

The Individualist-Reductionist Approach in Nature Conservation

The individualist approach has quite different implications due to its understanding of the reality constituted by "community" (in the sense of "Gesellschaft"). What should count as a community depends on the theories, definitions and questions brought to bear by the scientist. Statements about a community being harmed or destroyed are valid only in relation to the scientist's definitions of what is to count as a community in the first place. Similarly, the phenomenon of maldevelopment is meaningful only in relation to these definitions. There is no state that could be formulated as a point of reference on the basis of the requirements of the community itself and in relation to which one might judge changes to be either positive or negative. The community does not maintain *itself* against disturbances from outside itself; rather, changes in the environmental conditions affecting individual species bring about a rearrangement of species into a new group. There is no reason to describe certain combinations as "intact" communities and others not. Since this approach has no concept of a group of species as a community, as in the model of an organism, it makes no sense to ascribe a "perspective" to a group of organisms,

---

[53] Nothing should be added into the organic community, either. This results in a rejection of foreign species (e.g. Disko 1996; for the debate about the role of foreign species in nature conservation, see Eser and Potthast 1999, Körner 2000).

[54] See. e.g. Rapport (1989, 1995); Costanza et al. (1992); Ferguson (1994); Rapport et al. (1998).

[55] "bizarre und paradoxe rhetorische Mischungen".

let alone one in which something may be "expedient" or "valuable" for the group itself. Instead, in making such an ascription, one is necessarily relying on expectations directed at the system in question from the outside – the system may be useful for particular purposes, but the purpose cannot lie in the system itself.[56] It is not possible, from the perspective of "communities", to demand that they should not be changed, but that they should be protected instead.[57] This is because they do not exist as living entities: one cannot meaningfully say that something is either good or bad for them.

In the individualist perspective, species combinations are arbitrary as far as further development is concerned; in contrast to organicism, they are not necessary life phases in the development of a community towards the climax state.[58] However, they are *not* arbitrary in the sense that there are no causal laws governing the arrangement of individual organisms into communities, because the abiotic environment effects a selection and individuals act upon one another in a causal way (competition, predator-prey relationships, etc.). If we knew the causal laws and the parameters at work here, we could create specific species combinations, including completely new ones. The technology appropriate to individualistically interpreted communities is therefore one based on *construction* and not on nurture and healing.[59]

Transcending the Dualism Between Holism and Reductionism?

Even if there is rarely any explicit mention of these classical positions in controversies over the correct ecological theory and research practice, they can often be reconstructed nonetheless as the axis around which the debate takes place. There have been numerous attempts to adopt positions in between the two (e.g. Levins and Lewontin 1980). Organicist holism is clearly discernible in those studies that are commonly described under the rubrics of "ecosystem health" or "the Gaia hypothesis" (e.g. Lovelock 1979). Apart from the latter, no one in today's ecology talks anymore about communities being superorganisms. Whenever this term is used, it does not refer to the essence of communities in general, but only to certain specific communities that are assumed to be units of selection (e.g. Wilson and Sober 1989, Sober and Wilson 1994). Today's ecology rarely contains such extreme organicist-holist positions as those that were common during the first few decades of the twentieth century. Changes have taken place since that time, including various

---

[56] On the movements that have prompted such trains of thought in the theoretical development of German environmental planning, see Eckebrecht (2002) among others.

[57] The purpose we determine for them can also be, of course, to preserve a particular species.

[58] This is why foreign species are not rejected in the corresponding conceptions of nature conservation (e.g. Reichholf 1993).

[59] The image of nature that comes to the fore here is hardly ever the individualistic one of a "community of autonomous individuals", but that of the *machine*, as we shall see below.

modernisations of the classical theories. For example, at the time when Clements's theory was highly influential in the USA, it was not so much its holistic character in general that was criticised but rather the point that, even if different sets of circumstances existed in a climatic region, there could only be *one single* climax community ("monoclimax"). The so-called polyclimax theory accorded local conditions a much bigger role in the development of vegetation. Thus, it is possible for different climax communities to come about in a single climax region, influenced, for example, by small-scale varying edaphic factors (e.g. Tansley 1935, Whittaker 1953). The structure of the classical holistic approach can be seen more or less clearly during the second half of the twentieth century, for example in the work of E.P. Odum (1953, 1969), Margalef (1958), Patten (1978), Trojan (1984), Ulanowicz (1997) and Jørgensen (2000), even if the organicist aspect usually remains secondary to a more *technical* perspective in studies oriented towards systems theory (see Chap. 27, as well as Chap. 15).

Individualist theories were initially paid scant attention.[60] It was only during the 1950s that they assumed a more prominent role (e.g. Curtis and McIntosh 1950; Egler 1951; Whittaker 1953), once certain approaches – sometimes described as "population-oriented" – had been developed and began to be used in community ecology. Ever since then, ecology has seen itself more within this tradition.[61] Some modern theories do exist that are very similar to the classical formulation of the individualist concept (e.g. Hubbell 2001). However, rather than discussing the history of these approaches in any further detail, we intend in the following to look instead at one important change to which holistic approaches were subject.

One factor that proved decisive for the further development of holism in ecology was the coining of the ecosystem concept (Tansley 1935). Another equally important factor – if not more so – was the attempt undertaken in other parts of biology (especially in physiology) to accommodate the extreme reductionist and holistic positions – encountered here predominantly in the form of an opposition between mechanism and vitalism – within the framework of a systems theoretical conception of the organism, and later within General Systems Theory (Bertalanffy 1932, 1949, 1968)[62]. Both these factors laid the groundwork for the development in the 1940s and 1950s of a strand of research known as the ecosystem approach (see also Chap. 15). Let it simply be noted here that ecosystem theories entail different sorts of ideas about systems or, to put it another way, there is an apparent ambiguity within the ecosystem concept, an ambiguity that is also found in General Systems Theory with regard to the holism-reductionism issue (cf. Müller 1996).

Nowadays ecosystem theory comes in a bewildering array of variants. Ecosystems are generally defined as open, self-regulating systems in a state of flow

---

[60] For Anglo-American ecology, cf. McIntosh (1975, 1985, 1995); Simberloff (1980); for Europe, see the fascinating debate around the essay by Peus (1954) in Schwerdtfeger et al. (1960/1961).

[61] See also the debate recounted in Saarinen (1982).

[62] For Bertalanffy's early systems theory see Müller (1996); Schwarz (1996); Voigt (2001).

equilibrium, their defining characteristic being the uptake and output of matter and energy (e.g. Lindeman 1942) and information (e.g. Margalef 1958).

Even if these ecosystem theories explicitly see themselves as holistic (e.g. Odum 1953; Jørgensen 2000; Jørgensen and Müller 2000) and even if they *are* holistic in the sense that they take the whole of the system as given and see the parts as existing in a functional relationship to the latter, they are nonetheless reductionist in the sense that they entail a considerable degree of abstraction and are resolutely scientistic. Rather than focusing on communities ("Gesellschaften") described as combinations of particular species, their object of study typically consists instead of organisms or groups of organisms understood as "compartments" and viewed in the same way as other components of the ecosystem, including abiotic ones, which fulfil the same function within it.[63]

Those ecological theories from the second half of the twentieth century referred to as holistic generally have this systems theoretical scientific character (e.g. Patten 1978; Jørgensen and Müller 2000) and it is this that distinguishes them fundamentally from the old organicist holism.[64] There are, it is true, plenty of ecologists who use the concept of ecosystem and associate it with certain features of communities described by the old organicist holism, such as that ecosystems are real units which – like living beings – are self-enclosed, exist in relationship to their environment, and develop like a unit towards individuality, so that in principle it is possible to speak of their destruction in the same way we speak of the destruction of an organism etc. There is also the view that the endogenous development of an ecosystem leads to a predictable increase in biomass, diversity and stability in the course of succession (particularly influential here: Odum 1969). To a certain extent such theories involve reformulating the holistic-organicist concept of wholeness within the framework of systems thinking (cf. McIntosh 1980). The systems concept, which General Systems Theory takes as its starting point, proves attractive insofar as it offers the possibility of conceptualising this (cf. Müller 1996). However, the transition from organicism to ecosystem theory entailed a process of scientization, the logic of which contains a tendency towards fundamentally changing the focus of research and, with it, the kind of objects with which that research is concerned (see Chap. 27).

The old organicist holism attempted to describe the entire range of organisms occurring in a specific space as a higher-order organism, or at least to identify "superorganisms" among the more or less close associations of individual organisms. In other words, it sought at the synecological level those objects in nature whose mode of functioning is aimed at self-maintenance. Ecosystem theory, by contrast, demarcates certain systems, largely for instrumental reasons. These reasons may be of a technical, practical nature or they may be theoretical. If they are the latter, however, then they are also aimed at potential technical mastery: ecosystem

---

[63] With regard to the laws that are formulated, it is often unclear, as in systems theory in general, whether they are laws of nature – that is (usually physical) laws that relate to a particular domain of the empirical world – or whether they are formal laws of the mathematical kind (cf. Müller 1996).

[64] See the firm rejection of "ecosystems research" by such circles in Detering and Schwabe (1978).

theory regards ecosystems from the point of view of specific functions which the system can fulfil for something outside of it, e.g. producing biomass, purifying liquid wastes, stabilising the climate etc. ("ecosystem services").[65] In principle the number of functions an ecosystem can fulfil is unlimited, as is the number of ecosystems that can be formed by stating the functions from whose perspective they have been defined. Thus, it is not the case that *there are* a certain number of (varieties of) ecosystems (whose function lies in their self-maintenance), and which one can discover and describe; rather, a fundamentally arbitrary number of (varieties of) ecosystems are constructed (in theory or in reality), depending on particular interests, just like the communities ("Gesellschaften") of the individualist approach. Ecosystems are therefore *artefacts*,[66] (super)organisms constitutively not. In moving from the concept of the superorganism to the concept of the ecosystem, the purpose of ascribing value to certain states of the parts and the whole becomes a slightly different one: the characteristics and behaviour of an ecosystem's components that are "good" for the ecosystem itself are those which make it possible for them to fulfil the functions defined *by us* and by our interests. The structure of (super)organisms is reflexive: the parts are self-generating, a process mediated via the whole, and *that* is the function of both the parts *and* the whole. Superorganisms decide for themselves, as it were, what is "good" for them. With ecosystems, however, *we* set an end point to what in principle is an endless cycle of functional utility, and we do so at a point that seems interesting to *us*. By contrast, the chains of function in an organism reconnect to themselves, and this is why an organism is an end in itself (cf. McLaughlin 2001). Despite the fact that the old organicist rhetoric continues to be used in ecosystem theory (e.g. through terms such as self-regulation and self-preservation as, for example, in Odum 1969), the meaning of those terms has changed (even if they are often intended to convey their old meaning). One refers thus to processes which, without any help from the outside, contribute towards maintaining a state that is useful for the purpose *defined by us.*

These ecosystem theories can be understood and criticised as being both reductionistic and holistic. They are addressed towards the ecosystem as a whole but reduce the diversity of its characteristics to a very few which, moreover, are amenable to causal-mechanical analysis. The old problem of the relationship between parts and wholes, or between individual and community, as it appears in the holism-reductionism debate, is not addressed in these ecosystem theories, insofar as their focus is largely on the material, energetic and possibly informational aspects of interactions in a system, in which the difference between abiotic and biotic components is irrelevant: the organisms actually involved in the interactions are not of interest (cf. Bergandi 1995; Voigt 2009). The idea abstracted from this is that the system's elements (or the biotic part of the system's elements) are individual organisms of different species – they are regarded merely in terms of their function (e.g. for the energy and material

---

[65] Cf. e.g. the articles in Costanza et al. (1997); Daily (1997); de Groot et al. (2002); Farber et al. (2002).

[66] On this concept in relation to the concept of function in organisms, see McLaughlin (2001).

cycle of the system) which in turn is defined with regard to functions for some utilitarian interest. This suggests that a different basic model has come to take the place of the "organic community", namely: the machine. cf. Voigt 2009.

Summary

Holism-reductionism controversies have long played a significant role in various sciences, but especially so in philosophy and political ideologies. It is only against this background that the holism-reductionism debate in ecology can be understood, because here the issue is more than simply one of whether scientific theories of a certain type describe certain natural phenomena correctly: it is a question of a conflict of ideas about the correct relationship between individual and community, human beings and nature, progress and tradition. Taking this as our starting point, we reconstruct the forms this controversy has assumed in ecology and which *conceptual structures* it is based on; there is not an arbitrary range of these forms, but rather only certain structurally feasible possibilities for conceptualising the relationship between parts and wholes.

Many different things are referred under the heading of holism and reductionism. For example, it is possible to establish systematically those aspects of an object of inquiry or of a scientist's relationship to it to which holistic or reductionist principles are being applied.

In ecology the whole is a *community* and the parts are *individuals*. The contrast between reductionism and holism takes the form of an opposition between *individualism* and *organicism*. These two approaches are presented here first as ideal typical constructions in the classical form in which they emerged during the opening decades of the twentieth century, using examples by way of illustration.

In organicist holism, "community" is conceived of specifically as an *organic* community, or as a superorganism, that is, the relationship between parts and wholes is conceptualised in analogy to the relationship between organ and organism. The key point is that the organicist-holistic theory is a theory of development. The development of the community is a process – conceived of in teleological terms – of adaptation to environmental conditions; at the same time it is a process of breaking free from certain constraints. This corresponds closely to the structure of conservative political philosophies: for the latter, too, society is an organic community, or a higher-order organism. Its development consists in the (active) perfecting of the qualities inherent in "nations" ("Völker") – their "character" – and in the perfection of the qualities inherent in the nature that exists in their "Lebensraum". It is in adapting to the dictates of actual nature within their "Lebensraum" – which simultaneously involves breaking free from the constraints of nature – that "culture" develops. Rarely is any attention paid to the fact that holistic theory is also a theory of development; the consequence is that holistic ecology's image of nature is falsely regarded as a static one, just like the image of society found in conservative political philosophy.

As far as individualistic ecological theories are concerned, the fundamental unit is the individual organism. They accord "reality" to it alone, not to the community. It is the needs of the individuals, not the functional necessities of a community, which force the individual organisms to establish relationships to others. Social change is without a goal, its direction dependent on chance factors, and if there is such a thing as development to a higher level, then this does not refer to a community coming closer its prescribed goal, but is instead an improvement from the perspective of the individuals – those who win through in the competition. Correspondingly, liberal political philosophies view society as a system of interactions between individuals, where interaction results from its degree of usefulness for the latter; it is also a system in which interests are reconciled in utilitarian fashion on the premiss of a "struggle for survival". History is an open-ended process shaped by autonomous subjects and determined by the forces of chance.

Starting out from an approach based on a theory of constitution that allows us to understand the two types of ecological theory as being "inspired" by political world views, it is possible to understand better both the differences between the two types and their immanent logic as well as the utterly divergent practical consequences entailed by holistic ecological theories on the one hand and reductionist theories on the other. "Practice" may be a reference to technology (e.g. "ecosystem management") or to "knowledge for orientation", in which nature is viewed within a context of values.

In conclusion, the dynamic that gives rise to both approaches and causes them to change is described using the example of the transformation of the classical holistic position. The old organicist holism sought at a synecological level those objects in nature whose mode of functioning had the purpose of self-preservation. In contrast to this, ecosystem theory – even when it explicitly calls itself holistic – tends to demarcate certain systems in the interests of instrumental control. Ecosystems are not superorganisms found in nature but rather artefacts. Through this change, the purpose of attributing value terms to certain states of the parts and the whole becomes a somewhat different one. A different basic model comes to take the place of the "organic community", namely: the machine.

References

Agazzi E (ed) (1991) The problem of reductionism in science. (Colloquium of the Swiss Society of Logic and Philosophy of Science, Zürich, May 18–19, 1990. Zürich, Schweiz: Schweizerische Gesellschaft für Logik und Philosophie der Wissenschaften). Kluwer Academic, Dordrecht

Anker P (2001) Imperial ecology: environmental order in the British empire. Harvard University Press, Cambridge, pp 1895–1945

Arneson RJ (ed) (1992) Liberalism 3 Vol. Schools of thought in politics 2 An Elgar reference collection. - Hants: Edward Elgar, Aldershot (UK)

Ayala FJ (1974) Introduction. In: Ayala FJ, Dobzhansky T (eds) Studies in the philosophy of biology: reductionism and related problems. Macmillan, London, pp 7–16

Ayala FJ, Dobzhansky T (eds) (1974) Studies in the philosophy of biology: reductionism and related problems. Macmillan, London

Bertalanffy von L (1932) Theoretische Biologie, vol I, Allgemeine Theorie, Physikochemie, Aufbau und Entwicklung des Organismus. Verlag Gebrüder Bornträger, Berlin, p 349

Bertalanffy von L (1949) Das biologische Weltbild, vol I, Die Stellung des Lebens in Natur und Wissenschaft. Francke, Bern

Bertalanffy von L (1968) General system theory: foundations, development, applications. Braziller, New York

Bergandi D (1995) "Reductionist holism": an oxymoron or a philosophical chimaera of E. P. Odum's systems ecology? Ludus vitalis 3(5):145–180

Bergandi D, Blandin P (1998) Holism vs. reductionism: do ecosystem ecology and landscape ecology clarify the debate? Acta Biotheor 46(3):185–206

Bews JW (1935) Human ecology. Oxford University Press, London

Bock GR, Goode JA (eds) (1998) The limits of reductionism in biology (Novartis Foundation symposium 213, held at the Novartis Foundation, London, May 13–15 1997). Wiley, Chichester

Botkin DB (1990) Discordant harmonies: a new ecology for the twenty-first century. Oxford University Press, New York

Braun-Blanquet J (1928) Pflanzensoziologie: Grundzüge der Vegetationskunde. J. Springer, Berlin

Brenner A (1996) Ökologie-Ethik. Reclam, Leipzig

Bueno G (1990) Holismus. In: Sandkühler HJ (ed) Europäische Enzyklopädie zu Philosophie und Wissenschaften. Felix Meiner Verlag, Hamburg, pp 552–559

Callicott JB (1989) In defense of the land ethic: essays in environmental philosophy. State University of New York Press, Albany

Capra F (1982) The turning point: science, society, and the rising culture. Simon & Schuster, New York

Cassirer E (1921) Kants Leben und Lehre. Verlag Bruno Cassirer, Berlin

Clements FE (1916) Plant succession: an analysis of the development of vegetation. Carnegie Institution of Washington, Washington

Clements FE (1936) Nature and structure of the climax. In: The Journal of Ecology 24: pp 252–284 (– reprinted in: Allred B W & Edith S Clements (eds.) (1945) Dynamics of Vegetation: selections from the writings of Frederic E. Clements. – New York: The H. W. Wilson Company, pp. 1–21)

Clements FE, Shelford VE (1939) Bio-ecology. Wiley, New York

Costanza R, Norton BG, Haskell BD (eds) (1992) Ecosystem health: new goals for environmental management. Island Press, Washington D.C.

Costanza R, d'Arge R, de Groot R, Farber S, Grasso M, Hannon B, Limburg K, Naeem S, O'Neill RV, Paruelo J, Raskin RG, Sutton P, van den Belt M (1997) The value of the world's ecosystem services and natural capital. Nature 387(6230):253–260

Crick F (1966) Of molecules and men. (The John Danz lectures). University of Washington Press, Seattle

Curtis JT, McIntosh RP (1950) The interrelations of certain analytic and synthetic phytosociological characters. Ecology 31:434–455

Daily GC (ed) (1997) Nature's services: societal dependence on natural ecosystems. Island Press, Washington D.C.

Dawkins R (1976) The selfish gene. Oxford University Press, Oxford

de Groot RS, Wilson MA, Boumans RMJ (2002) A typology for the classification, description and valuation of ecosystem functions, goods and services. Ecol Econ 41:393–408

Desmond A, Moore J (1991) Darwin. Michael Joseph, London, pp 21–807

Detering K, Schwabe GH (1978) System, Natur und Sprache. Scheidewege 8(1):104–132

Dilthey W (1883) Einleitung in die Geisteswissenschaften: Versuch einer Grundlegung für das Studium der Gesellschaft und der Geschichte (1). Duncker & Humblot, Leipzig

Disko R (1996) Mehr Intoleranz gegen fremde Arten. Nationalpark 93(4):38–42

Drengson A, Inoue Y (eds) (1995) The deep ecology movement: an introductory anthology. North Atlantic Books, Berkeley

Driesch H (1935) Die Maschine und der Organismus. In: Meyer-Abich A (ed) Bios 4. Barth, Leipzig

Drury WH, Ian Nisbet CT (1973) Succession. J Arnold Arboretum 54(3):331–368

Eckebrecht B (2002) Das Naturraumpotential. Zur Rekonstruktion einer geographischen Fachprogrammatik in der Landschaftsplanung (Beiträge zur Kulturgeschichte der Natur 4. In: Eisel U, Trepl L (eds) Freising: TU München, Lehrstuhl für Landschaftsökologie

Egler FE (1951) A commentary on American plant ecology based on the textbooks of 1947 – 1949. Ecology 32:673–695

Ehrenfels von C (1890) Über Gestaltqualitäten. Vierteljahrsschrift für wissenschaftliche Philosophie 14:249–292

Ehrenfels von C (1916) Kosmogonie. Eugen Diederichs, Jena

Eisel U (1980) Die Entwicklung der Anthropogeographie von einer "Raumwissenschaft" zur Gesellschaftswissenschaft, vol 17, Urbs et Regio. Kasseler Schriften zur Geographie und Planung, Kassel

Eisel U (1982) Die schöne Landschaft als kritische Utopie oder als konservatives Relikt. Soziale Welt 38(2):157–168

Eisel U (1991) Warnung vor dem Leben. Gesellschaftstheorie als "Kritik der Politischen Biologie". In: Hassenpflug D (ed) Industrialismus und Ökoromantik: Geschichte und Perspektiven der Ökologisierung. Deutscher Universitäts.-Verlag, Wiesbaden, pp 159–192

Eisel U (2002) Das Leben ist nicht einfach wegzudenken. In: Lotz A, Gnädinger J (eds) Wie kommt die Ökologie zu ihren Gegenständen? Gegenstandskonstitution und Modellierung in den ökologischen Wissenschaften. (Beiträge zur Jahrestagung des AK Theorie in der Ökologie. – Theorie in der Ökologie 7). Peter Lang, Frankfurt a. M, pp 129–151

Engels, F (1886) Dialektik der Natur. In: Marx Karl and Friedrich Engels (1962) Werke, vol 20. Dietz Verlag, Berlin, pp 305–570

Eser U, Potthast T (1997) Bewertungsproblem und Normbegriff in Ökologie und Naturschutz aus wissenschaftsethischer Perspektive. Zeitschrift für Ökologie und Naturschutz 6:181–189

Eser U, Potthast T (1999) Naturschutzethik – Eine Einführung für die Praxis. Nomos-Verlagsgesellschaft, Baden-Baden

Farber SC, Costanza R, Wilson MA (2002) Economic and ecological concepts for valuing ecosystem services. Ecol Econ 4(3):375–392

Ferguson BK (1994) The concept of landscape health. J Environ Manage 40:129–137

Forbes SA (1887) In: The lake as a microcosm. Bulletin of the Scientific Association, Peoria, pp. 77–87 (reprinted in 1925: Illinois Nat Hist Survey Bull. 15, 9: pp 537–550)

Foucault M (1969) L' archéologie du savoir. Gallimard, Paris

Friederichs K (1927) Grundsätzliches über die Lebenseinheiten höherer Ordnung und den ökologischen Einheitsfaktor. Naturwissenschaften 15(7):153–157, 182–186

Friederichs K (1934) Vom Wesen der Ökologie. Sudhoffs Arch Gesch Med Naturwiss 27(3):277–285

Friederichs K (1937) Ökologie als Wissenschaft von der Natur oder biologische Raumforschung. In: Bios 7. J. A. Barth, Leipzig

Friederichs K (1957) Der Gegenstand der Ökologie. Stud Gen 10(2):112–124, 10:3: 125–144

Gams H (1918) Prinzipienfragen der Vegetationsforschung Ein Beitrag zur Begriffsklärung und Methodik der Biocoenologie. Vierteljahresschrift Naturforschende Gesellschaft Zürich 63:293–493

Gleason HA (1917) The structure and development of the plant association. Bull Torrey Botanical Club 44:463–481

Gleason HA (1926) The individualistic concept of the plant association. Bull Torrey Botanical Club 53:7–26

Gleason HA (1927) Further views on the succession-concept. Ecology 8(3):299–326

Gnädinger J (2002) Organismenzentrierte Rekonstruktion funktionaler Grenzen von synökologischen Einheiten. In: Lotz A and Gnädinger J (eds) Wie kommt die Ökologie zu ihren Gegenständen? Gegenstandskonstitution und Modellierung in den ökologischen Wissenschaften. (Beiträge zur Jahrestagung des AK Theorie in der Ökologie. – Theorie in der Ökologie 7). Peter Lang Verlag, Frankfurt a.M, pp 195–209

Goldstein K (1934) Der Aufbau des Organismus: Einführung in die Biologie unter besonderer Berücksichtigung der Erfahrungen am kranken Menschen. M. Nijhoff, Haag

Golley FB (1993) A history of the ecosystem concept in ecology: more than the sum of the parts. Yale University Press New Haven, New Haven
Greiffenhagen M (1971) Das Dilemma des Konservatismus in Deutschland. Piper, München
Grisebach A (1838) Über den Einfluß des Klimas auf die Begrenzung der natürlichen Floren. Linnaea 12:159–200
Habermas J (1968) Erkenntnis und Interesse. In: Habermas, Jürgen: Technik und Wissenschaft als "Ideologie". Suhrkamp, Frankfurt a.M, pp 146–168
Hagen JB (1992) An entangled bank: the origins of ecosystem ecology. Rutgers University Press, New Brunswick
Haldane JS (1931) The philosophical basis of biology (Donnellan lectures, University of Dublin 1930). Hodder and Stoughton, London
Hard G (1994) Die Natur, die Stadt und die Ökologie. Reflexionen über "Stadtnatur"und „Stadtökologie". In: Ernste H (ed) Pathways to human ecology. Lang, Bern, pp 161–180
Harrington A (1996) Reenchanted science: holism in German culture from Wilhelm II to Hitler. Princeton University Press, Princeton
Horn HS (1976) Succession. In: May RM (ed) Theoretical ecology: principles and applications. Blackwell Scientific Pub. Ltd., Oxford, pp 187–204
Hubbell SP (2001) The unified neutral theory of biodiversity and biogeography. Princeton University Press, Princeton
Hull DL, Ruse M (eds) (1998) The philosophy of biology. Oxford University Press, Oxford
Humboldt von A (1806) Ideen zu einer Physiognomik der Gewächse. Cotta, Tübingen
Jax K (1998) Holocoen and ecosystem on the origin and historical consequences of two concepts. J Hist Biol 31:113–142
Jax K (2002) Die Einheiten der Ökologie: Analyse, Methodenentwicklung und Anwendung in Ökologie und Naturschutz. Theorie in der Ökologie, 5th edn. Peter Lang, Frankfurt/M
Jørgensen SE (2000) A general outline of thermodynamic approaches to ecosystem theory. In: Jørgensen SE, Felix M (eds) Handbook of ecosystem theories and management. Lewis Publishers, London, pp 113–135
Jørgensen SE, Müller F (2000) Ecosystems as complex systems. In: Jørgensen SE, Felix M (eds) Handbook of ecosystem theories and management. Lewis Publishers, London, pp 5–21
Kant I (1970) Kritik der Urteilskraft, edition 1995. Suhrkamp, Frankfurt/M
Keller DR, Golley FB (eds) (2000) The philosophy of ecology: from science to synthesis. University of Georgia Press, Athens
Kirchhoff T (2005) Kultur als individuelles Mensch-Natur-Verhältnis. Herders Theorie kultureller Eigenart und Vielfalt. In: Weingarten M (ed) Strukturierung von Raum und Landschaft. Konzepte in Ökologie und der Theorie gesellschaftlicher Naturverhältnisse. Westfälisches Dampfboot, Münster, pp 63–106
Kirchhoff T (2007) Systemauffassungen und biologische Theorien. Zur Herkunft von Individualitätskonzeptionen und ihrer Bedeutung für die Theorie ökologischer Einheiten. (= Beiträge zur Kulturgeschichte der Natur, Band 16). Freising
Köchy K (1997) Ganzheit und Wissenschaft: das historische Fallbeispiel der romantischen Naturforschung, vol 180, Epistemata, Reihe Philosophie. Königshausen & Neumann, Würzburg
Köchy K (2003) Perspektiven des Organischen: Biophilosophie zwischen Natur- und Wissenschaftsphilosophie. Schöningh, Paderborn
Köhler W (1920) Die physischen Gestalten in Ruhe und im stationären Zustand: eine naturphilosophische Untersuchung. Vieweg, Braunschweig
Körner S (2000) Das Heimische und das Fremde: Die Werte Vielfalt, Eigenart und Schönheit in der konservativen und in der Liberal-progressiven Naturschutzauffassung. (Fremde Nähe Beiträge zur interkulturellen Diskussion 14). LIT, Münster
Kuhn TS (1962) The structure of scientific revolutions. University of Chicago Press, Chicago
Kühnl R (1982) Das liberale Modell politischer Herrschaft. In: Abendroth W (ed) Einführung in die politische Wissenschaft, 6th edn. Francke, München, pp 57–85
Langthaler R (1992) Organismus und Umwelt: die biologische Umweltlehre im Spiegel traditioneller Naturphilosophie, vol 34, Studien und Materialien zur Geschichte der Philosophie. Georg Olms Verlag, Zürich

Lenoble F (1926) À propos des associations végétales. Bulletin de la Société Botanique de France 73:873–893

Leopold A (1949) A Sand County almanac: and Sketches here and there. Oxford University Press, New York

Levins R, Lewontin RC (1980) Dialectics and reductionism in ecology. Synthese 43:47–78

Levins R, Lewontin RC (1994) Holism and reductionism in ecology. CNS 5(4):33–40

Lindeman RL (1942) The trophic-dynamic aspect of ecology. Ecology 23(4):399–418

Loeb J (1916) The organism as a whole: from a physicochemical viewpoint. Putnam's Sons, New York

Looijen RC (2000) Holism and reductionism in biology and ecology: the mutual dependence of higher and lower level research programmes, vol 23, Episteme. Kluwer Academic Publisher, Dordrecht

Lovelock JE (1979) Gaia: a new look at life on earth. Oxford University Press, Oxford

MacMahon JA, Schimpf DJ, Andersen DC, Smith KG, Bayn RLJ (1981) An organism-centered approach to some community and ecosystem concepts. J Theor Biol 88(2):287–307

Margalef R (1958) Information theory in ecology. Gen Syst 3:36–71

Mayr E (1982) The growth of biological thought: diversity, evolution and inheritance. Belknap, Cambridge

Mayr E (1988) The multiple meanings of teleological. In: Toward a new philosophy of biology: observations of an evolutionist. Belknap Press of Harvard University Press, Cambridge, pp 38–66

McIntosh RP (1975) H. A. Gleason. "Individualistic Ecologist" 1882–1975: his contributions to ecological theory. Bull Torrey Botanical Club 102(5):253–273

McIntosh RP (1980) The background and some current problems of theoretical ecology. Synthese 43:195–255

McIntosh RP (1985) The background of ecology: concept and theory. Cambridge University Press, Cambridge

McIntosh RP (1995) H. A. Gleason´s "individualistic concept" and theory of animal communties: a continuing controversy. Biol Rev Camb Philos Soc 70:317–357

McLaughlin P (2001) What functions explain: functional explanation and self-reproducing systems. Cambridge University Press, Cambridge

Meyer-Abich A (1934) Ideen und Ideale der biologischen Erkenntnis. Beiträge zur Theorie und Geschichte der biologischen Ideologien, vol 1, Bios. Barth, Leipzig

Meyer-Abich A (1948) Naturphilosophie auf neuen Wegen. Hippokrates, Stuttgart

Meyer-Abich KM (1984) Wege zum Frieden mit der Natur: praktische Naturphilosophie für die Umweltpolitik. Hanser, München

Mittelstraß J (1995) Holismus. In: Mittelstraß J (ed) Enzyklopädie Philosophie und Wissenschaftstheorie. Metzler, Stuttgart, pp 123–124

Mocek R (1974) Wilhelm Roux, Hans Driesch: Zur Geschichte der Entwicklungsphysiologie der Tiere, Entwicklungsmechanik. Fischer, Jenas

Mocek R (1998) Die werdende Form: eine Geschichte der kausalen Morphologie, vol 3, Acta biohistorica. Basilisken-Presse, Marburg

Müller K (1996) Allgemeine Systemtheorie. Geschichte, Methodologie und sozialwissenschaftliche Heuristik eines Wissenschaftsprogramms, vol 164, Studien zur Sozialwissenschaft. Westdeutscher Verlag, Opladen

Naess A (1973) The shallow and the deep, long-range ecology movement: a summary. Inquiry 16:95–100

Nagel E (1949) The meaning of reduction in the natural sciences. In: Stauffer RC (ed) Science and Civilization. University of Wisconsin Press, Madison

Nagel E (1961) The structure of science: problems in the logic of scientific explanation. Harcourt, Brace & World, New York

Nagel E (1979) Teleology revisited. In: Nagel E (ed) Teleology revisited and other essays in the philosophy and history of science. Columbia University Press, New York, pp 275–316

Needham J (1932) Thoughts on the problem of biological organization. Scientia 52:84–92

Negri G (1928) Popolamento vegetale ed animale delle alte montagne: relazione illustrativa delle proposte presentate dal Comitato Geografico Nazionale Italiano al Congresso internazionale di Cambridge. Istituto geografico, Florenz militare
Nordenskiöld E (1928) The history of biology: a survey. (Originally issued as Biologins historia, in three volumes, (1920–1924) Translated from the Swedish). Alfred A Knopf, New York
Odum EP (1953) Fundamentals of ecology. W. B. Saunders, Philadelphia
Odum EP (1969) The strategy of ecosystem development: an understanding of ecological succession provides a basis for resolving man's conflict with nature. Science 164:262–270
Oppenheim P, Putnam HW (1958) Unity of science as a working hypothesis. In: Feigl H, Scriven M, Grover M (eds) Concepts, theories and the mind-body problem. Minnesota studies in the philosophy of science, 2nd edn. University of Minnesota Press, Minneapolis, pp 3–36
Patten BC (1978) Systems approach to the concept of environment. Ohio J Sci 78:206–222
Peus F (1954) Auflösung der Begriffe "Biotop" und "Biozönose". Deutsche Entomologische Zeitschrift 1:271–308
Phillips J (1934) Succession, development, the climax and the complex organism: an analysis of concepts. Part I. J Ecol 22:554–571
Phillips J (1935) Succession, development, the climax and the complex organism: an analysis of concepts Part II & III. J Ecol 23:210–246, 488–508
Pickett STA, Parker VT, Fiedler PL (1992) The new paradigm in ecology: implications for conservation biology above the species level. In: Fiedler PL, Jain SK (eds) Conservation biology. The theory and practice of conservation, preservation and management. Chapman & Hall, New York, pp 65–88
Portmann A (1948) Die Tiergestalt Studien über die Bedeutung der tierischen Erscheinung. Reinhardt, Basel
Putnam H (1987) The many faces of realism. Open Court Publishing, La Salle
Ramensky LG (1926) Die Gesetzmäßigkeiten im Aufbau der Pflanzendecke. Botanisches Centralblatt 7:453–455
Rapport DJ (1989) What constitutes ecosystem health? Perspect Biol Med 33(1):120–132
Rapport DJ (1995) Ecosystem health: more than a metaphor. Environ Values 4:287–309
Rapport DJ, Costanza R, McMichael AJ (1998) Assesing ecosystem health. Trends Ecol Evol 13(10):397–402
Reichholf JH (1993) Comeback der Biber: ökologische Überraschungen. Beck, München
Rosenberg A (1985) The structure of biological science. Cambridge University Press, Cambridge
Roux W (1895) Gesammelte Abhandlungen über Entwickelungsmechanik der Organismen, vol 2. Wilhelm Engelmann, Leipzig
Ruse M (ed) (1973) Philosophy of biology. Hutchinson, London
Saarinen E (ed) (1982) Conceptual issues in ecology. Pallas paperback 23, Reidel
Scherzinger W (1995) Blickfang – Mitesser – Störenfriede. Nationalpark 88(3):52–56
Scherzinger W (1996) Naturschutz im Wald Qualitätsziele einer dynamischen Waldentwicklung. Ulmer, Stuttgart
Schoenichen W (1942) Naturschutz als völkische und internationale Kulturaufgabe. Eine Übersicht über die allgemeinen, die geologischen, botanischen, zoologischen und anthropologischen Probleme des heimatlichen wie des Weltnaturschutzes. Fischer, Jena
Schwarz AE (1996) Aus Gestalten werden Systeme: Frühe Systemtheorie in der Biologie. In: Mathes K, Broder B, Klemens E (eds) Systemtheorie in der Ökologie. Beiträge zu einer Tagung des Arbeitskreises "Theorie" in der Gesellschaft für Ökologie: Zur Entwicklung und aktuellen Bedeutung der Systemtheorie in der Ökologie. ecomed, Landsberg, pp 35–45
Schwarz AE (2003) Wasserwüste Mikrokosmos Ökosystem. Eine Geschichte der 'Eroberung' des Wasserraumes. Rombach, Freiburg im Breisgau
Schwerdtfeger F, Friederichs K, Kühnelt W, Illies JB, Schwenke W (1960/1961) Kolloquium über Biozönose-Fragen. Z Angew Entomol 47:90–116
Simberloff DS (1980) A succession of paradigms in ecology: Essentialism to materialism and probabilism. Synthese 43:3–39

Smuts JC (1926) Holism and evolution. Macmillan, New York
Sober E, Wilson DS (1994) A critical review of philosophical work on the units of selection problem. Philos Sci 61:534–555
Stöckler M (1992) Reduktionismus. In: Ritter J, Gründer K (eds) Historisches Wörterbuch der Philosophie 8. Schwabe, Basel, pp 378–383
Sukachev VN (1958) On the principles of genetic classification in biocenologie. Ecology 39:364–367
Tansley AG (1935) The use and abuse of vegetational concepts and terms. Ecology 16(3):284–307
Thienemann A (1941) Vom Wesen der Ökologie. Biologia Generalis 15:312–331
Thienemann A (1944) Der Mensch als Glied und Gestalter der Natur. Wilhelm Gronau, Jena
Thienemann A (1954) Ein drittes biozönotisches Grundprinzip. Arch Hydrobiol 49(3):421–422
Thienemann A, Kieffer JJ (1916) Schwedische Chironomiden. Arch Hydrobiol 2(Suppl):483–553
Tobey RC (1981) Saving the prairies the life cycle of the founding school of American plant ecology, 1895–1955. University of Carlifonia Press, Berkley
Tönnies F (1887) Gemeinschaft und Gesellschaft: Abhandlung des Communismus und des Socialismus als empirischer Culturformen. Fues, Leipzig
Trepl L (1987) Geschichte der Ökologie. Vom 17. Jahrhundert bis zur Gegenwart. Athenäum, Frankfurt/M
Trepl L (1993) Was sich aus ökologischen Konzepten von "Gesellschaften" über die Gesellschaft lernen läßt. Loccumer Protokolle 75(92):51–64
Trepl L (1994) Holism and reductionism in ecology: technical, political, and ideological implications. CNS 5(4):13–31
Trepl L (1997) Ökologie als konservative Naturwissenschaft. Von der schönen Landschaft zum funktionierenden Ökosystem. In: Eisel U, Schultz H-D (eds) Eographisches Denken, vol 65, Urbs et Regio, pp 467–492
Trojan P (1984) Ecosystem homeostasis. Dr. W. Junk Publishers, the Hague
Troll W (1941) Gestalt und Urbild: Gesammelte Aufsätze zu Grundfragen der organischen Morphologie. Akademische Verlagsgesellschaft, Leipzig
Ulanowicz RE (1997) Ecology, the ascendent perspective. Columbia University Press, New York
Uexküll von J (1920) Staatsbiologie: Anatomie, Physiologie. Pathologie des Staates. Berlin, Paetel
Voigt A (2001) Ludwig von Bertalanffy: Die Verwissenschaftlichung des Holismus in der Systemtheorie. Verhandlungen zur Geschichte und Theorie der Biologie 7:33–47
Voigt A (2009) Die Konstruktion der Natur. Ökologische Theorien und politische Philosophien der Vergesellschaftung. Franz Steiner, Stuttgart
Weber M (1904) Die "Objektivität" sozialwissenschaftlicher und sozialpolitischer Erkenntnis. In: Weber M (1988) Gesammelte Aufsätze zur Wissenschaftslehre. 7. (ed) Tübingen: Mohr, Siebeck, pp 146–214
Weil A (1999) Über den Begriff des Gleichgewichts in der Ökologie Ein Typisierungsvorschlag. In: Trepl L (ed) Gleichgewicht – Funktion der Biodiversität. Landschaftsentwicklung und Umweltforschung, vol 112. TU Berlin, Berlin, pp 7–97
Weil A (2005) Das Modell "Organismus" in der Ökologie: Möglichkeiten und Grenzen der Beschreibung synökologischer Einheiten, vol 11, Theorie in der Ökologie. Peter Lang, Frankfurt/M
Wertheimer M (1912) Experimentelle Studien über das Sehen von Bewegung. Barth, Leipzig
Whittaker RH (1953) A consideration of climax theory: the climax as a population and pattern. Ecol Monogr 23:41–78
Wilson DS, Sober E (1989) Reviving the superorganism. J Theor Biol 136:337–356
Wolf J (1996) Die Monoklimaxtheorie: Das biologische Konzept vom Superorganismus als Entwicklungstheorie von Individualität und Eigenart durch Expansion. In: Naturalismus. Projektbericht in zwei Bänden. S. 231–308. Bd. I. TU Berlin, Fachbereich 7

Wolf KL, Troll W (1940) Goethes morphologischer Auftrag. Versuch einer naturwissenschaftlichen Morphologie. Akademische Verlagsgesellschaft, Leipzig

Worster D (1985) Nature's economy: a history of ecological ideas. Studies in environment and history. Cambridge University Press, Cambridge

Wright L (1973) Functions. In: Conceptual issues in evolutionary biology, 2 edn. MIT Press, Cambridge, pp 27–49 (reprinted in: Sober E (ed) (1994))

Part III
About the Inner Structure of Ecology – Some Theses

Chapter 6
Conceptualizing the Heterogeneity, Embeddedness, and Ongoing Restructuring That Make Ecological Complexity 'Unruly'

Peter Taylor

Introduction

My thesis is that, although ecologists have not named the concept as such, they are always dealing with unruly complexity, that is, with ongoing change in the structure of situations that have built up over time from heterogeneous components and are embedded or situated within wider dynamics. Ecology tends to suppress such complexity by mimicking the physical sciences in constructing – materially and conceptually – well-bounded systems, which have clearly defined boundaries, coherent internal dynamics, and simply mediated relations with their external context. Ecologists can envisage themselves positioned outside the systems and seek generalizations and principles that afford a natural or economical reduction of complexity. If researchers want, in contrast, to discipline unruly complexity without suppressing it, they need to recognize that control and generalization are difficult and no privileged standpoint exists; that ongoing assessment is needed, and this requires engagement in the changing situations. The inner structure of ecology rests, therefore, on the tension between unruliness and attempts to discipline it.

This article, which builds on Taylor and Haila (2001), reviews the recent history of ecological theory with a view to highlighting the challenges of conceptualizing heterogeneity, embeddedness, and ongoing restructuring. The subject matter of this review is not well-bounded; at places I point well beyond the terrain of ecology proper. Indeed, the HOEK project of illuminating ecological concepts by positioning them in relation to the socio-historical context in which they are produced and deployed invites us to consider embeddedness of other kinds: ecological situations within socio-environmental processes; natural science within the systematic study of social change; conceptual work within scientific practice; and interpretation of science within engagement in scientific and social change. All these interconnected realms pose analogous challenges of dealing with unruly complexities (Taylor 2005). Although the unruliness-system tension can be seen in other disciplines,

P. Taylor (✉)
University of Massachusetts Boston, USA
e-mail: peter.taylor@umb.edu

such as evolutionary biology, epidemiology, and developmental psychology, ecology provides a fruitful entry point for exploring this epistemic type.

A Brief History of Recent Ecological Theory

Let me acknowledge at the outset that my conceptual map is centred in the United States and needs to be balanced by reference to other HOEK entries (see also Chaps. 7 and 8). During the 1960s and 1970s many U.S. ecologists sought theories of ecological structure and function that would be general and not dependent on historical particularities (Kingsland 1995, pp. 176–205). Systems ecologists, through extensions of thermodynamics and information theory to open biological systems, sought to explain complexity in terms of the nutrient, energy, and information flows within entire ecosystems (Taylor 2005). Community ecologists made theoretical propositions, often expressed as mathematical models, which focused on the regulation of population sizes and distributions through competition for limiting resources and other interactions. The two schools mapped broadly onto a series of conceptual-methodological contrasts: function and process vs. structure and demography; properties of wholes vs. explaining parts and building up from there; field measurements vs. mathematical modeling (Hagen 1989; see Taylor 1992 for a more complex map of commonalities and distinctions).

The status of model building was somewhat ambiguous: were models idealized representations of ecological reality (see, e.g., the "perfect crystals" of May 1973) or heuristic devices to generate further theoretical questions (Levins 1966)? Moreover, by the early 1980s ecologists of a particularistic bent were questioning many of community ecology's models, rejecting them when their fit to data was no better than alternative "null" hypotheses or "random" models (Strong et al. 1984). Scepticism about the possibility of general ecological theory became widely expressed. As Simberloff (1982) argued: Many factors operate in nature and in any particular case at least some of them will be significant. A model cannot capture the relevant factors and still have general application. Instead, ecologists should intensively investigate the natural history of particular situations and test specific hypotheses about these situations experimentally. They may be guided by knowledge about similar cases and may add to that knowledge, but they should not expect their results to be extrapolated readily to many other situations.

To some extent the particularistic current of the 1980s had been prefigured in plant ecology's shift from predictable stages of succession to shifting associations of individual species determined by their particular life histories and environmental requirements (McIntosh 1985; Taylor 1992 - or even earlier in the ecologically-rich third chapter of Darwin's 1859 "On the Origin of Species"). In other ways, however, vegetation ecology's long tradition of descriptive studies was leading to a different understanding of the difficulty of theorizing about ecological processes. Multivariate statistical techniques (or pattern analysis) could be used to cluster ecological sites into distinct communities (classification) or position them along continuous axes (ordination). The patterns exposed could then be used to generate hypotheses about causal factors or underlying environmental gradients. By the 1980s, however, vegetation ecologists (especially in Australia) had shown that the results of pattern analyses were sensitive to the models underlying the technique used and the sampling sites from the space of environmental possibilities. Popular techniques, such as principal components analysis and detrended correspondence analysis, when tested on simulated data, did not recover well the simulated environmental gradients. Techniques that reduce this model-dependence also tend to produce degenerate patterns (Faith et al. 1987; Minchin 1987). The Catch-22 is that one needs to know a lot about the causal factors behind the data in order to design efficient and distortion-free multivariate techniques that would expose those factors (Austin 1987). Inferring process from pattern is a problem that I remark on later, after further discussion of particularism versus modeling and theory-building.

The scepticism about theory and a one-sided emphasis on hypothesis testing that gained attention in the 1980s have been resisted from several angles: Observation and experiment can contribute to the generation of theory in ways other than through crucial hypothesis tests. Indeed, observations constructed for testing of a specific hypothesis may not be useful for thinking about anything beyond the local configuration observed. Theory generation draws on the many other faces of data: initial category-generating generalizations from observations, comparisons, analytic redescriptions (Haila 1988; see also Chap. 7 for an analytic description of ecological concepts and theories). Moreover, exploration of verbal and mathematical models has a valuable role in generating new concepts, framings, questions, and hypotheses.

An example of exploration of models, which is relevant to unruly complexity, concerns investigation of how complexity of communities is related to their persistence or stability. Originally, ecological theory implied that ecological complexity was able to persist because of the enhanced stability of complex ecological systems. However, mathematical analysis during the 1970s and 1980s showed that complexity works strongly against stability unless the complexity is nearly decomposable, i.e., consists of loosely linked subsystems. Subsequently, a "landscape" view arose, which holds that a community may persist in a landscape of interconnected patches even though the community is transient in each of the patches (DeAngelis and Waterhouse 1987). Meta-population theory, an actively explored variant, examines the persistence not of communities, but of populations (or phoretic associations of communities on carrier species) in such a landscape (Hastings and Harrison 1994). Another variant of the landscape view emerges from construction of model systems by addition and elimination of populations. This exploration shows that complexity can persist—at levels far greater than found in decomposable systems – even when any particular system is transient. Investigations of ecological complexity should incorporate continuing turnover of species, not only analysis of the stability and structure of the current configuration (Nee 1990; Taylor 2005). (Notice, however, that the investigation of assembly rules in community ecology – see Weiher and Keddy (1999) – tends to conflate pattern and process. In the absence of information about historical trajectories, assembly rules are better thought of as patterns of co-occurrence that are statistically significantly different from patterns that are produced by randomly sampling – "assembling" – species from the appropriately delimited species pool; Kelt and Brown 1999.)

By reintroducing historical contingency, transient or non-equilibrium situations, and embeddedness in larger contexts, exploratory modeling of the construction and turnover of systems, among other factors, contributes to undermining the aspirations of earlier decades for identifying general principles about systems and communities (Kingsland 1995, pp. 213–251). Since the 1980s ecologists in general have become increasingly aware that situations may vary according to historical trajectories that have led to them; that particularities of place and connections among places matter; that time and place is a matter of scales that differ among co-occurring species; that variation among individuals can qualitatively alter the ecological process; that this variation is a result of ongoing differentiation occurring within populations – which are specifically located and inter-connected – and that interactions among the species under study can be artifacts of the indirect effects of other "hidden" species.

In patch dynamic studies, for example, the scale and frequency of disturbances that create open "patches" is now emphasized as much as species interactions in the periods between disturbances (Pickett and White 1985). Studies of succession and of the immigration and extinction dynamics for habitat patches pay attention to the particulars of species dispersal and the habitat being colonized, and how these determine successful colonization for different species (Gray et al. 1987). On a larger scale such a shift in focus is supported by biogeographic comparisons that show that continental floras and faunas are not necessarily in equilibrium with the extant environmental conditions (Haila and Järvinen 1990). From a different angle, models that distinguish among individual organisms (in their characteristics and spatial location) have been shown to generate certain observed ecological patterns, such as patterns of change in size distribution of individuals in a population over time, where large scale, aggregated models have not (Huston et al. 1988; Lomnicki 1988). And, the effects mediated through the dynamics of populations not immediately in focus, or, more generally, through "hidden variables," upset the methodology of observing the direct interactions among populations and confound many principles, such as the competitive exclusion principle, derived on that basis (Wootton 1994).

Hidden variables and indirect effects have potentially profound consequences for conceptualizing ecology. Consider the strategy of scientific simplification in which models refer only to a few populations, even though those populations are embedded in naturally variable and complex ecological situations. Unless ecologists know that the full community has been specified, their "simple" models – including "null" models – are primarily redescriptions of the particular observations that do not provide, through their fit or lack of fit, sure or general insight about actual ecological relationships. It should be noted that progress in the physical sciences depends greatly on controlled experiments, in which systems are isolated from their context. Yet this model of science is not appropriate for understanding organisms embedded in a dynamic ecological context and responding to resources and hazards that are unevenly distributed across place and time (Taylor 2005).

Embeddedness and the confounding effects of hidden variables should prod theoreticians to scrutinize the analogies and conceptual borrowings drawn from work on well-bounded systems. Similarly, the heterogeneity of units in ecology and their disparate temporal and spatial scales of activity limits the relevance of complexity theory in which iterations of simple rules over time and space lead to complex behaviors for which parallels in real-life are suggested (Waldrop 1992). Long-standing physical and chemical theories in which macro-regularities arise statistically from large numbers of similar entities also seem problematic for linking patterns of ecological complexity and corresponding processes.

Extending thermodynamics to open "systems of man and nature" was, indeed, a central motivation for H. T. Odum's pioneering contributions to systems ecology (Taylor 2005). Although the search for theoretical principles became less important in subsequent systems ecology (during the 1960s and 1970s, especially during the International Biological Programme), it reemerged in the 1980s in accounts of ecological complexity as a hierarchy of systems embedded within larger systems with complementary processes and patterns at each level or scale. One hope of hierarchy theory was that, if the right measure were found for extracting patterns from data, a natural reduction of complexity might be achieved (Allen and Starr 1982).

To some extent the problem of inferring process from pattern can be overcome through the use of analysis of variance and related statistical techniques on data from replicated, multi-factorial field experiments (Underwood 1997). Strictly speaking, however, such results are local, that is, contingent on the configuration of other factors held experimentally or statistically constant for the experiment (Lewontin 1974). Localization poses few problems when ecological engineering affords control over conditions and isolates the system from any surrounding dynamics. But these are special cases.

Moving from Systems to Intersecting Processes

For some ecologists the growing emphasis since the 1980s on situated, scale-crossing processes means that ecology needs to be reconceived as an "historical" science (Schluter and Ricklefs 1993). Like epidemiologists, paleontologists, and historians, ecologists face the challenge of historical explanation. That is, they have to assemble a composite of past conditions sufficient for the subsequent outcomes to have followed and not some other, while, at the same time, they must not obscure the provisionality of such accounts in the face of competition from other plausibly sufficient accounts (Miller 1991). This "composite of past conditions" would include considerable historical and geographical contingency (such as which organisms survived in pockets when Mt. St. Helens erupted, Franklin and MacMahon 2000) and the evolutionary particularity or "individuality" of species (Sterelny and Griffiths 1999, pp. 253 ff.). Yet historicity need not eliminate ideas about regularities or structuredness of ecological patterns and processes. To say that ecological structure has a history could be to say that it changes in structure and is subject to contingent, spatially located events, while at the same time the structure constrains and facilitates the living activity that constitute any ecological phenomenon in its particular place.

Whether or not the label "historical" is used, a key challenge for conceptualizing ecology is to allow for particularity and contingency intersecting with structure, and for that structure to change, be internally differentiated, and, because of overlapping scales of different species' activities, have problematic boundaries. Systems that are well bounded or have simple relations with their external context, when they are encountered, could be viewed not as simple situations, but as special cases whose existence requires explanation.

It is here that conceptual clarification may benefit from viewing ecological situations as special cases of intersecting social-environmental processes, which would mean giving less weight to cases in which human disturbance is minimal or constant, and allowing the natural science of ecology to be informed by debates in social science. In particular, there is a "political ecological" current of anthropological and geographical research that focuses on situated, scale-crossing socio-environmental processes. This research analyzes environmental problems in terms of intersecting economic, social and ecological processes, which operate across various spatial and temporal scales and are mutually implicated in the production of any outcome and in their own ongoing transformation (Taylor and García-Barrios 1995; Peet and Watts 1996). Accounts of soil erosion or collapse of fish stocks, for example, may tie together the local and regional ecological characteristics, local institutions of production and associated agro- or aqua-ecologies, the social differentiation in a given community and its social psychology of norms and reciprocal expectations, and national and international political economic changes (Little 1987; García-Barrios and García-Barrios 1990; Taylor 2005).

Researchers who analyze "intersecting processes" have not articulated a mature conceptual or methodological framework, but explanations that preserve heterogeneity of causes and complexity of their inter-linkages warrant much more attention from philosophers and ecologists. Conceptual work in this area would require, among other things, attention to researchers' practice and engagement with the complexity studied (Haila and Levins 1992; Goldman et al. 2011). Moreover, intersecting processes accounts expose multiple points of potential engagement – each one partial in the sense of being insufficient to overcome the focal problem, and thus needing to be inter-linked within the ongoing intersecting processes (Taylor 2005).

Partiality is pertinent even when researchers do not focus on socio-environmental dynamics, but confine themselves to natural ecology. The exploratory use of models, mentioned earlier, retains support, in part, because of an unstated implication that, if the different exploratory models could be combined, they would yield an understanding of ecological phenomena that could not be achieved through the construction of all-encompassing systems models. For example, the idea that there is a limit to the similarity of co-existing species might be combined with the ideas that spatial heterogeneity or an intermediate level of disturbance promote diversity, and so on. But how? The means of weaving together or synthesizing necessarily partial models, or heuristics, has yet to be articulated. On the reasonable assumption that few ecologists can juggle more than a few heuristics, new approaches to conceptual work need to be developed that bring different types of ecologists into sustained interaction (Lee 1993; Walters 1997; Wondolleck and Yaffee 2000; Resilience Alliance, http://www.resalliance.org (04.06.2007)).

Social Interactions in the Production of Ecological Knowledge

Self-consciousness about social interactions involved in producing knowledge lay behind Levins' (1966) strategy of modeling, which distinguished his perspective from contemporaries in mathematical ecology. Levins has been concerned with the vitality of the modeling process – with a never-ending process of disturbing the provisional validity of models (Taylor 2000). His interest in the circumstances under which theoretical principles might be overthrown – circumstances that are not always apparent to scientists – has led him to consider the social conditions in which knowledge is produced (Haila and Levins 1992). For example, under a research and development system geared to firms making profits, pesticides have been favored over biological control of pests (Levins and Lewontin 1985, pp. 238–241).

Socio-historical contextualization should, as a matter of logical consistency, apply to any HOEK-style interpretation of ecological concepts. Indeed, I locate the origins of my interest in unruly complexity at the turbulent intersection of ecology-as-science and ecology-as-social-action during the 1970s. During this period ecology-as-social-action involved a serious critique of the scientific enterprise. This challenged researchers not only to attend to environmental concerns, but also to shape scientific practices and products self-consciously so as to contribute to transforming the dominant structure of social and environmental relations. This context led me to engage with the complexities of environmental, scientific, and social change together, as part of one project. The intersections of these three kinds of change conditioned me to emphasize, both conceptually and in practice, problematic boundaries, heterogeneity of ecological and social agents, and continuing process over competed product (Taylor 2005).

It is relevant to describe here one aspect of the subsequent conceptual evolution up to my current status as a HOEK contributor. I first attempted to interpret science in its socio-historical context during the 1980s when I examined H. T. Odum's efforts to reduce the complexity of social and ecological relations to a single currency – energy – whose flows could be adjusted or redesigned. I associated this with the post-war climate of "technocratic optimism" and proposed Odum had found in nature a special role for systems engineers, such as himself, working in the service of society (Taylor 2005). Although I had shown that the social embeddedness of science can have systematic effects on the content of scientific knowledge, the scientist in me wanted to develop ways to bring such interpretations to bear productively on subsequent research. To provide insights about how that might be achieved, a finer-grained analysis than the broad historical interpretation of Odum seemed to be ca yze shorter-term projects of socio-environmental assessment likely to be governed by more complex and contested pragmatics (Bocking 1997). During this interpretive work, my image of scientists working in a social context evolved from "social-personal-scientific correlations" into one where scientists have to deal with diverse considerations in practice. This allowed me to emphasize a range of different points at which researchers, interpreters of science as well as scientists, could engage differently in scientific practice and try to modify its outcomes. Whether any specific modifications are do-able depends on the position and resources of the specific researchers – myself included – as they enter into negotiations with other relevant social agents. The ability of researchers in practice to make knowledge is distributed beyond their persons, not concentrated mentally inside them; it depends on intersecting processes (Taylor 2005).

How do we get to know ecological complexity? The answers depend on paying more attention to who "we" are, to the associations different people make as they position themselves – in practical as well as conceptual terms - in relation to life's complex, changing, and rarely well-bounded ecological context. The inner structure and process of ecology is surely a matter of heterogeneity, embeddedness, and ongoing restructuring.

References

Allen TFH, Starr TB (1982) Hierarchy. perspectives for ecological theory. University of Chicago Press, Chicago
Austin MP (1987) Models for the analysis of species' response to environmental gradients. Vegetatio 69:35–45
Bocking S (1997) Ecologists and environmental politics. A history of contemporary ecology. Yale University Press, New Haven
Darwin C (1964/1859) On the origin of species. Harvard University Press, Cambridge
DeAngelis DL, Waterhouse JC (1987) Equilibrium and non-equilibrium concepts in ecological models. Ecol Monogr 57:1–21
Faith DP, Minchin PR, Belbin L (1987) Compositional dissimilarity as a robust measure of ecological distance. Vegetatio 69:57–68
Franklin JF, MacMahon JA (2000) Messages from a mountain. Science 288:1183–1185
García-Barrios R, García-Barrios L (1990) Environmental and technological degradation in peasant agriculture. A consequence of development in Mexico. World Dev 18(11):1569–1585
Goldman MJ, Nadasdy P, Turner MD (eds) (2011) Knowing Nature: Conversations between Political Ecology and Science Studies, University of Chicago Press, Chicago
Gray AJ, Crawley MJ, Edwards PJ (eds) (1987). Colonization, succession and stability. 26th Symposium of the British ecological society. Blackwell, Oxford
Hagen JB (1989) Research perspectives and the anomalous status of modern ecology. Biol Philos 4:433–455
Haila Y (1988) The multiple faces of ecological theory and data. Oikos 53:408–411
Haila Y, Järvinen O (1990) Northern conifer forests and their bird species assemblages. In: Keast A (ed) Biogeography and ecology of forest bird communities. SPB Academic, the Hague, pp 61–85
Haila Y, Levins R (1992) Humanity and nature. Ecology, science and society. Pluto, London
Hastings A, Harrison S (1994) Metapopulation dynamics and genetics. Annu Rev Ecol Syst 25:167–188
Huston M, DeAngelis D, Post W (1988) From individuals to ecosystems. A new approach to ecological theory. Bioscience 38:682–691
Kelt DA, Brown JH (1999) Community structure and assembly rules. Confronting conceptual and statistical issues with data on desert rodents. In: Weiher E, Keddy P (eds) Ecological assembly rules, perspectives, advances, retreats. Cambridge University Press, Cambridge, pp 75–107
Kingsland S (1995) Modeling nature: Episodes in the history of population biology. University of Chicago Press, Chicago

Lee K (1993) Compass and gyroscope. Integrating science and politics for the environment. Island Press, Washington DC

Levins R (1966) The strategy of model building in population biology. Am Sci 54:421–431

Levins R, Lewontin R (1985) The dialectical biologist. Harvard University Press, Cambridge

Lewontin RC (1974) The analysis of variance and the analysis of causes. Am J Hum Genet 26:400–411

Little P (1987) Land use conflicts in the agricultural/pastoral borderlands. The case of Kenya. In: Little P, Horowitz M, Nyerges A (eds) Lands at risk in the third world. Local-level perspectives. Westview Press, Boulder, pp 195–212

Lomnicki A (1988) Population ecology of individuals. Princeton University Press, Princeton

May RM (1973) Stability and complexity in model ecosystems. Princeton University Press, Princeton

McIntosh RP (1985) The background of ecology. Concept and theory. Cambridge University Press, Cambridge

Miller RW (1991) Fact and method in the social sciences. In: Boyd R, Gasper P, Trout JD (eds) The philosophy of science. MIT Press, Cambridge, pp 743–762

Minchin PR (1987) An evaluation of the relative robustness of techniques for ecological ordination. Vegetatio 69:89–107

Nee S (1990) Community construction. Trends Ecol Evol 2:337–343

Peet R, Watts M (eds) (1996) Liberation ecologies. Environment, development, social movements. Routledge, London

Pickett STA, White PS (eds) (1985) The ecology of natural disturbance and patch dynamics. Academic, Orlando

Schluter D, Ricklefs R (1993) Species diversity. An introduction to the problem. In: Ricklefs R, Schluter D (eds) Species diversity in ecological communities. University of Chicago Press, Chicago, pp 1–10

Simberloff D (1982) The status of competition theory in ecology. Ann Zool Fenn 19:241–253

Sterelny K, Griffiths P (1999) Adaptation, ecology, and the environment. Sex and death. An introduction to the philosophy of biology. University of Chicago Press, Chicago

Strong DR, Simberloff D, Abele LG, Thistle AB (eds) (1984) Ecological communities, conceptual issues and the evidence. Princeton University Press, Princeton

Taylor PJ (1992) Community. In: Keller EF, Lloyd E (eds) Keywords in evolutionary biology. Harvard University Press, Cambridge, pp 52–60

Taylor PJ (2000) Socio-ecological webs and sites of sociality. Levins' strategy of model building revisited. Biol Philos 15(2):197–210

Taylor PJ (2005) Unruly complexity: Ecology, interpretation, engagement. University of Chicago Press, Chicago

Taylor PJ, García-Barrios R (1995) The social analysis of ecological change. From systems to intersecting processes. Soc Sc Info 34(1):5–30

Taylor PJ, Haila Y (2001) Situatedness and problematic boundaries. Conceptualizing life's complex ecological context. Biol Philos 16(4):521–532

Underwood AJ (1997) Experiments in ecology. Their logical design and interpretation using analysis of variance. Cambridge University Press, Cambridge

Waldrop MM (1992) Complexity. The emerging science at the edge of order and chaos. Simon and Schuster, New York

Walters C (1997) Challenges in adaptive management of riparian and coastal ecosystems. Conserv Ecol 1(2):1, http://www.consecol.org/vol1/iss2/art1

Weiher E, Keddy P (eds) (1999) Ecological assembly rules. Perspectives, advances, retreats. Cambridge University Press, Cambridge

Wondolleck JM, Yaffee SL (2000) Making collaboration work. Lessons from innovation in natural resource management. Island Press, Washington DC

Wootton JT (1994) The nature and consequences of indirect effects in ecological communities. Annu Rev Ecol Syst 25:443–466

Chapter 7
A Few Theses Regarding the Inner Structure of Ecology

Gerhard Wiegleb

Introduction

Ecology is a pluralistic science (McIntosh 1985, 1987; Cherrett 1989; Dodson 1998). Pluralism can be explained by the following hypotheses:

It is a result of recent diversification. Applications in environmental sciences lead to differentiation within an accepted theoretical framework (centrifugal model).
It is a result of a long historic development. Incorporation of disparate branches of sciences were accompanied by incomplete theory reduction and unification (attractor model).
It is a result of a complete lack of coherent theory formation. Disagreement on aims, methods and accepted theories prevails (anarchy model).

I assume in this context that the third possibility holds. Any overview of ecology requires a delimitation towards nonecology. Here I define the current set of accepted facts, theories and laws in ecology according to the contents of the textbooks Begon et al. (1986) and Krebs (1995). This circumscription will help us to trace back ecological approaches before the rise of self-conscious ecology. In accordance with Begon et al. I define ecology as the science that deals with the description, explanation and prediction of individuals, populations, and communities in space and time. Variables of interest are distribution, amount (biomass), number (abundance, diversity) and composition (similarity) of biotic entities (Peters 1991; Krebs 1995).

First I will focus on the internal structure of ecology with special reference to its historical roots. I present a short outline of my personal approach towards history of science. I then give an overview of the main phases of the development of ecology including an investigation of the importance of benchmark papers and books since 1920. In the second part I investigate the fields in which current ecologists work. Those are defined by systematic groups, habitat types, observational levels, spatial scales, research programmes, schools and traditions, key concepts, and applications. Criteria such as the distinction between reductionistic vs. holistic

G. Wiegleb (✉)
Chair General Ecology, Brandenburg University of Technology, Cottbus, Germany
e-mail: wiegleb@tu-cottbus.de

ecology (Wilson 1988), between organismic vs. individualistic ecology (Simberloff 1980; Trepl 1987), competing methodological approaches (experimental, comparative, exploratory, simulation, Grime 1979), or competing aims (explanation, prediction, description; Wiegleb 1989, Peters 1991) are used for distinguishing research programmes, schools, or preferences for key concepts.

A Short History of Ecology

General Aspects

History of ecology is regarded as the history of ecological ideas (May and Seger 1986; Trepl 1987). The theory of Kuhn (1976), assuming scientific revolutions and subsequent change of paradigms (= complete replacement of valid ideas), is not applicable to ecology. According to Trepl (1987) ecology never developed any paradigm comparable to physics, chemistry, or molecular biology. Simberloff (1980) distinguished two paradigms in ecology, the organismic and the individualistic one. However, both sets of ideas cannot be regarded as paradigms in a Kuhnian sense. Or we have to assume the coexistence of two different, noncommunicating sciences. Neither can the "paradigms" of Regier and Rapport (1978) be seriously considered as Kuhnian paradigms.

I am inclined to follow Lakatos (1978) in assuming coexistence of competing research programmes and stepwise replacement of degenerative approaches by more promising ones. In ecology, research programmes were not deliberately formulated but developed as a casual consequence of benchmark papers or textbooks. Thus they are only recognizable in retrospective. According to Toulmin (1978) scientific continuity can be recognized on the levels of theory, content, sociology, and psychology. Questions such as the following deserve an answer: Which were the factors of innovation in the past which produced preliminary variants of ideas? What was the context of discovery? Which were the factors of selection fixing final variants of ideas? What was the context of agreement? Which factors prevented the break-through of ideas? Additionally, the approach of Mayr (1984) contributes significantly to gaining insight into historical processes. He describes the growth of ideas in individual persons, asking: Which scientists were exceptional, what did they think? How did researchers proceed to achieve their results? Which were the individual influences on their work on a biographical level? He thoroughly investigates the psychological and sociological hindrances of agreement upon theories and facts.

In this brief historical account the interplay between the development of ecology, the progress in philosophy of science, and the societal context cannot be analysed in detail. Progress in epistemology and philosophy of science between Paracelsus, Bacon and Descartes (empirism vs. rationalism) and Leibniz, Hume and Kant played an important role in the early separation of biology from natural history (until 1790). Later inner-biological developments and the accumulation of facts played a great role (until 1940). More recently, external factors relating ecological ideas to the development of social groups, institutions, politics, and technology have become more important. To date those chapters of the history of ecology are mostly unwritten (except for Küppers et al. 1978; Golley 1993; Anker 2001).

Early Phases of Ecology Development

Here I distinguish between seven phases of ecology, which have left their traces in recent ecological theory and practise. Besides autochthonous development, external scientific traditions were incorporated in the course of time.

Phase 1. During ancient to medieval times (600 BC to AD 1300) science was part of philosophy. Not much genuine ecological knowledge was accumulated. It is restricted to medicinal botany (from Dioscorides to Hildegard von Bingen 1150) and population studies (since 1200, L. Pisano; P. de Crescenzi; Egerton 1973, 1983).

Phase 2. During early modern times (from 15th to middle of eighteenth century) several traditions developed which paved the way for the achievements of Linnaeus. The most prominent are:

- Around 1,520 botany was revived by so-called "herbalists". Several nicely illustrated herbal books appeared between 1,520 and 1,540. Authors such as O. Brunsfels, H. Bock, L. Fuchs, A. Mattioli, R. Dodonaeus, Ch. de L'Écluse, and V. Cordus partly included pedological observations and descriptions of the growth place of plants (Mägdefrau 1992).
- Encyclopedic natural history started in the sixteenth century with its most famous representatives K. Gesner, U. Aldrovandi, P. Belon and G. Rondelet (Jahn et al. 1985).
- Early phytogeography of the seventeenth century borrowed both from herbalism (Fuchs) and natural history sources (Gesner). Important representatives are B. Varenius and J.P. de Tournefort (Egerton 1977).
- Harmonistic viewpoints of nature date back to ancient times, e.g. Pythagoras (Egerton 1973). They were revived at the end of the seventeenth century by J. Ray, who in his "Natural history of animals" advanced an explicit application of the "Balance of Nature" concept to natural history (Jansen 1972; Egerton 1973). W. Derham, a student of Ray's, developed "Physicotheology" into a comprehensive system which was discussed as a serious theory until 1850, in particular in England, later seeking confrontation with Darwinism. As a part of religious and environmentalist ways of thinking it is still recognizable in the current discussion on environmental protection.
- Realistic paintings of plants and insects (still life) influenced natural history during the seventeenth and early eighteenth century. Based on earlier Dutch examples A.S.M. Merian and J. Rösel von Rosenhof edited coloured books showing animals feeding on plants. Linnaeus used such material for describing species (Jahn et al. 1985).
- Starting around 1,580 the work on population development by F. Botero, D. Peteau, Th. Browne, J. Graunt, M. Hale and D. Dodart culminated 1798 in T.R. Malthus'

treatment "Essay on the Principle of Population". He postulated geometric growth of human population leading to famine and disease ("misery and vice"). The mathematical-theoretical population ecology was important for the origin of evolution theory, but it remained an independent line of thought until the beginning of the twentieth century (R. Pearl, A. Lotka, V. Volterra). Only around 1,970 an amalgamation to the mainstream of ecology occurred (Egerton 1976, 1977).

Phase 3. The scientific revolution during the age of enlightenment and the origin of science in a modern sense (around 1,750) lead to the protoecological work of C. von Linnaeus and some predecessors in the framework of natural history.

- Important predecessors of the early eighteenth century were A. van Leeuwenhoek, studying dynamics of microbes, R. Bradley, founding production biology, and R.A.F. de Reaumur, founding animal ecophysiology and ecophysiology of plants (Abbot 1983; Egerton 1969, 1977).
- Linnaeus contributed to various disciplines of ecology, e.g. floristics, vegetation geography (description of Skandinavian vegetation in terms of altitudinal stages, zonal stages and environmental gradients), mire science (description of mire types), lake science (description of lake types according to the vegetation, speculations on nutrient content of the lakes), indicator plants (formulation of the indicator principle), succession of plant communities (in bog waters), food chains (relationships between animals and their host plants, observations on the function of carnivores), experimental ecology (planting of plants of foreign countries in Sweden imported by his students), dispersal ecology (speculations on creation centres in disjunct species), and phenology (Bremekamp 1952; DuRietz 1957).
- The general importance of Linnaeus might be seen in his distinction between religion and science. He treated the "Balance of Nature" concept as a scientific theory rather than as a proof of divine wisdom (Egerton 1973; Querner 1980). Further achievements of Linnaeus were the foundation of biological systematics and the delimitation of the science from other cultural enterprises such as medicine, pharmacy, agriculture, and cooking. In his work, Linnaeus combined observation, theory formation and experiment.

Phase 4. Development of "self-conscious ecology" in the nineteenth century was based on the influence of two main lines of scientific development (Worster 1977; Trepl 1987; see also Chap. 4).

- The development of biogeography, in particular plant geography is based on the seminal work of Alexander von Humboldt (1806), later followed by De Candolle (père et fils), Schouw, Meyen, Grisebach, Kerner von Marilaun, Drude, and Schimper in plant geography, and Latreille and Wallace in animal geography. (Nelson 1978; Jahn et al. 1985). Animal geography was trailing behind and reached the mainstream only around 1920 (Shelford 1913).
- Evolutionary biology as developed by J.B. de Lamarck and C. Darwin between 1800 and 1860 paved the way for ecological thought. While Lamarck's theory was physiologically oriented, Darwin presented a theory in which external factors were responsible for organism change (Stauffer 1960; Vorzimmer 1965; Egerton 1968; Wuketits 1995).

– Theoretical considerations by E. Haeckel and practical contributions by K. Möbius (Reise 1980) and V. Hensen (Lussenhop 1974) between 1866 and 1878 lead to the development of the first ecological research programme by Warming (1895); (Goodland 1975).

Phase 5. Soon after the publication of Warming's book (between the edition of the English translation in 1909 and 1925), ecology diversified in various directions. Already existing lines of research discovered the ecological nature of their work. This is true for vegetation science (Clements 1916; Gleason 1926; Braun-Blanquet 1927; Whittaker 1962), marine ecology (Petersen 1913; Zauke 1989) and limnology (Hagen 1992; Golley 1993), all of which had been practised since around 1840 (see for more details Chaps. 19 and 26), and theoretical population ecology (Scudo 1971; Simberloff 1980; Kingsland 1985), which had a longer history. Shortly thereafter, major conceptual progress was marked by the work of C. Elton (1927) and Tansley (1935).

Phase 6. A second phase of scientific progress occurred after World War II. The development of "New Ecology" was accompanied by the publication of E.P. Odum's textbook in 1953 and the development of the systems concept in ecology (Worster 1977; Golley 1993). Simultaneously, the work of R. MacArthur supported the progress of community ecology. Both lines of development are connected by the contribution of G.E. Hutchinson, who was likewise the supervisor of R.L. Lindeman (whose work strongly influenced Odum's work) and R. MacArthur (Hagen 1992). Both lines culminated in the development of key concepts of modern ecology (Cherrett 1989). The phase is characterized by growing mathematization and formalization, turning from descriptive to process and causality orientated research.

Phase 7. Since the beginning of the 1980s various attempts have been made to unify ecology based on partly iconoclastic ecology critique (Harper 1982; Peters 1991), hierarchy theory (Allen et al. 1984) or neutral theories (Hubbell 2001).

Benchmark Papers and Influential Books

Benchmark papers imposed a certain direction on further research (see also Keller and Golley 2000) since 1920 when, after the foundations of the British and American ecological societies, indications of professionalization in ecology became visible (Lowe 1976; Cittadino 1980). Table 7.1 shows an idiosyncratic overview of benchmark papers.

There are different types of benchmark papers. Gleason (1926) was a concept paper or rather a self-review. It had no immediate consequences, revival of his ideas only took place after the publication of Bray and Curtis (1957) and other papers of the Wisconsin school. Lindeman (1942) published an original research paper executing conceptual ideas of others (Tansley 1935), subsequently stimulating further research. A similar relationship can be found between the work of Volterra (1926) and Gause's (1934) book. The subsequent papers of Fisher et al. (1944),

Table 7.1 List of benchmark papers

Author	Year	Keywords
Arrhenius	1921	Species-area relation
Gleason	1926	Individualistic concept
Volterra	1926	Lotka-Volterra equation
Tansley	1935	Ecosystem
Lindeman	1942	Trophic dynamic approach
Egler	1942	Relay floristics
Fisher, Corbet and Williams	1944	Diversity
Novikoff	1945	Levels of organisation
Leslie	1945	Leslie matrix
Watt	1947	Mosaic cycle
Skellam	1951	Random dispersal
Hutchinson	1957	Niche
Bray and Curtis	1957	Ordination
Huffaker	1958	Predator–prey systems
Slobodkin	1960	Energy relations
Connell	1961	Competition
Preston	1962	Species individual relation
MacArthur and Wilson	1963	Island biogeography
Margalef	1963	Ecosystem theory
Yoda et al.	1963	Self-thinning
Paine	1966	Food webs
Bormann and Likens	1967	Biogeochemistry
Porter and Gates	1969	Biophysical ecology
Simberloff and Wilson	1969	Zoogeography
May	1974	Stability
Connell and Slatyer	1977	Succession models
Connor and McCoy	1979	Passive sampling
Harper	1982	After description
Juhász-Nagy and Podani	1983	Spatial processes
Allen, O'Neill and Hoekstra	1984	Hierarchy theory
Kolasa and Pickett	1989	Scaling theory
Legendre	1993	Spatial autocorrelation

Preston (1962) and Connor and McCoy (1979) still form the backbone of community ecology. Connell and Slatyer (1977) are still unrivalled in the field of succession theory. Interestingly, since 1980 no important original paper appeared, while hierarchy theory, scaling theory and theoretical treatment of heterogeneity and spatial autocorrelation made some progress (see Kolasa and Pickett 1991; Wiegleb 1992; Palmer and White 1994; Jax et al. 1996).

Many important ideas were first published in books. Table 7.2 shows another idiosyncratic overview of important ecology books. The first textbook by Warming (1895: English edition 1909) was replaced by a new textbook only 50 years later (Odum 1953). A new generation of textbooks (Krebs 1975; Begon et al. 1986) replaced Odum's work, which is no longer edited.

Table 7.2 List of important books

Author	Year	Keywords
Warming	1895	Community
Clements	1916	Succession
Elton	1927	Animal ecology
Gause	1934	Competition
Odum	1953	General textbook
Andrewartha and Birch	1954	Population ecology
MacArthur	1972	Geographic ecology
Harper	1975	Population ecology
Krebs	1975	General textbook
Green	1979	Research method
Grime	1979	Plant strategies
Gates	1980	Biophysical ecology
Box	1981	Growth form and climate
Tilman	1982	Competition
Begon, Harper and Townsend	1986	General textbook
Allen and Hoekstra	1992	Unification
Hubbell	2001	Neutral theory

Elton's (1927) can likewise be regarded as the necessary counterpart to Warming in the field of animal ecology. Beyond that it presented conceptual novelties and led to enhanced empirical research. Small books on seemingly restricted subjects such as Grime (1979); Box (1981); Tilman (1982) or Hubbell (2001) triggered empirical research and had impact far beyond their original intensions.

Current Structure of Ecology

Organisms (Systematic and Functional Groups)

Ecology of systematic groups distinguishes between microorganisms (microbial ecology, geomicrobiology), plants (geobotany, plant ecology; mostly restricted to vascular plants); and animals (animal ecology, concentrating on vertebrates and insects). Ecology of man (human ecology) is excluded here as a special case, as it is rather a hybrid of ecology with social sciences. Classification by Wikipedia (2004) deliberately includes human ecology in the treatment. All subgroups have their own textbooks and journals. Departments at universities and funding organisations are orientated towards the taxonomic approach. Justification of the taxonomic approach is derived from objective differences between major taxa, e.g. trophic status or mobility. Dodson's (1998) "organism approach" also allows functional classifications according to growth form (tree ecology), mobility type (zooplankton ecology), trophic relations (parasite ecology) or life span (ecology of long lived organisms). Life form classifications (e.g. Raunkiaer 1934) played an important role in plant ecology.

Habitat Types

Ecology of habitat types or locations (Dodson 1998) is divided along the land-water borderline. Terrestrial ecology includes further subdivisions such as tropical to polar ecology, desert to forest ecology, or agro-, forest and urban ecology. Soil is often treated as a separate habitat type (soil ecology). Aquatic ecology deals with the sea (as does marine ecology) or inland waters (as limnology, hydrobiology, or freshwater ecology). The latter is, furthermore, strongly divided into lake and river ecology, sometimes including, sometimes deliberately excluding wetland ecology. This subdivision can be encountered in denominations of ecological departments, in the organisation of chapters in ecology textbooks, and in specific journal and societies devoted to the study of habitat centred ecology. Justification of the habitat approach is derived from objective differences between habitat types, including main limiting factors, homogeneity of resource distribution, or driving forces in community and ecosystem organisation.

Observational Levels Versus Scales

The idea of observational levels is derived from hierarchy theory (Novikoff 1945). It was introduced by Eugene P. Odum (1959) in the organisation of an ecology textbook. Observational levels serve for the constitution of ecological units (Jax et al. 1998). Different levels are treated in own textbooks: Individuals (autecology), populations (demecology, population ecology), communities (community ecology, biocoenology, often called "synecology"), ecosystems (ecosystem ecology, system ecology), and landscapes (landscape ecology). Dodson (1998), following Allen and Hoekstra (1992) is speaking of perspectives instead of observational levels. Thus "individual ecology" is separated into physiological ecology and behavioural ecology. Physiological ecology may include chemical, molecular or biophysical ecology. Exotoxicology, on the other hand, is no subdiscipline of ecology. Even though dealing with the same substances as physiological or habitat ecology the centre of interest is the chemical substance and its fate and not any biological entity.

The range of observational levels covered by ecology is shown in Table 7.3. Levels range from the individual to the biome. Only Allen and Hoekstra (1992) recognize all levels as ecologically relevant. Some authors (Shelford 1913), in the tradition of Warming or Klötzli (1989, being strongly influenced by systems thinking) regarded one of the levels as truly ecological. One might argue that any statement on the appropriate level of observation in ecology has paradigmatic character.

While some of the observational levels refer to a defined scale (e.g. landscape ecology is always large-scale, Forman and Godron 1996), others such as population or community ecology can be carried out on different scales (Allen and Hoekstra 1990). A distinction of ecological approaches according to spatial scales or spatio-temporal domains was introduced by Delcourt et al. (1983). Different spatial scales and coinciding temporal scales require different research strategies. In plant ecology

7 A Few Theses Regarding the Inner Structure of Ecology

Table 7.3 Observational levels of ecology

Author	Individual	Population	Community	Ecosystem	Landscape, biome
Begon et al. (1986)	X	X	X	–	–
Southwood (1977)	–	X	X	–	–
Shelford (1913)	–	–	X	–	–
Allen and Hoekstra (1992)	X	X	X	X	X
Rowe (1961)	X	–	–	X	–
Odum (1971a)	–	X	X	X	X
Klötzli (1989)	–	–	–	X	–

Table 7.4 Research programmes in ecology (revised after Wiegleb 1996)

Level Year	Individuals and population	Community	Ecosystem and landscape
1895	–	Warming (causal ecology)	–
1942	–	–	Lindeman (trophic-dynamic approach)
1954	Andrewartha and Birch (distribution and abundance)	–	–
1967	Harper (Darwinian ecology)	–	Borman and Likens (bio-geochemical cycles)
1972	–	MacArthur (geographical ecology)	–
1984	Allen, O'Neill and Hoekstra (hierarchy theory)	Allen, O'Neill and Hoekstra (hierarchy theory)	Allen, O'Neill and Hoekstra (hierarchy theory)
1988	–	Carpenter and Kitchell (trophic cascades)	Carpenter and Kitchell (trophic cascades)
1996	–	Mooney et al. (biodiversity and ecosystem function)	Mooney et al. (biodiversity and ecosystem function)

this was reflected by the distinction between small-scale "vegetation science" and large-scale "phytogeography". Recently, large scale ecology was linked to community ecology as "macroecology" (Gaston and Blackburn 1999).

Key Concepts

Key concepts had a major influence on the course of research in ecology (see Chap. 2). Based on an interview of 500 British ecologists on the occasion of the 75th anniversary of the British Ecological Society (in 1986), the following key concepts of ecology, ordered according to organisational levels, were identified. In brackets, ranking positions are given. Some concepts may appear twice or more (e.g. diversity includes both as "diversity" and "diversity-stability-relation").

- Population and evolutionary ecology: Life History and Strategies (9; also 12, Adaptation), various aspects of Population Dynamics (15, 17, 19), Coevolution (24, also 34), r and K selection (33).
- *Community ecology*: Succession (2; also 41, Climax), Competition (5; also 30, Competitive Exclusion), Niche (6), the Community (8), Species diversity (14; incl. 35, Diversity-stability Relation), Limiting factors (16), Predation (20, also 21 and 38), Island Biogeography and Species-area Relation (22, 39), Natural Disturbance (26), Indicator Organisms (29).
- *Ecosystem and landscape ecology*: The Ecosystem (1), Energy Fluxes (3), Material Cycling (7), Ecosystem Fragility (10), Food Webs (11, also 31), Heterogenity (13), Maximum Sustainable Yield (18).
- *Nature conservation and environmental protection*: Resource protection (4), Bioaccumulation (23), Habitat Restoration (27), Management of nature reserves (28).

Key concepts are almost equally distributed among the population, community and ecosystem level. In contrast to the community and the ecosystem, the population is not regarded as a genuine ecological concept. Ecosystem and community concepts take the highest ranks (ecosystem, succession, energy flux, competition, niche, material flux, community and diversity). Some concepts are related to dynamic processes (succession, material and energy cycling), others to important driving forces (competition, disturbance). The only applied concept in the Top 10 is resource protection. Today "resource protection" would surely be replaced by "sustainability".

Today, the ecosystem is still important as a general framework for ecological studies. However, most attempts to discover nontrivial ecosystem properties (emergent properties such as goal functions) failed (see Jørgensen 1992; Müller et al. 1996; Gnauck 2002). Energy and material flux belong to the "trivial" (reductionistic) properties of ecosystems, which need to be studied for the description of the system but which don't seem to have any interesting relationship to non-obvious biotic parameters. Succession is a concept dating back to pre-Linnean times (Clements 1916). Despite being a prominent concept, succession theory is still immature. This is due to the fact that long-term observations are lacking for most community types and many scientific ideas are based on indirect inference (short-term observations, chrono-sequences, spatial gradients).

Competition is one of the most disputed concepts in ecology. Competition is easy to produce under experimental circumstances (De Wit 1962), but almost impossible to observe under natural conditions. Again, indirect inference prevails (e.g. null models, Harvey et al. 1983; Gotelli and Graves 1996). The importance of the niche concept is clearly declining and niche theory might be regarded as a part of a competition theory. From my personal point of view, the community itself is and will be the central concept of ecology, despite the fact that the dispute between the functional community view and the statistical community view is still unsettled.

The concept of diversity has been recently transformed into "biodiversity". Most probably it would be voted No. 1 in a current poll. However, inclusion of nonbiological aspects has created some communication and research problems.

The biological part of biodiversity is intimately related to other important key-concepts such as the species-area relationship. Another concept of growing importance is disturbance. Disturbance is a more recent concept, even though ideas of catastrophic events can be traced back to the times of Linnaeus and Warming (Worster 1977). Main hindrance for the development of disturbance theory is the fundamental disagreement on the nature of disturbance as defined by Grime (1979) versus Pickett and White (1985).

Research Programmes

Research programmes have played an important role in ecology (Wiegleb 1996). Nine research programmes are distinguished in Table 7.3. Partly they were published in books, partly in conceptual papers. The first ecological research programme advanced by E. Warming referred to the level of the plant community. Community ecology soon reached zoology (Petersen 1913; Shelford 1913). The community ecology research programme was later refined and generalized by MacArthur (1972) treating almost any important community ecology concept. As the treatment of Hubbell (2001) shows, the process of integrating disparate view on community ecology and relevant aspects of population and evolutionary ecology is still unfinished. Hubbelll's book does not include a truly new research programme.

Lindeman (1942) created a new "trophic-dynamic approach" as a counterpart to the "static species distribution approach" (comprising classification of communities, habitat ecology, life and growth form analysis) and the "dynamic species distribution approach" (including succession research). Both refer to Warming's approach and Clements' dynamic extension. Further ramifications of the trophic dynamic approach have been described by Golley (1993). The most original one is that of Bormann and Likens (1967) which might be regarded as a research programme in its own right. Some reductionistic approaches (e.g. Odum 1971b) fullfill the requirement of being a true research programme, but not necessarily one in ecology. The integration of ecosystem ecology and community ecology, which had been lost in the times of hard-boiled systems ecology was re-established by Carpenter and Kitchell (1988), based on an idea of Hairston et al. (1960); Mooney et al. (1996) made another integrative attempt. Also their idea was not new, but can be found in many papers since the end of the 1980s. Mooney et al. (1996) reached popularization, as the time was ripe for a revival of a more elaborate version of the diversity-stability-hypothesis.

Population ecology for a long time remained divided into studies on animal (Andrewartha and Birch 1954) and plant populations (Harper 1975). Yet, in marine ecology a separate line of population and life-history-centred ecology prevailed (Zauke 1989). Despite the obvious effort of Hubbell (2001) the lose ends of "individualistic" community ecology (Gleason 1926), Darwinian ecology (Harper 1967), individual based modelling (Wissel 1989), theoretical population ecology (Volterra 1926; Pielou 1969), and life history theory (Stearns 1976) still need further integration.

At first glance, hierarchy theory of Allen et al. (1984) does not relate to any of the preceding traditions. Instead, it tries to unify all possible research programmes in ecology under one common headline, hierarchy theory (see also Chap. 6). But it is clearly an offspring of the Wisconsin school of plant ecology, trying to integrate achievements of modern science and systems theory into community ecology in a wider sense.

Schools and Traditions

The distinction between research programmes and schools (or "traditions") assumes that there are schools within larger research programmes, or even schools outside of recognizable research programmes. Schools characterise a lower level of scientific integration. Differentiation into schools is well described for systems ecology (Golley 1993), community ecology (Hubbell 2001) and vegetation science (Whittaker 1962).

Ecosystems ecology followed quite divers directions, e.g. thermodynamics, exergy, networks, cybernetics, automaton theory etc. One might distinguish a Margalef school, an H.T. Odum school, a Jørgensen school etc. (Regier and Rapport 1978; Jørgensen 1992; Golley 1993; Gnauck 2002). In contrast to the assumptions of Regier and Rapport (1978) none of these schools ever reached the paradigmatic stage. Ulanowicz (1990) paper can be regarded as the last attempt to save systems ecology and render Harper's (1982) and Fenchel's (1987) criticism obsolete.

Community ecology, differentiated according to the interest in competition as a major driving force of structuring natural communities (the ghost of competition past, assembly rules; Diamond 1975; Connor and Simberloff 1979) and to methodological preferences (e.g. Pielou 1969; Hurlbert 1990; mathematical ecology). Today the situation of community ecology is more relaxed, even though approaches centred around null models (Strong 1980; Gotelli and Graves 1996) or neutral theories (Hubbell 2001) tend to undermine the conventional wisdom of community ecology.

Vegetation science has been analyzed by Whittaker (1962). At an early stage, vegetation science split into conflicting schools based on floristic traditions (emphasis on species composition: north European, south European, Russian, British and an American tradition) or physiognomic traditions (emphazising vegetation structure or architecture; found in Scandinavia, North America, also in tropical ecology). Simberloff (1980) assumes that vegetation science differentiated on a paradigmatic level. However, scientists such as Gleason, Clements and Tansley can still be seen as members of Warming's research programme. This is not true for Central European phytosociology (Braun-Blanquet 1927; see Chap. 21 and also Chap. 19). The Wisconsin school (Bray and Curtis 1957) founded a new tradition within an old research programme. Major progress in computer technology and programming (Gauch 1982, Ter Braak 1988) lead to a relaxion of tensions between schools and eventually to an intrusion of vegetation scientific methods into the core of community ecology.

Applicability

Distinctions according to applicability can be drawn between theoretical ecology or basic ecology on the one hand and applied ecology on the other (Dodson 1998). The terms "theoretical" and "basic" have different connotations. Theoretical ecology is mostly equated to mathematical ecology (Pielou 1969; Wissel 1989), even though mathematical models can be applied to practical questions. Basic ecology refers to any scientific research, the results of which can be transferred to applied sciences such as conservation biology, agriculture, wildlife management or restoration ecology (Simberloff 1999). Ecology is a pure or basic natural science. For the application of ecological knowledge additional normative elements are needed (see Chap. 26). For the relationship between ecology and its applied sciences such as nature conservation, the same relationship holds as between physiology and medicine, or physics and electrical engineering. This distinction is confused in many ecological writings. As a recent example, Wikipedia (2004) is mentioned. The distinction between basic and applied is, however, dynamic. Genecology and palaeoecology have changed recently from theoretical exercises to applied disciplines.

Crossover of ideas lead to the introduction of "biodiversity research" (CBD 1992) including both a basic part (systematics and ecology) and an applied part (planning and socioeconomy; Wiegleb 2003). This is a reaction to the fact that basic ecological research is not funded in most countries. In Germany, most funding was spent on applied ecological research in the context of explaining forest die-back, mitigating consequences of intensive agriculture, restoration of rivers, lakes, flood plains, or drainage areas, and the like. Many of these studies yielded interesting results concerning the case studied. Casually, these studies also yielded concepts of theoretical interest (e.g. Müller et al. 1996; Hauhs and Lange 1996). An attempt to proceed in a different way (study applied questions on a solid theoretical basis, Zauke 1989; Vareschi and Zauke 1993) never received adequate funding and was closed down before having really started.

Conclusions

Is there really "more to ecological science in the postdescriptive phase than acquiring the ability to handle unique anecdotal management problems" (Harper 1982)? The above discussion shows that there are few general laws or rules available in ecology, that there is no unification in ecology as to aims, methods and conceptual approaches, and that all attempts to reduce the separate ecologies to one consistent bulk of knowledge have failed so far. So far, no general agreement as to the question of Dice (1955) "What is ecology?" has been reached. Progress has been made in the description of single species or single habitats. However, most of the knowledge collected cannot be transferred to other species of different life form types or other habitats under different climatic conditions.

We are confronted with the unusual situation that ecology is not characterized by paradigm formation. "Paradigms" in ecology such as thermodynamics, stoechiometry, or evolution are borrowed from other sciences. We observe a separation of ecology into simultaneous schools or traditions that behave like paradigms. Communication among schools does not take place either because it is not intended (lacking respect for other sub-disciplines) or impossible (language barriers). I assume that scientific communication has been possible since 1800, even though the sheer number of new publications nowadays may impose physical boundaries on information transfer. In the 1930s in the USA, adherents of Clements would not read any paper by Gleason. Followers of a "physicalistic" school (Odum 1971b; Jørgensen 1992) would never read papers such as Harper (1967, 1982); Walter (1973); Stearns (1976) or Krebs and Davies (1993). Followers of the Braun-Blanquet approach deliberately neglected the whole bulk of Anglo-American community ecology for decades. Likewise, the important paper of Petersen (1913) was neglected by terrestrial ecologists. Attempts to bridge the gap between observational levels (e.g. Gates 1980; Carpenter and Kitchell 1988; Turner 1989; Ulanowicz 1990; Jones 1995; Bartha et al. 1998) are usually not recognized outside a specialized group of scientists. Other people might think that Carpenter and Kitchell (1988) wrote on "lakes", but their topic is much wider.

Even within its traditions, ecological knowledge is far from being cumulative. Are we really wiser with respect to the life history of plants than Salisbury (1932) has been? Good old wisdom is forgotten (e.g. the nonvalidity of crude forms of the stability-diversity hypothesis, Goodman 1975; Trepl 1995). Money is still being spent to prove this obsolete hypothesis. Nowadays it is difficult to distinguish between a proper research programme and ephemeral fashions. Research in ecology has become opportunistic. Philosophers, economists, politicians, and jurists have adopted ecological terms or rather developed their own "creole" or "pidgin" languages on ecology. The ecologist, e.g. when reading the advertisements of funding organisations, has to understand these texts. The main reason for this situation is that scientific standards such as recognition of the stochastic and historic nature of ecological systems (Simberloff 1980; Levins and Lewontin 1980), the necessity of predictive ecology (Harper 1982; Peters 1991), or the necessity of inter-level reduction of phenomena (Allen et al. 1984; Kolasa and Pickett 1989; Shrader-Frechette and McCoy 1990) have not yet become common sense in ecology.

Acknowledgment I thank U. Böring for the critical reading of various versions of the text and help with the technical preparations of the manuscript.

References

Abbot D (1983) The biographical dictionary of scientists. Biologists. Blond Educational, London
Allen TFH, Hoekstra TW (1990) The confusion between scale-defined levels and conventional levels of organisation. J Vegetable Sci 1:5–12

Allen TFH, Hoekstra TW (1992) Toward a unified ecology. Columbia University Press, Columbia
Allen TFH, O'Neill RV et al (1984) Interlevel relations in ecological research and management: some working principles from hierarchy theory. USDA, Forest Service, General Technical Report RM-110
Andrewartha HG, Birch LC (1954) The distribution and abundance of animals. University of Chicago Press, Chicago
Anker P (2001) Imperial ecology: the environmental order of the British Empire, 1895-1945. Harvard University Press, Cambridge
Arrhenius O (1921) Species and area. J Ecol 9:95–99
Bartha S, Czaran T et al (1998) Exploring plant community dynamics in abstract coenostate spaces. Abstracta Bot 22:49–66
Begon M, Harper JL et al (1986) Ecology: individuals, populations and communities. Blackwell, Sunderland
Bormann FH, Likens G (1967) Nutrient cycling. Science 155:424–429
Box EO (1981) Macroclimate and plant form: an introduction to predictive modelling in phytogeography. Junk, the Hague
Braun-Blanquet J (1927) Pflanzensoziologie. Bornträger, Berlin
Bray JR, Curtis JT (1957) An ordination of the Upland Forest Communities of Southern Wisconsin. Ecol Monogr 27:325–349
Bremekamp CEB (1952) Linné's significance for the development of phytogeography. Taxon 2:47–54
Carpenter S, Kitchell JF (1988) Strong manipulations and complex interactions: consumer control of lake productivity. Bioscience 38:764–769
CBD (1992). from www.biodiv.org.
Cherrett JM (1989) Key concepts: the result of a survey of our members' opinions. In: Cherrett JM (ed) Ecological concepts: the contribution of ecology in understanding of the natural world. Blackwell, Oxford, pp 1–16
Cittadino E (1980) Ecology and the professionalization of botany in America, 1890-1905. Stud Hist Biol 4:171–198
Clements FE (1916) Plant succession. An analysis of the development of vegetation. Carnegie Institution of Washington, Washington, DC
Connell JH (1961) The influence of interspecific competition and other factors on the distribution of the Barnacle Chthamalus stellatus. Ecology 42:710–723
Connell JH, Slatyer RO (1977) Mechnisms of succession in natural communities and their role in community stability and organization. Am Nat 111:1119–1144
Connor EF, McCoy ED (1979) The statistics and biology of the species-area relationship. Am Nat 113:791–833
Connor EF, Simberloff D (1979) The assembly of species communities: chance or competition? Ecology 60:1132–1140
De Wit C (1962) On competition. Pudoc, Wageningen
Delcourt HR, Delcourt PA et al (1983) Dynamic plant ecology: the spectrum of vegetational change in space and time. Q Sci Rev 1:153–175
Diamond JM (1975) Assembly rules of species communities. In: Cody ML, Diamond JM (eds) Ecology and evolution of communities. Harvard University Press, Cambridge, pp 342–444
Dice LR (1955) What is ecology? Sci Monogr 80:346–351
Dodson S (ed) (1998) Ecology. Oxford University Press, Oxford
DuRietz GE (1957) Linnaeus as a phytogeographer. Vegetatio 6:161–168
Egerton FN (1968) Studies on animal populations from Lamarck to Darwin. J Hist Biol 3:225–259
Egerton FN (1969) Ricardo Bradley's understanding of biological productivity: a study of eighteenth-century ecological ideas. J Hist Biol 2:391–410
Egerton FN (1973) Changing concepts of the balance of nature. Q Rev Biol 48:322–350
Egerton FN (1976) Ecological studies and observations before 1900. In: Taylor BJ, White TJ (eds) Issues and ideas in America. University of Oklahoma Press, Oklahoma, pp 311–351
Egerton FN (1977) A bibliographical guide to the history of general ecology and population biology. Hist Sci 15:189–215

Egerton FN (1983) The history of ecology: achievements and opportunities, part one. J Hist Biol 16:259–310
Egler FE (1942) Vegetation as an object of study. Philos Sci 9:245–260
Elton C (1927) Animal ecology. Methuen, London
Fenchel T (1987) Ecology potentials and limitations. International Ecology Institute, Oldendorf
Fisher RA, Corbet AS et al (1944) The relation between the number of species and the number of individuals in a random sample of an animal population. J Anim Ecol 12:42–58
Forman RTT, Godron M (1996) Landscape ecology. Wiley, New York
Gaston KJ, Blackburn TM (1999) A critique for macroecology. Oikos 84:353–368
Gates D (1980) Biophysical ecology. Springer, New York
Gauch HG (1982) Multivariate analysis in community ecology. Cambridge University Press, Cambridge
Gause GF (1934) The struggle for existence. Williams & Wilkins, Baltimore
Gleason HA (1926) The individualistic concept of the plant associations. Bull Torrey Bot Club 53:7–26
Gnauck A (2002) Automatentheorie in der Ökologie. In: Gnauck A (ed) Systemtheorie und Modellierung von Ökosystemen. Physica, Berlin, pp 32–48
Golley FB (1993) A history of the ecosystem concept in ecology: more than the sum of the parts. Yale University Press, New Haven
Goodland RJ (1975) The tropical origin of ecology: Eugen Warming´s Jubilee. Oikos 26:240–245
Goodman D (1975) The theory of diversity-stability relationships in ecology. Q Rev Biol 50:237–266
Gotelli NJ, Graves GR (1996) Null models in ecology. Smithsonian Institution Press, Washington, DC
Green RH (1979) Sampling design and statistical methods for environmental biologists. Wiley-Interscience, New York
Grime JP (1979) Plant strategies and vegetation processes. Wiley, Chichester
Hagen JB (1992) An entangled bank: the origins of ecosystem ecology. Rutgers University Press, New Brunswick
Hairston N, Frederick G et al (1960) Community structure, population control, and competition. Am Nat 94:421–425
Harper JL (1967) A Darwinian approach to plant ecology. J Ecol 55:247–270
Harper JL (1975) Population ecology of plants. Academic, New York
Harper JL (1982) After Description. In: Newmann EI (ed) The plant community as a working mechanism. Blackwell, London, pp 11–25
Harvey PH, Colwell RK et al (1983) Null models in ecology. Annu Rev Ecol Syst 14:189–211
Hauhs M, Lange H (1996) Ecosystem dynamics viewed from an endoperspective. Sci Total Environ 183:125–136
Hubbell SP (2001) The Unified Neutral Theory of biodiversity and biogeography. Princeton University Press, Princeton
Huffaker CB (1958) Experimentals studies on predation: dispersion factors and Predator-Prey Oscillations. Hilgardia 27:343–383
Hurlbert SH (1990) Spatial distribution of the montane unicorn. Oikos 58:257–271
Hutchinson GE (1957) Concluding remarks, population studies: animal ecology and demography. Cold Spring Harb Symp Quant Biol 22:415–427
Jahn I, Löther R et al (eds) (1985) Geschichte der Biologie. Theorien, Methoden, Institutionen und Kurzbiographien. Fischer, Jena
Jansen AJ (1972) An analysis of "Balance of Nature" as an ecological concept. Acta Biotheor 21:86–114
Jax K, Jones CG et al (1998) The self-identity of ecological units. Oikos 82:253–264
Jax K, Potthast T et al (1996) Skalierung und Prognoseunsicherheit bei ökologischen Systemen. Verhandlungen der Gesellschaft für Ökologie 26:527–535
Jones CG (ed) (1995) Linking species and ecosystems. Chapman & Hall, New York

Jørgensen SE (1992) Integration of ecosystems theories: a pattern. Kluwer, Dordrecht
Juhász-Nagy P, Podani J (1983) Information theory methods for the study of spatial processes and succession. Vegetatio 51:129–140
Keller DR, Golley FB (eds) (2000) From science to synthesis. Readings in the foundational concepts of the science of ecology. University of Georgia Press, Athens
Kingsland SE (1985) Modelling nature. Episodes in the history of population ecology. University of Chicago Press, Chicago
Klötzli F (1989) Ökosysteme. Fischer, Stuttgart
Kolasa J, Pickett STA (1989) Ecological systems and the concept of biological organisation. Proc Natl Acad Sci USA 86:8837–8841
Kolasa J, Pickett STA (eds) (1991) Ecological heterogeneity. Ecological studies. Springer, New York
Krebs CJ (1975) Ecology. The experimental analysis of distribution and abundance. Harper Collins, New York
Krebs CJ, Davies NB (1993) An introduction to behavioural ecology. Blackwell, London
Krebs CJ (1995) Ecology. The experimental analysis of distribution and abundance. Harper Collins, New York
Kuhn TS (1976) Die Struktur wissenschaftlicher Revolutionen. Suhrkamp, Frankfurt
Küppers G, Lundgreen P et al (1978) Umweltforschung - die gesteuerte Wissenschaft? Eine empirische Studie zum Verhältnis von Wissenschaftsentwicklung und Wissenschaftspolitik. Suhrkamp, Frankfurt
Lakatos I (1978) Die Geschichte der Wissenschaft und ihre rationale Rekonstruktion. In: Diederich W (ed) Theorien der Wissenschaftsgeschichte. Suhrkamp, Frankfurt, pp 55–119
Legendre P (1993) Spatial autocorrelation: trouble or new paradigm? Ecology 74:1659–1673
Leslie PH (1945) On the use matrices in certain population mathematics. Biometrika 33:183–212
Levins R, Lewontin R (1980) Dialectics and reductionism in ecology. Synthese 43:47–78
Lindeman RL (1942) The trophic-dynamic aspect of ecology. Ecology 23:339–418
Lowe PD (1976) Amateurs and professionals: the institutional emergence of British plant ecology. J Soc Bibliogr Nat Hist 7:517–535
Lussenhop J (1974) Victor Hensen and the development of sampling methods in ecology. J Hist Biol 7:319–337
MacArthur RH (1972) Geographical ecology. Harper & Row, New York
MacArthur RH, Wilson EO (1963) An equilibrium theory of insular zoogeography. Evolution 17:373–387
Mägdefrau K (1992) Geschichte der Botanik. Leben und Leistung großer Forscher. Fischer, Stuttgart
Margalef R (1963) On certain unifying principles in ecology. Am Nat 97:357–374
May RM (1974) Biological populations with non-overlapping generation: stable cycles, and chaos. Science 186:645–647
May RM, Seger J (1986) Ideas in ecology. Am Sci 74:256–267
Mayr E (1984) Die Entwicklung der biologischen Gedankenwelt. Vielfalt, Evolution und Vererbung. Springer, Berlin
McIntosh RP (1985) The background of ecology: concept and theory. Cambridge University Press, Cambridge
McIntosh RP (1987) Pluralism in ecology. Annu Rev Ecol Syst 18:321–341
Mooney HA, Cushman JH et al (eds) (1996) Functional roles of biodiversity – A global perspective. Wiley, Chichester
Müller F, Fränzle O et al (1996) Modellbildung in der Ökosystemanalyse als Integrationsmittel von Empirie, Theorie und Anwendung – eine Einführung. EcoSys 4:1–16
Nelson G (1978) From candolle to croizat: comments on the history of biogeography. J Hist Biol 11:269–305
Novikoff AB (1945) The concept of integrative levels and biology. Science 101:209–215
Odum EP (1953) Fundamentals of ecology. Saunders, Philadelphia
Odum EP (1959) Fundamentals of ecology. Saunders, Philadelphia

Odum EP (1971a) Fundamentals of ecology. Saunders, Philadelphia
Odum EP, Odum HT (1953) Fundamentals of ecology. Saunders, Philadelphia
Odum HT (1971b) Environment, power and society. John Wiley & Sons, London
Paine RT (1966) Food web complexity and species diversity. Am Nat 100:65–75
Palmer MW, White PS (1994) Scale dependence and the species-area relationships. Am Nat 144:717–740
Peters RH (1991) A critique for ecology. Cambridge University Press, Cambridge
Petersen CGJ (1913) Valuation of the sea II. The animal communities of the sea bottom and their importance for marine zoogeography. Rep Danish Biol Stat 21:1–44
Pickett STA, White PS (1985) Patch dynamics – a synthesis. In: Pickett STA, White PS (eds) The ecology of natural disturbance and patch dynamics. Academic, San Diego, pp 371–384
Pielou EC (1969) An introduction to mathematical ecology. Wiley Interscience, New York
Porter WP, Gates DM (1969) Thermodynamic equilibria of animals with environment. Ecol Monogr 39:224–244
Preston FW (1962) The canonical distribution of commonness and rarity. Ecology 43(185–215):431–432
Querner H (1980) Das teleologische Weltbild Linne's - Observationes, Oeconomia, Politia. Veröff Joachim Jungius-Ges Wiss Hamburg 43:25–49
Raunkiaer C (1934) The life forms of plants and statistical plant geography. Clarendon, Oxford
Regier HA, Rapport DJ (1978) Ecological paradigms, once again. Bull Ecol Soc Am 59:2–6
Reise K (1980) Hundert Jahre Biozönose. Die Evolution eines ökologischen Begriffes. Naturwissenschaftliche Rundschau 33:328–335
Rowe JS (1961) The level-of-integration concept and ecology. Ecology 42:420–427
Salisbury EJ (1932) The East Anglian flora. Trans Norfolk Norwich Natl Soc 13:191–263
Scudo F (1971) Vito Volterra and theoretical ecology. Theor Popul Biol 2:1–23
Shelford VE (1913) Animal communities in temperate America as Illustrated in the Chicago region. Bulletin of the Geographical Society of Chicago, Chicago
Shrader-Frechette K, McCoy E (1990) Theory reduction and explanation in ecology. Oikos 58:109–114
Simberloff D (1980) A succession of paradigms in ecology: from essentialism to materialism and probabilism. Synthese 43:3–39
Simberloff D (1999) The role of science in the preservation of forest biodiversity. Forest Ecol Manage 115:101–111
Simberloff D, Wilson EO (1969) Experimental zoogeography of islands: the colonization of empty islands. Ecology 50:278–296
Skellam JG (1951) Random dispersal in theoretical populations. Biometrika 38:96–218
Slobodkin LB (1960) Ecological energy relationships at the population level. Am Nat 44:213–236
Southwood TRE (1977) Ecological methods. Chapman & Hall, London
Stauffer RC (1960) Ecology in the long manuscript version of Darwin's origin of species and Linnaeus' Oeconomy of nature. Proc Am Philos Soc 104:235–241
Stearns SC (1976) Life history tactics: a review of ideas. Q Rev Biol 51:3–47
Strong D (1980) Null hypothesis in ecology. Synthese 43:271–285
Tansley AG (1935) The use and abuse of vegetational concepts and terms. Ecology 16:284–307
Ter Braak CJF (1988) CANOCO – A FORTRAN program for canonical community ordination by [Partial] [Detrended] [Canonical] correspondence analysis and redundancy analysis (Version 2.1). GLW, Wageningen
Tilman D (1982) Resource competition and community structure. Princeton University Press, Princeton
Toulmin S (1978) Kritik der kollektiven Vernunft. Suhrkamp, Frankfurt
Trepl L (1987) Geschichte der Ökologie. Athenäum, Frankfurt
Trepl L (1995) Die Diversitäts-Stabilitäts-Diskussion in der Ökologie. Berichte ANL, Beiheft 12:35–49

Turner MG (1989) Landscape ecology: the effect of pattern on process. Annu Rev Ecol Syst 20:171–197
Ulanowicz RE (1990) Aristotelian causalities in ecosystem development. Oikos 57:42–48
Vareschi E, Zauke GP (1993) Entwicklung eines theoretischen Konzepts zur Ökosystemforschung im Wattenmeer. UBA-Texte 47/39:1–142
Volterra V (1926) Fluctuations in the abundance of a species considered mathematically. Nature 118:558–560
Vorzimmer P (1965) Darwin´s ecology and its influence upon his theory. Isis 56:148–155
Walter H (1973) Allgemeine geobotanik. Ulmer, Stuttgart
Warming E (1895) Plantesamfund, grundträk af den Ökologiske plantegeografi. Philipsens Forlag, Copenhagen
Watt AS (1947) Pattern and process in the plant community. J Ecol 35:1–22
Whittaker RH (1962) Classification of natural communities. Bot Rev 28:1–239
Wiegleb G (1989) Explanation and prediction in vegetation science. Vegetatio 83:17–34
Wiegleb G (1992) Explorative Datenanalyse und räumliche Skalierung – eine kritische Evaluation. Verhandlungen der Gesellschaft für Ökologie 21:327–338
Wiegleb G (1996) Konzepte der Hierarchietheorie in der Ökologie. In: Mathes K, Breckling B, Ekschmitt K (eds) Systemtheorie in der Ökologie. Ecomed, Marburg, pp 7–24
Wiegleb G (2003) Was sollten wir über Biodiversität wissen? Aspekte einer angewandten Biodiversitätsforschung. In: Weimann J, Hoffmann A, Hoffmann S (eds) Messung und Bewertung von Biodiversität: mission impossible? Metropolis, Marburg, pp 151–178
Wilson DS (1988) Holism and reductionism in evolutionary ecology. Oikos 53:269–273
Wissel C (1989) Theoretische Ökologie. Springer, Berlin
Worster D (1977) Nature´s economy. The roots of ecology. Sierra Club Books, San Francisco
Wuketits FM (1995) Evolutiontheorie. Historische Vorraussetzungen, Positionen, Kritik. Wiss. Buchgesellschaft, Darmstadt
Yoda K, Kira T et al (1963) Self-thinning in overcrowded pure stands under cultivated and natural conditions. J Biol Osaka City Univ 14:107–129
Zauke GP (1989) Konzeptionelle Überlegungen für einen Forschungsschwerpunkt "Wattenmeer" aus der Sicht der theoretischen Ökologie. UBA-Texte: 253–264

Chapter 8
Dynamics in the Formation of Ecological Knowledge

Astrid Schwarz

The Field of Knowledge "Ecology"

In many societies a growing consensus has arisen about the importance of ecological knowledge: It helps us to address some of the most pressing problems we face at both global and regional level by providing research and effective management options to deal with global warming, diminishing natural resources and the deterioration of soils and water resources. Debates about natural disasters, about the purity of natural things, and about the perceived crisis of the nature-culture relationship in general all revolve around the role and the importance of ecological knowledge in social processes and in negotiations about the kind of nature with and in which we wish to live. For this reason it seems not only worthwhile but also essential, in a sense, to take a closer look at the logical and disciplinary construction of ecological knowledge from a philosophy of science perspective. Inversely, for the philosophy of science ecology is an interesting field having identified as one future perspective that it is important to link general philosophy of science with special philosophies of science in a more fruitful way. This is the framing I have in mind when I propose in the following to explore a distinct epistemological approach for ecological knowledge.

First of all, when "philosophy of science deals with the foundations and the methods of science" its attention is most often focused on the foundations and methods of a discipline.[1] This already entails the expectation that scientific knowledge needs to be developed in a distinct institutional setting, in one that is in some sense a "closed society", with its own language and customs. This is certainly true of scientific disciplines such as physics or astronomy, which have an age-long tradition and a similarly long tradition of philosophical reflection. But there are other epistemic domains whose foundations and methods cannot be described adequately by framing them as a discipline in this sense. They should rather be conceived as a field of

---

[1] This is a widespread characterization of philosophy of science – for instance in the call for paper for a conference in Tilburg in 2010 organized by Stefan Hartmann and Paul Griffiths on *The Future of Philosophy of Science*.

A. Schwarz (✉)
Institute of Philosophy, Technische Universität Darmstadt, Schloss, 64283, Darmstadt, Germany
e-mail: schwarz@phil.tu-darmstadt.de

knowledge that is scattered between different academic disciplines and ultimately even beyond the academic context itself. I propose, first, that ecology be understood in the sense of a field of knowledge rather than as a discipline; and second, I propose an epistemological description to encompass this patchwork field. The theories in this field move back and forth between three basic conceptions and it is this productive movement, I argue, that stabilizes ecological knowledge.

At the end of this chapter will be offered another more general perspective that is a discussion on the relationship between the philosophy of nature and the philosophy of science in view of the field of ecological knowledge. I will provide some reflections on how the two might be linked and how this could help to better understand the plurality in the field.

The starting point for all these considerations was the finding that one and the same ecological research object is "seen" in very different ways and that it may be described simultaneously in different theories and narratives: Organisms in a lake, for example, can be communities, societies or merely assemblages, depending on how strongly their mutual interconnection is seen to be and how necessary their incorporation into their environment. These organisms may have predominantly friendly or indifferent "neighbourly" relations, or they are hostile towards one another; the resources available to the organisms exist in unlimited quantity, or they are described as being permanently scarce; and, finally, organisms themselves can be conceived of principally as a unit of production, as a storage container or as a unit of selection.

A good example of an assemblage in terms of a community might be the recent www-model of a forest (Wiemken and Boller 2002), which describes the rhizomic system of plants and fungi in a forest. The wood wide web connects all individuals to one another in a mainly cooperative way and thus enables an exchange of energy, nutrients and even genetic information in the community. To illustrate an assemblage in terms of a society, I have chosen one of the very first descriptions of a lacustrian system, the microcosm lake, given in 1897 by Stephen Forbes, an American zoologist who originally started out working in the field of entomology. Forbes wonders if the "system of life is such that a harmonious balance of conflicting interests has been reached where every element is either hostile or indifferent to every other, may we not trust much to the outcome where, as in human affairs, the spontaneous adjustments of nature are aided by intelligent effort, by sympathy, and by self-sacrifice?" He then describes certain pelagic forms in a lake "often exquisitely transparent, and hence almost invisible in their native element" which seems to "protect them (perfectly) against their enemies". But, "with an ingenuity in which one may almost detect the flavor of sarcastic humor, Nature has turned upon these favored children and endowed their most deadly enemies with a like transparency, so that wherever the towing net brings to light a host of these crystalline Cladocera, there it discovers also swimming, invisible, among them, a lovely pair of robbers and beasts of prey" (1897, p. 545).

Clearly, we are confronted here with different stages of precision of scientific concepts. Following Rudolf Carnap, one might say that Forbes' system of life is still in the classificatory stage, whereas the wood wide web model has already encompassed the comparative stage and is now in a qualitative stage. This is not terribly surprising, as there are roughly 100 years of conceptual evolution between the two. But what might seem surprising is that the conceptual scheme of describing an assemblage of individuals of plants or animals appears to be quite robust: from the very beginning of ecological thinking, there have been communities with more competitive and hostile relations among one another, held together by a contract; and at the same time, there have been communities where more friendly relations dominate and/or the reference to a superordinate unity is of utmost importance (Clements' superorganism). The idea of the basic conceptions is to capture these different orientations by describing a connection between the philosophy of nature involved in a basic conception and the formation of scientific concepts, thus in a philosophy of science that is interested in the semantic, pragmatic and cognitive evolution of concepts.

Why Basic Conceptions?

To presuppose this kind of "orientation" is not new. Certainly one could say that Imre Lakatos's "hard core" contains also implicit ideas about nature, and this is certainly the case with the "Naturcharakter" the nature character -, a rather phenomenological concept proposed by Gernot Böhme, which is at work behind the scenes of the epistemic operations which ascribe meaning to nature. Thus, this paper argues that ecological plurality is shaped in a distinct way and can be conceptualized as a structure consisting of three so-called basic conceptions. Each of the three characterizes a particular historical field of knowledge that embraces practices and theories about living beings in their environments. Over time, basic conceptions are flexible, they show a dynamic behaviour that is described as an oscillation. This triadic conceptual system is a suggestion for a dynamic conceptualization of ecological knowledge.

A basic conception is basical in the sense that most ecological theories can be integrated in one of the three basic conceptions. The basic conceptions embody an implicit idea about nature that is embedded in the structure and formation of theories and thus also in the structure and formation of knowledge. The basic conception operates as a kind of vantage point that organises concepts and theories.

Why Three of Them?

The seductive power of triads is well-known especially when it comes to Hegelian-Marxist dialectic. It promises a kind of logical and also historical stability that has a dangerous flipside. Obviously the danger of a dialectical approach lies in the authoritarian attitude and also in the inexorability of the process. However, it seems possible to escape these dangers as Lakatos has demonstrated, who was claimed by his friend Paul Feyerabend to be a "big bastard – a Pop-Hegelian Philosopher born from a Popperian father and an Hegelian mother." (Motterlini 2001, p. 1).

The triade presented here is rather based historically, but nevertheless claims to give an analytical tool at hand to describe the plurality of ecological theories and concepts in a heuristically productive and adequate way. The basic conceptions do not follow one another in the sense of a dialectical scheme, there is no synthetic position and the dynamic of the basic conception is rather driven by competition to advance theories or research programmes. These multiple research programmes are not necessarily related to each other; thus concepts and theories used in the field might be incommensurable.

Basic Conceptions and Pluralism

The observation that ecology is a science characterized by plurality has become almost commonplace (Cooper 1996; Kiester 1980; McIntosh 1987; Shrader-Frechette and McCoy 1994; Haila and Taylor 2001). The reasons given for this pluralism are manifold: Ecology encompasses a variety of epistemological positions and grapples with a multiplicity of ontological relations. It has often been assumed by philosophers – and indeed by ecologists themselves – that ecology is an immature and impure science that ought better to be broken down into separate fields of knowledge. Even worse, some assume that ecology is just bad science and the idea of describing ecology as a plural science has frequently been judged as misguided. This is the case in philosophy[2] as well as in ecology itself (see, for example, Roughgarden 1984; Peters 1991). If we wish to reject such assumptions, it follows that we need to recognize the partiality of knowledge as important and useful and, subsequently, to feed it into a different framework that appreciates scientific plurality.

From the point of view of philosophy, this prompts the question of how such a framework – one that is open to pluralism and is not constrained by a strong commitment to unified science or even monism – can be adequately conceptualized. "The multiplicity of approaches is usefully addressed not by comparative evaluations directed at selecting the uniquely correct one, but by appreciating the partiality of each. […] In concert, they [the approaches] constitute a nonunifiable plurality of partial knowledges" (Longino 2006, p. 127). According to this conceptualization, it is not only admissable to produce partial knowledge, it is also acknowledged and accepted that these partial knowledges do not operate in parallel, as in a neat division of labour where every question is approached with its own distinctive set of methods.

---

[2] See for instance Haila and Taylor (2001, p. 93): "Ecology has received relatively slight attention among non-biologist scholars, including philosophers, who interpret and comment the life sciences."

The present study aligns itself with Longino's comments and with arguments for greater tolerance of different knowledge claims and interrelations within disciplines (conceptual, causal, model-driven, data-driven etc.). It also represents an appeal for acceptance of different linguistic forms in the sciences and for the permissibility of different philosophical accounts of scientific methodology. It thus adopts the pragmatic approach proposed by Rudolf Carnap to learn

> from the lessons of history. Let us grant to those who work in any special field of investigation the freedom to use any form of expression which seems useful to them; the work in the field will sooner or later lead to the elimination of those forms which have no useful function. Let us be cautious in making assertions and critical in examining them, but tolerant in permitting linguistic forms. (Carnap 1956, p. 40)

Plurality in Ecology

The present approach argues that the existence of a plurality of theories (or programmes) can have a positive impact because it allows for greater logical flexibility and thus for more explanatory power as well. On this point, Nancy Cartwright (1999) has shown that, if anything, the search for explanatory unity detracts from the search for truth. In a similar vein, Shrader-Frechette and McCoy (1994) argue in the course of their attempt to develop a philosophical vocabulary adequate to partial knowledges that, in ecology, the main method used for linking data with a hypothesis is not the classical deductive scheme but rather (in case studies, for instance) a variety of logic they call "informal inferences". Cooper (1998) suggests a three-fold scheme in which theoretical principles, phenomenological patterns and causal generalizations are the basic forms of generalization in ecology. This constitutes a philosophical taxonomy which, as Cooper points out, should not be taken as a rigorous or categorical classification; instead, it should function as an aid to distinguishing the different modes of investigation (model-driven or data-driven ecology, for example) and acknowledging their varied generalizations while not dismissing the possibility that laws may exist in ecology.

> The various regions in the taxonomic space [...] are all more or less occupied. There is a great deal of variation in scope and reliability among ecological generalizations. [...] (I)f laws are what philosophy of science has tended to take them to be, then there are no laws in biology. But that does not mean that everything in biology is equally contingent. (Cooper 1998, p. 582, 584)

Thus all these authors share an unease toward any logical or methodological unity in science. Instead, they support the idea of different epistemological strategies that can be described in a philosophically sound way. In ecology these epistemological strategies range from experimental studies in the lab, through real-world experiments, quasi-experimental studies and case studies, to purely observational studies in the field.

Taking for granted a plurality of methods and knowledge in ecology, this study argues that ecological plurality is shaped in a distinct way and can be conceptualized as a structure consisting of three basic conceptions, namely "energy", "niche" and "microcosm". These three are conceived of as existing in a rather antagonistic relation to one another, each of them typifying a particular historical field of historical knowledge which embraces experimental practices and theories about living beings in their environments. Over time, all three basic conceptions are flexible; they display a dynamic behaviour that is described in the following as "oscillation". The triadic structure is additionally seen as having systematic implications, in that it functions as a kind of pattern. This pattern is informed by a distinctive conceptualization of modernity as described from a philosophy of history perspective.

Depending on what political or societal exigencies are at stake (though also as a result of its internal dynamic), a research programme from the sphere of the "microcosm" conception may move into the domain of "energy". It is assumed, then, that there are relations of competition and dominance between the basic conceptions, such that any one basic conception might push another into the background but is never able to eliminate it entirely. Instead, a permanent oscillation between the conceptions enables stability to be attained, without the conceptions ever becoming unified.[3]

This may sound rather schematic for now, but this monochrome blueprint will acquire more colour as examples are provided from the context of theory building in early ecology. This will be preceded, though, by a section in which a few aspects of the philosophy of science debate about unity versus plurality in the sciences are presented. The expectation is that the conceptualization of the basic conceptions may benefit from this debate – and perhaps, conversely, the pluralism debate may acquire an additional facet by way of a relatively obscure field of study in the philosophy of science.

Excursus on Plurality and Lawlikeness in Ecology and Biology

Viewing the unity of "science" (in the singular) as a measure of its goodness has long since ceased to be a commonplace topos in the philosophy of science. The capacity to think of the sciences in terms of a plurality in both theory and practice is not only acknowledged as useful, it is now encouraged and indeed required in this field. An increasing number of proposals – themselves pluralistic – relating to scientific pluralism have been put forward over the last 20 years. Among their more well-known exponents are Nancy Cartwright, Ian Hacking, Helen Longino, Alan Richardson, while their forerunners are often said to include Patrick Suppes, Alfred North Whitehead, and William Whewell; even Karl Popper and Paul Feyerabend agreed about this point that "'theoretical pluralism' is better than 'theoretical monism'" (Lakatos 1972, p. 135).

---

[3] This is explicated in detail later in this essay in the language of the Lakatosian research programmes.

In the philosophy of the biological sciences (including ecology) special attention is given to plurality in the sense of the difference and diversity of research objects and interrelationships. It is this that makes biological research so attractive. This applies above all to those biological disciplines which are concerned less with nomothetic than with idiographically constituted knowledge, that is, those areas of knowledge that are not so much about theory or physical laws in general (as in molecular biology or evolutionary biology) as about providing an appropriate (in philosophy of science terms) description of case studies. The issue is thus one of understanding real world experiments and quasi-experiments, of model-based versus data-based methods and modes of representation.

The struggle to formulate an appropriate description of this difference and diversity of objects and interrelations has been a part of biological sciences since they emerged as a field of academic knowledge. What distinguishes the law of gravity from talk of a law governing the cycle of organic substance, from the basic law of biocoenosis or from Mendel's laws? This question of what distinguishes ecological or biological regularities from physical or chemical regularities is one of determining and evaluating exceptions and limits, the single and the individual. Also at stake in this question, though, is a natural philosophy, one which carries with it a claim to universality and laws in nature. As both ecology and biology grew stronger in science and society – particularly during the second half of the twentieth century – and as philosophy of science began to be applied to biological disciplines, controversies arose as to the existence and status of biological (or ecological) laws.

The possibility of biological laws is ruled out from the start when the concept of (a physical, natural) law is conceived in terms of traditional philosophy of science and is identified with physics as the ideal science. In this scheme, a natural law must satisfy three conditions: it must apply at all times and in every place; it must not contain any reference to individual names; and finally, there must be no exceptions. Smart (1963, p. 52) names a fourth condition which amounts to a fourth argument against the law-based character of generalizations in biology, namely the role of coincidence. This, so he asserts, can never be completely eliminated in biology, which is why "biology is physics and chemistry plus natural history".[4]

John Beatty (1995) pursues a similar line of argument to the extent that he, too, believes there can be no laws in biology on account of the evolutionary contingency thesis. Wherever it is possible to identify laws in biology, these are not "biological" laws but ultimately physical or chemical laws. The evolutionary process generates a situation of "high-level contingency" which ultimately prevents a species from behaving exactly the same way if "the experiment" were to be repeated – even if the same environmental conditions prevailed.

This has been countered by the argument that there certainly can be laws in biology – or at least law-like structures. Martin Carrier, for instance, argues that the extent to which laws underlie facts is more a matter of degree than of principle: the concept of fitness, for example, enables us to explain certain aspects of the

---

[4] Smart also stresses that generalizations in biology should be understood in a similar way to those in the field of technology: both are at the mercy of historical contingency (Schweitzer 2000, p. 369).

evolutionary process, and the so-called Lotka-Volterra model in ecology makes it possible to predict the development of a population – within a well-defined range. The crucial point is to tolerate the idea that these laws

> apply to a variety of physically distinct cases. They express features that are quite differently realized physically. For this reason supervenient concepts are apt to capture general traits inaccessible on the purely physical level [...]. I maintain, biology and the physical sciences are in the same ballpark. They both contain laws. (Carrier 1995, pp. 92, 97)

What is now virtually uncontested is that this debate about the concept of laws is inextricably linked with a struggle over hierarchies and the power of knowledge within the sciences. Similarly, most would agree that the debate about realism went hand-in-hand with a tendency to confuse the level of description with that being described and thus to "naturalize" the concept of laws. Philosophically, however, this is not a legitimate move. On this issue, philosopher of science Michael Hampe notes:

> The search for syntactic, semantic or 'architectural' criteria for law-like regularity is always a search within the domain of instruments of description, not in that of the aspect of nature being described. [...] One might well say that no philosophically serious attempts are nowadays made anymore to establish the natural regularity of nature on the basis of an understanding of nature that is somehow 'above' the sciences – which, after all, have always postulated certain laws of nature.[5]

This remark concerning the naivety and ignorance of essentialist reasoning is echoed in a rather laconic remark made by theorist of biology Tim Allen and colleagues, who call on ecologists to adapt their narratives to the postmodern world in which we already live: "The material world will not tell you what decisions you must make as a scientist ... Instead of retreating to naïve objectivism, scientists need to adapt to a postmodern age by becoming conscious of the significance of their narratives" (Allen et al. 2001, p. 484).[6]

To seek to establish laws of nature on the basis of nature itself is philosophically impossible; as a result, laws lose their normative power as unifying and indeed necessary instruments of description to certify the scientific character of a discipline. This opens up a space for other concepts and "narratives" and, not least, for a different philosophy of nature.

In the 1970s, philosopher of biology Michael Ruse argues against identifying the concept of laws with physics, as this, he says, relegates biology to the second

---

[5] Hampe 2000, p. 250 f. He also draws attention to the fact that in the "strategy of legitimizing laws through more general laws, it is not possible to eliminate contingency. Any legitimation of natural laws has to appeal to something other than laws." The question of why laws of nature apply, though, essentially depends, as Hampe points out, on the question of legitimation. God as lawmaker is an inadmissable option to date; and analytical philosophy of science has no general criterion to offer either for laws that would enable a distinction to be made between "legitimate" laws and universal statements which "make an illegitimate claim to the mantle of laws" (see also Giere (1995) for an extended discussion from a skeptical perspective on the laws of nature). If not otherwise marked, translation of citations was done by Kathleen Cross.

[6] Allen et al. are not only taking ecologists to task here on account of their alleged naive realism; they are also drawing attention to the problem of underdetermination.

league of science from the start. He also argues, though, that physical laws, themselves contain exceptions and proper nouns and that Mendel's Laws most certainly can be subjected to a process of independent examination (Ruse 1973). Kenneth Waters (1998) is also wary of the concept of law. He explores a conceptualization of biological generalizations in which a distinction is drawn between two types of generalization: The first refers to the distribution of traits among biological entities such as populations or groups, while the second describes dispositions of causal regularities.[7] The trade-off in this distinction is that the evolutionary contingency hypothesis only applies to some biological generalizations, mainly the first type, that of distribution. However, generalizations about biological regularities can be made for distinct system classes which are independent in time and space.[8]

The fact that contingency does not always play a role in ecology is a point also emphasized by Gregory Cooper. His concern is to find a mechanism for recognizing degrees of contingency in ecological generalizations. In the end, he suggests that "the attempt to partition generalizations into the two categories of laws and non-laws should be abandoned in favor of the concept of nomic force, […] which recognizes that nomicity in biology comes in degrees and over restricted domains" (Cooper 1996, p. 33 f.). Similarly, Sandra Mitchell argues that the dichotomous oppositions "law vs accident" and "necessity vs contingency" are, if anything, an obstacle that impoverishes the conceptual framework. Such an approach, she says, "obscures much interesting variation in both the types of causal structures studied by the sciences and the types of representations used by scientists" (Mitchell 2000, p. 243). She defends a "multidimensional account" of scientific knowledge, proposing a multi-dimensional conceptual space that spans the axes of abstraction, stability and strength. This scheme allows for a more comprehensive conception of "law" in which both the law of conservation of mass and Mendel's law can be represented in the same conceptual space, with the latter just being less stable. Thus the advantage of the model is that "the strength of the determination can also vary from low probability relations to full-fledged determinism, from unique to multiple outcomes" (Mitchell 2000, p. 263).

Once the multidimensionality of scientific knowledge is acknowledged, one might even go further and admit that scientific knowledge is not only a matter of concepts and theories but also a matter of practice. This is an issue raised by Shreder-Frechette and McCoy when they argue that the "logic" of case studies consists in a certain pragmatic procedure which ultimately makes them a reliable basis for comparison as well as generalization. This is because the rationality of practice relies on rules and on the reference to a community of individuals – just as scientific concepts themselves do. "Hence the 'logic' of case studies may be

---

[7] Waters's oft-quoted example of law-like causal regularity is the following: "Blood vessels with a high content of elastin expand as internal fluid pressure increases and contract as the pressure decreases." (Waters 1998, p. 19).

[8] For further details, see Weber (1999) and Schweitzer (2000). Drawing on Waters's analysis, Weber eventually draws the not so surprising conclusion that "ecology knows evolutionarily invariant generalizations which are law-like and at the same time distinctively biological." (1999, p. 71).

appropriate to science if one conceives of scientific justification and objectivity in terms of method, in terms of practices that are unbiased – rather than in terms merely of a set of inferences, propositions that are impersonal" (Shrader-Frechette and McCoy 1994, p. 243). Thus the uniqueness of case studies must be seen to lie not only in their enactment of purely subjective rules and untestable principles. Instead, ecological case studies should be perceived and appreciated epistemologically for the specific form of knowledge they manifest, which accommodates practical as well as conceptual and methodological analysis. It is likely that this epistemic form is closer to the ideal of a proper epistemic model in ecology.

Thus there are a range of promising approaches available within the philosophy of science to enable an adequate description of a corpus of knowledge as plural in terms of its methods, practices, concepts and objects as that generated in scientific ecology. The approaches discussed,[9] be they Cooper's concept of nomic force, Carrier's concept of supervenience, or Mitchell's multidimensional space – all these expounding a more differentiated gradation of scientific generalizations – or again Shrader-Frechette and McCoy's proposal of a "logic" of case studies based on pragmatic procedures: all these approaches provide appropriate instruments for describing ecology as a powerful scientific discipline.

A Rational Reconstruction of Ecology

The question now is how the conceptual triad grounded in natural philosophy can be linked with the dynamics of theory development: How do the basic conceptions and thus ecological theories take over where the others leave off, what is the connection between the natural philosophical content – the "character" – of a basic conception and the theories developed within each one? Which philosophy of science is capable – to borrow a meta-history of science norm from Lakatos as a criterion of quality – of integrating "history of science" the most, that is, of reconstructing the most concepts and theories in a rational manner? The philosophy of science method of rational reconstruction proposed by Lakatos certainly does seem best suited to describe adequately the dynamic development of theory in ecology. This is because, first, the method does not prescribe its methodological construction according to criteria set by some external authority but rather orients itself towards the judgment of the community itself regarding the extent to which a theory is regarded as progressive or degenerative. This provides a good depiction of ecological relationships, because it is precisely here that these latent simultaneities of research programmes are encountered. Second, methodological strategies are proposed in order to tie the dynamics of theory building back to a natural philosophical substance, or "natural character". An additional third argument might be that Lakatos holds not only that "the history of science is the history of

---

[9] To rule out the possibility of misunderstandings at this point: none of the approaches mentioned lays claim to completeness.

research programmes rather than of theories" but beyond that this "may therefore be seen as a partial vindication of the view that the history of science is the history of conceptual frameworks or of scientific languages." (Lakatos 1972, p. 132).

Research programmes consist of a "hard core" which is adhered to even when difficulties with experiments arise, and a "protection belt", in which auxiliary hypotheses are developed and adjustments made in order to protect the core. The adjustments are guided by a positive heuristic, which defines problems and makes methods available. It is the "protective belt of auxiliary hypotheses which has to bear the brunt of tests and get adjusted and re-adjusted, or even completely replaced, to defend the thus-hardened core" (loc cit., p. 133). The heuristic power of a research programme is the capacity to anticipate novel facts in its growth. Thus, a research programme is successful as long as its progressive problemshift continues. "Creative imagination is likely to find corroborating novel evidence even for the most 'absurd' programme, if the search has sufficient drive." (loc cit., p. 187). Lakatos even admits that the increase in content needs to prove its worth only occasionally in retrospect in order to have enough scope for giving a rational explanation for dogmatically clinging to a research programme (even in the face of refutations): how many new facts did it produce, how great was the capacity of research programmes "to explain their refutations in the course of their growth" (loc cit., p. 137).

This idea makes it possible to say, then, that the refutations are able to have no impact on the hard core: Thus research programmes are scientific achievements which can be evaluated on the basis of progressive and degenerating problemshifts. The basic unit of evaluation for progressiveness or degeneration is not the individual theory[10] but a series of theories – the research programme. The function of the research programme consists in predicting anomalies through its positive explanatory potential and integrating them by force of the dynamics of persuasion. What is decisive is that the research programmes are simultaneously characterised by a sluggishness which guarantees their continuity. This continuity can not simply be interrupted by a "crucial experiment" or an individual theory, either. Lakatos states that in this sense no experiment can ever be regarded as decisive, neither at the time it was conducted, and certainly not beforehand.

The replacement of one research programme by another occurs when it "loses its empirical power", proposes Lakatos by paraphrasing Popper (loc cit., p. 154). A new programme already developed in nuce during the lifetime of the old programme comes to compete with the older one and eventually outstrips it by its greater problem-solving power. "While the old programme increasingly requires theoretical adaptations to uncooperative experiential data (degenerating problem shifts), theoretical development in the new one maintains the upper hand (progressive problem shift): empiricism becomes not the task of the new programme but a successful test of it" (Diederich 1974, p. 14). A research programme continues to advance as long

---

[10]Diederich (1974) and other authors (e.g. Wolfgang Stegmüller) complain that Lakatos's use of the term "theory" is not as unequivocal, as the powerful weighting of theory towards research programmes would suggest with some justification.

as new facts can be predicted with a certain degree of success, and it stagnates if its theoretical growth is retarded – if it is able to offer only post-hoc explanations. The point at which it is completely cancelled is when a new research programme is capable of demonstrating greater explanatory potential compared with its predecessor.

However, what is key here is that Lakatos admits of competition between two research programmes as a "long extended process, during which it is possible to work rationally on each of the two programmes (*or, if possible, on both*)" (Lakatos 1974, p. 282; emphasis in original). The difficulty of deciding when a research programme can finally be regarded as having been superseded could be made further acute insofar as the extended process of competition between two research programmes can not, in principle, be contained at all. "Even when a research programme is seen to be swept away by its predecessor, it is not swept away by some 'crucial' experiment; and even if some crucial experiment is later called in doubt, the new research programme cannot be stopped without a powerful progressive upsurge of the old programme." (Lakatos 1972, p. 163). Thus research programmes would exist in perpetuity and – depending on the one hand on the internal development of the degenerating problem shift and on the other on external societal developments – alternately put the other "rival to one side", which then lurks in the background, awaiting its opportunity to reconquer the field. However, Lakatos points out that there lies no necessity in the series of alternation of speculative conjectures and empirical refutations. Rather, the "dialectic of research programmes" is characterized by a broad variety: "which pattern is actually realized depends only on historical accident." (loc. cit., p. 151).

This "dialectic" and the variety of interactions between the evolution of the programme and the empirical checks as well as the simultaneity of co-existing research programmes seems to be a fitting description of the plural situation in ecology. This is the sense in which the three ecological basic conceptions of "microcosm", "energy" and "niche" will be discussed in the following. In the descriptive mode of the Lakatosian research programme, the basic conceptions's natural philosophical content (or natural character) merges and becomes a part of the hard core.

Three Basic Conceptions in Ecology

The purpose of the term "basic conception" is to describe the possibility of three conceptions of nature and to render them operational in such a way that, by providing a normative pattern, they in turn render the discipline of ecology comprehensible. This pattern opens up and simultaneously limits the space of possibilities for concept formation and theory building. The partial knowledge areas and the narratives present in ecology are structured along these lines. The basic conception of microcosm, for example, typically contains a romantic narrative, whereas the basic conception of "niche" is characterized more by a drive economic narrative based on the economy of drives. This pattern of basic conceptions turns out to be quite robust; in a certain sense it might even be called archaic: it is linked to the very beginnings of ecology as a scientific discipline, has been reproduced time and again ever since, and serves as a rather intuitive meta-narrative of the discipline. While the triadic pattern is systematically robust, it is historically flexible insofar as the power of representation of the basic conceptions among each other can be switched and repositioned, depending on which character of nature fits best with the narrative or heuristics concerned and eventually helps to get a research programme accepted. Which of the three conceptions of nature is able to become established in a specific historical situation is subject not only to conditions within science itself but is also influenced by societal projections and expectations of ecological concepts or models. In order to get an impression of the success, historically speaking, of a particular basic conception, and to assess the joint oscillation of all three basic conceptions over a longer period of time, it is important not only to pay heed to the dynamics between the three characters of nature and those of the formation of concepts and hypotheses, but also to consider which socio-political expectations and prevailing circumstances[11] the ecological description of nature refers to and builds upon (and, conversely, which ones it influences).

Basic Conception of "Niche"

Within the basic conception of "niche" organisms are imagined as "self-determined" individuals in early ecology. In "the lake as a microcosm", a paper that was influential in the historical development of ecology, American zoologist Stephen Alfred Forbes (1887) describes organisms as "remarkably isolated" and indifferent or even hostile towards one another. There is an abundantly clear reference in Forbes to a economic driven character of nature and therefore to "niche", and this is confirmed by closer analysis at the conceptual level.[12] Predator-prey relationships are included in the field of view here; they appear to be ubiquitous and the relationships between the competing individuals are influenced by the principle that every animal has its enemies, and "mercy and welfare are completely unknown" (Forbes 1887). These organisms come together in societies in which the distribution of roles are determined solely according to the purpose of the association: the organisms build a society based on what might be called a "social contract". The organisms'

---

[11] Particularly Ludwig Trepl and collaborators underline the socio-political influence on ecological theories; beyond this rather general claim they see a close relationship between a distinct political constellation and the concepts of ecological society or community (see Trepl 1987, 1994, and also Trepl & Voigt Chap. 5, this volume).

[12] It may at first seem confusing that the very publication in which the lake is proclaimed – for the first time – to be a microcosm is not enlisted for the basic conception of "microcosm". Here, though, microcosm refers above all to the idea of a system per se, a combined view of individual parts which at that point could not yet be integrated conceptually. But this new perspective is in some sense the ecosystem perspective and can occur in all three basic conceptions. So the phrase "lake as microcosm" is not on its own sufficient for deciding to which of the basic conceptions it should be ascribed.

environment – biotic and abiotic – is at most a neutral, but more usually hostile affair, which intervenes in the life of the individuals for purposes of regulation.

The Darwinian principle of natural selection is omnipresent here, so that the evolutionary biological model of the niche is able to become quite well established as a result. One of the most successful ecological theories is developed in this context, namely the "competitive exclusion principle" (CEP). The CEP states that two species coexisting in the same space compete for limited resources. The niche is regarded as a spatially organized structural unit which the competing species endeavour to gain possession of. The structure of the niche lies to a certain extent at the bottom of the society, whose stability is an outcome of the regulation that occurs through competition between the species. The CEP is held to be a model which is so well developed mathematically that it is also considered capable of making predictions. It is conceived as a key to explaining the composition of animal and plant societies, leading to the dominance of the basic conception of "niche" in theory building in ecology (Gause 1934).

When Charles Elton (1927) introduced the functional concept of niche, the CEP was temporarily in trouble. What Elton wants to achieve with his conception of niche is to describe the position of an animal in the food chain and therefore its economic significance. This new function-based niche did not entirely supplant the old spatially-conceived niche. The old meaning of niche as habitat and not as a position in the food chain crops up at various points (Kingsland 1991, p. 6). So the CEP also plays a role in Elton's conception, albeit not a dominant one; the principles of competition and selection receded into the background in Elton's work. Ultimately, the heuristic of the spatial concept of niche becomes weakened as a result. In Elton's work the concept of niche is transformed: the spatial niche becomes the functional niche and at the same time is "transferred from the evolutionary biological context into an ecological context" (Trepl 1987, p. 170). This "ecological context" fits with the basic conception of "energy" in the triadic conception, where relational conditions and a systemic perspective play a role above all.

However, this is by no means the end of the evolutionary biological concept of niche: it experiences a renaissance in the 1950s, especially in plant ecology, and is then brought successfully to bear in opposition to the so-called climax theory (basic conception "microcosm"). Eventually, the functional and the spatial concept of niche are integrated in a single theory: the niche concept is conceptualized as a geometric, "n-dimensional hyperspace" (Hutchinson 1957) which makes it possible to depict all ecological factors, i.e. the sum total of a species' conditions of existence – upon which the building of successful theory takes place in the context of the basic conception of "energy".

Basic Conception of "Microcosm"

The "microcosm" relates to romantic nature. As a more or less well thought-through position in terms of philosophy of nature, this nature can currently be found, for example, in nature conservation ecology, but is also quite a widespread narrative in theoretical ecology. This applies all the more to early ecology at the close of the nineteenth century. Descriptions of a lake as the "arena of life"[13] or as the "self-enclosed, clearly demarcated whole" (Forel 1901), or again as a "mirror image of processes in the larger whole" (Zacharias 1905) in which an "extremely intricate living hustle and bustle" (ibid.) is found, are not unusual. The principle of selection at the centre of the basic conception of "niche", that of the control of individual organisms from the outside, is replaced in the basic conception of the "microcosm" by principles of reciprocal and (usually) highly hierarchical control. As a result, the biological community is viewed not according to the premise of competition between individuals but according to the premise of fitting in or adaptation to the community, along with an emphasis on the interplay and the mutual dependency among all individuals or species. Here, the local conditions are not an expression of relentless natural laws by which the plants and animals occurring more or less by chance in a certain place are externally controlled. Instead, the community actively adapts to the local conditions, and in the process of adaptation, these local conditions are "internalized".

In plant ecology, for example, this position has been represented by the "superorganism" theory, in which the plant community is conceptualized as a superorganismic individual, whose individual organisms exist in intra-societally organized functional interrelationships. The superorganism successively works to develop its individuality through the interrelations between organisms and general local conditions, which at the same time are a specific place. In this process of adaptation the plant community increasingly detaches itself from the immediate and specific constraint of nature and simultaneously adapts to the region and climate. The total organism "plant community" develops the specific aspects of the place, which distinguish it from all other places and the organisms adapted to them. Since the plant community contributes to shaping the space surrounding it, though, its relationship to this space is not exclusively one of dependence but rather highlights the particularity of the organized individual.

Basic Conception of "Energy"

The introduction of the conception of energy in physics established a unity between different sub-fields of physics.[14] Qualitatively different natural phenomena can not only be "reduced to a common one in the conception of energy"; more than this,

---

[13] Forbes 1887; Forel 1891, 1901; Zacharias 1904, 1905, 1907, 1909.

[14] "Unwittingly perhaps, Helmholtz was the first great bourgois philosopher of labor power precisely because in his essays on Kraft he does not distinguish between natural, mechanical, or human labor power. For him all expenditure of energy produced work, and conversely all work involved the consumption of energy. His conception of labor power reveals no self-moving power, no social labor that is not at the same time a natural force. With Helmholtz work was reduced to a quantitative phenomenon subject to a system of mathematical equivalents" (Rabinbach 1990, p. 61).

"energy is seen in some sense as the origin of all changes in nature and as the measure of all impacts" (Breger 1982, p. 41). It just seemed to be a matter of consistency that, after the conception of energy had been introduced in physics, within a short time it became common practice to describe all natural phenomena by giving their energy distribution.

In the physiology of the 1850s this idea of the standardization of the forces of nature using the concept of energy was taken up in something of a vulgar materialistic variation. The "eternal cycle of material-bound life", in fact, life per se, was declared to be "nothing but material", which in turn was supposed to be nothing other than force or energy.[15] A material characterized in this way can quite easily be non-living or living nature, and could now be conveyed on a material basis via the cycle.

This is taken up in early aquatic ecology, for instance with the notion of the circulating "organic substance" in a lake. This organic substance was not characterized by the property of a particular material or of a chemical element but by its function of connecting the organisms in the lake with one another and with their inorganic environment. Organic substances are materials with which work is done on an organism, i.e. ones which contribute to its maintenance or are brought forth by it. The crucial point is that these functions of organic substance are always related to the "lake as a system", in which the cycle of organic substance takes place. By analogy to the physical conception of energy, the reference here, too, is to an image of nature in which the idea of balance and of unity are of central significance – we are looking at a nature of control.

Provisional Result

– The simultaneity of co-existing research programmes seems to be a fitting description of the plural situation in ecology. This is the sense in which the three ecological basic conceptions of "microcosm", "energy" and "niche" were discussed.
– The three basic conceptions provide a normative pattern that opens up and simultaneously limits the space of possibilities for concept formation and theory building. Concepts and narratives present in ecology are structured along these lines.
– This pattern of basic conceptions turns out to be quite robust; it has been reproduced time and again ever since the scientific field of ecological knowledge exists, and seems to serve as a meta-narrative of the discipline.

---

[15] In physics the concept "energy" is shaped not before the 1850s. However, a unifying force that goes far beyond Newton's concept of force, is already in discussion much earlier (see for instance Schelling) (Breger 1882, p. 98 f.).

- The triadic pattern is flexible over time in so far as the power of representation of the basic conceptions among each other can be switched and repositioned, depending on which character of nature fits best with the narrative or heuristics concerned and eventually helps to get a research programme accepted.
- Which of the three basic conceptions is able to become established in a specific historical situation is subject not only to conditions within science itself but is also influenced by societal projections and expectations of ecological concepts or models.

Oscillating Basic Conceptions

A crucial part of the theoretical dynamics described here is that the basic conceptions are able to coexist. Which of the basic conceptions is able to predominate over the others seems to be less a question of the internal theoretical dynamics and more one of the degree of fit between the structural core, or character, of the basic conception and the socio-political context.

In early ecology it was the basic conception of "niche" that came powerfully to the fore.[16] Towards the end of the nineteenth century plant ecologist Eugen Warming (1841–1924) emphasized that competition should be regarded as the most important relationship between the members of a plant community, thus aligning himself with a prevailing "Darwinian paradigm". "Community" was conceived of as a collection of individuals that confront each other in the competition over scarce resources in either an indifferent or a hostile way. From this perspective, individuals and species in a given location live in a biological community, but are largely determined by their individual biological capacities. In liberal social scientific theories, the contradiction between the active component in this model – the active individual fighting for survival – and the passive component – the external exigencies of the environment – is resolved in an endless and undirected evolutionary process, that is, in progress. This heavy emphasis on endlessness and progress in Darwinian theory and later in Gleason's theory of succession (Gleason 1926) became a point of widespread criticism in ecological community, ringing in – as historian and theorist of ecology Ludwig Trepl argues, for example – a conservative phase. From the conservative perspective the biological community does not function on the premise of competition between individuals but on the premise of adaptation and an emphasis on the interplay and the mutual dependency among all individuals or species. By analogy to the conservative social philosophy, this construction corresponds to the idea of the "superorganism": the whole comes before the parts, which fulfil their task in a functionalist, teleological way. The "superorganism" contains and combines

---

[16] This was discussed in more detail in Schwarz and Trepl (1998, p. 305 f).

the cooperating elements into a harmonious structure of a higher order (Clements 1936, but also already 1916). The conservative idea of the state is also characterized by the endeavour to eliminate the liberal distinction between state and society or at least to weaken it, so that the community (which ultimately means an indivisible unity of state and society) acquires a more important role than the one assigned it by the liberal position:

> The societal organism which is recognized as relatively higher must be not only that in which greater scope is given to the political freedom of the individual citizens or social groups, but also that in which state power in its greatest concentration also possesses the greatest independence, self-determination, and freedom of action. (Lilienfeld 1873, p. 84)

In ecological theories of the romantic-holistic type, similar structures which arose at roughly the same time can be identified:

> Every oyster bank is in some sense a community of living creatures, a selection of species and a sum of individuals which find in this very place all the conditions for their emergence and maintenance, that is, the right soil, sufficient food, the appropriate salt content and tolerable temperatures favourable for development. ... All living members of a living community keep [...] the balance with the physical conditions of their biocoenosis because they maintain themselves and reproduce themselves in the face of all the influences of external irritations and against all attacks on the continued existence of their individuality. (Möbius 1986 (1877), p. 74 f.)

In the 1920s and 30s the idea of superindividual units became dominant in aquatic ecology as well. Richard Woltereck (1877–1944) concerned himself extensively with the "spatial structures of biosystems" (title of chapter eight; Woltereck 1940 (1932), p. 208) and developed an intricate terminology in which individuals were conceived of as "components of multi-sectional systems (collective structure)". The so-called "multi-person systems" were able to be "physically unified edifices" which "belong together only in spatial terms". Edifices are structures used to describe an "ordered multiplicity", and between whose "components interactions take place which are merged together by superordinate relationships into a whole or gestalt. This applies to the individual organism as a self- or individual edifice [...], secondly to the organism in its environment [...], thirdly to multi-sectional, but self-enclosed systems of organisms" (ibid.). All "three kinds of system" are to be conceived of as "edifices of relationships".

Woltereck distances himself vigorously from the "niche position" and thus simultaneously from political liberalism: Chance, according to Woltereck, was "nothing but a shabby word, a non-concept and a nonsense (...), despite the supposedly creative power which the term 'selection' was to infuse into this chance." (l.c., p. 230).

In the years after 1930 this holistic image of nature as "romantic nature" was gradually replaced by the basic conception of "energy" and thus by a synthetic position between so-called reductionism and organicism (or vitalism). However, Ludwig von Bertalanffy writes already in 1929 (p. 95):

> We understand on the basis of the maintenance of gestalts how the organism preserves itself in metabolism. ...Our view is a theory of a system remaining steadfast in its condition and overcomes both machine theory, which is not adequate, as well as the scientifically impossible vitalist view.

This position became highly successful in ecology; the early ecosystem theories were developed out of the basic conception of "energy" (Schwarz 1996). Ecosystem wholeness promised controllability and the technical reconstruction of nature.

What Holds Ecology Together?

The question remains, however, why a plural ecology does not fall apart but instead remains a reference system to which research programmes and concepts are bound and a particular corpus of knowledge refers, even when the disciplinary edges become blurred. Why does a system – be it an experimental system,[17] a discipline, or a research programme – peters out (or not)?

A variety of epistemological proposals have been presented, each referring to different systemic levels: The experimental system becomes exhausted when there are too many irreconcilable, conflicting answers available; the heuristic of a research programme becomes negative when its theories are no longer able to explain the phenomena observed; and a discipline falls apart when its instruments and methods become progressively diverse while other epistemic objects, or "question machines" (Rheinberger 1992, p. 72), emerge.

The approach proposed in the following is based above all on an analysis of concepts as constitutive elements of ecology as a knowledge system. It therefore argues that it is mainly the conceptual architecture, rather than theories, methods or instruments that guarantees disciplinary and logical robustness. This does not mean, of course, that in the triadic design of the basic conceptions of energy, niche and microcosm theories or narratives are not equally at work. After all, concepts only function as such when they are appropriately contextualized – in distinct theories or narratives. Thus there is typically a romantic narrative contained in the basic conception of microcosm, for example. The basic conception "niche" tends to be influenced by an economy of drives, while the conception "energy" relies on narratives in which functional und systemic relations play a prominent role. Based on the triadic approach, then, the temporal and epistemic stability of ecology can be construed as follows:

It is generated (a) by the interrrelational context in which the concepts are embedded, represented by distinct theories and narratives; (b) by the dynamic which links them, and finally (c) by being tied to the historical genesis of a triadic concept of nature.

In order to strengthen the latter claim, a philosophy of history approach is mobilized which is aimed at a pluralistic concept of nature. This is where the triadic construction of the basic conceptions finds a philosophical rationale.

---

[17] According to a conception developed by Hagner et al. an "experimental system" in laboratory science is one in which differences are generated (Hagner, Rheinberger, Wahrig-Schmidt 1994, p. 10).

Historical Genesis a Rationale for Pluralism in Ecology Based on the History of Philosophy

Based on the above, three "characters of nature" can be ascribed to the three ecological basic conceptions. Working on the basis of Marquard's ideas, the status of these characters of nature is important in systematic terms. The basic conception of "niche" is supported by drive nature, that of "microcosm" by romantic nature, and the basic conception of "energy" is closest to controlling nature.

The historical genesis of the ecological triad can be localized around the turn of the nineteenth century. In the first instance this is not especially surprising because it coincides with the general perception of a historical break at that time, particularly for the sciences of biology and chemistry. A myriad of studies has been published probing various issues in this regard; however, most of them share the following story: The relationship between nature and culture changes more or less radically, this being apparent in the reorganisation of conceptual frameworks in philosophy, science and literature, in the situatedness of ontology, and also in the narrative forms. These changes permeate through to all scientific disciplines and societal discourses and practices in equal measure. The accelerated secularisation of the order of Nature is reflected in a gradual shift of traits from the outside of bodies to inside them, from concrete and visible signs to abstract and unseen ones. A mathematically describable and law-like Nature steps up to compete with the perception and representation of Nature as a hierarchical chain of beings – the visible signs of the hand of God (allegorically speaking). At the "end of natural history" (Lepenies 1976, Latour 2005) we are left with a Nature which is split off from everything, which is considered to be culturally conditioned – a split which is also reflected in the distinction between two concepts of history: the history of Nature and the history of humans.[18] Historians and philosophers place temporal and spatial limits on this historical break have it lasting for a longer or shorter time, and consider it to be more rhetorically than pragmatically effective – or vice versa. Usually, however, it is identified with the beginnings of modernity and of a construction of objectivity in scientific knowledge production, which relies on what Feyerabend has called the "separability thesis".

In environmental philosophy, debates abound concerning the shift that occurred in the concept of Nature at that time and also concerning the traces of these concepts in contemporary theories and practices of perceiving and representing Nature. Much less common is the effort to link these philosophical conceptions together in order to substantiate the epistemology of ecology, as proposed in the following.

---

[18] In Schelling the latter is merely "inauthentic" history which, moreover, is inferior to the authentic history of nature, a history of the whole of Nature (natura naturans), of Nature that is constantly reproducing itself. This conception of a processual Nature is an oft-used figure of thought in environmental philosophy. (For a review of Schelling's work in the field of ecocriticism, especially with respect to wilderness, see Wilke 2008 from a perspective of literary criticism).

Philosopher Odo Marquard conceptualizes the historical break in which the hitherto fixed relationships between reason, humanity and nature have to be reformulated as a period of transition. The modern constitution of nature as "the sphere opposite to reason" (das Andere der Vernunft) (Böhme and Böhme 1983) is not yet consolidated conceptually in this historical phase, manifesting in a "triadisation" of the concept of nature. The conceptual triad is a snapshot of relationships that remain blurred, within which the new relations emerge.

The thesis put forward in this paper, then, is that these three concepts of nature and their relations and references to and among one another contribute towards stabilising the plural local and temporary development of concepts and theories in ecology, up to and including the present day. Latour's "we have never been modern" applies to ecology to the extent that the latter never obeyed the dominant narrative of a "work of purification" identified with modernity, either historically or systematically. This narrative is replaced in the triadisation of the nature concept by a constellation in permanent transition.

A Triadic Conceptualization of a Philosophically Delocalized Nature

Marquard goes along with the common mode of reading which sees nature, when the historical break[19] occurs, become the estranged and split off counterpart, the other of reason. At that moment, the concept of nature is no longer bound up as part of a metaphysically conceptualized world; instead, it is constructed through scientific experience, which is oriented towards control and imagines nature as a subject to making and remaking ("machbare Natur"). As a result, statements about the essence of nature become "dispensable, or rather impossible".[20] This constellation gives rise to a dilemma, which leads to a shift in the concept of nature along with a process of differentiation. The philosophical reformulation of the concept of nature is challenged, according to Marquard, by a reason that has "begun to waver". Reason threatens to depart from human nature – something that cannot actually happen, as the very realization of human nature itself is identified with reason.[21]

---

[19] On the question of a "reasonable" construction of an epochal break see Nordmann (2006).

[20] Mittelstraß 1981, p. 64. In philosophy the historical break is characterised by the transition from transcendental philosophy to philosophy of history. This proves to be a dilemma in the context of the "anthropological reorientation towards nature" regarding the "question of essence", which in the context of the philosophy of history can only be a false question: the transcendental philosophical framework has been dissolved.

[21] In Marquard's words, human nature is "tied to reason and to that which makes reason what it is, what enables it to flourish" (Marquard 1987, p. 54). Drawing on the Aristotelian concept of nature, M. describes this as "actual" nature. If we attempt to subtract human reason from human nature, it will continue to strive "in itself" towards a realization of reason, without being able to find it – a state which, as "potential nature", is in opposition to actual nature. To assert that this potential nature is one that is completely emancipated from reason is to accept that it is a "mere possibility".

Philosophically, he argues, nature in this situation can no longer be conceptualized within a transcendental context and cannot yet be conceptualized through the philosophy of history. This situation of a delocalized nature explained by Marquard through reference to the interplay of three concepts of nature: controlling nature (Kontrollnatur), romantic nature (Romantiknatur) and drive nature (Triebnatur).

Drive nature refers to lack of control, to the satisfaction of individual desires, and to the realm of the senses in general. Hobbes is called as the crown witness in this context, with his famous dictum: "Hereby it is manifest, that during the time men live without a common Power to keep them all in awe, they are in that condition which is called Warre; and such warre, as is of every man, against every man" (Hobbes 1651, p. 62). When there is no common, binding set of rules to channel passions, the result, says Hobbes, is a civil war in which everyone fights against everyone else. Only fear in the face of a superordinate power can keep humans from hurtling headlong into violence and lack of restraint. This natural state of human beings – of humanity per se – is ever present, even if it is not always equally powerful. It gives a society (though not a community), a particular character that is opposed to the state: "drive nature is present (in tempered form) as society" (Marquard 1987, p. 55).

Romantic nature is above all nature perceived as organism; it can be exalted and beautiful. Whether manifesting as landscape or as fertile wilderness, the "condition of innocence and naivety" (Marquard 1987, p. 57) is ever present in it. Contained in romantic nature are notions of a harmonious cosmos, a symmetrical relationship between microcosm and macrocosm, as well as that of a complete, whole primeval state. Nature conceptualised as romantic also refers to inwardness and temperament, to feeling and longing, to the realms of imagination in general. As a consequence it is especially the domain of literary fiction. "The poet understands nature better than the scientist" (Novalis, quoted in Marquard 1987, p. 57). Only in the presence of the poet does it show itself and reveal its secrets. This is why – in Novalis's interesting reversal of the argument – genuine "physics [is] nothing other than the study of the imagination": "the more poetic, the more true". Romantic nature is above all nature perceived and interpreted aesthetically.

Finally, Marquard identifies controlling nature as that of the exact sciences. It is that which, through observation, experiment and mathematics, can be recorded, manipulated, predicted and therefore controlled. Above all, however, this nature is not "merely sensual" nature but the nature of rational rules, the nature of the rational mind. It is not only derived from the laws of the possibility of experience: it is identical with them. For Marquard, Kant is the source for controlling nature with his famous statement that the general laws of nature must be settled independently (that is before every experience). According to Marquard then, controlling nature is the nature of science and technology, present as "design, method, result and application" (1987, p. 56).

Thus only the last of these three concepts of nature, controlling nature, is one relevant to the natural sciences. This constitutes a pronounced difference in comparison with the ecological basic conceptions, all three of which aim to provide an opening into science. What Marquard's scheme certainly does do, however, is provide a historical rationale for the pluralization of the concept of nature and a philosophical rationale for the triadisation. Both are useful here in order to fashion a systematically sound foundation for the triadic conceptualization of ecology. In this sense delocalized nature – that is, the triadization of the concept of nature – is viewed as a scheme rooted in natural philosophy, which underlie the basic conceptions, acting as a framework which, from a philosophy of science perspective, cannot be circumvented. Unlike hypotheses, such frameworks are exempted from critical examination, being used instead to guide the activity and orientation of hypothesis formation.

This interconnection between hypothesis formation exposed to empirical verification protected from it has been described in the context of various approaches in the philosophy of science. The "metaphysical core" in Imre Lakatos's research programmes[22] as already discussed and the "themes" in scientific thoughts as described by Gerald Holton are examples. Similarly, Gernot Böhme proposes the notion of physiognomically deducible "characters", namely "characters of nature" (Naturcharaktere) which are at work behind the epistemic operations ascribing meaning to nature. Characters of nature are

> sentences in which certain character traits are ascribed to nature, [they] are an expression of highly aggregated experiences. They name an overall impression of nature that is grasped intuitively and formulate it in physiognomic manner: As nature shows certain basic traits, it is ascribed a character, as it were; it is said what is basically expected from it. Statements about the character traits of nature are made not on the basis of scientific experiences alone. For science, they perform the function of heuristics, that is, of instructions for what should be sought in nature and what is to be expected of it. Statements about characters of nature are thus still related to science and its transformation, but they are not scientific statements per se. They can be understood as genuine statements of a philosophy of nature to the extent that the latter is an endeavour to understand nature as a whole and as such. (Böhme 1992, p. 211)

In this sense the three basic conceptions provide a theoretical foundation for ecology based on the philosophy of nature: they make explicit three conceptions of nature which are identified historically with the beginning of modernity and are capable systematically of stabilizing the plural representation of this science.

References

Allen TFA, Tainter JA, Chris Pires J, Hoekstra T (2001) Dragnet ecology "Just the facts Ma'm": The privilege of science in a postmodern world. Bioscience 51:475–484

Beatty J (1995) The evolutionary contingency thesis. In: Wolters G, Lennox JG, McLaughlin P (eds) Concepts, theories, and rationality in biological sciences. The Second Pittsburgh-Konstanz Colloquium in the Philosophy of Science. University of Pittsburgh Press, Pittsburgh, pp 45–81

---

[22] It is important to note here that Lakatos makes no difference between the methodology of a research programme with a "metaphysical" core and "one with a 'refutable' core except perhaps for the logical level of the inconsistencies which are the driving force of the programme." (Lakatos 1972, p. 127).

Böhme G (1992) Wissenschaft-Technik-Gesellschaft. 10 Semester interdisziplinäres Kolloquium an der THD. TH Darmstadt, Darmstadt

Böhme H, Böhme G (1983) Das Andere der Vernunft: zur Entwicklung von Rationalitätsstrukturen am Beispiel Kants. Suhrkamp, Frankfurt/M

Breger H (1982) Die Natur als arbeitende Maschine. Zur Entstehung des Energiebegriffs in der Physik 1840–1850. Campus, Frankfurt/M

Carnap R (1956) Meaning and necessity: a study in semantics and modal logic. (Reprinted of Revue Internationale de Philosophie 4 (1950): 20-40). University of Chicago Press, Chicago

Carrier M (1995) Evolutionary change and lawlikeness. Beatty on biological generalizations. In: Wolters G, Lennox JG (eds) Concepts, theories, and rationality in the biological sciences. Universitätsverlag Konstanz/University of Pittsburgh Press, Konstanz/Pittsburgh

Cartwright N (1999) The dappled world. A study of the boundaries of science. Cambridge University Press, Cambridge

Clements FE (1936) Nature and structure of climax. J Ecol 24:252–284

Cooper GJ (1996) Theoretical modeling and biological laws. Philos Sci 63:28–35

Cooper GJ (1998) Generalizations in ecology: a philosophical taxonomy. Biol Philos 13:555–586

Diederich W (ed) (1974) Theorien der Wissenschaftsgeschichte. Beiträge zur diachronischen Wissenschaftstheorie. Suhrkamp, Frankfurt/M

Elton C (1927) Animal ecology. Sidgwick & Jackson, London

Forbes SA (1887) The lake as a microcosm. Bull Peoria Sci Assoc 111:77–87. (*Reprinted* Bull Nat Hist Surv 15:537-550, Nov 1925)

Forel F-A (1891) Allgemeine Biologie eines Süßwassersees. In: Zacharias O (ed) Die Tier- und Pflanzenwelt des Süßwassers. Weber, Leipzig, pp 1–26

Forel F-A (1901) Handbuch der Seenkunde. Allgemeine Limnologie. Engelhorn, Stuttgart

Gause GF (1934) The struggle for existence. Williams and Wilkins, Baltimore

Giere R (1995) The skeptical perspective: science without laws of nature. In: Weinert F (ed) Laws of nature. Essays on the philosophical, scientific and historical dimension. de Gruyter, Berlin/New York

Gleason HA (1926) The individualistic concept of the plant associations. Bull Torrey Bot Club 53:7–26

Haila Y, Taylor P (2001) The philosophical dullness of classical ecology, and a Levinsian alternative. Biology and Philosophy 16:93–102

Hagner M, Rheinberger H-J, Wahrig-Schmidt B (1994) Objekte, Differenzen, Konjunkturen. In: Hagner M, Rheinberger H-J, Wahrig-Schmidt B (eds) Objekte-Differenzen-Konjunkturen: Experimentalsysteme im historischen Kontext. Akademie Verlag, Berlin, pp 7–22

Hampe M (2000) Gesetz, Natur, Geltung. Historische Anmerkungen. Philos Nat 37:241–254

Hobbes T (1651) Leviathan, or, The matter, forme, and power of a common wealth, ecclesiasticall and civil. Printed for Andrew Crooke, London

Hutchinson GE (1957) Concluding remarks. Population studies: Animal ecology and Demography. Cold Spring Harbor Symposia on Quantitative Biology. T. b. L. C. S. Harbor. New York, Long Island Biological Association. 22:415–422

Kiester RA (1980) Natural kinds, natural history and ecology. Synthese 43:331–342

Kingsland SE (1991) Defining ecology as a science. In: Real L, Brown JH (eds) Foundations of Ecology. The University of Chicago Press, Chicago/London, pp 1–13

Lakatos I (1972) Falsification and the methodology of scientific research programmes. In: Lakatos I, Musgrave A (eds) Criticism and the growth of knowledge. Cambridge University Press, Cambridge

Lakatos I (1974) Die Geschichte der Wissenschaft und ihre rationalen Rekonstruktionen. In: Lakatos I, Musgrave A (eds) Kritik und Erkenntnisfortschritt. Vieweg, Braunschweig, pp 271–312

Latour B (2005) From Realpolitik to Dingpolitik or how to make things public. In: Latour B, Weibel P (eds) Making things public. Atmospheres of democracy. MIT Press, Cambridge, pp 14–43

Lepenies W (1976) Das Ende der Naturgeschichte. Wandel kultureller Selbstverständlichkeiten in den Wissenschaften des 18. und 19. Jahrhunderts. Hanser, München

Lilienfeld Pv (1873) Gedanken über die Socialwissenschaft der Zukunft. Behre, Mitau
Longino HE (2006) Theoretical pluralism and the scientific study of behaviour. In: Kellert SH, Longino HE, Kenneth C (eds) Scientific pluralism. Minnesota studies in the philosophy of Science. University of Minnesota Press, Minneapolis/London, pp 102–131
Marquard O (1987) Transzendentaler Idealismus, Romantische Naturphilosophie, Psychoanalyse. Verlag Jürgen Dinter, Köln
McIntosh RP (1987) Pluralism in ecology. Annu Rev Ecol Syst 18:321–341
Mitchell S (2000) Dimensions of scientific law. Philos Sci 67:242–265
Mittelstraß J (1981) Das Wirken der Natur. In: Rapp F (ed) Naturverständnis und Naturbeherrschung: philosophiegeschichtliche Entwicklung und gegenwärtiger Kontext. Wilhelm Fink, München, pp 36–69
Möbius KA (2006) Zum Biozönose-Begriff. Die Auster und die Austernwirtschaft 1877 (2nd ed. by T Potthast; 1st edition and comment by G Leps 1986). Harri Deutsch, Frankfurt/M
Motterlini M (2001) Reconstructing Lakatos. A reassessment of Lakatos' philosophical project and debates with Feyerabend in light of the Lakatos archive. London school of economics and political science. Centre for philosophy of natural and social science. Discussion paper series LSE (DP 56/01), pp 1–48
Nordmann A (2006) Collapse of distance: epistemic strategies of science and technoscience. Dan Yearb Philos 41:7–34
Peters RH (1991) A critique for ecology. Cambridge University Press, Cambridge
Rabinbach A (1990) The human motor. Energy, fatigue, and the origins of modernity. University of California Press, Berkeley
Rheinberger H-J (1992) Experiment, Differenz, Schrift: zur Geschichte epistemischer Dinge. Basilisken-Presse, Marburg/L
Roughgarden J (1984) Competition and theory in community ecology. In: Salt G (ed) Ecology and evolutionary biology: a round table on research. University of Chicago Press, Chicago
Ruse M (1973) Philosophy of biology. Hutchinson, London
Schwarz AE (1996) Gestalten werden Systeme: Frühe Systemtheorie in der Ökologie. In: Mathes K, Breckling B, Ekschmidt K (eds) Systemtheorie in der Ökologie. Ecomed, Landsberg, pp 35–45
Schwarz AE, Trepl L (1998) The relativity of orientors: interdependence of potential goal functions and political and social developments. In: Leupelt M, Müller F (eds) Eco targets, goal functions, and orientors. Springer, Berlin, pp 298–311
Schweitzer B (2000) Naturgesetze in der Biologie. Philos Nat 37:367–374
Shrader-Frechette K, McCoy ED (1994) Applied ecology and the logic of case studies. Philos Sci 61:228–249
Smart JJC (1963) Philosophy and scientific realism. Routledge and Kegan Paul, London
Trepl L (1994) Holism and reductionism in ecology: technical, political, and ideological implications. CNS 5:13–40
Trepl L (1987) Geschichte der Ökologie. Athenäum, Frankfurt/M
Waters CK (1998) Causal regularities in the biological world of contingent distributions. Biology and Philosophy 13:5–36
Weber M (1999) The aim and structure of ecological theory. Philosophy of Science 66:71–93
Wiemken V, Boller T (2002) Ectomycorrhiza: gene expression, metabolism and the wood-wide web. Curr Opin Plant Biol 5:355–361
Wilke S (2008) From 'natura naturata' to 'natura naturans': 'Naturphilosophie' and the concept of a performing nature. Interculture 4:1–23
Woltereck R (1940) Ontologie des Lebendigen. Stuttgart, Ferdinand Enke
Zacharias O (1904) Skizze eines Spezial-Programms für Fischereiwissenschaftliche Forschungen. Fischerei-Zeitung 7:112–115
Zacharias O (1905) Über die systematische Durchforschung der Binnengewässer und ihre Beziehung zu den Aufgaben der allgemeinen Wissenschaft vom Leben. Forschungsberichte aus der biologischen Station Plön 12:1–39
Zacharias O (1907) Das Süsswasserplankton. Teubner, Leipzig
Zacharias O (1909) Das Plankton als Lebensgemeinschaft. Unsere Welt 1:5–14

Part IV
Main Phases of the History of the Concept "Ecology"

Chapter 9
Etymology and Original Sources of the Term "Ecology"

Astrid Schwarz and Kurt Jax

Literal Translation in German: *Ökologie*; in French: *Écologie*

The term "Oecologie" was coined by the German zoologist Ernst Haeckel in 1866 in his book *Generelle Morphologie der Organismen*.[1] It derives from the Greek "οικοσ" (oikos; house, household, also dwelling place, family) and "λογοσ" (logos; word, language, language of reason). "Oecologie", later appearing as "Ökologie" (from around 1890),[2] was used to refer to the "whole science of the relations of the organism to its surrounding outside world"[3] or, to put it differently, the science of the household of nature or the economy of organisms.[4] The term was taken up rather rapidly by some authors (e.g. by Semper 1868, p. 229), but more than 20 years passed until it became widely used. It was not until 1885 that it was

---

[1] Haeckel 1866, Vol.1, p. 8 (footnote), 237, 238 (table); Vol.2, p. 286. There is a persisting tale that Thoreau used the word prior to this (see, for instance, Michel Serres in *Revisiting the natural contract* 2006). According to Walter Harding, editor of "The correspondence of Henry David Thoreau", this interpretation is a misreading that confused geology with ecology: "I must assume that, since *geology* makes as much sense in the context as *ecology* does, *geology* must have been the word that Thoreau intended". (Harding 1965, p. 707 (emphasis in the original); see also Egerton 1977 or Acot 1982).

[2] In the 1896 German version of his book on Ecological Plant Geography, Warming wrote *Ökologie* while Dahl (1898), for example, still used the spelling *Oekologie*.

[3] "Gesammte *Wissenschaft von den Beziehungen des Organismus zur umgebenden Aussenwelt*", Haeckel 1866 (p. 286, emphasis in original).

[4] Haeckel 1870 (p. 365) originally referred only to animals: "[der] Haushalt der thierischen Organismen" ("the household of animals"). This paper was a written version of his inaugural lecture as a professor in Jena.

A. Schwarz (✉)
Institute of Philosophy, Technische Universität Darmstadt, Schloss, 64283 Darmstadt, Germany
e-mail: schwarz@phil.tu-darmstadt.de

K. Jax
Department of Conservation Biology, Helmholtz Centre for Environmental Research (UFZ),
Permoserstr. 15, 04318 Leipzig, Germany
e-mail: kurt.jax@ufz.de

used for the first time in a book title, namely in Reiter's *Die Consolidation der Physiognomik als Versuch einer Oekologie der Gewaechse*.

In English the term "Oecologie" was initially translated as "œcology". Its first use occurred, it seems, in the English translation of Haeckel's book *Natürliche Schöpfungsgeschichte (The History of Creation[5])* in 1876 (cf. Bather 1902,[6] p. 748; Benson 2000, p. 60). In the early 1890s, not least following a recommendation of the Madison Botanical Congress in 1893 (Madison Botanical Congress 1894, pp. 35–38), the double-letter was dropped in favour of the final English form "ecology". From that time on, the term "ecological" was widely used in publications[7] and from 1904 the *Botanical Gazette*, a University of Chicago publication, produced a column entitled *Ecological notes*.

However, the old wording still persisted for some years in the writings of a number of authors and even more in dictionaries (Bessey et al. 1902; Bather 1902). For example, the English translation of Warming's (1895) seminal book *Plantesamfund* was published in 1909 as *Oecology of plants*.

In French the word was also introduced as "oecologie" via the translation of Haeckels "Natürliche Schöpfungsgeschichte" (Histoire de la création des êtres organisés) in 1874. Its first use by French scientists (now as "écologie") can be traced definitively to the year 1900 in the context of plant ecology. Charles Flahault, a botanist loosely associated with the Montpellier school, used the word in its adjectival form in his "projet de nomenclature phytogéographique".[8] However, despite being introduced by a famous zoologist and Darwinian apologist, the word was not very successful in the French community of biologists; this might be due to the fact that resistance to Darwinian ideas was much greater in France than in the English-speaking world.[9]

In Russian the term "ekologia" was first introduced through an abridged translation of Haeckel's *Generelle Morphologie der Organismen* in 1869. In similar fashion to its impact on the French and the English-speaking world, it was not influential in the sense of leading to the founding of an ecological research programme or institutions. In the late 1890s, this situation changed with the translation of Eugenius Warming's *Plantesamfund* into *Oikologicheskaya Geografia Rastenii* (1901) which was highly influential, together with the translation of Grisebach's work (1874, 1877): researchers now began to focus on groups of organisms and to develop a synecological approach.

---

[5] Haeckel 1876, Vol. II, p. 354.

[6] Bather erroneously assumes that Haeckel also *coined* the word in the *Natürliche Schöpfungsgeschichte*. Haeckel's *Generelle Morphologie der Organismen* (1866), where the word was in fact used first, was never published in English translation.

[7] For instance A. S. Hitchcock, *Ecological Plant Geography of Kansas (1898)* or H. C. Cowles, *The ecological relations of the vegetation on the sand dunes of Lake Michigan* (1899).

[8] C. Flahault, Projet de nomenclature phytogéographique, Actes du 1er Congrès international de botanique tenu à Paris à l'occasion de l'exposition universelle de 1900, Lons-le-Saunier, 1900, p. 440, 445 (cf. Matagne 1999, p. 107).

[9] Matagne 1999, p. 109; also Acot 1982, p. 106 ff.

Brief Overview of Part IV to VII

The term "Oecologie" was coined in the second half of the nineteenth century. However, ANTECEDENTS OF AN ECOLOGICAL IDEA in the sense of a modern science and in contrast to natural history existed before it was described as "ecology". Such ideas were present, for example, in Alexander von Humboldt's "physiognomic system of plant forms", Alfred R. Wallace's "geography" of animal species, and Charles Darwin's "entangled bank"; they also existed in Louis Agassiz' studies on lakes and oceans. In 1866 "Oekologie" was introduced by the German zoologist Ernst Haeckel. From the beginning he referred to the meaning of the Greek "oikos" as "household", suggesting that "ecology" is the science of the household of organisms, i.e. of their relation to their biotic and abiotic surroundings. None of the definitions he offered denoted an existing research programme, nor was Haeckel's aim to develop such a programme. The term primarily filled an empty place within his disciplinary system of zoology: that of "external physiology". This search for an order of the study of living beings was in line with contemporary efforts to find a general system of biology. However, the new name "Oekologie" gave way to a focus on the field as a self-conscious enterprise, despite varying LOCAL CONDITIONS OF EARLY ECOLOGY according to nation-state and language area.

At the time a number of COMPETING TERMS existed. "Ecology" was used in a wide range of ways to refer to different domains of objects and phenomena. During the late nineteenth and early twentieth century, what is nowadays called "ecology" was also described as "ethology" or "biology" in a narrower sense. Likewise, different approaches were taken towards delimiting the term "ecological" to terms such as "physiological" and/or "sociological".

In the first half of the twentieth century a STABILISING OF THE CONCEPT occurred. At the same time the FORMATION OF SCIENTIFIC SOCIETIES, academic institutions and publishing bodies accelerated, and the later SUBDIVISIONS OF ECOLOGY appeared on the horizon. With the RISE OF SYSTEMS THEORY in the 1940s and THE ARRIVAL OF THE ENVIRONMENTAL MOVEMENT in the 1960s the concept broadened and ecology came to be described as a "super-science". "Ecology" in this sense served to blur the boundaries between scientific, philosophical and political knowledge, and at the methodological level there was a merging of facts and values, the epistemic and the social. The BORDER ZONES BETWEEN SCIENTIFIC ECOLOGY AND OTHER FIELDS became the subject of highly controversial discussions. From this time on, "ecology" can refer to a variety of completely disparate ideological doctrines and political stances.

The struggle to define ecology and its sub-disciplines has thus always been both one of structuring and assessing the complex subject matter of "the interdependence of living and non-living nature" and a debate about the delimitation of institutional and social groups – in academia and beyond. This process is still ongoing and will be of central importance in LOCATING ECOLOGY IN THE CONTEXT OF 21ST CENTURY ENVIRONMENTAL SCIENCES. It also has importance with respect to the question of whether THE DOMAIN OF ECOLOGY IS DETERMINED BY METHOD, OBJECT OR INSTITUTION.

Chapter 10
The Early Period of Word and Concept Formation

Kurt Jax and Astrid Schwarz

The neologism "Oecologie" was coined by German zoologist Ernst Haeckel in 1866. His intention in doing so, however, was not to establish a discipline of "Oecologie" along with its own concepts, theories and practices. Haeckel himself never engaged in "ecological" research, but rather invented the word to identify a hitherto unnamed branch in his system of zoology.[1] It was not until around the 1890s that ecology became a "self-conscious"[2] enterprise. Prior to that, the term served more as a focal point to denote certain activities that had been undertaken in disciplines such as zoology, botany, physiology, geography and oceanography, which in turn constituted the diverse roots of what would later be known as "ecology".

Haeckel presented the term "Oekologie" in several publications, providing a variety of definitions for it.[3] The elements contained in these definitions, along with his characterization of the place of ecology within biology, are still in evidence in debates about ecology today.

In Haeckels's system of zoology, ecology refers to the *external physiology* of organisms:

> Die Physiologie theilen wir ebenfalls in zwei Disciplinen: I. Die Physiologie der Conservation oder Selbsterhaltung (a. Ernährung, b. Fortpflanzung), II. die Physiologie der Relationen oder Beziehungen (a. Physiologie der Beziehungen der einzelnen Theile des

---

[1] This is highlighted by the fact that Haeckel first used the term "Oecologie" in a diagrammatic representation before setting out to explicate it in words (Haeckel 1866, vol.1, p. 238).
[2] Allee et al. 1949, pp. 19, 42.
[3] Haeckel 1866, 1868, 1870.

K. Jax
Department of Conservation Biology, Helmholtz Centre for Environmental Research (UFZ), Permoserstr. 15, 04318 Leipzig, Germany
e-mail: kurt.jax@ufz.de

A. Schwarz (✉)
Institute of Philosophy, Technische Universität Darmstadt, Schloss, 64283 Darmstadt, Germany
e-mail: schwarz@phil.tu-darmstadt.de

Organismus zu einander (beim Thiere Physiologie der Nerven und Muskeln); b. Oecologie und Geographie des Organismus oder Physiologie der Beziehungen zur Aussenwelt).[4]

One perspective, then, places the focus on the *interrelations between an organism and its physical surroundings*, including interactions with other organisms:

> Unter *Oecologie* verstehen wir die gesammte *Wissenschaft von den Beziehungen des Organismus zur umgebenden Aussenwelt,* wohin wir im weiteren Sinne alle *‚Existenz-Bedingungen'* rechnen können. Diese sind theils organischer, theils anorganischer Natur.[5]

An alternative perspective focuses on ecology as an "*economy of nature*", implying a reference to liberal economic theories as supposedly advocated – at least in Haeckel's view – by Darwin:

> Die *Oecologie der Organismen,* die Wissenschaft von den *gesammten Beziehungen der Organismen zur umgebenden Außenwelt,* zu den organischen und anorganischen Existenzbedingungen; die sogenannte *‚Oekonomie der Natur',* die Wechselbeziehungen aller Organismen, welche an einem und demselben Orte mit einander leben, ihre Anpassung an die Umgebung, ihre Umbildung durch den Kampf um's Dasein.[6]

> Unter *Oecologie* verstehen wir die Lehre von der Oeconomie, von dem Haushalt der thierischen Organismen. Diese hat die gesammten Beziehungen des Thieres sowohl zu seiner anorganischen, als zu seiner organischen Umgebung zu untersuchen, vor allem die freundlichen und feindlichen Beziehungen zu denjenigen Pflanzen und Thieren, mit denen es in directe oder indirecte Berührung kommt; oder mit einem Worte alle diejenigen verwickelten Wechselbeziehungen, welche Darwin als die Bedingungen des Kampfes um´s Dasein bezeichnet.[7]

Haeckel's definitions explicitly excluded the spatial (topographical and geographical) dimension, which he reserved for what he called "chorology" (*Arealkunde*,

---

[4] "We divide physiology also into two disciplines: I. the physiology of conservation or self-preservation (a. nutrition, b. reproduction), II. the physiology of relations (a. physiology of the relations of parts of the organism to each other (meaning, for animals, the physiology of nerves and muscles); b. ecology and geography of the organism or physiology of the relations with the external world)" (Haeckel 1866, vol. 1, p. 237).

[5] "By ecology we mean the *whole science of the relations of the organism to its surrounding outside world*, which we may consider in a broader sense to mean all *'conditions of existence'*. These are partly of an organic nature and partly of an inorganic nature" (Haeckel 1866, vol. 2, p. 286 (emphasis in original)).

[6] "The *ecology of the organisms*, the science of the whole *relations of organisms to their surrounding world*, towards the organic and inorganic conditions of existence; the so-called '*economy of nature*', the interrelations of all organisms which live in one and the same place, their adaptations to their environment, their transformation through the struggle for existence" (Haeckel 1868, p. 539 (emphasis in original)).

[7] "By *ecology*, we mean the science of the economy, of the household of animal organisms. This has to study the entirety of relations of the animal both to its inorganic and its organic environment, in particular the benign and hostile relations with those plants and animals with which it comes directly into contact; or, to be concise, all those intricate interrelations which Darwin calls the struggle for existence" (Haeckel 1870, p. 365).

10 The Early Period of Word and Concept Formation

biogeography).[8] This spatial dimension, however, became one of the basic pillars of the concept of ecology.

The positioning of ecology as part of physiology was part of Haeckel's search for a structure of biology that was logically related to Darwin's theory of evolution, from which he also drew insights about the importance of those relations of organisms which he called "ecological".[9]

Figure 10.1a shows the scheme of the main branches of zoology, in which Haeckel placed ecology and chorology in 1866: ecology and chorology are situated within the "physiology of the relations of organisms to their outside world". They are part of the "Relations-Physiologie der Thiere" (physiology of relations of animals), the other part of which is devoted to the relations of the parts of the (animal) body to each other. This physiology of relations was set in contrast to the "Conservations-Physiologie" (physiology of conservation, meaning the physiology by which the organism keeps itself alive and reproduces). As already noted critically by Tschulok (1910, pp. 141 ff) the specific placement of ecology and, especially, chorology (geography and topography as a part of physiology) does not follow any clear logic or explicit criteria. It is not obvious, for example, why the physiology of muscles and nerves (which he provides as examples of the "relations of the individual parts of the animal body to each other") is closer to the physiology of the organism's relations to its surroundings than they are to the physiology of internal metabolism or reproduction. It seems that Haeckel's desire to find a consistent scheme of clear-cut *dichotomies* into which the different parts of biology (or here, specifically zoology) could be ordered, was a dominant force in the design of his scheme. This is evident also from the ease with which Haeckel rearranged parts of the scheme. Figure 10.1b shows a second visualization of the divisions of zoology, published in 1902.[10] Here, parts of what had previously appeared as "relations physiology" (Relations-Physiologie), in particular the "relations of the individual parts of the animal body to each other", have been shifted to the other part of physiology, the physiology of working functions (Physiologie der Arbeitsleistungen), without any clear indication as to why – or indeed to which specific category – they were moved. While keeping the overall number of branches of zoology constant (eight at the lowest level), chorology and ecology now become more

---

[8] "Unter *Chorologie* verstehen wir die gesammte *Wissenschaft von der räumlichen Verbreitung der Organismen*, von ihrer geographischen und topographischen Ausdehnung über die Erdoberfläche" (Haeckel 1866, Vol. 2, p. 287, emphasis in original). (By *chorology* we mean the whole *science of the spatial distributions of the organisms*, of their geographical and topographical extension over the surface of the earth).

[9] See Stauffer (1957). During the nineteenth and the first half of the twentieth century several publications attempted to link the different branches of biology in a consistent scheme; in addition to that of Haeckel (1866), these included in particular Burdon-Sanderson (1893), Wasmann (1901); Tschulok (1910), Gams (1918), and Du Rietz (1921).

[10] Although the figure is taken from an 1870 "reprint" of Haeckel's paper in a collection of his popular writings, it was not included in the original paper. Also, the terminology and the ordering used there (1870) is intermediate between those of the two schemes displayed in (Fig. 10.1).

Fig. 10.1 Ernst Haeckel's ideas about the subdivisions of zoology. (**a**) From Haeckel (1866) (Vol. 1, p. 238), (**b**) from Haeckel (1902), p. 29. See text for explanation

separate as the *only* parts of the physiology of relations (with chorology now also including the subject species' migrations). Obviously, in both representations the field "Oecologie" followed (at most) the internal logic of Haeckel's system and was never intended as a vehicle to develop a consistent concept of ecology in the sense of a new research programme.

Nevertheless, the concept of ecology as the "external physiology" of organisms was taken up early on,[11] and by the turn of the twentieth century at the latest, the

---

[11] This is the case especially in the German-speaking world, where the reference to the physiological tradition was more important than in the Anglo-Saxon world (e.g. Trepl 1987, p. 26). One of the first authors to develop a research programme in this sense was Karl August Möbius, with his case study of the economy of oyster beds (Möbius 1877).

early protagonists of ecology such as Carl Semper, A.F. Wilhelm Schimper and Eugenius Warming, as well as Frederic Clements and Henry Chandler Cowles, succeeded in developing it into a research programme. The emphasis on the biotic interrelations of organisms as well as a consideration of the whole "economy of nature" were part of these and other research programmes. Stabilizing the range of meanings of the concept of ecology and its varied transformations was one of the key concerns of ecologists in the early stages of the discipline's formation.[12]

References

Allee WC, Emerson AE, Park O, Park T, Schmidt KP (1949) Principles of animal ecology. Saunders, Philadelphia
Burdon-Sanderson JS (1893) Inaugural address. Nature 48:464–472
Du Rietz GE (1921) Zur methodologischen Grundlage der modernen Pflanzensoziologie. Adolf Holzhausen, Wien
Gams H (1918) Prinzipienfragen der Vegetationsforschung. Ein Beitrag zur Begriffsklärung und Methodik der Biocoenologie. Vierteljahresschr. Naturf Gesellsch Zürich 63:293–493
Haeckel E (1866) Generelle Morphologie der Organismen. Georg Reimer, Berlin
Haeckel E (1868) Natürliche Schöpfungsgeschichte: gemeinverständliche wissenschaftliche Vorträge über die Entwickelungslehre im Allgemeinen und diejenige von Darwin, Goethe und Lamarck im Besonderen, über die Anwendung derselben auf den Ursprung des Menschen und andere damit zusammenhängende Grundfragen der Naturwissenschaft. Reimer, Berlin
Haeckel E (1870) Über Entwicklungsgang und Aufgabe der Zoologie. Jenaische Z Med Naturwiss 5:353–370
Haeckel E (1902) Gemeinverständliche Vorträge und Abhandlungen aus dem Gebiete der Entwickelungslehre. Emil Strauß, Bonn
Möbius KA (1877) Die Auster und die Austernwirtschaft. Wiegandt, Hempel & Parey, Berlin
Stauffer RC (1957) Haeckel, Darwin, and ecology. Q Rev Biol 32:138–144
Tschulok S (1910) Das System der Biologie in Forschung und Lehre. Eine historisch-kritische Studie. Gustav Fischer, Jena
Wasmann E SJ (1901) Biologie oder Ethologie? Biols Zentralbl 21:391–400

---

[12] See Jax, Chap. 12, this volume.

Chapter 11
Competing Terms

Kurt Jax and Astrid Schwarz

Up until the early twentieth century, competing terms for "ecology" were "natural history", "biology", "bionomics", and "ethology".

Natural history is one of the roots of ecology and played an important role in the emergence of ecology. Haeckel noted in 1870 (p. 365): "Ecology (often also inappropriately called biology in the narrower sense) has, up to now, constituted the main component of so-called 'natural history' in the usual sense of this word."[1]

Although he places ecology (as "external physiology") in the context of a modern system of biology,[2] he nevertheless acknowledges the importance of natural history as a fundamental root of ecology. The traces of a natural history in the sense of "structuring visible nature"[3] never disappeared entirely in ecology. This is despite the fact that, by the turn of the twentieth century, natural history was largely identified with a rather indiscriminate and "unscientific" way of collecting data about natural phenomena. However, this diagnosis was already a result of the steady boundary work going on in the emerging laboratory-based biological sciences. In fact, the natural history of the nineteenth century never was merely a matter of haphazard sampling and describing; rather, it had a very specific methodology and a very specific set of theoretical questions.[4] Such questions included the issue of

---

[1] "Oecologie (oft unpassend auch als Biologie im engeren Sinne bezeichnet) bildete bisher den Hauptbestandtheil der sogenannten ‚Naturgeschichte' in dem gewöhnlichen Sinne des Wortes."

[2] In his writings Haeckel always refers specifically to zoology.

[3] Translated from Foucault 1974, p. 177.

[4] A detailed discussion of the changing rationality of natural history through the centuries is given in *Cultures of Natural History* (1996), edited by Jardine, N., E.C. Spary and J.A. Secord; R. Kohler (2006) presented a detailed study on specific scientific practices in natural history in the nineteenth century through to the 1950s, while D. Takacs shows in *The Idea of Biodiversity. Philosophies of Paradise* (1996) that natural history is very present in the modern biosciences.

K. Jax
Department of Conservation Biology, Helmholtz Centre for Environmental Research (UFZ),
Permoserstr. 15, 04318 Leipzig, Germany
e-mail: kurt.jax@ufz.de

A. Schwarz (✉)
Institute of Philosophy, Technische Universität Darmstadt, Schloss, 64283 Darmstadt, Germany
e-mail: schwarz@phil.tu-darmstadt.de

explaining the distribution and origin of natural variety and the mechanisms of nature's perceived order.[5]

The identification of ecology with "classical" natural history was strongly rejected by most early ecologists as being too broad (e.g. Wheeler 1902). In response, however, to what he perceived as an overemphasis of laboratory biology at the cost of observing animals in the field, Charles Elton wrote of ecology in 1927 in terms of it being "scientific natural history" (Elton 1927, p. 1). Shelford (1937 p. 32, FN 1) also noted that "[t]he term is applied to those phases of natural history and physiology which are organized into a science, but does not include all the unorganizable data of natural history".

In the second half of the nineteenth century and into the early twentieth century, "*biology*" was used (especially in Germany) both in the broad sense common today and in a narrower sense, designating a concept that was close to, if not identical with, Haeckel's definition of ecology.[6] This dual use of the word and its application to "ecology" (or sometimes also "ethology") was criticised for several reasons, not least because it was seen as leading to confusion (Haeckel 1866, 1870; Wasmann 1901;[7] Dahl 1898[8]). Wasmann, who on the one hand criticised this dual use, tried on the other to save and refine the notion of "biology" in its narrower sense. For him, it included "the external activities of life which pertain to organisms as individuals and at the same time regulate their relations to other organisms and to the inorganic conditions of existence".[9] It thus comprised the external habits of organisms (such as feeding and reproduction), their interactions, and their conditions of existence. In his view, this concept was broader than that of "ethology" and even more so than that of Haeckel's "ecology".

---

[5] The questions posed here were also "why" and not only "how". The methodology of natural history, as the term denotes, was a heavily historical one, explaining the specific patterns and processes found in nature in a nevertheless systematic way (Trepl 1987, p. 46). Farber (1982, p. 150 f.) sees natural history and "scientific" physiology as parallel traditions in the nineteenth century, with natural history guided by theoretical questions (e.g. the relations between classification, morphology and history) and culminating, by the middle of the century, in evolutionary theory.

[6] Haeckel saw ecology as a substitute for what he perceived as the meaning of "biology in the narrow sense".

[7] In spite of the criticism that he himself also raised with regard to this issue, Wasmann argued in favour of retaining the term "biology" in the narrower sense – rejecting both "ethology" and "ecology".

[8] Dahl (1898, p. 121 f.) argued: "Hat man doch bisher nicht einmal einen Namen fur dieses Gebiet gefunden, der allgemein anerkannt wurde. Man nannte es fruher Biologie. Nachdem aber diese Bezeichnung im weitesten Sinne auf die Erforschung aller Lebewesen in Anwendung gekommen ist und die Zellforschung im Speziellen sich Biologie nennt, müssen wir als die minder Bekannten und Geachteten das Feld räumen". ("Thus far, a name that might meet with general agreement has not even been found for the field. It used to be called biology. But since this name has now been applied in the broadest sense to the investigation of all living beings and since cytological research in particular calls itself biology, we, as those who are less well-known and respected, are compelled to beat a retreat").

[9] "…[die] äußeren Lebensthätigkeiten, die den Organismen als Individuen zukommen, und die zugleich auch ihr Verhältnis zu den übrigen Organismen und zu den anorganischen Existenzbedingungen regeln" (Wasmann 1901, p. 397).

11 Competing Terms

The use of "biology" rather than "ecology" persisted in the writings of the community, a fact that was noted and criticised by, among others, Sinai Tschulok in 1910 (p. 211 f.):

"Es ist sehr zu bedauern, daß selbst in wissenschaftlichen Werken die Ökologie noch sehr häufig als Biologie bezeichnet wird. Denn einmal ist Biologie die Gesamtwissenschaft von den Lebewesen, ein anderes Mal ein Teil davon, d.h. eine bestimmte Betrachtungsweise [...]. Noch schlimmer ist es, wenn unter Biologie etwas mehr als Ökologie verstanden werden will und damit doch nicht die gesamte Lehre von den Organismen gemeint ist".[10]

In the second half of the nineteenth century another, narrower meaning of biology in the sense of "cytology" (later relating also to "microscopic knowledge of organisms") had been established, especially in France and Belgium (Dahl 1898; Wasmann 1901; Wheeler 1926). This in turn led to the necessity to coin a new term for the "science of the ways of living of animals and plants" (Wasmann 1901, p. 394), namely, *"ethology"*.[11] Ethology was a term preferred especially by zoologists (e.g. Dahl 1898; Wheeler 1902). Dahl saw ecology as part of ethology, together with what he called "trophology" (pertaining to the food of animals). Both Dahl and Wheeler saw "ecology" as too narrow, arguing that it referred (at least through the direct connotation of the word "oikos") mainly to "dwelling" ("Aufenthalt", Dahl 1898, p. 122) or "habitat" (Wheeler 1902, p. 973). Wheeler also opted for the term "ethology" to describe the whole complexity of animal life, including (in contrast to plant life) social and psychological phenomena of the organisms' relations, while this emphasis was not a major interest of Dahl's. Like Dahl, Wheeler too saw ecology as a *part* of ethology. Attempts to substitute "ecology" for "ethology", however, were not a great success. By the 1920s, if not earlier, the word "ecology" had already become the dominant term within zoology as well, while ethology had soon become restricted to the behaviour of animals in terms of their individual social interactions. Carpenter's "Ecological Glossary" from 1938 doesn't even mention "ethology". However, in the same period the German biologist Jakob von Uexküll propagated a "theory of environment" (Umweltlehre), managing to ignore both the labels "ecology" and "ethology" in the process.[12]

---

[10] "It is highly regrettable that even in scientific works ecology is still very frequently referred to as biology. For biology is at once the overall science of living being as well as a part of it, i.e. a particular perspective... The matter becomes even worse when biology is intended to denote something more than ecology and thus not the general study of organisms".

[11] The first author to use "ethology" in this sense was Isidore de Geoffrey St. Hilaire (1854) (Wheeler 1926; van der Klaauw 1936a; Jahn and Sucker 2000). See van der Klaauw 1936a, p. 140 for the precise wording of this first description of "ethology".

[12] "Die Umweltlehre besteht (...) aus zwei Hauptpunkten. Neben der Anerkennung der planerzeugten Umwelten fordert sie die Anerkennung des Zusammenhanges aller Umwelten in einer allumfassenden Planmäßigkeit." ("The theory of the environment consists of (...) two main points. In addition to acknowledging plan-based environments it demands acknowledgement of the interrelatedness of all environments in an all-encompassing orderliness". Uexküll 1929, p. 45). Uexküll's specific concept of *Umwelt* (not really captured in today's colloquial term "environment") was at once nomothetic and idiographic, in the sense of depending on the specific individual (biological subject).

"*Bionomics*", was another, less commonly used concept that "competed" with ecology,[13] and still appears to be the least defined of these concepts even today. In 1910 Dahl – in contrast to his earlier writings – saw ethology as being on an *equal level with ecology* which, together with psychology, constitutes animal bionomics, while he considered plant bionomics as being identical with plant ecology (Dahl 1910, p. 3). For Friederichs (1930) bionomics was the science of the specific "laws of life" (Lebensgesetzlichkeit; p. 11) of a species as these are manifested externally, but was still closely related to ecology. Carpenter (1938) defines bionomics as "autecology (used more loosely)" (p. 44); for Schaefer (2003) it is "the science of the way of living of a species, also called 'biology in the narrower sense'[14]; and for Lincoln et al. (1998, p. 41) it is a synonym of ecology: "Ecology; the study of organisms in relation to their environment". Thus, the naming competition seems to be ongoing even now, despite the fact that ecology nowadays is well institutionalised.

References

Carpenter RJ (1938) An ecological glossary. Kegan Paul, Trench, Trubner & Co., London
Dahl F (1898) Experimentell-statistische Ethologie. Verhandlungen der Deutschen Zoologischen Gesellschaft, Jahresversammlung 1898 in Heidelberg:121–131
Dahl F (1910) Anleitung zu zoologischen Beobachtungen. Quelle & Meyer, Leipzig
Elton C (1927) Animal ecology. Sidgwick & Jackson, London
Farber PA (1982) The transformation of natural history in the nineteenth century. J. Hist. Biol. 15:145–152
Foucault M (1974) Die Ordnung der Dinge. Suhrkamp, Frankfurt/Main
Friederichs K (1930) Die Grundfragen und Gesetzmäßigkeiten der land- und forstwirtschaftlichen Zoologie, insbesondere der Entomologie. Erster Band: Ökologischer Teil. Paul Parey, Berlin
Haeckel E (1866) Generelle Morphologie der Organismen. Georg Reimer, Berlin
Haeckel E (1870) Über Entwicklungsgang und Aufgabe der Zoologie. Jenaische Zeitschrift für Medizin und Naturwissenschaften 5:353–370
Jahn I, Sucker U (2000) Die Herausbildung der Verhaltensbiologie. In: Jahn I (ed) Geschichte der Biologie, 3 edn. Spektrum, Akademischer Verlag, Jena, pp 580–600
Jardine N, Secord JA, Spary EC (eds) (1996) Cultures of natural history. Cambridge University Press, Cambridge
Klaauw C, J. van der (1936) Zur Geschichte der Definitionen der Ökologie, insbesonders aufgrund der Systeme der zoologischen Disziplinen. Sudhoffs Archiv für Geschichte der Medzin und der Naturwissenschaften 29:136–177
Kohler RE (2006) All Creatures: Naturalists, Collectors and Biodiversity 1850-1950 Princeton University Press, Princeton, New Jersey
Lincoln R, Boxshall G, Clark P (1998) A dictionary of ecology, evolution and systematics. Cambridge University Press, Cambridge

---

[13]Wheeler (1926) mentions E. Ray Lankester as one who used the term in 1889, while Jahn & Sucker (2000) refer to J. Wilhelm Haake as the first to use it, Lankester and Haake being students of Haeckel. "Zoonomie", describing the (natural) history of animals, can already be found in the early nineteenth century (see van der Klaauw 1936a, p. 137 f.).

[14]"Lehre von der Lebensweise einer Art, wird auch als 'Biologie im engeren Sinne' bezeichnet" (Schaefer 2003, p. 50).

Schaefer M (2003) Wörterbuch der Ökologie, 4 edn. Spektrum Akademischer Verlag, Heidelberg

Shelford VE (1937) Animal communities in temperate America, 2 edn. University of Chicago Press, Chicago

Takacs D (1996) The idea of biodiversity: philosophies of paradise. John Hopkins University Press, Baltimore & London

Trepl L (1987) Geschichte der Ökologie. Vom 17. Jahrhundert bis zur Gegenwart. Athenäum, Frankfurt/Main

Tschulok S (1910) Das System der Biologie in Forschung und Lehre. Eine historisch-kritische Studie. Gustav Fischer, Jena

Uexküll Jv (1929) Welt und Umwelt. Aus deutscher Geistesarbeit 5:20–26, 36–46

Wasmann ESJ (1901) Biologie oder Ethologie? Biologisches Zentralblatt 21:391–400

Wheeler WM (1902) 'Natural history', 'oecology' or 'ethology'? Science 15:971–976

Wheeler WM (1926) A new word for an old thing. Q. Rev. Biol. 1:439–443

Chapter 12
Stabilizing a Concept

Kurt Jax

The concept of ecology emerged well before Haeckel's coining of the word in the nineteenth century, when the logical possibility – and with it the quest – arose to understand the patterns of plant and animal distribution as a consequence of their immediate environmental relations. However, the formation of ecology into a self-conscious discipline[1] – that is, a systematic enterprise engaged in by researchers who described themselves as doing "ecological" work – postdated Haeckel's definition of ecology by some two decades. In between lay the struggle to mark the boundaries of the concept of ecology, and this phase of stabilizing ecology lasted well into the first decade of the twentieth century.

Henry Chandler Cowles, for example, was asked to report to the American Association for the Advancement of Science (AAAS) about "[t]he work of the year 1903 in ecology". He began his report by stating: "It is more than impossible to do such a task for ecology, since the field of ecology is chaos. Ecologists are not agreed even as to fundamental principles or motives; indeed, no one at this time, least of all the present speaker, is prepared to define or delimit ecology". Cowles (1904, p. 879). Ecology at the turn of the twentieth century was, in fact, still struggling to find its place within the field of tension between its basic roots and constituent elements, namely natural history and biogeography on the one hand, and modern biology (specifically, physiology) on the other – to name only the most prominent ones.[2]

---

[1] An expression used first by Allee et al. 1949 and then by McIntosh 1985.
[2] Oceanography and limnology also contributed to the formation of ecology, albeit often through a somewhat separate discourse. See, e.g. Schwarz 2003.

K. Jax (✉)
Department of Conservation Biology, Helmholtz Centre for Environmental Research (UFZ), Permoserstr. 15, 04318 Leipzig, Germany
e-mail: kurt.jax@ufz.de

The tension between the more descriptive, heavily context-dependent approaches of natural history and those of physiology as a science based largely on the (controlled) experiment has remained in evidence throughout the entire history of ecology. This tension proved (and still proves) to be both productive and problematic. It transformed ecology into an approach which not only united different ways of looking at living nature but also developed powerful centrifugal forces, resulting in attempts to "purify" the new discipline and rid it of particular strands of its tradition, particular that of natural history.[3]

The strong emphasis placed on physiology by many early ecologists was also a means to professionalize a certain part of biology, one which lay beyond the mainstream of laboratory biology and was predominant at the end of the nineteenth and the beginning of the twentieth century. The upcoming field of ecology at that time was frequently seen as a passing fad by many scientists. Bringing physiology from the lab to the field, not only made use of the established reputation of exact and precise biological methodology but also served to separate the "mere naturalists" from the "professional ecologists" (Hagen 1986; Cittadino 1980). Major proponents of the young science (e.g. Carl Semper, Andreas Schimper, Henry C. Cowles, Frederic Clements, Victor Shelford, Richard Hesse) subscribed to this view.[4]

The Madison Botanical Congress, struggling with the proper definition (and orthography) of "ecology", dealt with this issue in the context of "plant physiology" – or, as it was sometimes also named at that time, "vegetable physiology".[5] Frederic Clements (1905, p. 2) stated: "There can be little question in regard to the essential identity of physiology and ecology. (…) Ecology has been largely the descriptive study of vegetation; physiology has concerned itself with function", urging a merger of the two, which "will combine the good in each, and at the same time eliminate superficial and extreme tendencies" (ibid. p. 3).

Many early animal ecologists also perceived ecology in this way. For the German biologist Carl Semper, ecology was part of a research programme to validate Darwin's evolutionary theory by investigating "the influence of temperature, light, heat, humidity, nutrition, etc. on the living animal" (Semper 1868, p. 228f.).[6] Victor Shelford, quoting approvingly from Semper's book on animal ecology (Semper 1880 – the first book on animal ecology)[7] wrote in 1913: "At present *ecology is that part of general physiology which deals with the organism as whole, with its general life processes, as distinguished from the more specialized physiology of organs, and which also considers the organism with particular reference to its usual environment*".[8]

---

[3] See Simberloff (1980) for a discussion of these endeavours.

[4] See also Kohler (2002) for a historical account of the shifting lab-field boundary in biology.

[5] Madison Botanical Congress (1894, pp. 35ff).

[6] "… die Einflüsse des Lichtes und der Wärme, des Feuchtigkeitsgrades, der Nahrung etc. auf die lebenden Thiere". See also Chap. 19; for a discussion of Semper and the physiological tradition in early German ecology.

[7] The English versions of Semper's book – with slightly different titles in the UK and the USA – were published in 1881.

[8] Shelford (1913, p. 1 (emphasis in original)).

And in 1927, Richard Hesse wrote: "[E]cology [is] merely a continuation of and a supplement to physiological anatomy [...]; the conditions of the environment are incorporated into the theoretical linking of the individual processes".[9]

The physiological tradition of ecology (and, in particular, of what was later termed "autecology") had already been prepared by a school of German plant physiologists around Simon Schwendener, Gottlieb Haberlandt and Ernst Stahl, who – using an evolutionary perspective – sought to link morphology with physiology in "Physiologische Pflanzenanatomie" (physiological plant anatomy).[10] This approach investigated the morphological features of plants with respect to their functional (adaptive) significance, bringing laboratory science to the field.[11] An important step, however, which had a decisive influence on the stabilization of the concept of ecology, was the merging of the physiological perspective in ecology with a biogeographical one. This entailed, at the same time, a shift from the environmental relations (and adaptedness) of single species or two-species interactions (such as parasitism and mutualism) to that of *communities*. It was a step first taken systematically by Eugenius Warming in Denmark and Andreas Schimper in Germany (Warming 1896; Schimper 1898). They completed the transformation of Alexander von Humboldt's physiognomic plant geography (1807) from a partly aesthetic, partly scientific perspective to an exclusively scientific one, a transformation that had already been prefigured some decades earlier by the work of August Grisebach (1838). With Warming and Schimper, the programme of making physiology the basis of an *ecological* plant (and animal) geography (as opposed to a merely taxonomic or historical one) thus became a major aim of early ecology. Warming (1896) wrote: "A purely physiognomic system has no scientific significance; only when physiognomy is explained physiologically and ecologically does it gain this significance".[12] and a few pages prior to this: "*Why* do plants unite into communities and *why* do they have the physiognomy they have?".[13]

Warming, who can more plausibly be considered a founder of ecology than Haeckel, wrote this in his well received book entitled *Lehrbuch der ökologischen Pflanzengeographie* (Textbook of ecological plant geography). Here he distinguished "ökologische Pflanzengeographie" ("ecological plant geography") from "floristischer Pflanzengeographie" ("floristic plant geography"):

> "Die *ökologische Pflanzengeographie* [...] belehrt uns darüber, wie die Pflanze und die Pflanzenvereine ihre Gestalt und ihre Haushaltung nach den auf sie einwirkenden Faktoren,

---

[9] Hesse (1927, p. 944): "[...D]ie Ökologie [ist] nur eine Fortführung und Ergänzung der physiologischen Anatomie [...]; es werden die Bedingungen der Umwelt mit einbezogen in die gedankliche Verknüpfung der Einzelvorgänge."

[10] This was also the title of Haberlandt's (1884) seminal book on the new approach.

[11] See especially Cittadino (1990) for a description of these schools.

[12] Warming (1896, p. 5): "Ein rein physiognomisches System hat keine wissenschaftliche Bedeutung: erst wenn die Physiognomie physiologisch und ökologisch begründet wird, erhält sie eine solche".

[13] Ibid., p. 3 (emphasis in the original): "*Weshalb* schließen sich Arten zu bestimmten Gesellschaften zusammen und *weshalb* haben diese die Physiognomie, die sie besitzen?". For the transformation of Humboldts's plant physiognomy into ecology see (Trepl 1987, pp. 104ff; Schwarz 2003, pp. 28ff).

z.B. nach der ihnen zur Verfügung stehenden Menge von Wärme, Licht, Nahrung und Wasser u.a. einrichten".[14]

Extending the notion of "external physiology" from the physiology of organisms to the physiology of communities proved to be a decisive step for ecology. It was this programme that Schimper, Clements, Shelford and Hesse also pursued (Schimper 1898; Clements 1905; Shelford 1913; Hesse 1924).

In its stringent version, however, the programme of a "purely" physiological and experimental ecology never fulfilled its promises, at least not within terrestrial ecology and particularly with regard to its emphasis on the experimental method (Hagen 1988). Description, comparison and classification – the basic tools of natural history – remained essential for ecological work and theory building. Indeed Warming's book itself (and many of Clements' empirical writings) deals not only with abiotic factors and experimental approaches to investigating how plants relate to these factors, but also with interactions among organisms and the dynamics and classification of whole plant communities.[15] Warming integrated a physiognomic perspective (derived from earlier plant geography), a physiological perspective concerned with the "household" of plants (and plant societies) in relation to their abiotic environment, and a strong emphasis on the interactions among organisms as a structuring element of communities.

Thus around 1900 the main "ingredients" of the concept of ecology were all in place – not just as mere definitions but as *practice*:[16]

- The external physiology of organisms as an undisputed core element of ecology, involving
- The interactions of organisms with abiotic factors as well as their interactions with each other
- The set of relations between external physiology and the local and global (or at least regional) distribution of organisms
- An extension of the basic objects of ecology from the individual organism to whole communities[17]

---

[14] Ibid., p. 2, emphasis in original: "*Ecological plant geography* [...] teaches us how plants and plant communities adjust their forms and their housekeeping to the factors that affect them, e.g. the available amount of heat, light, nutriment, water, and so forth".

[15] The original Danish title (Warming 1895) was "Plantesamfund", meaning "Plant society" or "Plant community"; see Chap. 23 for a detailed account of Warming's notion of ecology.

[16] In terms of his pure verbal *definition* of ecology, Warming simply refers to Haeckel's description of ecology as the "Wissenschaft von den Beziehungen der Organismen zur Aussenwelt" ("the science of the organisms' relations to the exterior world"; Warming 1896, p. 2, footnote 1). However, as described above, his *concept* of ecology is much more elaborate.

[17] Different notions arose concerning the nature of "wholes" (or units) as objects of ecology, based partly on physiognomic and "statistical" concepts and partly on more relational or even functional ones (see Jax 2006 and also Chap. 4). Warming's position, for example, came close to that formulated later on by H.A. Gleason as the "individualistic concept of the plant community" (Gleason 1917, 1926). "Populations" and "ecosystems" were later arrivals as units of ecological investigation (see also Chap. 14).

- Natural history remaining an indispensable part of ecology[18]

Despite this, the exact boundaries of the concept of "ecology" continued to be contested and to "oscillate"[19]; to a lesser extent, this remains the case today. There was considerable variation in the degree of emphasis placed on the many different "ingredients" of the ecology concept. This became particularly apparent in the many debates about the contexts in which the words "ecology" and "ecological" are applied and in disagreements over how to demarcate ecology from other scientific fields.[20] A few important strands of these debates are outlined in the remainder of this chapter.

British plant ecologist Arthur Tansley embraced a view of ecology similar to Warming when he stated in his presidential address as the first president of the new British Ecological Society in 1914: "We claim for ecology that it is before all things a way of regarding the plant world, that it is *par excellence* the study of plants for their own sakes as living beings in their natural surroundings, of their vital relations to these surroundings and to one another, of their social life as well as their individual life." Tansley (1914, p. 195 (emphasis in original)). Others, however, like botanist Paul Jaccard, attempted to introduce a distinction between ecology, chorology and the sociology of plants and to restrict the meaning of "ecology". He proposed to "... use ecology only for the edaphic, physiographic, and climatic factors" (Jaccard in Flahault and Schröter 1910, p. 21) and suggested using "chorology" for the geographic relations of plants and "sociology" for the biotic interactions between plants. His definition, entailing a narrower use of "ecology" than that proposed by Haeckel and others, was not, however, widely accepted.

At the same time, the "plant *sociology*", which Jaccard sought to restrict to biotic relations alone, was often made synonymous with ecology – in particular with *synecology* (as in Nichols 1923[21]; Tansley 1920) or was conceived of even more broadly, including chorology.[22] Josias Braun-Blanquet wrote in the introduction to his textbook on plant sociology: "Plant sociology, the science of communities or the knowledge of vegetation in the widest sense, includes all phenomena which

---

[18] Albeit only its "organized parts", as Victor Shelford emphasized, among others: "The term ecology is applied to those phases of natural history and physiology which are organized into a science, but does not include all the unorganizable data of natural history". Shelford (1913, p. 32, footnote 1). Similarly, more than a decade later, Charles Elton defines ecology as "scientific natural history" (Elton 1927, p. 1).

[19] Schwarz (2003, pp. 254ff), focusing mainly on aquatic ecology, describes ecology as oscillating between three basic concepts, which she labels "microcosm", "niche" and "energy", representing in turn a physiognomic, a physiological, and a functional approach.

[20] See contributions in Part VII where the border zones between ecology and other scientific fields are discussed in more detail.

[21] "The study of plant communities in their relation to environment comprises the field of what might be called *ecological plant sociology*; more commonly it has been called synecology". Nichols (1923, p. 11 (emphasis in original)).

[22] Which is in fact also contained in Warming's concept of ecology.

touch upon the life of plants in social units".[23] The "life of plants in social units" meant for him: "the organization or structure of the community (...), synecology: the study of the dependence of plant communities upon one another and upon the environment, synergetics (development of communities) (...), synchorology (geographic distribution of communities) (...), sociological classification (systematics) (...)".[24] Some later authors, however, used the word "plant sociology" (even more common: "phytosociology") for the *classification* of plant assemblages only (see Friederichs 1963). Thus "phytosociology" is defined in a British dictionary of ecology as: "The classification of plant communities based on floristic rather than life-form or other considerations".[25]

"Sociological" is used in a much more restricted way in animal ecology, where it is usually (though not always) used to denote interactions between animals but not between animals and their inanimate environment (see also Friederichs 1963). This view thus comes closer to the original meaning of the word "sociology" as coined in the nineteenth century by Auguste Comte to describe the relations between human individuals and society. The equivalent to what Braun-Blanquet described as plant sociology is thus not "animal sociology" but "animal ecology" in the broadest sense.[26]

The notion of ecology as a science that studies "wholes" which extend beyond the biological community – the entire "household of nature" – was developed mainly by scientists working in aquatic ecology, e.g. August Thienemann and Richard Woltereck (e.g. Thienemann 1939). The way in which these "wholes" were conceived, as well as the metaphors used, were highly varied (see Jax 1998; Schwarz 2003 and Chap. 4), even more so after the emergence of systems theories. With the rise of systems theory in ecology (see Chap. 15) around the middle of the twentieth century, some authors felt it necessary to modify the traditional – organism-focused – definitions of ecology and to bring the importance of systems theory and of ecosystems as the basic objects of ecology to the fore. The web of interrelations and the emergence of whole systems (including abiotic factors) were given priority here.

---

[23] Braun-Blanquet (1932, p. 1): "Die Pflanzensoziologie, die Lehre von den Pflanzengesellschaften, auch Vegetationskunde im weitesten Sinn, umfaßt alle das soziale Zusammenleben der Pflanzen berührenden Erscheinungen". Braun-Blanquet (1928, p. 1) Instead of providing a translation of the 1928 text of Braun-Blanquet myself, I am here using the text of the English translation of his book from 1932.

[24] Braun Blanquet (1932, p. 1f): "...das Gesellschaftsgefüge (Organisation oder Struktur)(...), der Gesellschaftshaushalt (Synökologie)(...), die Gesellschaftsentwicklung (Syngenetik)(...), die Gesellschaftsverbreitung (Synchorologie) (...) und die Klassifikation oder Systematik der Pflanzengesellschaften". Braun-Blanquet (1928, p. 2). This definition of plant sociology was later adopted by the International Congress of Botany. in Paris in 1954 (according to Braun-Blanquet 1964, p. 22).

[25] Allaby (1994, p. 302); a similar definition can be found in Calow (1998).

[26] In fact the conceptual development of plant and animal ecology, as of aquatic and terrestrial ecology, has only partially overlapped, but instead, especially in the early years of ecology, occurred in rather separated circles (see also McIntosh 1985).

Thus Eugene Odum remarked: "As you know ecology is often defined as: The study of interrelationships between organisms and environment. I feel that this conventional definition is not suitable; it is too vague and too broad. Personally I prefer to define ecology as: The study of the structure and function of ecosystems. Or we might say in a less technical way: The study of the structure and function of nature". Odum (1962, p. 108). In a similar way, Spanish ecologist Ramon Margalef defined ecology as the "biology of ecosystems" (Margalef 1968, p. 4).

A more recent view, which attempted to integrate the different aspects described above, is the definition of ecology given by Gene Likens: "Ecology is the scientific study of the processes influencing the distribution and abundance of organisms, the interactions among organisms, and the interaction between organisms and the transformation and flux of energy and matter" (Likens 1992, p. 8). This definition represents a continuation of developments in twentieth century ecology: while focusing on organisms as the core of all ecology (paraphrasing Andrewartha and Birch 1961; Krebs 1985[27]) it also places emphasis on the "transformation and flux of energy and matter", thus alluding to those aspects most frequently seen as constituting the core of ecosystem ecology.

Another point of controversy regarding the delineation of the ecology concept should be mentioned again here, having been touched upon only briefly above. This is the relationship between ecology and biogeography, which was and is perceived in very different ways and has been discussed extensively time and again (e.g. McMillan 1956; Major 1958; Müller 1980; Browne 2000), not only as a conceptual issue but also as a matter of marking off (and occupying) professional fields of activity. For Haeckel both ecology and biogeography ("chorology") were biological (physiological) sub-disciplines of equal status (see Chap. 10). As quoted above, Warming (1896, p. 2) distinguished "ecological plant geography" (i.e. ecology in his sense) from traditional "floristic (and at the same time historical) plant geography".[28] In doing so, he included some issues of plant biogeography in the domain of ecology, as did many other authors (such as H.C. Cowles[29]). Biogeography is thus frequently understood as a kind of "border science", one side belonging to biology, the other to geography (Hesse 1924; Major 1958; Illies 1971; Leser et al. 1993). For some authors, by contrast, biogeography is an umbrella science, extending far beyond ecology. For these authors, ecology is a subsidiary branch of biogeography.[30] At the same time, biogeography is

---

[27] Krebs 1985, p. 4: "Ecology is the scientific study of the interactions that determine the distribution and abundance of organisms".

[28] This approach is similar to that of Krebs (1985, p. 41). The historical dimension of biogeographical explanations is emphasized by many authors, including Simberloff (1983) and Ricklefs (2004).

[29] Cowles (1901, p. 74) made a further distinction between those parts of community-related ecology that deal with the regional distribution of organisms, which he called "biogeographical ecology", and those that deal with local distribution, or local communities, which he called "physiographic ecology".

[30] Dansereau, for example, writes in the preface of his book *Biogeography*: "The scope of this book extends across the fields of plant and animal ecology and geography, with many overlaps into genetics, human geography, anthropology, and the social sciences. All of these together form the domain of biogeography". Dansereau (1957, p. v).

sometimes perceived as a branch of ecology (this includes, in my view, Walter and Breckle 1983, even though they do not explicitly use the word biogeography), or else the two disciplines are considered as being indistinguishable. MacArthur & Wilson commented in their book *Island Biogeography*: "Now we both call ourselves biogeographers and are unable to see any real distinction between biogeography and ecology". MacArthur and Wilson (1967, p. v.).

A more recent – albeit not very successful – tendency is the endeavour to create a terminological distinction between concepts and research approaches of a branch of "bioecology"[31] (as a part of biology) and "geoecology" (as a part of geography), which are then both seen as constituting the interdisciplinary "umbrella science" of ecology. This is promoted mainly by Hartmut Leser in Switzerland.[32] Whether or not this is a useful distinction is questionable. Although geographical aspects are important in ecology, it is always characterized by a concern with living beings and thus is always an essentially biological discipline. There can be – as Walter and Breckle (1983, p. 1) remarked – a geography of the moon, but never an ecology of the moon.

More recently, the meaning of "ecology" has been further complicated by its usage in the environmental movement and within the realm of "political ecology" (see Chap. 16). Here, ecology has become a buzzword, which is extended to almost everything (for example, see McIntosh 1985, pp. 6ff). In some quarters it represents a worldview, while in others it simply stands for "systemic" or for "related to interactions and connections". However, this is almost exclusively the case when the term is used outside the narrower scientific community of ecologists.

The meanings of "ecological" and the explicit definitions of "ecology" within science varied and still vary in the emphasis they give to organism-environment relations, interactions between organisms, patterns of distribution and abundance, and a view of the "functioning" of ecological wholes (units) beyond the individual organism. Struggles over the definition of ecology and its sub-disciplines have thus always been both about structuring and assessing the complex subject matter of the interdependence of living and non-living nature as well as about debating the boundaries of institutional and social groups – in academia and beyond. This process is still ongoing.

---

[31] Walter Taylor used the term "bio-ecology" as far back as 1927. An American textbook from 1939 (Clements and Shelford 1939) also bore this title. In neither case, however, was the term used as a means of demarcation from geoscience. It was used instead to emphasize the integration of the two subdisciplines plant ecology and animal ecology, which had commonly been treated in separate textbooks prior to this, in order to create an ecology of all living beings. The term "bioecology" has not met with broad acceptance in ecology.

[32] Leser (1984, 1991); see also the Diercke Dictionary Ecology and Environment, Leser et al. (1993); also Rowe and Barnes (1994).

References

Allaby M (ed) (1994) The concise Oxford dictionary of ecology. Oxford University Press, Oxford
Allee WC, Emerson AE, Park O, Park T, Schmidt KP (1949) Principles of animal ecology. Saunders, Philadelphia
Andrewartha HG, Birch LC (1961) The distribution and abundance of animals. University of Chicago Press, Chicago
Braun-Blanquet J (1928) Pflanzensoziologie. Springer, Berlin
Braun-Blanquet J (1932) Plant sociology: the study of plant communities. McGraw-Hill, New York
Braun-Blanquet J (1964) Pflanzensoziologie. Grundzüge der Vegetationskunde, 3 edn. Springer, Berlin, Wien, New York
Browne J (2000). History of biogeography. Encyclopedia of Life Sciences [online]. Wiley & Blackwell. URL: www.els.net
Calow P (ed) (1998) The encyclopedia of ecology and environmental management. Blackwell, Oxford
Cittadino E (1980) Ecology and the professionalization of botany in America, 1890-1905. Stud Hist Biol 4:171–198
Cittadino E (1990) Nature as the laboratory: Darwinian plant ecology in the German Empire, 1880–1900. Cambridge University Press, Cambridge
Clements FE (1905) Research methods in ecology. The University Publishing Company, Lincoln
Clements FE, Shelford VE (1939) Bio-ecology. Wiley, New York
Cowles HC (1901) The physiographic ecology of Chicago and vicinity; a study of the origin, development, and classification of plant societies. Bot Gaz 31:73–108, 145–182
Cowles HC (1904) The work of the year 1903 in ecology. Science 19:879–885
Dansereau P (1957) Biogeography: an ecological perspective. Ronald Press, New York
Elton C (1927) Animal ecology. Sidgwick & Jackson, London
Flahault C, Schröter C (eds) (1910) Phytogeographische Nomenklatur. III. Internationaler Botanischer Kongress, Brüssel 1910. Zürcher & Furrer, Zürich
Friederichs K (1963) Über den Gebrauch der Worte und Begriffe "Gesellschaft" und "Soziologie" in verschiedenen Sparten der Wissenschaft. KZfSS 15:449–461
Gleason HA (1917) The structure and development of the plant association. Bull Torrey Bot Club 44:463–481
Gleason HA (1926) The individualistic concept of the plant association. Bull Torrey Bot Club 53:7–26
Grisebach (1838) Über den Einfluß des Klimas auf die Begrenzung der natürlichen Floren. In: Grisebach A (ed) Gesammelte Abhandlungen und kleinere Schriften zur Pflanzengeographie. Verlag von Wilhelm Engelmann, Leipzig, pp 1–29
Haberlandt G (1884) Physiologische Pflanzenanatomie. Engelmann, Leipzig
Hagen JB (1988) Organism and environment: Frederic Clements´s vision of a unified physiological ecology. In: Rainger, Ronald, Benson KR, Jane Maienschein (eds) The American development of biology. University of Pennsylvania Press, Philadelphia, pp 257–280
Hagen JB, (1986) Ecologists and taxonomists: divergent traditions in twentieth-century plant geography. J Hist Biol 19:197–214
Hesse R (1924) Tiergeographie auf ökologischer Grundlage. Gustav Fischer, Jena
Hesse R (1927) Die Ökologie der Tiere, ihre Wege und Ziele. Naturwissenschaften 15:942–946
Alexander von Humboldt (1807) Ideen zu einer Geographie der Pflanzen. In: Dittrich M (1957) (ed) Ideen zu einer Geographie der Pflanzen, Akademische Verlagsgesellschaft Geest & Portig, Leipzig, pp 29–50
Illies J (1971) Einführung in die Tiergeographie. Gustav Fischer, Stuttgart
Jax K (1998) Holocoen and ecosystem: on the origin and historical consequences of two concepts. J Hist Biol 31:113–142
Jax K (2006) The units of ecology: definitions and application. Q Rev Biol 81:237–258

Kohler RE (2002) Landscapes and labscapes: exploring the lab-field border in biology. University of Chicago Press, Chicago/London

Krebs CJ (1985) Ecology: the experimental analysis of distribution and abundance. Harper & Row, New York

Leser H (1984) Zum Ökologie-, Ökosystem- und Ökotopbegriff. Nat Land 59:351–357

Leser H (1991) Ökologie wozu? Der graue Regenbogen oder Ökologie ohne Natur. Springer, Berlin

Leser H, Streit B, Haas H-D, Huber-Fröhli J, Mosimann T, Paesler R (1993) Diercke Wörterbuch Ökologie und Umwelt. Band 2 N-Z. dtv/Westermann, München, Braunschweig

Likens GE (1992) The ecosystem approach: its use and abuse. Ecology Institute, Oldendorf/Luhe

MacArthur RH, Wilson EO (1967) The theory of island biogeography. Princeton University Press, Princeton

Madison Botanical Congress (1894) Proceedings of the Madison Botanical Congress, Madison, 23–24 Aug 1893

Major J (1958) Plant ecology as a branch of botany. Ecology 38:352–363

Margalef R (1968) Perspectives in ecological theory. The University of Chicago Press, Chicago/London

McIntosh RP (1985) The background of ecology: concept and theory. Cambridge University Press, Cambridge

McMillan C (1956) The status of plant ecology and plant geography. Ecology 37:600–602

Müller P (1980) Biogeographie. Eugen Ulmer, Stuttgart

Nichols GE (1923) A working basis for the ecological classification of plant communities. Ecology 4:11–23, 154–179

Odum EP (1962) Relationship between structure and function in the ecosystem. Jpn J Ecol 12:108–118

Ricklefs RE (2004) A comprehensive framework for global patterns in biodiversity. Ecol Lett 7:1–15

Rowe JS, Barnes BV (1994) Geo-ecosystems and bio-ecosystems. Bull Ecol Soc Am 75:40–41

Schimper AFW (1898) Pflanzengeographie auf physiologischer Grundlage. Gustav Fischer, Jena

Schwarz AE (2003) Wasserwüste - Mikrokosmos - Ökosystem. Eine Geschichte der "Eroberung" des Wasserraums. Rombach-Verlag, Freiburg

Semper K (1868) Reisen im Archipel der Phillipinen. Zweiter Teil: Wissenschaftliche Resultate. Erster Band: Holothurien. Verlag von Wilhelm Engelmann, Leipzig

Semper K (1880) Die natürlichen Existenzbedingungen der Thiere. Brockhaus, Leipzig

Semper K (1881) Animal life as affected by the natural conditions of existence. D. Appleton, New York

Shelford VE (1913) Animal communities in temperate America. University of Chicago Press, Chicago

Simberloff DS (1980) A succession of paradigms in ecology: essentialism to materialism and probabilism. Synthese 43:3–39

Simberloff DS (1983) Biogeography: the unification and maturation of a science. In: Brush AH, Jr Clark GA (eds) Perspectives in ornithology. Cambridge University Press, Cambridge, pp 411–455

Tansley AG (1914) Presidential address to the first annual general meeting of the British ecological society. J Ecol 2:194–202

Tansley AG (1920) The classification of vegetation and the concept of development. J Ecol 8:118–149

Taylor WP (1927) Ecology or bio-ecology. Ecology 8:280–281

Thienemann (1939) Grundzüge einer allgemeinen Ökologie. Arch Hydrobiol 35:267–285

Trepl L (1987) Geschichte der Ökologie. Vom 17. Jahrhundert bis zur Gegenwart. Athenäum, Frankfurt/M

Walter H, Breckle SW (1983) Ökologie der Erde. Ökologische Grundlagen in globaler Sicht, vol 1. Gustav Fischer, Stuttgart

Warming E (1895) Plantesamfund. Grundtræk af den økologiske plantegeografi. Philipsen, Kjøbenhavn

Warming E (1896) Lehrbuch der ökologischen Pflanzengeographie: Eine Einführung in die Kenntnis der Pflanzenvereine. Gebrüder Bornträger, Berlin

Chapter 13
Formation of Scientific Societies

Kurt Jax

An important step in consolidating ecology as a "self conscious" science was the formation of scientific societies explicitly devoted to ecology or particular parts of it.

The first ecological societies were formed in Great Britain (1913) and the USA (1915). From the outset, these societies intended to accommodate both botanists and zoologists, although this did not mean that close co-operation between the two fields became common in practice. The forerunner of the British Ecological Society (BES) was the British Vegetation Committee, initiated by Arthur Tansley in 1904, a body established with the primary purpose of surveying and mapping Britain's vegetation.[1] The BES also published the *Journal of Ecology* which, like the society as a whole, was initially dominated by plant ecologists. Two decades later (in 1932) the society established a second journal on the initiative of Charles Elton (*Journal of Animal Ecology*), devoted to zoological and ecological studies. Two years after the founding of the BES, the Ecological Society of America (ESA) came into being along with its new journal "Ecology".[2] (see Part VI, this volume for further details of local traditions).[3]

In contrast to specialised branches of ecology and related research fields, international societies covering ecology as a whole were much less successful and are a rather late development. The International Association for Ecology (INTECOL) was founded in 1967 as part of the Section of the Environment of the International Union of Biological Sciences (IUBS), and has held several major conferences (one every 3 years since 1975). However, it has never achieved the renown of some of

---

[1] For the special role of this committee, see Fischedick (2000).
[2] The journal itself started in 1920, i.e. five years after the founding of the society.
[3] For a history of the British Ecological Society, see Salisbury (1964) and Sheail (1987), for that of the Ecological Society of America, Burgess (1977).

K. Jax (✉)
Department of Conservation Biology, Helmholtz Centre for Environmental Research (UFZ), Permoserstr. 15, 04318 Leipzig, Germany
e-mail: kurt.jax@ufz.de

the national societies or more specialised global organisations. At the European level, there is a Federation of Ecological Societies (European Ecological Federation, EFF), but no distinct Society in its own right.

Within the more specialised branches of ecology, the most successful international societies have been formed within aquatic ecology and vegetation science. The special role played within ecology by aquatic ecology is demonstrated by the fact that separate societies were formed for these disciplines. At the international level, the International Limnological Society (SIL) was founded by August Thienemann and Einar Naumann in 1922 (see Elster 1974; Rodhe 1974; Steleanu 1989; Schwarz 2003) and became a major focal point for research on freshwater ecology. In some countries, such as the USA,[4] national societies for limnology and/or marine ecology (and, even more broadly, oceanography) also developed, though mostly not until the second half of the twentieth century. Limnology and marine ecology were among those parts of ecology that were able to establish themselves rather early on in university departments and other institutions, not least because these fields were often closely related to fisheries and water treatment.[5] Aquatic research, especially in marine research societies and institutions, was and is generally much broader in scope than "mere" marine ecology or freshwater ecology, i.e. it also involves much hydrographic research.[6] A combined emphasis on or even the study of both fields (marine *and* freshwater ecology) in scientific societies, as in the American Society of Limnology and Oceanography, remains a very rare exception, however.

Focusing on theoretical and practical studies of vegetation, the International Association for Vegetation Science (IAVS) was founded in 1947 in Hilversum/ Netherlands. Its precursor was the (very short-lived, on account of World War II) International Phytosociological Society (IPS) created in 1939 with its headquarters in Montpellier/France. In fact, the IPS did not consider itself as being "only" devoted to ecological research. In the founding statement of the society, as published in the journal *Ecology* (Anonymous 1939, p. 110) the first aim of the society is stated as: "The development of phytosociology (and geobotany) by a closer collaboration between phytosociologists and ecologists", thus identifying phytosociology and (plant) ecology as overlapping but not identical fields.

No similar organisation exists in relation to animal ecology. In contrast to aquatic ecology, the establishment of academic institutions was much more difficult

---

[4] The Limnological Society of America was formed in 1936 and in 1948 merged with the Oceanographic Society of the Pacific to form the American Society of Limnology and Oceanography (ASLO).

[5] It is for this reason that marine biological stations and several national marine research commissions and associations, such as the Marine Biological Association of the United Kingdom, developed as early as the late nineteenth century (Hedgpeth 1957).

[6] The founders of the SIL emphasized that limnology is *"the science of freshwater as a whole* and includes everything that concerns freshwater (…), freshwater hydrography and freshwater biology". (*"die Wissenschaft vom Süßwasser im ganzen* umfaßt alles, was das Süßwasser betrifft (…), limnische Hydrographie und limnische Biologie") (Naumann and Thienemann 1922, p. 585; emphasis in original).

for terrestrial ecology.[7] This changed dramatically along with the increasing awareness of environmental problems from the 1960s onwards, with the creation of many new chairs in ecology and ecology departments.

References

Anonymous (1939) The international society of phytosociology. Ecology 20:110
Burgess RL (1977) The ecological society of America. Historical data and some preliminary analyses. In: Egerton FN (ed) History of American ecology. Arno Press, New York, pp 1–24
Crowcroft P (1991) Elton's ecologists. A history of the bureau of animal population. Chicago University Press, Chicago
Elster H-J (1974) History of limnology. Mitteilungen. Internationale Vereinigung für Limnologie 20:7–30
Fischedick KS (2000) From survey to ecology: the role of the British Vegetation Committee, 1904–1913. J Hist Biol 33:291–314
Godwin H (1977) Sir Arthur Tansley: the man and the Subject. J Ecol 65:1–26
Hedgpeth JW (1957) Introduction. Treatise on marine ecology and paleoecology. Geol Soc Am Mem 67:1–16
Naumann E, Thienemann A (1922) Vorschlag zur Gründung einer internationalen Vereinigung für theoretische und angewandte Limnologie. Arch Hydrobiol 13:585–605
Rodhe W (1974) The International Association of Limnology: creation and functions. Mitteilungen. Internationale Vereinigung für Limnologie 20:44–70
Salisbury E (1964) The origin and early years of the British Ecological Society. J Ecol 52(Suppl):13–18, Appendix p. 244
Schwarz AE (2003) Wasserwüste – Mikrokosmos - Ökosystem. Eine Geschichte der "Eroberung" des Wasserraums. Rombach-Verlag, Freiburg
Sheail J (1987) Seventy-five years in ecology. The British Ecological Society. Blackwell, Oxford
Steleanu A (1989) Geschichte der Limnologie und ihrer Grundlagen. Haag & Herchen, Frankfurt/Main

---

[7] The fate of the two most eminent British ecologists of the first half of the twentieth century, Charles Elton and Arthur Tansley, is a good example for these problems: Elton struggled – successfully – to establish a kind of semi-private institution, the Bureau of Animal Population at Oxford University, which, however, did not survive after his retirement (Crowcroft 1991). Tansley, one of the founders of the BES and an internationally renowned plant ecologist and Fellow of the Royal Society, only received a chair at Oxford at the age of 56 (in 1927; see Godwin 1977).

Chapter 14
The Fundamental Subdivisions of Ecology

Kurt Jax and Astrid Schwarz

The conceptual foundations of ecology were developed rather independently in different biological fields (McIntosh 1985; Jax 2000; Schwarz 2003), leading early on to an array of subdivisions within the emerging discipline. These subdivisions result in part from research traditions that go back beyond the formation of ecology as a science and in part from new distinctions arising out of specialisations and new emerging topics within ecology. One distinction that was especially important in shaping the character of ecology as a concept was that created between those scientific fields within ecology that dealt with individual organisms and those that dealt with groups of organisms, in particular the distinction between autecology and synecology and, later on, population ecology.

The first kind of subdivisions mentioned above refer to the division into fields dealing with particular taxonomic groups, namely animal and plant ecology (more recently also microbial ecology), and those focusing on the differences between different types of environments (terrestrial, freshwater, marine).

Plant and animal ecology as well as aquatic and terrestrial ecology at first developed quite independently from each other. Even more distinct fields were paleoecology and parasitology.[1] The existence of these separate branches of ecology was in part brought about by the traditional academic divisions of biology into, for example,

---

[1] Parasitology is generally addressed in the context of disease research in human and veterinary medicine or plant pathology. The term "paleoecology" (or "paleo-ecology") was coined by Frederic Clements (1916, p. 279), while studies on the relationship between fossil organisms and their environment had already begun to be conducted by Edward Forbes and others by at least 1840 (see Cloud 1959, p. 927f; Hecker 1965, pp. 1ff; McIntosh 1985, pp. 98ff). Paleoecology often remained equally or even more closely related to geology than to biology and ecology.

K. Jax (✉)
Department of Conservation Biology, Helmholtz Centre for Environmental Research (UFZ), Permoserstr. 15, 04318 Leipzig, Germany
e-mail: kurt.jax@ufz.de

A. Schwarz
Institute of Philosophy, Technische Universität Darmstadt, Schloss, 64283 Darmstadt, Germany
e-mail: schwarz@phil.tu-darmstadt.de

zoology and botany, but also by the specific characteristics of their objects and the different methods required for their investigation.

Botany (and plant geography), zoology, and the aquatic sciences (marine as well as freshwater) were the traditional fields in which ecology first developed. Plant ecology was the forerunner in shaping ecology and contributed crucially to its formation as a self-conscious science – an issue that was also acknowledged by zoologists (e.g. Shelford 1915, p. 1; Hesse 1927). A major step in the stabilisation of ecology as a distinct research field with its own concepts was the merging of the biogeographic and the physiological perspective on plants and, especially, plant communities, which occurred during the last decade of the nineteenth century (see Chap. 12). Animal ecology, by contrast, remained a science without a strong common conceptual core or a common sense of constituting a scientific community in its own right for at least another 2 decades (Trepl 1987, p. 160f).

In terms of their training and the objects they studied, aquatic ecologists were mainly zoologists. As with plant ecology, limnology and oceanography had their roots in a longer tradition of geographical research on seas and freshwater bodies (in particular lakes), which were systematically related to biological studies only towards the end of the nineteenth century. Aquatic ecologists were much more inclined than terrestrial ecologists to emphasize and investigate the "whole" web of interrelations in a particular place (e.g. a lake) and thus to consider larger wholes, which included the physical and chemical properties of the habitat they studied, as the basic objects of ecology.[2]

Beyond the question of which kinds of organisms constituted the object of study, another important subdivision was created by the delimitation of those scientific branches within biology (and thus also within ecology) that dealt with individual organisms from those that dealt with groups of organisms. Carl Schröter and Oskar Kirchner in 1902 distinguished between the ecology of individual organisms, or autecology (German: *Autökologie*), and that of species assemblages, or synecology (German: *Synökologie*). The term *Synökologie* was coined thus: "I propose to introduce for this important discpline [the science of plant communities] the name 'formation science' or 'synecology', from =with, and οικοσ=house, i.e., the science of the plants that live together, and at the same time the science of the plants that seek out analogous ecological conditions".[3] A similar division of ecology into two "phases" – as he called it – was also undertaken by Henry Chandler Cowles, although he did not provide specific names for these subdisciplines: "Whatever its limits may be, ecology is essentially a study of origins and life histories, having two well-marked phases; one phase is concerned with the origin and development of plant structures,

---

[2] See Schwarz (2003) for a history of the development of aquatic ecology.
[3] "Ich schlage vor, für diese wichtige Disziplin [die Lehre von den Pflanzengesellschaften] den Namen ‚Formationslehre' oder ‚Synökologie' einzuführen, von συν=mit, zusammen wohnen und οικοσ=Haus, also die Lehre von den Pflanzen, welche zusammen wohnen, und zugleich die Lehre von den Pflanzen, welche analoge ökologische Bedingungen aufsuchen" (Schröter and Kirchner 1902, p. 63).

the other with the origin and development of plant societies or formations" (Cowles 1901, p. 73). Other authors substituted the words *Biocoenotik* (biocoenology) (Gams 1918) or *Biosoziologie* (biosociology) (Du Rietz 1921) for "synecology". This distinction, between the biology and ecology of individual organisms and that of groups of organisms, was emphasized in particular by Gams (1918)[4] and Schwenke (1953) and considered by them as being of greater significance than the distinctions, for example, between ecology and chorology, or physiology and morphology.[5]

The division of ecology into autecology, population ecology (also: demecology, in German: *Demökologie*), and synecology, which is currently common, is of more recent origin.[6] Population ecology is sometimes subsumed under synecology and sometimes considered a (sub)discipline in its own right.

Population ecology – despite also having far-reaching roots, especially in human demography (see e.g. Hutchinson 1978, pp. 5ff) – made a rather late appearance as a subdiscipline of ecology. Its growth, though rapid, only began to occur during the 1920s and 1930s. Despite its late appearance on the scene, population ecology was the first subdiscipline of ecology to enter the stage of mathematical theory building (Kingsland 1985/1995). Alongside autecology, population ecology was that part of ecology which pursued most stringently the "hard science" approach. Although it was claimed (by Thomas Park (1946), for example) that population ecology paid no heed to boundaries of habitat and taxonomy in the scope of its studies, it was in fact almost completely a zoological enterprise. Even though Clements and Tansley also devoted time to working experimentally on plant populations, a *population ecology* of plants was practically non-existent in the first half of the twentieth century and was systematically developed only during the 1960s and 1970s in Britain by John Harper.[7]

The investigation of "ecosystems" is most commonly subsumed under synecology. While it is common in English speaking countries to distinguish between population ecology, community ecology and ecosystem ecology – especially after Eugene Odum and his highly influential textbook[8] – the subdivision of ecology based on "individuals, populations and communities"[9] found in the most popular contemporary English textbook of ecology (Begon et al. 1990, 1996, 2005) – is again close to that used for

---

[4]For Gams (1918, p. 297), this distinction was a fundamental subdivision of the whole field of *biology*, which he divided into the two major branches of *Idiobiologie* (idiobiology: the science of individual organisms) and *Synbiologie* (synbiology: the science of biological communities).

[5]See especially van der Klaauw (1936) for a detailed discussion of this subdivision.

[6]We managed to find this systematic division in the German-language literature first in Schwerdtfeger 1968, who apparently coined the term *Demökologie* as an alternative word for "population ecology". Schwerdtfeger himself points out that population ecology had already been established previously as a separate branch between autecology and synecology in the English-language literature; see, e.g. Park (1946).

[7]The seminal textbook of plant population biology was Harper (1977).

[8]First edition 1953; revised and extended editions followed in 1959 and 1971.

[9]In the 4th edition of the book (2005) the part formerly headed "Communities" is now entitled "Communities and Ecosystems".

a long time in German speaking ecological circles. Depending on author and viewpoint one the "levels" is sometimes omitted in one or the other textbook. A uniform scheme does not exist here. More recently, other authors (e.g. Jones and Lawton 1995) emphasize the distinction between an ecology of populations and communities on the one hand (focusing on specific species' interactions) and ecosystem ecology on the other, the latter seen as dealing mainly with flows of energy and matter.

During the 1980s, landscape ecology became yet another subdiscipline of ecology, for the most part emphasizing large-scale topological aspects of ecological phenomena.[10] Distinctions vis-à-vis the other subdisciplines mentioned above, especially ecosystem ecology, are not always clear-cut.

Since the 1970s and 1980s, when ecology became a more established and popular science, numerous additional branches of ecology have been constituted, each one focusing on specific methods, perspectives, and objects. They include, for example, behavioral ecology, molecular ecology, functional ecology, microbial ecology, and evolutionary ecology. Some of these new ecological specialisms and their scientific communities overlap with the broader, more traditional branches, while others (such as microbial ecology) are addressed by rather distinct and separate research communities.

References

Begon M, Harper JL, Townsend CR (1990) Ecology: individuals, populations and communities, 2nd edn. Blackwell, Oxford
Begon M, Harper JL, Townsend CR (1996) Ecology. Individuals, populations and communities, 3 edn. Blackwell, Oxford
Begon M, Townsend CR, Harper JL (2005) Ecology: from individuals to ecosystems, 4th edn. Blackwell, Oxford
Clements FE (1916) Plant succession: an analysis of the development of vegetation. Carnegie Institution of Washington, Washington, DC, Publication No. 242
Cloud PE Jr (1959) Paleoecology: retrospect and prospect. J Paleontol 33:926–962
Cowles HC (1901) The physiographic ecology of Chicago and vicinity; a study of the origin, development, and classification of plant societies. Bot Gaz 31:73–108, 145–182
Du Rietz E G (1921) Zur methodologischen Grundlage der modernen Pflanzensoziologie. Adolf Holzhausen, Wien
Gams H (1918) Prinzipienfragen der Vegetationsforschung. Ein Beitrag zur Begriffsklärung und Methodik der Biocoenologie. Vierteljahresschrift der Naturforschenden Gesellschaft in Zürich 63:293–493

---

[10] In contrast to Anglo-Saxon tradition, landscape has been an object of German ecology already since the first half of the twentieth century. However, these traditions derived from a different context, connecting cultural and natural aspects of human living spaces, and not from a purely scientific approach (see Trepl 1995; Klink et al. 2002, and Chap. 25).

Harper JL (1977) Population biology of plants. Academic Press, London
Hesse R (1927) Die Ökologie der Tiere, ihre Wege und Ziele. Naturwissenschaften 15:942–946
Hecker [Gekker] RF (1965) Introduction to paleoecology, American Elsevier, New York
Hutchinson GE (1978) An introduction to population ecology. Yale University Press, New Haven/London
Jax K (2000) History of ecology. In: Encyclopedia of Life Sciences. [online]. Wiley. URL: www.els.net
Jones CG, Lawton JH (eds) (1995) Linking species and ecosystems. Chapman & Hall, New York
Kingsland SE (1985/1995) Modeling nature: episodes in the history of population ecology. University of Chicago Press, Chicago
van der Klaauw CJ (1936) Zur Aufteilung der Ökologie in Autökologie und Synökologie, im Lichte der Ideen als Grundlage der Systematik der zoologischen Disziplinen. Acta Biotheor 2:195–241
Klink H-J, Marion P, Bärbel T, Gunther T, Martin V, Uta S (2002) Landscape and landscape ecology. In: Bastian O, Uta S (eds) Developments and perspectives of landscape ecology. Kluwer, Dordrecht, pp 1–47
McIntosh RP (1985) The background of ecology: concept and theory. Cambridge University Press, Cambridge
Odum EP (1953) Fundamentals of ecology, 1st edn. W.B. Saunders, Philadelphia
Park T (1946) Some observations on the history and scope of population ecology. Ecol Monogr 16:313–320
Schröter Carl, Oskar Kirchner (1902) Die Vegetation des Bodensees, 2. Teil. Lindau i. B.: Kommissionsverlag der Schriften des Vereins der Geschichte des Bodensees und seiner Umgebung von Joh. Thom. Stettner
Schwarz AE (2003) Wasserwüste – Mikrokosmos – Ökosystem. Eine Geschichte der "Eroberung" des Wasserraums. Rombach-Verlag, Freiburg
Schwenke W 1953. Biozönotik und angewandte Entomologie. Beiträge zur Entomologie 3, Beiheft: 86–162
Schwerdtfeger F (1968) Ökologie der Tiere. Band II: Demökologie: Struktur und Dynamik tierischer Populationen. Paul Parey, Hamburg/Berlin
Shelford VE (1915) Principles and problems of ecology as illustrated by animals. J Ecol 3:1–23
Trepl L (1987) Geschichte der Ökologie. Vom 17. Jahrhundert bis zur Gegenwart. Athenäum, Frankfurt/Main
Trepl L (1995) Die Landschaft und die Wissenschaft. In: Erdmann K-H, Kastenholz HG (eds) Umwelt- und Naturschutz am Ende des 20. Jahrhunderts. Probleme, Aufgaben und Lösungen. Springer, Berlin/Heidelberg/New York, pp 11–26

Part V
"Ecology", Society and the Systems View in the Twentieth and Twenty-first Century

Chapter 15
The Rise of Systems Theory in Ecology

Annette Voigt

The emergence of systems theory in ecology, particularly during the 1950s and 1960s, was accompanied by the hope that ecology might turn into an exact science with prognostic potential and a set of uniform theoretical foundations. The impact of systems theory on ecology was manifested mainly in the formulation and development of ecosystem theory. The widely-held view is that ecosystem theory is concerned primarily with units comprising communities of organisms of various species and the abiotic environment of these communities. The components of systems are seen to interact with one another.

The main elements in any historical reconstruction of the emergence of ecosystem theory include the establishment of general systems theory and its associated theories (including cybernetics, information theory etc.), the introduction of the term "ecosystem" and early ecosystem theories.

General Systems Movement

During the 1940s a variety of approaches were developed in different parts of Europe, the USA and the USSR, which were later to become united under the rubric of "systems theory". Most of these theories and practices came about as a result of encounters between scientists from different disciplines. Perhaps the one motivation they can all be said to have shared was an interest in the scientific description of "gestalt", a term found throughout the scientific literature of the first few decades of the twentieth century, in a wide range of disciplines.[1] Otherwise,

---

[1] The most prominent is probably the gestalt concept in psychology (Köhler 1920), but see also the field concept(s) in physics, and the "gestalt laws" (Bertalanffy 1926, 1929). The role of the latter with respect to ecology is discussed in Schwarz (1996).

A. Voigt (✉)
Urban and Landscape Ecology Group, University of Salzburg, Hellbrunnerstraße 34,
5020 Salzburg, Austria
e-mail: annette.voigt@sbg.ac.at

these approaches all differ in terms of their aims, their problem focus and their institutional background. Approaches with more of an engineering, mathematical or physical background include cybernetics (Wiener 1948) and information theory (Shannon and Weaver 1949); other theories, such as game theory (Neumann and Morgenstern 1944), operations research (Churchman et al. 1957), action theory (Parsons 1937) and general systems theory (GST), emphasize more psychological, physiological and philosophical aspects.

The major figures regarded as the founders of GST are economist Kenneth E. Boulding, neurophysiologist Ralph W. Gerard, biomathematician Anatol Rapoport and, in particular, biologist Ludwig von Bertalanffy.[2] Bertalanffy started out from the problem of how to provide a scientific explanation of "life". His theory of the *organism* as a hierarchically organised, open system was intended in part as a means of overcoming the dispute in biology between mechanism and vitalism (Bertalanffy 1932, 1949). Thus, organisms were to be described in non-reductionist terms, as living wholes, yet still within a scientific framework. Taking this organismic systems theory as a starting point, and having observed certain parallels between the concepts and models found in different scientific disciplines, Bertalanffy, from the 1950s onwards, began to define a *generalised* systems concept, related to all "sets of elements standing in interrelation"[3]: technical systems, organisms, social systems, and so on.[4] According to Bertalanffy GST is a new scientific logico-mathematical discipline: 'Its subject matter is formulation of principles that are valid for 'systems' in general, whatever the nature of their component elements and the relation or 'forces' between them' (Bertalanffy 1968, p. 36).

The new thing about GST that set it apart from prevailing understandings of science up until then was that its aim was to reach a "new level" of theory making: "Developing unifying principles running 'vertically' through the universe of the individual sciences, this theory brings us nearer to the goal of the unity of science" (Bertalanffy 1968, p. 37). As such, it ignored the division between the human and the natural sciences, which had become institutionalised in the nineteenth century, and applied mathematical procedures to areas where, up until then, hermeneutical methods had seemed appropriate (Lilienfeld 1978; Müller 1996). The aim, however, was not to seek explanations that reduced phenomena to their elemental "fundamental units" (molecules and so on), but rather to provide a mathematical description of the "system as a whole", one in which the system's elements were to be regarded primarily from the point of view of their function in relation to other elements or to "the whole". Approaches adopted within information theory and cybernetics in the 1950s extended systems theory by adding an explicitly social scientific dimension, since it is possible to apply a mathematical concept of

---

[2] Boulding 1941, 1953, 1956; Bertalanffy 1950, 1951, 1955; Gerard 1940, 1953; Rapoport 1947, 1950. Also cf. Davidson 1983; Müller 1996; Hammond 2003.

[3] "A system can be defined as a complex of interacting elements p1, p2 ... pn. Interaction means that the elements stand in a certain relation, R, so that their behaviour in R is different from their behaviour in another relation, R'." (Bertalanffy 1950, p. 143).

[4] Bertalanffy 1950, 1955, 1968; also cf. Müller 1996; Schwarz 1996; Voigt 2001.

information to the sphere of the social without opening oneself up to the accusation of naturalisation. In addition, cybernetic concepts enable a connection to be made to mathematical constructions and technological issues.

"Systems theory" thus encompasses various approaches that have proven their worth in describing, monitoring and constructing complex systems.[5] Systems are classified according to their characteristics, while the specific focus may be on their relationships to their environment, their complexity, their mode of self-organisation or the capacity of a system for feedback. Systems theoretical approaches exist today in all those sciences that see their objects of interest as having, at least in part, a systemic character; these include – to name just a few – sociology, psychology, geography, physics, cognitive and neurosciences and ecology.

Historical Overview of Early Ecosystem Theory[6]

Ecology developed its own systemic theories early on, later applying systems theoretical approaches of different kinds.[7]

In 1935 vegetation ecologist A.G. Tansley introduced the term "ecosystem" as a "fundamental concept in ecology". Tansley's new concept represented a response to the debate about the structure and organisational form of units of vegetation in ecology.[8] On one side of this debate was the holistic-organicist approach (e.g. Clements 1916, 1936; Friederichs 1927, 1934, 1937; Phillips 1934, 1935; Clements and Shelford 1939), according to which a biotic community ("Lebensgemeinschaft") is essentially determined by internal, functional relationships of dependency between individual organisms. As a whole, it has the character of a superorganism and is usually conceived of as a real unit. At the opposite end of the debate was the reductionist-individualist approach (e.g. Gleason 1917, 1926; Gams 1918; Ramensky 1926), which held that the term "association" relates to temporary combinations of species determined both by needs that either coincide or are complementary and by the random character of immigration. It is accepted that this "association" is only of heuristic use; no real, natural units exist above the level of the organism.[9] Tansley's ecosystem is neither a superorganism nor a chance

---

[5]More modern systems theoretical variants include non-equilibrium thermodynamics (Prigogine 1955) and theories about adaptive, self-organised and self-referential systems, e.g. autopoiesis (Maturana and Varela 1987).

[6]Hagen (1992) and Golley (1993) provide a detailed account of the history of ecosystem theory. Also cf. McIntosh 1985.

[7]On the theory of the transfer of ideas from ecology to systems theory, Chap. 27.

[8]On this debate, see e.g. Tobey 1981, p. 76–109; Worster 1985, p. 205–220; McIntosh 1995, p. 76–85, 1995; Trepl 1987, p. 139–158; Hagen 1992, p. 15–32; Golley 1993, p. 8–34; Botkin 1990; Jax 2002; Chaps. 19 and 20.

[9]The radical individualist position (Peus 1954) rejects the notion of associations as an object of science because it sees them as "fictions".

combination: it contains not only organisms, but also their environment,[10] and these components and the interactions that exist between them are viewed in *physical* terms. Tansley's formulation, "The whole method of science [...] is to isolate systems mentally for the purposes of study [...]" (1935, p. 299f.) suggests the view that the system being studied by the scientist is not a real object but rather an idealisation. No claim is made to be studying all the variables of a phenomenon; instead, abstractions are made only in relation to those issues that are of interest to the scientist. Even though the distinction between living and non-living parts is secondary in Tansley's system concept, he does refer explicitly to ecological objects and does not pursue the high level of abstraction found in GST. Tansley distanced himself from Bertalanffy's systems theory and, while not doing ecosystem research in today's sense (Golley 1993, p. 34), his concept of ecosystem nonetheless contributed greatly to a physically oriented ecosystem theory.

The period after 1935 saw, on the one hand, the rise of theories oriented towards physics, which described the transfer of matter, energy and information within the ecosystem and, on the other, more biologically-oriented positions, which saw the key to understanding ecosystems in the characteristics of the populations and individuals that constitute them (e.g. Lamotte and Bourliere 1978).[11]

The ecosystem concept was first applied in 1942 by the limnologist Raymond L. Lindeman. He described a lake as an energetically open ecosystem consisting of biotic and abiotic components. Organisms are significant only insofar as they fulfil specific functions within the system. They are arranged in trophic levels (producers, primary and secondary consumers, decomposers). "The basic process in trophic dynamics is the transfer of energy from one part of the ecosystem to another". (Lindeman 1942, p. 400). Energy from the sun is accumulated in the producers by means of photosynthesis, and only a portion of this energy is transferred via consumption to the next level – everything else is lost (by respiration and decomposition). It is possible to quantify both the productivity of each level and the degree of efficiency of the energy transfer, as well as that of both changes in the succession. This approach enables ecosystems to be described in thermodynamic terms.

One key figure in the history of systems theory in ecology is Lindeman's teacher George Evelyn Hutchinson. Hutchinson, an ecologist, was part of the core group that organised the ten "Macy Conferences" (1946–1953) to explore the possibility of using scientific ideas that had emerged in the war years as a basis for both interdisciplinary research alliances and solving peacefully the complex problems facing

---

[10] Previous approaches to conceptualising communities along with their environment include Thienemann's concept of the biosystem (Thienemann and Kieffer 1916) and Friederichs' concept of the holocoen (1927). However, these differ from the concept of ecosystem on account of their holistic-morphological and/or holistic-organicist orientation (Chap. 4).

[11] In addition, the concept of ecosystem research is related to the fact that everything is considered that is relevant ecologically in a specific site. That is, not only all the organisms are considered but all edaphic and climatic factors as well.

the postwar world.[12] The first meeting of this "Cybernetics Group" was called "Feedback Mechanisms and Circular Causal Systems in Biological and Social Systems". It was here, where the different kinds of systems theories discussed above finally found their way into ecology. With his paper on "Circular Causal Systems in Ecology" Hutchinson presented a theory for describing a community using the cybernetic terms feedback and circular causality (Hutchinson 1948). Within certain boundaries, ecosystems are "self-correcting" by means of "circular causal paths", so that conditions of equilibrium prevail. The assumption of regulating feedback systems forms the basis of both his biogeochemical approach (following V. Vernadsky) – in which the transfer of substances through the systems is described quantitatively and without any specifically biological terms – and of his biodemographic approach, which describes population developments with reference to quantitative theories of ecology (e.g. Lotka 1925; Volterra 1926). Abiotic and biotic factors alike are looked at from the point of view of the extent to which their effect is to stabilise the equilibrium. The carbon cycle, for example, is corrected by the regulating effects of the oceans and the biological cycle, while the size of a population is regulated by purely physical conditions (e.g. size of area available) or by the behaviour of (groups of) organisms (e.g. competition).

The ecosystem approach was promulgated above all, however, by Howard T. and Eugene P. Odum. E.P. Odum placed at the centre of his holistic ecology a systems concept whose lack of clarity was thoroughly characteristic of early ecosystem theory: on the one hand, ecosystem is a concept that describes any unit consisting of a living component (including, for example, cells) and their environment (and whose energy and matter transfer one is studying); on the other hand, it is also a term that refers to a specific ecological unit (in addition to organism and population, for example) which, as a concrete object, contains *all* the organisms and their abiotic environment located in a specific area (E.P. Odum 1953, 1971; also cf. Golley 1993). Odum postulates the necessity of a "whole-before-the-parts" approach when studying ecosystems, because they possess emergent characteristics. He emphasises the organismic attributes of the ecosystem and draws parallels between succession controlled by the organic community and the development of the individual organism; both, according to him, are oriented towards achieving homeostasis (Odum 1969; also cf. Worster 1994; Hagen 1992, p. 128). Looking at successive editions of their influential book "Fundamentals of Ecology" (1953, 1959, 1971) we can see how the Odum brothers increasingly built on the energetic approach as the basis of ecology. It was Howard T. Odum in particular – likewise a student of Hutchinson – who developed it further. He depicted the energy transfer

---

[12] The Macy Conferences, in which figures such as N. Wiener, J. von Neumann, R. Gerard, G. Bateson, A. Rosenblueth, M. Mead, J. von Foerster and G.E. Hutchinson participated, contributed decisively towards the dissemination of cybernetic approaches in the 1940s and 1950s far beyond the sphere of their technical application, into areas such as the social sciences, psychology, biology and the human and life sciences (cf. Taylor 1988; Heims 1993; Pias and Foerster 2003).

in ecosystems using energy flow diagrams and electrical circuits (Odum 1956). By "transforming" everything into energy he used the energetic approach as the sole basis for researching both natural and social systems. Their energy balance serves as well as a basis on which to evaluate them and as the starting point for technocratic concepts of control ("ecological engineering") – of social systems among others (H.T. Odum 1971; for more detail, cf. Taylor 1988). Claims of this kind to be able to explain everything and to exercise control are also found in other systems theoretical approaches.

Ecosystem theory became the leading paradigm in ecology through various large-scale research programmes, including studies on the distribution of radioactivity, funded by the US Atomic Energy Commission, the "International Biological Program" (IBP), undertaken in the USA from the 1960s onwards (cf. Kwa 1987; Hagen 1992; Golley 1993) and later in Europe as well e.g. the Solling Project (Ellenberg 1971, 1986), and the Hubbard Brook Project (Bormann and Likens 1967; Likens et al. 1977). With the emergence of the environmental movement in the 1960s and 1970s, ecosystem research took on the task of analysing environmental problems such as the impact of pesticides, the eutrophication of lakes etc., assessing and combating its consequences (e.g. in the context of UNESCO's "Man and the Biosphere Program" (MAB). Ecosystems were to be not only studied but managed as well. In addition, it was hoped that ecosystem theory would offer a deeper understanding of the effects of human action on "nature", as well as providing, on the one hand, an ultimately technocratic solution to the environmental crisis and, on the other, a new, holistic human-nature relationship.

A number of factors – including the reception of different variants of systems theory, the associated transfer of modern physical and mathematical theories into ecology, as well as developments within the discipline – led to the flowering of a bewildering array of ecosystem theoretical views and applications. Only a few examples can be named here.[13] Ecosystems are often described as systems characterised by an open energy flow and matter cycle, and they are usually looked at on the basis of theories of thermodynamics (e.g. Jørgensen 2000; Kay 2000). According to this view, organisms assimilate energy in order to counter "entropic decline". Systems are frequently seen as being cybernetic: they can regulate themselves physically or biologically within certain boundaries via feedback loops (e.g. Patten 1959; Patten and Odum 1981).[14] The succession of ecological communities can be described using concepts from information theory.[15] The application of the hierarchy theory in ecology (Allen and Starr 1982; O'Neill et al. 1986) means that the ecosystem can be seen as a hierarchically organised system – neither cybernetic models (as ultimately mechanistic models) nor the processing of ever more detailed

---

[13] An overview of more recent developments in ecosystem theory can be found in Frontier and Leprêtre 1998, also cf. articles in Pomeroy and Alberts 1988; Higashi and Burns 1991; Vogt et al. 1997; Pace and Groffman 1998; Jørgensen and Müller 2000; see also Chap. 27.

[14] For the opposing position, cf. Engelberg and Boyarsky 1979.

[15] E.g. Margalef 1958, 1968; on more modern information theoretical ecosystem approaches, see Ulanowicz 1997; Nielsen 2000; see also Hauhs and Lange 2003.

data are considered appropriate to complex systems. In order to understand a phenomenon at any level of a particular hierarchy, such as that of the "ecosystem", it is necessary, according to this view, to look at its relationship to both higher levels of the hierarchy (e.g. the biosphere) and lower ones (e.g. organisms). The issue of the unpredictability of ecosystem dynamics is taken account of by referring to catastrophe and chaos theories. A large section of the ecosystem research community is concerned with computer modelling and simulation of ecosystems, e.g. on the basis of fuzzy logic and artificial neural networks (cf. diverse articles in Hall and Day 1977; Recknagel 2003). Attempts at combining population and ecosystem ecology can be found in Jones and Lawton (1995).

The Scientization of Organicism by the Systems Concept

The different kinds of systems theories contributed to transform earlier organicist notions of ecological units into scientific concepts, as had been the original intention of Tansley in coining the term "ecosystem". But there is an apparent ambiguity within the ecosystem concept.

The conception of synecological units as spatial (super)organisms (Clements 1916, 1936; Clements and Shelford 1939; Phillips 1934, 1935) implies transferring the idea of individuality to a larger entity of organisms, to a community or organic community. In this community individual organisms are looked at with regard to the contribution they make towards the community's ability to function and its overall maintenance; its existence, characteristics and external relations are explained by reference to the function they have for the community. In this respect, the latter appears to be a "higher-level individual". The ecosystem concept scientizes this individual wholeness: (1) the system is "holistic" because, apart from the "sum of its parts", it comprises the relationships between the parts (and processes) that, in principle, can be explained in causal terms. The ecosystem is a physical object and is looked at as being analogous to a *machine*.[16] The systems approach is driven by an interest in technical knowledge to the extent that it promises the possibility of managing complex natural systems, optimising them and making them available for use. (2) Ecosystem approaches in many cases call themselves holistic (e.g. Odum 1953) and *are* indeed such, in that they take the whole of the system as their starting point. But even so, they are reductionistic in the sense that, to a large extent, they develop abstractions on the basis of actual objects (it was this in particular that prompted the accusation of reductionism). In the (physical) ecosystem perspective, the main concern is with the material-energetic (and possibly informational) aspects of interactions; the actual species involved are only of interest insofar as their specific features are relevant to the transformation of matter and energy. Systems concepts based on set theory (e.g. Hall and Fagen 1956)[17] go further still: they

---

[16]Cf. Taylor 1988; Hagen 1992; Golley 1993.

[17]"A system is a set of objects together with relationships between the objects and between their attributes" (Hall and Fagen 1956, p. 18).

assert that the system's components are not real objects but rather are defined by the characteristics of similar classes.[18] (3) With this constructivist concept of system, "wholes" come about through a scientific operation, i.e. they are not real entities, as in organicism.[19] Ecosystems are models – theoretical constructions of the mind – for bringing order to diverse phenomena. They are constructed according to specific functions of the ecosystem defined by the *observer* (e.g. biomass production). As such, their components are looked at from the point of view of how they contribute towards fulfilling this function. The whole is now considered to be anything necessary for fulfilling the function defined by the observer.

Even so, ecosystem theories may be viewed within the anti-mechanistic, organicist tradition if, for example, it is assumed that systems have real spatial boundaries: systems are (spatial) entities demarcated from one another by *their* processes; they are subject to a succession that leads towards dynamic equilibrium, and they can be destroyed. As such they can be depicted in terms of their relevant components, i.e. this *realistic* approach assumes that the abstractions undertaken in the model correspond to actual relationships. If the mode of functioning of an ecosystem is conceptualised in such a way that its purpose is self-preservation, i.e. it is an end in itself ("Selbstzweck"), then the ecosystem is being conceptualised in analogy to the organism – in other words, a circular dependency exists between its parts; it produces, develops and maintains itself, and it can be destroyed. This view of ecosystems is also widely found in nature conservation and environmental ethics.

Conclusion

In conclusion, systems theoretical views and their mathematical formulations – therein lies both an integrating function as well as a certain set of problems – generally allow constructivist and realist interpretations.[20] A self-regulating ecosystem can be seen as both a machine and an organism. The term "equilibrium" can be seen in relation to something that is defined in physical-mechanistic terms, but can also be applied to a state that is actively and intentionally maintained.[21] Moreover, the idea of the system as an organism does not stand in contradiction to the claim to capture, manage or control it with regard to energy flow. Organicist and technocratic perspectives may well be combined. It is between the extremes of constructivism – realism, mechanism – organicism and reductionism – holism that the theoretical debates within ecosystem research can be located.

---

[18] Müller 1996.
[19] Cf. Tobey 1981; Jax 1998.
[20] Müller 1996.
[21] Cf. Weil 1999.

References

Allen TFH, Starr TB (1982) Hierarchy: perspectives in ecological complexity. University of Chicago Press, Chicago
Bertalanffy L (1926) Zur Theorie der organischen 'Gestalt'. Roux' Archiv: 413–416
Bertalanffy L (1929) Vorschlag zweier sehr allgemeiner biologischer Gesetze. Biol. Zentralbl. 49: 83–111
Bertalanffy L (1932) Theoretische Biologie, Bd. I: Allgemeine Theorie, Physikochemie, Aufbau und Entwicklung des Organismus. Borntraeger, Berlin
Bertalanffy L (1949) Das biologische Weltbild. Die Stellung des Lebens in Natur und Wissenschaft. Francke, Bern
Bertalanffy L (1950) An Outline of General System Theory. Brit. J. Philos. Sci. 1:134–165
Bertalanffy L (1951) General System Theory: A New Approach to Unity of Science. Problems of General System Theory. Human Biology 23/4:302–312
Bertalanffy L (1955) General System Theory. Main Currents in Modern Thought 11:75–83
Bertalanffy L (1968) General system theory: foundations, development applications. George Braziller, New York
Botkin DB (1990) Discordant harmonies: a new ecology for the twenty-first century. Oxford Univ. Pr., New York
Bormann FH & Likens GE (1967) Nutrient cycling. Science 155(3461): 424–429
Boulding KE (1941) Economic analysis. Harper & Brothers, New York
Boulding KE (1953) Toward a general theory of growth. Canadian J. o. Economics and Political Science 19/3:326–340
Boulding KE (1956) Generals systems theory. The skeleton of science. Management Science 2:197–208
Churchman CW, Ackoff RL, Arnoff EL (1957) Introduction to operations research. Wiley, New York
Clements FE (1916) Plant succession: an analysis of the development of vegetation. Carnegie Institution of Washington, Washington, DC
Clements FE (1936) Nature and structure of the climax. J. of Ecology 24:252–284
Clements FE, Shelford VE (1939) Bio-ecology. Wiley, New York
Davidson M (1983) Uncommon sense: the life and thought of Ludwig von Bertalanffy, father of general system theory. JP Tarcher, Los Angeles
Ellenberg H (ed) (1971) Integrated experimental ecology: methods and results of ecosystem research in the German Solling Project. Springer, Berlin
Ellenberg H (ed) (1986) Ökosystemforschung. Ergebnisse des Sollingprojektes, 1966–1986. Ulmer, Stuttgart
Engelberg J, Boyarsky LL (1979) The noncybernetic nature of ecosystems. Am Nat 114(3): 317–324
Friederichs K (1927) Grundsätzliches über die Lebenseinheiten höherer Ordnung und den ökologischen Einheitsfaktor. Naturwissenschaften 8:153–157, 182–186
Friederichs K (1934) Vom Wesen der Ökologie. – Sudhoffs Arch. Gesch. d. Medizin u. Naturwissens 27 (3): 277–285
Friederichs K (1937) Ökologie als Wissenschaft von der Natur oder biologische Raumforschung. Barth, Leipzig
Frontier S, Leprêtre A (1998) Développements récents en théorie des écosystèmes. Ann. Inst. océanogr. Paris 74(1): 43–87
Gams H (1918) Prinzipienfragen der Vegetationsforschung. Ein Beitrag zur Begriffsklärung und Methodik der Biocoenologie. Naturf. Gesellschaft Zürich. Vierteljahresschr. 63:293–493
Gerard RW (1940) Unresting Cells. Harper & Brothers, New York
Gerard RW (1953) The Organismic view of society. Chicago Behavioral Science Publications 1: 12–18

Gleason HA (1917) The structure and development of the plant association. Bull Torrey Bot Club 44:463–481

Gleason HA (1926) The individualistic concept of the plant association. Bull Torrey Bot Club 53:7–26

Golley FB (1993) A history of the ecosystem concept in ecology: more than the sum of the parts. Yale University Press, New Haven/London

Hagen JB (1992) An entangled bank: the origins of ecosystems. Chapman & Hall, New York

Hall CAS, Day J (eds) (1977) Ecosystem modeling in theory and practice. Wiley, New York

Hall AD, Fagen RE (1956) Definition of System. General System, 118–28

Hammond D (2003) The science of synthesis: exploring the social implications of General Systems Theory. Univ. Pr. of Col., Colorado

Hauhs M, Lange H (2003) Informationstheorie und Ökosysteme. Handbuch der Umweltwissenschaften. Ecomed, München: 1–22

Heims SJ (1993) Constructing a social science for postwar America: the cybernetics group, 1946 – 1953. MIT Press, Cambridge

Higashi M, Burns TP (eds) (1991) Theoretical studies of ecosystems. Cambridge University Press, Cambridge

Hutchinson GE (1948) Circular causal systems in ecology. Annals of the New York Academy of Sciences 50:221–246

Jax K (1998) Holocoen and ecosystem: on the origin and historical consequences of two concepts. J. Hist. Biology, 31:113–142

Jax K (2002) Die Einheiten der Ökologie. Analyse, Methodenentwicklung und Anwendung in Ökologie und Naturschutz. Lang, Frankfurt/M

Jones CG, Lawton JH (1995) Linking species and ecosystems. Chapman & Hall, New York

Jørgensen SE (2000) A general outline of thermodynamic approaches to ecosystem theory. In: Jørgensen S, Müller F (eds) Handbook of ecosystem theories and management. Lewis, London/New York/Washington, DC

Jørgensen SE, Müller F (2000) Handbook of ecosystem theories and management. Lewis, London/New York/Washington, DC

Kay JJ (2000) Ecosystems as self-organising holarchic open systems: narratives and the second law of thermodynamics. In: Jørgensen S, Müller F (eds) Handbook of ecosystem theories and management. Lewis, London/New York/Washington, DC

Köhler W (1920) Die physischen Gestalten in Ruhe und im stationären Zustand: eine naturphilosophische Untersuchung. Vieweg, Braunschweig

Kwa C (1987) Representations of nature mediating between ecology and science policy: the case of the International Biological Programme. Social Studies of Science 17, 3, 413–442

Lamotte M, Bourliere F (1978) Problemes d' écologie, structure et fonc-tionnement des écosystèmes terrestres. Masson, Paris

Lotka, AJ (1925) The elements of physical biology. Williams & Wilkins, Baltimore

Lindeman RL (1942) The trophic-dynamic aspect of ecology. Ecology 23:339–418

Likens GE, Bormann FH, Pierce RS, Eaton JS, Johnson NM (1977) Biogeochemistry of a forested ecosystem. Springer, New York

Lilienfeld R (1978) The rise of systems theory. Wiley, New York

Margalef R (1958) Information theory in ecology. YearB Soc Gen Syst Res 3:36–71

Margalef R (1968) Perspectives in ecological theory. University of Chicago Press, Chicago, pp 1–25

Maturana HR & Varela FJ (1987) Der Baum der Erkenntnis: die biologischen Wurzeln des menschlichen Erkennens. Scherz Verlag, Bern

McIntosh RP (1985) The background of ecology: concept and theory. Cambridge University Press, Cambridge

McIntosh RP (1995) H. A. Gleason's 'Individualistic concept' and theory of animal communities: a continuing controversy. - Biol. Rev., 70:317–357

Müller K (1996) Allgemeine Systemtheorie. Studien zur Sozialwissenschaft 164. Opladen

Neumann J, Morgenstern O (1944) Theory of games and economic behavior. Princeton Univ. Press, Princeton, NJ
Nielsen SN (2000) Ecosystems as information systems. In: Jørgensen S, Müller F (eds) Handbook of ecosystem theories and management. Lewis, London/New York/Washington, DC
Odum E (1953, 1959, 1971) Fundamentals of ecology. Saunders, Philadelphia
Odum HT (1956) Primary production in flowing waters. Limnology and Oceanography 1:102–117
Odum EP (1969) The strategy of ecosystem development: an understanding of ecological succession provides a basis for resolving man's conflict with nature. Science 164:262–270
Odum HT (1971) Environment, power and society. Wiley, London
O'Neill RV, DeAngelis DL, Waide JB, Allen TFH (1986): A hierarchical concept of ecosystems. Princeton Univ. Pr., Princeton, NJ
Parsons T (1937) The structure of social action. McGraw-Hill, New York
Pace ML, Groffman PM (eds) (1998) Successes, limitations, and frontiersn in ecosystem science. Springer, New York
Patten BC (1959) An introduction to the cybernetics of the ecosystem: the trophic dynamic aspect. Ecology 40:221–231
Patten BC, Odum EP (1981) The cybernetic nature of ecosystems. Am Nat 118:886–895
Peus F (1954) Auflösung der Begriffe "Biotop" und "Biozönose". Deutsche Entomologische Zeitschrift N F 1:271–308
Phillips J (1934,1935) Succession, development, the climax, and the complex organism: an analysis of concepts. Part 1–3. J Ecol 22:554–571, 23: 210–246, 3: 488–508
Pias C & Foerster H (eds) (2003) Cybernetics: the Macy-Conferences 1946–1953. Diaphanes, Zürich
Pomeroy LR, Alberts JJ (eds) (1988) Concepts of ecosystem ecology. Springer New York
Prigogine I (1955) Introduction to thermodynamics of irreversible processes. Thomas, Springfield
Ramensky LG (1926) Die Gesetzmäßigkeiten im Aufbau der Pflanzendecke. Botanisches Centralblatt N F 7:453–455
Rapoport A (1947) Mathematical theory of motivation of interactions of two individuals. Bulletin of Mathematical Biophysics 9,1:17–27
Rapoport A (1950) Science and the goals of man: a study in semantic orientation. Harper, New York
Recknagel F (ed) (2003) Ecological informatics: understandig ecology by biologically-inspired computation. Springer, Berlin
Shannon CE, Weaver W (1949) The mathematical theory of communication. University of Illinois Press, Urbana, Illinois
Schwarz AE (1996) Aus Gestalten werden Systeme: Frühe Systemtheorie in der Biologie. In: Mathes K, Breckling B, Eckschmitt K (eds) Systemtheorie in der Ökologie. Landsberg, pp 35–45
Tansley AG (1935) The Use and abuse of vegetational concepts and terms. Ecology 16(3): 284–307
Taylor P (1988) Technocratic optimism, H.T. Odum, and the partial transformation of ecological metaphor after World War II. – J. Hist. Biol., 21(2):213–244
Thienemann A, Kieffer JJ (1916) Schwedische chironomiden. Arch. hydrobiol. 2(Suppl):489
Tobey RC (1981) Saving the prairies. University of Carlifonia, Berkeley
Trepl L (1987) Geschichte der Ökologie. Vom 17. Jahrhundert bis zur Gegenwart. Athenäum, Frankfurt a. M.
Ulanowicz RE (1997) Ecology, the ascendent perspective. Columbia University Press, New York
Vogt KA, Gordon JC, Wargo JP, Vogt DJ, Asbjorsen H, Palmiotto PA, Clark HJ, O'Hara JL, William S-K, Toral P-W, Larson B, Tortoriello D, Perez J, Marsh A, Corbett M, Kaneda K, Meyerson F, Smith D (1997) Ecosystems: balancing science with management. Springer, New York

Voigt A (2001) Ludwig von Bertalanffy: Die Verwissenschaftlichung des Holismus in der Systemtheorie. Verhandlungen zur Geschichte und Theorie der Biologie 7:33–47

Volterra V (1926) Variazioni e fluttuazioni del numero d'individui in specie animali conviventi. Mem. Accad. Lincei series 6, 2(36):31–113

Weil A (1999) Über den Begriff des Gleichgewichts in der Ökologie - ein Typisierungsvorschlag. Unversitätsverlag, TU Berlin, Berlin

Wiener N (1948) Cybernetics or control and communication in the animal and the machine. Wiley, New York

Worster D (1994) Nature's economy: a history of ecological ideas. Camb. Univ. Pr., Cambridge

Chapter 16
Ecology and the Environmental Movement

Andrew Jamison

Introduction

With the emergence of the environmental movement in the 1960s, the science of ecology was transformed from being a relatively minor sub-field of biology into an object of political engagement and public interest. For a brief historical moment, ecology became more than a mere science; it became a component part of what I have previously characterized as an emerging ecological culture (Jamison 2001). Even though many of the political struggles that brought it into being have faded into the past, the environmental movements of the 1960s and 1970s continue to influence scientific ideas and personal values, as well as broader socio-political discourses.

Ecology became a kind of "super-science" that was considered to have a crucial role to play in the newfound mission, or political project, of environmental protection. From various points along the political spectrum, it was felt that the science of ecology could provide concepts, theories, and methods that could help guide society into a more sustainable, environmentally-friendly direction. These developments were initiated in the United States but spread to Europe in the course of the 1970s as environmental movements took shape in many countries. The influence of a politicized ecology was particularly strong in the Scandinavian countries, where there were indigenous traditions of both ecological science and environmental politics dating back to the eighteenth century.

Opinions varied, however, in regard to just what the science of ecology had to offer to the broader politics of the environment. According to the distinction made by the Norwegian philosopher Arne Naess in 1972, there was both a "shallow" and a "deep" version of ecology that could be found in what was starting to be considered an environmental movement. Where the shallow ecologists went to the science primarily in search of operational concepts and administrative tools with which to carry out their political struggles, the deep ecologists took their point of departure in

---

A. Jamison (✉)
Department of Development and Planning, Institut for Samfundsudvikling og Planlægning, Aalborg University, Fibigerstræde 13, 9220 Aalborg, Denmark
e-mail: andy@plan.aau.dk

the science for the creation of a new world-view, or belief system, an ecological philosophy. In a similar vein, the American activist Murray Bookchin distinguished between what he termed "environmentalists" and "ecologists" within the emerging movement (e.g. Bookchin 1982). Bookchin's argument was that the environmentalists reacted to particular cases of environmental destruction, while the ecologists reacted to the underlying social and political conditions behind the particular cases. Others referred to a tension in the emerging environmental consciousness between ecocentrists and anthropocentrists, which was due, in large measure, to different meanings that were given to the science of ecology in its relation to environmental politics.

Wherever environmental movements developed, the science of ecology came to play a significant role in the collective identity, or cognitive praxis, of the activists and their activities. Ecology, in various ways, took on an ideological function, rather than, or in addition to, its more traditional scientific role in society. The "use" of ecology for political purposes proved to be problematic, however, and, in the course of the 1980s, the science of ecology and the environmental movement more or less parted company. The discourse of sustainable development tended to replace ecology as an overarching ideology, or political doctrine, for environmental activists and green party politicians, and most ecologists tended to disavow, or at least, disassociate themselves from the explicit political meanings that had been attributed to their science. But the links that were established in the 1970s have continued to affect both the science of ecology, as well as the broader politics of the environment. Indeed, in the recent challenges to the scientific understanding of climate change from self-declared skeptics such as the Danish political scientist Bjørn Lomborg, the views of environmental scientists and environmental activists are conflated with one another. For skeptics such as Lomborg, the statements of ecologists and other environmental scientists are not to be believed, at least in part because of their association with environmental organizations, such as Greenpeace (Lomborg 2001).

The aim of this article is to discuss the interactions between ecology and the environmental movement with particular reference to the way that ecology came to be a kind of "super-science" in the 1960s and 1970s.

The Traditions of Ecology

In an influential account, first written in the 1970s, Donald Worster pointed to two main streams of thought that had come together in the environmental movement, two opposing attitudes to nature that had led, through the centuries, to two different kinds of ecology, or ecological traditions (Worster 1979). He traced an "imperialist" tradition back to the writings of Francis Bacon in the early seventeenth century and his ideas about the human domination of nature. Carl Linneaus and Georges Buffon in the eighteenth century helped give this imperialist ecology a more systematic and scientific form. Nature was conceptualized in mechanical and instrumental terms, which helped make possible the effective utilization of nature for human exploitation. In the course of industrialization, this highly utilitarian view of non-human reality became the dominant discourse, or philosophy, of nature, especially in the natural sciences, as they took on a more professional and disciplined organizational identity. The Linnean system provided methods that were used both for scientific research and for more technical kinds of work. The various plants and animals were given names, which made it easier to understand and analyze their functions and interrelations, and they were also given structural characteristics, which proved useful for conducting practical experiments, such as breeding new plant varieties or exploiting new natural resources. The imperialist tradition represented the experimental, systemic approach to knowledge-making, which entered the science of ecology, when it was given a more formalized identity as a sub-field of biology around the turn of the century in both the United States and several European countries.

Opposed to the imperialists were the nature-lovers, to whom Worster gave the label "arcadian" in order to associate their version of ecology to the classical ideal of harmony between nature and society that had been depicted by Roman poets in the Greek region of Arcady. According to Worster, the back-to-nature folks began to articulate their counter-program at the dawning of the industrial era, as part of the Romantic movement. The arcadians shared many of the modernizing, scientific ambitions of the imperialists, but they came to develop a different way of investigating and understanding nature. Tracing arcadians back to the English pastor and writer Gilbert White, and especially to his work, "The Natural History of Selborne", originally published in 1789, Worster delineated a stream of experiential, or participatory, ecology that was perhaps most influentially developed further by Henry David Thoreau in the nineteenth century. In Germany and the Scandinavian countries, a related tradition of "*Naturphilosophie*" (or philosophy of nature) emerged in the late eighteenth century among academics and artists; this romantic tradition had an influence in both geology and geography, biology and chemistry, and even in physics, where the search for an underlying "spirit" in nature led the Danish scientist Hans Christian Ørsted to discover the connection between electricity and magnetism in 1820.

Worster's argument was that the two ecological traditions had both contributed to Charles Darwin's theory of natural evolution, but that they had subsequently given rise, in the course of the twentieth century, to two different ways of thinking about ecology and conducting ecological research. The one was systemic, while the other was individual in focus, and they fostered a large-scale, ecosystems oriented ecology, on the one hand, and a smaller-scale, population-oriented ecology on the other, the one taking its point of departure in the systemic relations that exist among species, and the other taking its point of departure in the dynamic relations of one species to its environment. The two traditions drew on different attitudes, or conceptions of nature, as well as different methodological and theoretical assumptions about how to investigate, or interrogate nature.

Worster's division into an imperialist and arcadian ecology captures a fundamental contradiction in the history of ecology. But it tends to disregard a third important source of inspiration for the environmental movement, as it developed in the 1960s and 1970s, namely the various "human ecologies" that had emerged in the nineteenth and early twentieth centuries, both in Europe and in the United States.

Table 16.1 Ecological traditions

	Imperialist	Arcadian	Human
Formative influences	Francis Bacon Carolus Linnaeus	Gilbert White Henry D Thoreau	George Marsh Lewis Mumford
Key mobilizers	Odum brothers Gro H Brundtland	Rachel Carson Arne Næss	Paul Ehrlich Barry Commoner
Type of sciencing	Systemic models experimentation	Natural history thick description	Mapping surveying
Relation to nature	Exploitation management	Participation harmony	Planning co-construction
Conception of nature	Ecosystem resource base	Community locality	Region landscape
Ideologies	Anthropocentric/modernism	Ecocentric/deep ecology	Pragmatic/postmodern

In part motivated by the journeys of exploration among biologists and geographers to South America and in the North American frontier, in part an outgrowth of infrastructural projects and urban planning and, in part, a sub-field of public medicine and public health, these human ecologies entered into the new social sciences of sociology and anthropology, of economics and political science, and of geography and planning as they grew into important fields in the course of the twentieth century.

To make the story somewhat more complete, it can therefore be useful to add a third tradition to Worster's two, and to distinguish three ideal-typical ecological traditions that have been mobilized in the making of the environmental movement. Each tradition – the imperialist, the arcadian, and the human – has its own characteristic conception of nature and its own preferred methods of investigation, as well as its own distinct version of an appropriate ecological practice or politics) Table 16.1).

The Mobilization of Traditions

It would be these three traditions and the various sub-sets thereof that came to be mobilized in the 1960s in the making of a new social movement. On the one hand, the imperialist tradition was reinvented, among other places, in the cybernetic language of ecosystems ecology and energy systems analysis (see Chap. 15). Systems ecology, as developed by Eugene and Howard Odum, became extremely influential among natural scientists, particularly during the International Biological Program, and, as a new approach to ecology, it would play a major role in the emergence of an environmental consciousness in the 1960s (Worster 1979, pp. 291ff.).

The Odums also illustrate the importance of established scientists in the articulation of the new environmentalism's collective identity, or cognitive praxis (Cramer et al. 1987). The environmental movement involved, at the outset, a kind of popularization of science, as well as a translation of concepts and terminology that had been developed in understanding non-human nature to human societies. Eugene and Howard Odum's popular writings provided a scientific legitimacy and authority to the new movement, as well as a powerful terminology and conceptual framework, while the movement helped provide ecological scientists with new opportunities for research, and a new political mission: ecologizing society (Söderqvist 1986).

In the World Wildlife Fund, founded in 1961, and then more scientifically in the International Biological Program, that was established in the mid-1960s, the "imperialist" tradition took on a more modern or contemporary manifestation. It became more explicitly international, as scientists and other conservationists came to take part in transnational research and development networks, particularly within the IBP projects (Kwa 1989). The imperialist tradition was also brought up to date technologically with the new cybernetic and computer-based approaches to research that were developed by the new breed of ecosystem ecologists, led by Eugene and Howard Odum. Particularly in the energy-flow schematizations of Howard Odum, the mathematical modeling of nature and society was presented in an ambitious and sophisticated manner that was to have a major importance on the emerging environmental consciousness.

It would be the biologist turned science writer Rachel Carson, whose eloquent writings would do most to give the arcadian tradition a contemporary resonance. "Over increasingly large areas of the United States", she wrote, "spring now comes unheralded by the return of the birds, and the early mornings are strangely silent where once they were filled with the beauty of bird song" (Carson 1962, p. 97). Her book "Silent Spring" served to awaken the industrial world from its postwar slumbers, and she was soon followed by other writers who, with their scientific pondus and more sober tone, had a somewhat different impact on the public consciousness.

After writing two best-selling nature books in the 1950s, Carson had grown concerned about the impact that the new chemical insecticides were having on the forests and on the animals that she loved so much. Her 4-year investigation of the environmental consequences of one of those pest-killers resulted in a new form of political broadside, a book of scientific poetry that brought the arcadian science of Thoreau, whose writings Carson admired so much, into the twentieth century. In any case, "Silent Spring" was to have a major influence on the cognitive praxis of the emerging environmental movement (Jamison and Eyerman 1994, pp. 92ff.). But it would also inspire a new generation of arcadian ecologists to reframe their message and challenge the more "technocratic" approaches of the systems ecologists. To a large extent, the historical dichotomy between the imperialists and the arcadians would be replayed in the tensions over direction and orientation in the fledgling environmental movement organizations that developed in the late 1960s and early 1970s.

The mobilization of human ecology came from many different directions. Some, like Murray Bookchin, who had been a labor activist in the 1930s, brought a socialist sensibility into the environmental movement. His book from 1963, "Our Synthetic Environment", was one of the first to present the wide range of new environmental problems – occupational health, chemical pollution, household risks, waste disposal – that were to gain increasing public attention in the years to come. Others, like the biochemist Barry Commoner, gave the environmental movement a more technical emphasis; Commoner depicted, in his first book, "Science and Survival" (1966), the subservient role that science was playing in society and production, and suggested a number of public service, or critical, activities for scientists to play in the emerging movement. The biologist Paul Ehrlich resurrected the Malthusian message of population pressures and resource limitations in his book, "The Population Bomb" (1968), and the different perspectives of Commoner and Ehrlich would subsequently combine in new activist organizations and environmental studies departments. Still others, like Lewis Mumford, would provide historical and philosophical perspectives to help understand the new environmental problems. As such, the human ecology tradition was also reinvented, or mobilized, in the 1960s.

Already then, however, the seeds were sewn for the differentiations that have followed. The mobilization of traditions did not lead to one coherent movement, but rather to different kinds of hybrid identities which, for the sake of simplicity, we can think of as practical, cultural, and political. The practical, or technical environmentalists, have reinvented the human ecological tradition with infusions of advanced technology, but also with influences from the other ecological traditions. And while the cultural and the political environmentalists have both identified most strongly with one of the earlier traditions, they too have combined the older idea and perspectives with new ingredients. For the culturalists, the hippie influence, and the more general critique of technocratic society and its "one-dimensional" thought, have been of fundamental importance; while, for the political environmentalists, the "globalizing" tendencies of capitalism, and the development of telecommunications and media technologies have been extremely significant factors in their professionalization.

The Age of Ecology

By the end of the 1960s, ecology had inspired both the emergence of new activist groups, such as Friends of the Earth, as well as a process of policy reform and institution building. In the early 1970s, most of the industrialized countries established new state agencies to deal with environmental protection, and environmental research and technological development were organized in new locations in both the private and public sectors – often in the name of ecology. Many national parliaments enacted more comprehensive environmental legislation and, at the United Nations Conference on the Human Environment in Stockholm in 1972, protecting the environment was recognized as a new area of international concern.

The manned landing on the moon in 1969 had provided the symbol for the conference, the blue planet viewed from space: small, fragile, and strikingly beautiful in its shape and color. A biologist, René Dubos, and an economist, Barbara Ward, collaborated on the book that would set the agenda for the conference. "Only One Earth" (1972), their book was called, and in it they made the case for a new kind of environmentalism, combining efficient management of resources with empathetic understanding: "Now that mankind is in the process of completing the colonization of the planet", they wrote, "learning to manage it intelligently is an urgent imperative.

Man must accept responsibility for the stewardship of the earth" (Ward and Dubos 1972, p. 25). They noted, in conclusion, that the reforms and policy proposals that they suggested in their book would not come easily: "The planet is not yet a centre of rational loyalty for all mankind. But possibly it is precisely this shift of loyalty that a profound and deepening sense of our shared and inter-dependent biosphere can stir to life in us" (Ward and Dubos 1972, p. 298).

Over the next few years, ecology would come to be drawn upon by almost all sides in the new environmental politics. There was also a range of "grass-roots" engineering initiatives that emerged in the fledgling environmental movement, a kind of ecological technology. In the United States, a group of self-proclaimed "new alchemists" moved from the university out to the country to experiment with ecological agriculture and renewable energy (Todd 1977). In many European countries, but perhaps especially in Britain, Denmark and the Netherlands, a number of research centers and projects in alternative or ecological technology were established. Also in Switzerland and in Germany, the so-called anthroposophic movement of Rudolf Steiner, that had been started in the 1920s with its biological-dynamic agri- and horticulture was reinvigorated, as were the organic-biological methods invented by the Swiss Hans Müller in the 1950s. At some of the "hippie" communes and production collectives that developed at the time, there was often an interest in energy and agriculture, and there were also, among architects and planners, attempts to develop more environmentally-friendly, or ecological approaches and techniques (Dickson 1974).

An interest in ecological science and technology developed as an integral part of the environmental movement in several countries, as ecological research was given greatly increased funding, and programs in environmental education were established at many universities. In retrospect, we can see that the environmental movement opened a public space for experimentation with a collective mode of science and engineering – or what Ivan Illich called at the time "tools for conviviality" (Illich 1973) – in relation to energy, agriculture, housing, and transportation. The particular technical interests have diffused widely into society – for are we not all a bit more "ecological" in the ways we garden, and decorate our homes, and move ourselves around? – while the collective creativity has largely dissipated. While new scientific institutions eventually emerged out of the movement, most of the attempts that were made in the early 1970s to "use" ecology directly for purposes of environmental decision making proved unsuccessful. But even more significant in the long run was the coming of the oil crisis and the central role that the struggle over nuclear energy came to play in the second half of the 1970s in both Europe and North America.

New kinds of disciplines, or sub-disciplines developed in many countries. Energy systems analysis became a recognized field for investigating the costs and benefits of different choices of energy supply and distribution. Human ecology took on the form of a recognized academic field, and developed its own theories, based on concepts of entropy and energy flow. Within ecology itself, a kind of bifurcation took place between ecosystem ecologists, on the one hand, who were often drawn into larger, multidisciplinary projects, and population, or evolutionary ecologists, on the other, who focused their attention on particular species or ecological communities. There was a still further specialization, due to the range of approaches that emerged in the established disciplines to take on the new environmental and energy issues.

The Politicization of the Movement

It was the first oil crisis, of 1973–74, that led to a major shift in environmental consciousness, as energy issues moved to the top of many national political agendas, especially in relation to nuclear energy. In many countries, the late 1970s were a period of intense political debate and social movement activity, as the pros and cons of nuclear energy, or "hard energy paths" in general, were contested (Lovins 1977). In certain countries, like Denmark, renewable energy experimentation became a social movement of its own, and led to new industries and government programs. Seen in retrospect, an important result of the energy debates of the 1970s was a professionalization of environmental concern and an incorporation by the established political structures of what had originally been a somewhat delimited political issue. As a result, there was both a specialization and institutionalization of knowledge production.

What, in retrospect, is most characteristic of the 1970s is the breadth, but also the unity and coherence, of the environmental movement. As a popular front, or campaign, against nuclear energy, the different traditions of ecology were combined into an integrative *cognitive praxis*, with a visionary ecological philosophy, or world-view guiding a range of practical experiments with alternative technology in settings that were largely autonomous, outside of the formalized rule systems and organizational frameworks of the larger society. In informal local groups, and movement-based workshops and study circles, technical projects and educational activities were conducted with participation of both "experts" and amateurs. The key point is that the movement, for a brief time, could provide an organized learning experience, in which theory and practice were combined in pursuit of a common, collective struggle. These settings would be difficult to maintain for very long, since they were, in many ways, too unstable for any kind of permanent institutionalization, and when the issues that inspired the movement were resolved, and taken off the political agenda, the different component parts split apart and fragmented (Cramer et al. 1987). The unity that had been achieved in struggle could simply not be sustained.

The challenge to the coherence of the movement was also, to a large extent, a result of the broadening and diversification of environmentalism in the late 1970s. While most activists in Europe were concerned with nuclear energy, which became an issue of major political importance in several countries, other issues were important and inspired new forms of activism in other parts of the world. In the United States, the discovery of toxic wastes buried under the neighbourhood of Love Canal in Buffalo, New York, inspired a new kind of locally-based, working class opposition to environmental pollution (Szasz 1994). It was also in the United States that the new techniques of genetic engineering were critically reviewed by activists for their risks and dangers to the communities in which the laboratory experiments were carried out. In opposing genetic engineering, environmentalists, such as Jeremy Rifkin, pointed to a new kind of futuristic challenge that made many of the actual environmental problems pale in significance. "Two futures beckon us",

Rifkin wrote, "[w]e can choose to engineer the life of the planet, creating a second nature in our image, or we can choose to participate with the rest of the living kingdom. Two futures, two choices. An engineering approach and an ecological approach" (Rifkin 1983, p. 252).

These new forms of environmentalism were difficult to contain within one unified movement; rather, in the intellectual and broader political traditions that they tapped into, and in the alliances that they made, sometimes with quite conservative and religiously fundamental groups, they were often articulating interests and strategies that were diametrically opposed to the positions of "modernist" anti-nuclear activists, as well as many of the professional environmentalists in the think tanks and the "mainstream organizations". As such, while the environmental movement was growing and expanding and diversifying, the seeds were sewn for a more explicit process of differentiation in the 1980s.

The tension between those who see the environmental crisis as fundamental and those who see it as merely one of many challenges facing modern society has been a defining feature of the environmental movement ever since, and it has exerted a strong influence on the way in which ecology has been used, or appropriated. The "fundamentalists" or deep ecologists tend to see ecology as an overarching philosophy or cosmology, while the pragmatists or realists often see ecology as a somewhat more limited scientific toolbox, providing methods and concepts for environmental managers and other scientists. Perhaps the most appropriate approach for environmentalists to follow in the future is to try to forge a new kind of hybrid identity that combines the passion, engagement and holistic thinking of the deep ecologists with the practical skills and professional rigor of the environmental managers.

Indeed, it might be suggested that it is in the hybridization of knowledge-making activities that the environmental movement, or environmental activism continues to have an impact on the development of ecology, and, more generally on the making of "green knowledge" (Jamison 2001). In recent years, in the quest for sustainable development, there have emerged a number of new cognitive combinations or transdisciplinary forms of knowledge production that make connections between the natural, social, and human sciences (Jamison 2005). As such, ecology has once again come to play a broader role in society, both in environmental politics and management (in such hybrid practices as industrial ecology and ecological economics), as well as within the larger culture (in relation to such hybrid "discourses" as ecological citizenship and ecological literacy).

References

Bookchin M (1982) The ecology of freedom. Cheshire Books, Palo Alto
Carson R (1962) Silent spring. Houghton Mifflin, Boston
Cramer J, Eyerman R, Jamison A (1987) The knowledge interests of the environmental movement and its potential or influencing the development of science. In: Blume S, Bunders J, Leydesdorff L, Whitley RP (eds) The social direction of the public sciences. Reidel, Dordrecht
Dickson D (1974) Alternative technology and the politics of technical change. Fontana, Glasgow

Illich I (1973) Tools for conviviality. Harper & Row, New York

Jamison A (2001) The making of green knowledge. Environmental politics and cultural transformation. Cambridge University Press, Cambridge

Jamison A (2005) Hybrid identities in the european quest for sustainable development. In: Paehlke R, Torgerson D (eds) Managing leviathan. Broadview Press, Peterborough

Jamison A, Eyerman R (1994) Seeds of the sixties. University of California Press, Berkeley

Kwa Chunglin (1989) Mimicking nature. The Development of Systems Ecology in the United States, 1950–1975. Dissertation, University of Amsterdam, Amsterdam

Lomborg B (2001) The skeptical environmentalist. Cambridge University Press, Cambridge

Lovins A (1977) Soft energy paths. Penguin, Harmondswoth

Rifkin J (1983) Algeny. Penguin, Harmondsworth

Szasz A (1994) Ecopopulism: toxic waste and the movement for environmental justice. University of Minnesota Press, Minneapolis

Söderqvist T (1986) The ecologists. From merry naturalists to saviours of the nation. Almqvist and Wiksell, Stockholm

Todd NJ (ed) (1977) The book of the new alchemists. E.P. Dutton, New York

Ward B, Dubos R (1972) Only one earth. The care and maintenance of a small planet. Andre Deutsch, London

Worster D (1979) Nature's economy. The roots of ecology. Anchor Books, Garden City

Chapter 17
Ecology and Biodiversity at the Beginning of the Twenty-first Century: Towards a New Paradigm?

Patrick Blandin

Introduction

Since the 1960s, Ecology is a matter of crucial concern, with people becoming increasingly aware of the increasing environmental crisis. In reality, as a scientific discipline, Ecology was first officially addressed when the International Union for the Protection of Nature (IUPN, now IUCN, with "C" for "Conservation") was created at the end of 1948: an agenda was drawn up, including the following topic: "the international cooperation for scientific research in the field of the Protection of Nature, especially concerning œcological research in the various fields of exact and natural sciences" (UIPN 1948, p. 15). The first concrete IUPN action was the organization, with UNESCO, of a technical conference, which took place at Lake Success (USA), in August, 1949. The Ecology Section of the conference was introduced by a French biologist, Georges Petit, who emphasized the fact that the relationships between the Protection of Nature and Ecology had been widely neglected, the Protection of Nature having been "considered for a long time only as the results of aesthetic or moral preoccupations" (Petit 1950, p. 304).

It is important to grasp the conceptual situation at this very moment when the conservationist movement met Ecology. The IUPN first General Secretary's viewpoint (Harroy 1949, p. 10) illustrates conservationists' expectations: "In order to efficiently protect the natural associations which are useful, Man must have them carefully studied beforehand. But to study these associations in the best conditions, I would say "in the state of a pure body", he must have protected them before, that is, in appropriate and sufficiently vast areas, to have them shielded from disturbing human influences which mask and distort the fundamental processes that the researcher attempts to observe and to order into laws". This is symptomatic of an ideology which considers man as an external factor, whose actions disturb the natural equilibrium – the so called "Balance of Nature": – an ideology which is the descendant of Bacon's and Descartes's dualist philosophies. Rapidly, this ideology

---

P. Blandin (✉)
Muséum National d'Histoire Naturelle, Départment Hommes-Natures-Sociétés, 57, rue Cuvier, 75005 Paris, France
e-mail: patrick.blandin@yahoo.fr

fit with the developing systemic and cybernetic approach of Ecology, which bloomed under the umbrella of the Odumian Ecosystem Paradigm (Bergandi 1995).

At the turn of the twentieth century, the environmental crisis – of which the Biodiversity accelerated loss and the climatic change are emblems – radically questions the comfortable dualist occidental ideology: new ways of living with Nature are to be invented. In this context, the Biodiversity challenge calls for the renewal of Ecology, which is in search of a paradigm for the ongoing twenty-first century. In this paper, I shall briefly analyse this evolution of Ecology, focussing on the issue of biological diversity.

The Balance of Nature Ideology and the Ecosystem Stability

At the Lake Success IUPN Conference, the term "ecosystem" (Tansley 1935) was not used: the prevailing concern was about the consequences of different human activities on the "Balance of Nature"; the ecosystem concept came into common use later, after the publication of Odum's "Fundamentals of Ecology" (1953). In fact, Ecosystem Ecology developed independently of conservation issues, but shared with the conservation world the Balance of Nature Ideology, which claims that equilibrium is the "normal" state of Nature.

Rapidly, in the USA, thanks to the availability of the first digital computers, the use of ecosystem analysis and cybernetics models became the core of ecological research for decades (Golley 1991). For many ecologists, the ecosystem being conceived as an organized, "cybernetic" entity, its biotic community could not be considered as a random assemblage of species. The community would present a "structure", determined by interactions between the coexisting species – the number of which being probably ecologically regulated, depending on available resources to be shared by competing species –, and resulting in global properties at the ecosystem level. The Odumian Ecosystem Paradigm implies that the interactions between the components of any natural ecosystem, playing under the control of external and internal constraints, result in the stability of the mature "climax" ecosystem. Little by little, the intuitive idea of some relationships between species diversity, structural and functional complexity, and ecosystem stability naturally emerged. Therefore, during the 1960s and the 1970s, much research focussed on the structure of ecological communities, and many papers dealt with the problem of the relationships between species diversity and ecosystem characteristics and stability (for example: Leigh 1965; Paine 1966; Margalef 1969; Loucks 1970 or Smith 1972).

In 1973, in his Presidential address to the British Ecological Society, Amian Macfadyen emphasized the fact that the question of species diversity – ecosystem stability relationships being controversial, Ecology was still in an early stage and lacked accepted paradigms (Macfadyen 1975). The question was reviewed by Daniel Goodman (1975), who observed that "It would seem, at first sight, to be plausible that the balance of nature is more readily balanced when there are more interacting species present. This functional relationship may be conceived in terms of spare parts, more links to take up the slack, or, nowadays, more opportunity for feedback loops." (p. 238). But Goodman concluded that "the expectations of the diversity-stability hypothesis are borne out neither by experiments, by observation, nor by models" (p. 261). Actually, at the end of the 1970s the interest in species diversity as a global characteristic of communities was decreasing, perhaps because it was actually impossible to develop "popperian" research on the supposed ecological functions of species diversity.

Biodiversity Appears on the Scene

Diversity began a new life at the USA National Forum on BioDiversity, held in 1986, which had an immediate impact on the public and the media. Edward O. Wilson, in his Editor's foreword of the resulting book, entitled "BioDiversity", (1988a, p. v), pinpointed two "more or less independent developments" to explain the increased focus, among scientists and portions of the public, on conservation of biodiversity: (i) – "the accumulation of *enough* data on deforestation, species extinction and tropical biology"; (ii) – "the growing awareness of the close linkage between conservation of biodiversity and economic development". With the feeling that a richness, which was unimaginable only a few years before, could be widely destroyed within a few decades, a radical change occurred. This was remarkably expressed by Terry L. Erwin (1988, p. 127) : "[…] we should think in terms of more than 30 million, or perhaps 50 million or more, species of insects on Earth.[…]. The extermination of 50% or more of the fauna and flora would mean that our generation will participate in an extinction process involving perhaps 20–30 million species. We are not talking about a few endangered species listed in the Red Data books […]. No matter what the number we are talking about, whether 1 million or 20 million, it is massive destruction of the biological richness of Earth".

Edward O. Wilson (1988b) gave no precise scientific definition of "Biodiversity", but simply stated (p. 3) that: "Biological diversity must be treated more seriously as a global resource, to be indexed, used, and above all, preserved". This is symptomatic of a conservationist and taxonomist conception, regarding biodiversity as a global collection of species and genes. The defenders of this approach missed the opportunity of giving genuine scientific meaning to the concept of biodiversity. After 1988, efforts were made to elaborate scientific definitions, even by conservationist circles; for example, Jeffrey A. McNeely et al. (1990, p. 17) made the following proposal: "Biological diversity" encompasses all species of plants, animals and micro-organisms and the ecosystems and ecological processes of which they are parts. It is an umbrella term for the degree of nature's variety, including both the number and frequency of ecosystems, species or genes in a given assemblage. It is usually considered at three different levels: genetic diversity, species diversity, and ecosystem diversity".

In 1994, the International Union for Biological Sciences (IUBS) organized in Paris an International Forum called "*Biodiversity, Science and Development*". In their introduction, Francesco di Castri and Talal Younès (1996) tried to provide a rigorous definition of biodiversity. Considering the "three levels approach", they pointed out the fact that previous definitions of biodiversity paid "little attention, if any, to the interactions within, between and among the various levels of biodiversity" (p. 1). Stressing the fact that "interaction is the main intrinsic mechanism to shape the characteristics and the functioning of biodiversity" (p. 2), and considering that the interactions between the three levels of biodiversity are of a hierarchical nature, forming the "unique trilogy of biodiversity" (p. 3), they called for a general theory and for the development of a transdisciplinary scientific field. Consequently, they proposed a hierarchical definition: "A more sophisticated definition of biodiversity could be, therefore, the ensemble and the hierarchical interactions of the genetic, taxonomic and ecological scales of organization, at different levels of integration" (p. 4). Moreover, they considered that: "from a practical viewpoint, structural and functional attributes of system stability, productivity and sustainability, as well as patterns of ecosystem functioning [...], can only be clarified if hierarchies and scales are considered in terms of their interactions" (p. 5). They concluded that: "The real challenge lies in the possibility of taking into account the emerging properties that appear by the interactions of the three diversities" (p. 9).

In this way, di Castri and Younès adopted a resolutely holistic approach. They considered evident the existence of emerging properties, resulting from the hierarchic organization of biological systems. For example, they considered that the major attributes of ecosystems (stability, productivity and sustainability) can be understood only by taking into account interactions between the different levels of integration. Thus, from an epistemological point of view, this functionalist approach proposes a scientific challenge which is much more stimulating than the taxonomists' claim to the total inventory of all forms of life.

Nevertheless, the risk still exists that the functionalist approach supports a static view of a balanced Nature. Demonstrating that we actually have to understand a *"Non-Equilibrium World"*, John A. Wiens (1984, p. 440) made the following statement: "Ecology has a long history of presuming that natural systems are orderly and equilibrial (the "balance of nature" notion; [...]), and the infusion of evolutionary thinking into ecology strengthened this view, providing a mechanism (natural selection) that may lead to the development of optimally structured communities". It may appear paradoxical that an evolutionary thinking favoured the idea of an Equilibrium Word, but it is a fact that the underlying Balance of Nature Ideology supported the view of the "climax" ecosystem as a regulated, stable system, optimally composed by a characteristic assemblage of co-adapted species, and showing the same organization – therefore being homogeneous – everywhere within its boundaries. Therefore, spatial and temporal heterogeneity has been for a long time rather an obstacle than a topic for ecological research.

The Decline of the Equilibrium World

The idea of a global "Balance of Nature", of an "Equilibrium World", has probably not only philosophical, but also psychological roots. They could explain the success of the stability-diversity hypothesis which, as suggested by Goodman (1975, p. 261), "may have caught the lay conservationists' fancy, not for the allure of its scientific embellishments, but for the more basic appeal of its underlying metaphor. It is the sort of thing that people like, and want, to believe". In the context of this ideology, perturbations were generally considered as catastrophic events.

Landscape ecology favoured an important conceptual shift, introducing a new way of looking at the spatial organization and dynamics of ecological systems. Heterogeneity was then considered as an attribute of these systems, and disturbance as the driving process producing mosaic landscapes (Blondel 1995). Moreover, it became obvious that any region is characterized by a specific disturbance regime. These ideas took shape progressively, with such papers as those by Loucks (1970), White (1979) or Sousa (1984), and the book "*The Ecology of Natural Disturbance and Patch Dynamics*" (Pickett and White 1985). Relationships between disturbance, patch formation, community structure and species diversity were also addressed, for example, by Levin and Paine (1974) or Sousa (1979).

At the landscape level, recurring perturbations result in a patchy structure, where ecological units at different successional states coexist. Consequently, the ecological diversity and the whole species diversity associated with the landscape are supposed to be sustained by the disturbance regime. Therefore, the mosaic landscape was substituted for the climatic ecosystem as the effective equilibrium system, called "*metaclimax*", the dynamics of which, driven by the regional disturbance regime, was recognized as the sustaining process for the regional biodiversity (Blondel 1986). This shift from "ecosystem equilibrium" to "landscape equilibrium" was important, as it allowed spatial and temporal heterogeneity to gain a conceptual status. But the Balance of Nature Ideology remained in the background, as it is suggested by the creation of the "metaclimax" concept. That's why the statement made by Baker (1995, p. 157), on the basis of simulation studies, is of major interest: "Landscapes with long rotation times may be in perpetual disequilibrium with their disturbance regimes, because climatic change may create a new disturbance regime before the landscape has fully adjusted to the old regime". Thus, different landscapes, following different trajectories, should be in different situations, and may be arrayed along a gradient of states ranging from non-equilibrium to equilibrium.

In such a new context, the interpretation of species diversity of ecosystems has to change. Around the 1960s–1980s, under the umbrella of the "equilibrium competitive community paradigm" (Wiens 1984, p. 456), many ecologists considered that the diversity of a local community is more or less fixed at a level – the saturation point – above which the addition of immigrant species is balanced by the extinction of pre-existing ones. As stated by Robert E. Ricklefs (1987): "Present-day ecological investigations are largely founded on the premise that local diversity – the number of species living in a small, ecologically homogeneous area – is the deterministic outcome of local processes within the biological community" (p. 167). But Ricklefs also observed that: "Ecologists are beginning to realize that local diversity bears the imprint of such global processes as dispersal and species production and of unique historical circumstance. These processes pose a challenge to community ecologists to expand the geographical and historical scope of their concepts and investigations" (p. 167). Ricklefs emphasized the necessity to consider the balance between local and regional processes, as well as the balance between short term events and long term processes, to understand the species diversity on the local scale: "The presence or absence of a species depends on the outcome of processes tending to increase or decrease its numbers. The latter are generally local in nature [...]. Most interactions between species are antagonistic, and selection favors increased competitive ability and predator efficiency. Thus, evolution, while fostering greater accommodation among coexisting species, ultimately tends to reduce species richness. Balancing these negative factors is [...] the immigration of individuals from other areas. The variety of immigrants to a particular place depends on such regional processes as the generation and dispersal of new species (speciation) and also on historical accidents and circumstances related to past climate history and geographical position of dispersal barriers and corridors. The stronger speciation and dispersal are, relative to local factors influencing adjustment of population size and adaptation of individuals, the deeper the imprint of history and geography on the local community" (p. 169). Moreover, Ricklefs underlined the fact that the historical dimension of any ecological system results in the diversity of local situations.

Therefore, everywhere, biodiversity is a stage of "a unique evolutionary play" (Ghilarov 2000, p. 411). In this perspective, the local species diversity can't be considered only as the result of past evolutionary processes, but also as a potential for further evolution. As early as 1959, Hutchinson suggested that a diversified community would have a higher aptitude to evolve than a community including a low number of species: doing so, he introduced the idea that not only the stability of an ecosystem, but also its adaptability, could depend on its species diversity. Blandin et al. (1976) developed such ideas. They proposed to consider, schematically, two different adaptive strategies for ecosystems. On one hand, the adaptability of ecosystems with low specific diversity would depend on the genetic diversity – and consequently on the adaptability – of a few species carrying out keystone functions. On the other hand, the adaptability of ecosystems with high specific diversity would depend on the existence of functionally redundant species with different ecological aptitudes. The more numerous are the coexisting species, the lower the individual number per species may be: within an ecosystem, the genetic diversity – and, therefore, the adaptability – of each species population can not be independent of the number of species sharing space and/or trophic resources. Obviously, the two ecosystem strategies are not exclusive, and any intermediate situation may exist, depending on the actual number of coexisting species, with a particular hierarchical interaction between genetic diversity and specific diversity (Blandin 1980).

Similar ideas have been expressed more recently. For example, Tisdell (1995) compared tropical ecosystems, owning rich biodiversity but being at considerable risk with any environmental change (because many species have little biological tolerance and little mobility), with temperate ecosystems showing less biodiversity but species with greater tolerance and mobility. He concluded that biodiversity as such is neither necessary nor sufficient to ensure the sustainability of ecosystems. In a more precise way, di Castri and Younès (1996, p. 5), underlined the possible role of redundant species, and the balance between species diversity and genetic diversity: "Not all species are equal when it comes to measuring the biodiversity of a system: a few species can play a keystone role in system functioning, while others may be redundant; some species are dominant and can embrace a very large number of individuals, thus decreasing the equitability of the system, and others, rare species, may be present in a very low number of individuals. Also, a lower number of species can be compensated – to a certain extent – by a very high genetic variability in some populations".

Undoubtedly, the issue of species diversity and genetic diversity interdependence should help to overcome the controversial hypothesis of a direct relationship between species diversity and the stability of ecosystems. Moreover, it allows us to address the question of a community's capacity to evolve, even if the "Life Paradox" – life continuity resulting from life change – remains difficult to overcome. Actually, in the perspective of conservation debates, it is not useful, it is necessary to consider that evolution is the background and the horizon.

Biodiversity Dynamics: Towards a New Paradigm

Francesco di Castri and Talal Younès (1996, p. 3) called for "a general theory integrating the hierarchical levels of biodiversity, how they come to be and interact". The consideration of the relative roles of global, regional and local processes in a historical perspective, the emphasis on heterogeneity on different spatial and temporal scales, the evidence of various patterns of interdependence between species diversity and genetic diversity – resulting from history and present ecological contexts – provide a possible framework for such a theory, which can be outlined as follows (Blandin 2004).

The ecosphere's trajectory is chaotic. At any time the future is unpredictable, at least for the long term, but the past can be explained by deterministic processes. An interacting web of global and local processes continuously produces new species and provokes the extinction of others. At the Earth scale, the balance between species origination and species extinction processes results in the global dynamics of biodiversity.

The number of species living within a region results, on one hand, from extra-regional species originations and subsequent immigrations, and from intra-regional species originations. On the other hand, these processes are counterbalanced by intra-regional extinctions and the shifting of species out of the region, for example in relation with climatic changes.

Within a regional heterogeneous landscape, the richness of a local assemblage of species depends first on the regional stock of the species which are capable of coexisting. Secondly, it depends on the landscape structure, which governs the migration flows between ecological units. The actual coexistence of populations of different species depends on the physical and chemical constraints of the environment, and on the local flows of resources that can be exploited by these species, according to their needs. It also depends on the variety and regime of disturbances. Lastly, catastrophic events may produce dramatic changes.

Taking into account these new insights, natural communities can be viewed as being arrayed along a gradient of situations resulting on one side from a history of stochastic events, on the other from long environmental stability, with intermediate situations resulting from more or less long periods characterized by more or less regular disturbance regimes. According to such different histories, the relative roles in patterns of selection, of environmental characteristics, of trophic flows and of inter-specific interactions may have changed widely: biotic coupling, allowing co-selection, is unlikely under stochastically changing conditions, but could be possible in recurrent contexts, giving rise to co-adapted species.

In each situation, each species is distributed through the landscape mosaic in a particular way, and participates in assemblages of species interacting in a more or less strict manner, according to the degree of co-adaptation reached through evolution. The genetic diversity of each species has a particular pattern, depending on the number and size of populations, on the local flows of individuals between populations and on the balance between immigration and emigration. The local sustainability of the species depends on this genetic diversity. The size of each species population is partially determined by the number and size of the populations of other species sharing space and resources: there are necessary interactions between the genetic diversity of species and the species diversity of assemblages, in the framework of the local ecological diversity. Therefore, the adaptative capacities of the biotic communities – on which depends their sustainability – are governed by the present pattern of interaction between the three biodiversity levels.

At any time, on any spatial scale, the global biodiversity is the unique heritage of past evolution, and the unique, limited potential for further evolution. At any time, many different trajectories are possible, but only one will be followed. As local, regional and global processes are continuously interacting, the evolution of ecological systems must be considered as a "web of interdependent trajectories": what we can call the "Transactional Trajectories Paradigm" – using "transactional" in the spirit of Dewey's epistemology (Dewey and Bentley 1949) – is taking the place of the Odumian Ecosystem Paradigm, deep-rooted in the Balance of Nature Ideology.

The Sustainable Adaptability of the Ecosphere

Evolution cannot be considered only in the limited sense of species originations and extinctions: it is a global process of ecological changes, co-evolutionary interactions and biodiversity transformation. Broadening the perspective at the Ecosphere level, the Israeli ecologist Zev Naveh (2000) has suggested that "The Total Human Ecosystem should be regarded as the highest coevolutionary ecological entity on Earth" and he maintained that "This conceptual approach enables us to view the evolution of Total Human Ecosystem landscapes in the light of new holistic and transdisciplinary insights into the dynamic process of self-organization and coevolution in nature and human societies" (p. 358). Then, the Life Paradox could be overcome saying that coevolution makes possible life's sustainability, with new species substituting others to perform continuously ecological functioning. This implies an ideological shift, "sustainability" being substituted for "conservation" as the fundamental aim.

This evolutionary perspective opens new insights. As the unique memory of the past evolution, with the scattered remains of the paleobiodiversity, the present biodiversity offers keys for understanding evolutionary processes, as well as opportunities for wonder. With the Russian ecologist Alexei M. Ghilarov (2000), we can say that it is "an example of evolutionary heritage that is probably worth protection no less than the heritage of our culture" (p. 411). Moreover, as the unique potential for further evolution, the present biodiversity is the unique available man's companion for the co-evolutionary adventure to come. To satisfy, through the generations, humans' needs implies the permanent availability of natural "resources" able to match the cultural diversity of humans' desires. Today, humans' aims and projects are diverse; tomorrow, they will be different; nobody can predict what future generations, living in new ecological and cultural contexts, will need and desire. Therefore, the main problem is to transmit, through generations, a "four-levels diversity" allowing the sustainability of "man-nature systems": this implies not only functional continuity, but also the maintenance of a capacity to evolve, which depends on the integration of climatic, specific, genetic and cultural diversities within man-nature systems. Obviously, research is needed to provide new insights into the interactions of these four diversities, on all spatial scales. Thus, no longer can Ecology be merely a "biological" or a "natural" science: Ecology must evolve in order to contribute to the construction of a transdisciplinary field, dealing with integrated natural and socio-cultural trajectories. This is necessary to provide the scientific bases that humans need to ensure the "sustainable adaptability of the Total Human Ecosystem".

References

Baker WL (1995) Longterm response of disturbance landscapes to human intervention and global change. Landscape Ecol 10(3):143–159

Bergandi D (1995) 'Reductionist holism': an oxymoron or a philosophical chimaera of E.P. Odum's systems ecology. Ludus Vitalis 3(5):145–180, reprinted in Keller, D.R. & F.B. Golley (eds.) 2000. The philosophy of ecology: from science to synthesis (abridged version), University of Georgia Press, Athens, pp 204–217

Blandin P (1980) Evolution des écosystèmes et stratégies cénotiques. In: Barbault R, Blandin P, Meyer JA (eds) Recherches d'écologie théorique. Les stratégies adaptatives. Maloine, Paris, pp 221–235

Blandin P (2004) Biodiversity, between science and ethics. In: Hanna S, Mikhail WZA (eds) Soil zoology for sustainable development in the 21st century. Egypte, Cairo, pp 3–35

Blandin P, Lecordier C, Barbault R (1976) Réflexions sur la notion d'écosystème: Le concept de stratégie cénotique. Ecol Bull 7:391–410

Blondel J (1986) Biogéographie évolutive. Masson, Paris

Blondel J (1995) Biogéographie. Approche écologique et évolutive. Masson, Paris

Dewey J, Bentley AF (1949) Knowing and the known. Beacon, Boston di Castri F, Younès T (1996) Introduction: Biodiversity, the Emergence of a New Scientific Field–Its Perspectives and Constraints. In: di Castri F, Younès T (eds) Biodiversity, science and development. Towards a new partnership. CAB International and IUBS, Paris, pp 1–11

Erwin TL (1988) The tropical forest canopy. The heart of biotic diversity. In: Wilson EO, Peter FM (eds) Biodiversity. National Academy Press, Washington, DC, pp 123–129

Ghilarov AM (2000) Ecosystem functioning and intrinsic value of biodiversity. Oikos 90:408–412

Golley FB (1991) The ecosystem concept: a search for order. Ecol Res 6:129–138

Goodman D (1975) The theory of diversity-stability relationships in ecology. Q Rev Biol 50(3):237–266

Harroy JP (1949) Définition de la protection de la nature. In: UIPN (ed) Documents préparatoires à la conférence technique internationale pour la protection de la nature. UNESCO, Paris, Bruxelles, pp 9–14

Hutchinson GE (1959) Homage to Santa Rosalia, or why are there so many kinds of animals? Am Nat 93:145–159

Leigh EG (1965) On the relationship between productivity, biomass, diversity, and stability of a community. Proc Natl Acad Sci USA 53:777–783

Levin SA, Paine RT (1974) Disturbance, patch formation, and community structure. Proc Natl Acad Sci USA 71:2744–2747

Loucks OL (1970) Evolution of diversity, efficiency and community stability. Am Zool 10:17–25

Macfadyen A (1975) Some thoughts on the behaviour of ecologists. J Anim Ecol 44:351–363

Margalef R (1969) Diversity and stability: a practical proposal and a model of interdependence. Brookhaven Symp Biol 22:25–37

McNeely JF et al (1990) Conserving the world's biological diversity. IUCN/WRI, CI, WWF-US, The World Bank, Gland, Switzerland, Washington, DC

Naveh Z (2000) The total human ecosystem: integrating ecology and economics. Bioscience 50(4):357–361

Odum EP (1953) Fundamentals of ecology. W.B. Saunders, Philadelphia

Paine RT (1966) Food web complexity and species diversity. Am Nat 100:65–75

Petit G (1950) Protection de la nature et écologie. In: IUPN (ed) International technical conference on the protection of nature, Lake Success, 22-29-VIII-1949, proceedings and papers. UNESCO, Paris, Bruxelles, pp 304–314

Pickett STA, White PS (eds) (1985) The ecology of natural disturbance and patch dynamics. Academic, New York

Ricklefs RE (1987) Community diversity: relative roles of local and regional processes. Science 235:167–171

Smith FE (1972) Spatial heterogeneity, stability and diversity in ecosystems. Trans Conn Acad Arts Sci 44:309–335

Sousa WP (1979) Disturbance in marine intertidal boulder fields: the nonequilibrium maintenance of species diversity. Ecology 60:1225–1239

Sousa WP (1984) The role of disturbance in natural communities. Annu Rev Ecol Syst 15:353–391

Tansley AG (1935) The use and abuse of vegetational concepts and terms. Ecology 16:284–307

Tisdell CA (1995) Issues in biodiversity conservation including the role of local communities. Environ Conserv 22(3):216–222

UIPN (1948) Union internationale pour la protection de la nature, créée à Fontainebleau le 5 octobre 1948. UIPN, Bruxelles

White PS (1979) Pattern, process and natural disturbance in vegetation. Bot Rev 45:229–299

Wiens JA (1984) On understanding a non-equilibrium world: myth and reality in community patterns and processes. In: Strong DR Jr, Simberloff D, Abele LG, Thistle AB (eds) Ecological communities: conceptual issues and the evidence. Princeton University Press Princeton, New York, pp 439–457

Wilson EO (1988a) Biodiversity. National Academy Press, Washington, DC, pp v–vii

Wilson EO (1988b) Biodiversity. National Academy Press, Washington, DC, pp 3–18

Chapter 18
An Ecosystem View into the Twenty-first Century

Wolfgang Haber

Introduction: Ecosystem Between Recognition and Disputation

The term "ecosystem", introduced by Arthur G. Tansley[1] in his epoch-making article of 1935, has provided us with a useful and promising concept to be applied for investigative understanding and solving the growing environmental problems of the twenty-first century. For both institutions and persons bearing responsibility for a sustainable development, usage of "ecosystem management" and "ecosystem services" has become their common language and means of communication. The "Millennium Ecosystem Assessment" launched by the United Nations at the turn of the millennia (MEA 2003) will further strengthen general attention for that rather abstract term and turn it into a self-evident component of the twenty-first century's everyday publicity.

Among environmental scientists, however, notably in the disciplines of ecology, biology, and geography, and their theorists and practitioners, "ecosystem" has been, and is being debated as a more or less contentious term. In order to bridge upcoming differences between scientific rigour and clarity on one side, and general, transdisciplinary usage of "ecosystem" in public affairs on the other, and to avoid further misunderstandings and errors, it appears useful to take a retrospective view of the history of the term and its varying meanings in theory and practice (cf. Allen et al. 2005; Peterson 2005; de Laplante 2005; Blandin 2006).

"Ecosystem" was the clear winner when in 1988, at the occasion of the 75 years' jubilee of the British Ecological Society the members were asked about the most important concepts in ecology. It was followed, but at a considerable distance, by "succession" (Cherrett 1989). "Ecosystem ecology" had become established as a

---

[1] Willis 1997 has pointed out that the term was originally coined by Clapham whom Tansley had asked "if he could think of a suitable word to denote the physical and biological components of an environment considered in relation to each other as a unit. When Clapham suggested 'ecosystem', Tansley, after some consideration, wholly approved of it" (p. 268). In his article of 1935, however, Tansley did not mention Clapham.

W. Haber (✉)
Technische Universität München, WZW, Lehrstuhl fur Landschaftsökologie,
Emil-Ramann-Strasse 6, D-85354 Freising, Germany
e-mail: WETHABER@aol.com

special branch of ecology (Pomeroy and Alberts 1988), and the popularity both of the term and the concept "ecosystem" literally reached their peak. E. O. Wilson (1996) even counted the ecosystem among the "metaphysical constructs" that would have proven more powerful and less vulnerable than ordinary scientific theories. Since 1998 it can also boast a special scientific periodical entitled "Ecosystems".

This seemed to be a triumphant success for the term Tansley had published – but not precisely defined–about 50 years earlier. Stations of this success were Lindeman's (1942) pioneering study of a small lake, and in particular E. P. Odum's ground-breaking textbook "Fundamentals of ecology" (1953) which was based on the ecosystem as its central concept; Bergandi (1995) called it one of the few really paradigmatic scientific texts. The ecosystem then became the core concept of the first great international research programmes launched since the 1960s, such as the International Biological Programme (IBP) and its follower "Man and the Biosphere" (MAB). Their worldwide success and reputation considerably strengthened and popularized the use of the ecosystem concept and were unthinkable without it.

In contrast to this success story, since the 1980s a growing number of ecologists began questioning the primacy of the ecosystem concept and started a controversial debate on its meaning and importance (cf. Reiners 1986, Likens 1992). In particular evolutionary and population biologists developed an aversion against ecosystem studies to which they reproached one-sided consideration of energy and matter transfers or trophic levels, neglecting or ignoring aspects of population and community ecology as well as biodiversity. In Holling's (1992, 1996) view, population and community ecology seemed to exist in their own world and ecosystem ecology in another one. O'Neill et al. (1986) confronted them, emphasizing that an integral view of an ecosystem requires a combination of the two approaches, because the many functions of an ecosystem cannot be understood without considering biotic interactions in all their variety.

As a matter of fact, E. P. Odum, the father of ecosystem ecology, did take well into account population and community ecology, which occupy special and important chapters both in the three editions of his above-mentioned textbook and in the various popular versions following it. Nevertheless, Bergandi (1995) called the Odumian ecosystem approach utterly reductionist. This holds, however, rather for the work of Howard T. Odum, Eugene's brother, who factually reduced all ecosystem processes to energy transformations. Eugene Odum, it is true, recognized and supported his brother's view, incorporating it into his own books, but always insisted upon his (Eugene's) approach as being truly "holistic".

Ecosystem Management and Ecosystem Services

The power of the ecosystem concept proved itself anew in America when in 1992 the U.S. administration, following a proposal of the then Vice President Gore, decided to change the management of the country's natural resources from a single-resource approach (forests, rangeland, wildlife etc.) to a comprehensive "wholistic" one for which the name "ecosystem management" was chosen. Although ecosystems as subjects of science were controversially debated among ecological scientists, they appeared to be rather easily utilized as subjects of environmental management. About 20 U.S. federal agencies formed an "Interagency Ecosystem Management Task Force" which in 1995 presented a comprehensive three-volume report entitled "Healthy ecosystems and sustainable economies – the Federal Interagency Ecosystem Management Initiative" (EMI). Its goal was a proactive approach to ensuring a sustainable economy and a sustainable environment through ecosystem management (Malone 1995). This initiative may be considered as important as that of the Biodiversity Convention of 1992 embracing ecosystem diversity–albeit on a national scale.

The EMI, however, enhanced the confusion about the ecosystem concept (Carpenter 1995). To cite from Stein and Gelburd (1998, p. 74; emphasis WH):

> The ecosystem approach is a method for sustaining or restoring *natural* systems and their functions and *values*. It is based on a *collaboratively* developed vision of desired future conditions that integrates ecological, *economic*, and *social* factors and is applied within a *geographic* framework defined primarily by *ecological* boundaries. The ecosystem approach was chosen to restore and sustain [reversed priority! W.H.] the *health*, productivity, and biological diversity of ecosystems and the overall quality of life – through a *natural resource* management approach that is fully integrated with social and economic goals.

This somewhat curious statement leaves one wondering whether the ecosystem approach is restricted to *natural* resources or *natural* systems, which is justified because ecosystem management *per se* does not include economic and social factors, or should include these aspects. Referring to the EMI, Szaro et al. broaden its vagueness by stating: "For practical purposes, ecosystem management is generally synonymous with sustainable development, sustainable management ... and a number of other terms being used to identify an ecological approach to land and resource management" (1998, p. 5).

The same authors state a few paragraphs further on that "[the] ecosystem approach emphasizes place- or region-based objectives" because ecosystem data are "inherently spatial" (Brussard et al. 1998), and that "Ecosystem planning must consider the dynamics of landscape scale patterns" (Szaro et al. 1998, pp. 2 f.). The seven case studies forming part of the EMI report included among others a large watershed, the Great Lakes basin, the whole of South Florida, and the Prince William Sound in Alaska (which had been devastated by the wreckage of the oil tanker "Exxon Valdez"). It seems that "ecosystem" was used here as a kind of umbrella term for quite a variety of management goals and measures, comparable to the German (untranslatable) term "Naturhaushalt". In a commentary on EMI, Sheifer (1996) consistently used the double term "ecosystems/ecoregions", suggesting a regional landscape dimension of ecosystem (cf. Blandin and Lamotte 1988).

The EMI, however, was not allowed to fulfil its objectives – not because of the confusion about the ecosystem concept, but for political reasons: the Republican majority in the US Congress, outcome of the 1996 federal elections, rejected the initiative and cut the funds for it. Nevertheless, the journal "Landscape and Urban Planning" devoted a 234-page special issue (Vol. 40, 1998) to the EMI which offers interesting reading for ecosystem discussions. Enhancing the confusion, a book entitled "Ecosystem Management", edited by the (American) authors Samson and

Knopf (1996), did not mention EMI in its final report at all but wrote in the preface "The fact that no one knows what ecosystem management means has not diminished enthusiasm for the concept" (Wilcove and Samson 1987, p. 322). And in 1995 the Ecological Society of America issued a document entitled "The scientific basis for ecosystem management", again without mentioning EMI – and as such criticized by Zeide (1996) for not being as scientific as intimated. A number of ecologists and other experts who had cooperated in the EMI worked out the results and put them together in a three-volume book entitled "Ecological stewardship: A common reference for ecosystem management" (Johnson et al. 1999). The emphasis had shifted from management to stewardship, with a bias towards both conservation biology and the biological features of the ecosystem, thus reflecting the growing interest in biodiversity.

Towards the end of the twentieth century, the notion of "ecosystem services" offered to people and to society in general attracted great public attention, in particular stimulated by Daily's (1997) book. These services linked ecosystems to what ecological economists had conceived of as "natural capital" in the then upcoming debate on "sustainable development". It is these ecosystem services which have served as the starting point for the newest and comprehensive global environmental programme called "Millennium Ecosystem Assessment" which, as mentioned at the beginning, was launched by the United Nations in June 2001 (MEA 2003). This scientific enterprise focuses on how changes in ecosystem services have affected and are affecting human well-beings, how ecosystem changes may affect people in future decades, and what types of response can be adopted at local, national, or global scales to improve ecosystem management and thereby contribute to human well-being and poverty alleviation. Ecosystem services, however, are not based on clearly delimitable environmental objects, but on functional units of variable size and composition. Regarding ecosystem management or stewardship, we are faced with a big problem of implementation requiring immense mental effort which is based on the ecosystem concept – both loading it with enormous expectations and implications and imposing a likewise huge responsibility on ecological science to develop a theoretically and practically reliable basis for the concept. This is clearly demonstrated in the (preliminary) synthesis report titled "Ecosystems and Human Well-Being" (MEA 2005).

The Ecosystem's Position in Hierarchies and Scales

How well are ecologists prepared to scientifically accompany, advise on and support the solving of this huge task of the twenty-first century? After all the controversies of the last decades of the twentieth century Mayr (1997) had argued that the ecosystem concept – after its great Odum-inspired popularity in the 1960s and 1970s – had lost its role of a dominant paradigm. O'Neill (2001), one of the protagonists of ecosystem ecology, had even asked "Is it time to bury the ecosystem concept? (With full military honours, of course!)", but gave the answer himself: "Probably not". The term which thinkers had earlier referred to as an epistemological nightmare, quite contrary to the view of Scheiner et al. (1993), has obviously become indispensable. How should it be handled? Again, a short retrospective appears helpful for finding an answer.

When Clapham suggested the term "ecosystem" to Tansley in the early 1930s (Willis 1997), he borrowed the word "system" from the then upcoming system studies. Of course "system" implies a holistic view, but at the same time emphasizes a physical character and a mechanical or machine-like approach – an analogy that, by the way, had appeared much earlier (1922) in Tansley's work. According to Golley (1993) and Jax (2002), the main motivation of Tansley, whom Evans (1976) and quite recently Sheail (2005) characterize as continuously preoccupied with philosophy and processes of thought, was to strengthen the recognition of ecology as a serious science and to keep it free from misleading philosophical abuses of holistic theories of his time.

There are two important aspects in Tansley's article that should be emphasized. The first is that he explicitly called the ecosystem a "mental isolate", i.e. a "metaphysical construct" to put it into E.O. Wilson's terms. The second one is that Tansley conceived of a sequence of organization levels of nature – he called them "physical systems of the universe" – "which range from the universe as a whole [!] down to the atom" (Tansley 1935, p. 299), and that he assigned to the ecosystem a specific level within this hierarchical sequence. It is astonishing that this second, no less important concept (Schultz (1967) even called it "far more important") has met with relatively little appreciation until today. It was Egler (1942) who took up this idea for his vegetation system and formulated a first but still incomplete hierarchical sequence of levels of organization of living nature, which was complemented by Novikoff (1945). Feibleman (1954) even propounded a "theory of integrative levels" based on twelve "laws", which were discussed in an ecosystem context by Schultz (1967). (These three authors, by the way, are never cited in more recent publications on hierarchy.)

Tansley's idea of the ecosystem as a mental construct means that the object does not exist as such in reality, but only in the researcher's mind, and it is here where it is defined regarding its content and especially its boundaries. This mental picture is then projected onto a real situation in nature, the mental boundaries being fitted to factual ones such as river banks and lake shores, forest edges, watershed boundaries or boundaries originating from substrate or soil differences, all of which can serve to delimit ecosystems. In today's landscapes, however, most boundaries result from human land use, although they may coincide with natural boundaries. Clearly, if the ecosystem is but a mental construct, its boundaries are set by the observer or investigator! Therefore, to liken an ecosystem to a Clementsian (super)organism is fundamentally wrong and misleading, because an organism has an outside boundary (skin, epidermis, cell wall etc.) produced by itself and accepted by the researcher.

Tansley's original concept – it was really only an idea – of the hierarchy of physical systems in nature ranked these from the smallest dimension to the largest, and for him the ecosystem was but one of these systems, apparently of an intermediate dimension (a "middle-number system" *sensu* Weinberg 1975). But this dimension must not be confounded with spatial scale, which would imply that, according to its position in the hierarchy, an ecosystem is on a larger spatial scale than a community or a population and should be treated accordingly – which lacks any logic. Moreover, many biotic processes like succession or migration occur on larger spatial or temporal scales than ecosystem processes such as nutrient cycling.

Roots of Confusion

In this context, we hit upon one of the roots of misunderstandings, semantic confusions, and contradictions encumbering the ecosystem concept. Lindeman's pioneering ecosystem study of a small lake (only 1 ha surface, 1 m deep) had assigned an explicit spatial dimension and a clear boundary (the shoreline) to Tansley's concept, so it fitted into his hierarchical order. Eugene Odum, however, in his 1953 book presented the ecosystem as a basic unit in ecology of *every* spatial dimension, from the population to the biosphere, a definition which was taken up and confirmed by Evans (1956). But in the second edition of his book (1959), Odum followed Tansley, restricting the ecosystem to a specific level in the hierarchy – but continued nevertheless to use the term in its broader sense too, and nobody appeared unduly disturbed by these contradictions (Golley 1993, p. 72).

This explains the ongoing confusion concerning the character of the levels in Tansley's (and his followers') hierarchy, i.e. between a scalar (spatial) level and an organizational level – which induced Wiegleb (1996) and Jax (2002), following Allen and Hoekstra (1990), to suggest replacing "organizational" with "observational". It also resulted in a proliferation of "branched" or "broken" hierarchies (see the discussion in Wiegleb 1996, chapter 3.4) to avoid this confusion. The book by Hölter (2002) may contribute to its elucidation. Anyway, the validity of hierarchy theory *above* the organism level – where *all* levels are "mental constructs" – is more strongly contested than at distinctly lower levels of organization (Fränzle 2001, p. 75). Thus the concept of ecosystem in the traditional hierarchy scheme runs counter to the concept of ecosystem as a perspective that can be studied as a whole, incorporating activities from different scales (Vogt et al. 1996).

A solution of this scale-related dilemma was proposed by Ellenberg (1973; see also Mueller-Dombois and Ellenberg 2002, p. 168–171) who, following Evans' ecosystem definition in principle, established a corresponding "system of ecosystems". He distinguished five classes of ecosystems ordered in a hierarchy of spatial dimensions, namely mega-, macro-, meso-, micro- and nano-ecosystems. He called the meso-ecosystems the "ecosystems in the strictest sense" (Ökoysteme im engeren Sinn, p. 237) and the basic types of ecosystem classification, thus approaching once again the ecosystem position in Tansley's (and the later Odum's) hierarchy. An example of a meso-ecosystem is the summer-green broadleaved forest which is subdivided into the different alliances and associations familiar to phytosociologists. However, Ellenberg's classification did not arouse much attention.

Ellenberg's understanding of ecosystem is reflected in the German IBP project in the Solling which he directed (Ellenberg et al. 1986) and which is regarded as a "milestone in the development of ecosystem research" (Fränzle 1998, p. 11). It was strongly influenced by the phytosociological approach of the Zurich-Montpellier school based on plant associations recognized by their floristic composition and character species. A plant community with its more or less distinct boundaries could be mapped and rather easily be likened to a (micro-)ecosystem. This equalization (more in practice than in theory) of plant associations or formations with ecosystems (meso- and micro-ecosystems in Ellenberg's terminology) is quite common in continental European ecosystem studies, especially when applied to nature conservation or landscape planning, and is considered a useful and proven approach to ecosystem studies. Both nature conservationists and environmentalists soon adopted the term "ecosystem", making liberal use of it and opened its way into everyday language.

It was during the IBP research that, almost imperceptibly, the "ecosystem" changed from Tansley's "mental isolate" to a real object in nature or, as Schultz (1967) put it, a "perceptible object of study". Thus ecosystem ecology owed its most important success to "a realist, even ontic perspective of the ecosystem as a real world entity" (Potthast 2002, p. 139).

The Ecosystem Concept and Theoretical Ecology

It was not until the 1990s that theoretical ecologists seriously tackled the epistemological background of the ecosystem concept. This Handbook is both a proof and a first essential result of this endeavour. At first they seemed to be caught in the "holism trap". Users of the concept are often criticized for favouring "organicist thinking" or adhering to Clementsian "superorganism belief" (cf. Trepl 1988, de Laplante 2005), often voiced with a hidden derogatory undertone. Such critique, however, is one-sided and leads one astray. No organism can exist without interacting with other organisms, be it in the form of symbiotic or antagonistic relationships. These constitute, especially when displaying a recurrent behaviour, a real network whose structure can be disclosed analytically without any organicist idea in mind, and which can of course be called a system. Such reasoning is recognized by Trepl (1988), without however curbing the tendency of lumping together all systems- or synthesis-minded ecologists into one organicist or "holist" box.

Therefore, ecological scientists (and also many biologists) remain seriously split on the issue of reductionism *versus* holism, instead of dealing simultaneously with the part and the whole (Blandin 2006). A reductionist approach is obligatory in any investigation of highly complex phenomena, as complexity can only be mastered by breaking it down into its (presumably) principal components. This analytical procedure, however, has to be followed, and completed by a re-synthesis of the components in order to comprehend the functioning of the complex whole or the "why" of complexity. Such a combination of a reductionist with a holistic approach, often done iteratively, is essential for ecosystem investigations, but is neglected by researchers who shun the "holism trap". According to Schultz (1967), this attitude goes back to the origins of science and the historical primacy of physics in the seventeenth and eighteenth century. The objects of the physicist's manner of scientific inquiry were inanimate and timeless matter and energy, enabling him to explain the universe by means of a purely reductionist approach and to rather easily find generalizations and laws. This became the "only justifiable" scientific method[2] which was imposed on late-comers such as the life sciences; these dealt with biological entities each of which is unique - an "event" or a "chunk of space-time" that is only typifiable and defies explanation in purely physical terms. Therefore, biology is essentially an "event science"; there are only rules, but no laws at the biological level, theories are essentially non-testable, and the life phenomena too indeterminate for assignment of causes and predictions. Moreover, one of the fundamental biological questions, that is even constitutive for ecology, addresses the assessment of environmental factors with regard to their suitability for, and promotion of, the self-maintenance of organisms. This is ultimately a teleological question (cf. Weil 2005) greatly exceeding "common" scientific laws and utterly reprehensible, if not abhorrent for physicists' minds; but it can be justified by heuristic necessity forgoing causal explanations. What would have been the development of modern science if biology instead of physics had achieved primacy? It might have been considerably delayed, because those early scientists of reductionist obsession would have been totally overcharged with tackling the huge diversity of life and living phenomena.

Schwarz (2003) has suggested a thoughtful and promising mode to bridge these abysmal differences. She places ecology as a kind of "third power", endowed with mediating abilities, between the two mainstreams of biology: (1) *physiology* as reductionist-analytical research pursuing explanatory goals, based on physicist reasoning and seeking general or generalisable laws, and (2) *physiognomy* founded upon integrative or synthetic comprehension (not explanation) of principally unique phenomena of "Gestalt" quality. Schwarz also gives full recognition to the indispensable importance of metaphors ("metaphysical constructs" *sensu* E. O. Wilson) and of heuristics. Such a mediating role of ecology, however, is never uniform, but varies or fluctuates according to changing preferences for central fundamental concepts of ecology, namely "energy", "niche" and "microcosm". The latter might be understood as "ecosystem" in its narrow definition, so it fits well into Schwarz's arguments. These are apt to break the unseemly haughtiness of the reductionist party and its unfounded superiority based on (mostly unconscious) incontestability of their scientific attitude.

The power, or pervading influence, of the ecosystem concept appears to be based on the deep-felt yearning of ecologists for "unifying ideas", which has continually been articulated since about 1960–it was the topic of the first International Congress

---

[2]Fränzle, however, points out that leading representatives of the scientific community around 1800, like Buffon, James Hutton, Georg Forster, Linné, Miraband, Kant were already primarily interested in defining interrelationships, and less so "in detailed generic analyses or unravelling specific microscale causalities" (2001, p. 60).

of Ecology in The Hague 1974. Ford and Ishii (2001), repeating the plea for a synthesis in ecology, consider "integrative concepts" an essential part of ecology because they enable scientists to organize and reorganize ecological knowledge as the science progresses. Who would contest the integrative notion of the ecosystem concept? Indeed, the two authors recommend, as a viable way to achieve synthesis, the construction [!] or development of integrative concepts regarding the internal organization of ecological systems. Fränzle argues that "the present theoretical background to ecosystem research has not yet reached the level of a comprehensive unified theory", but he sees "a commendable number of unifying concepts and integrative approaches with regard to ecosystem research" (2001, p. 83). In his opinion, all these concepts are derived from, or associated with, system theories in general, more specifically with those of self-organizing ecological systems. The latter integrate components of thermodynamics, dissipative processes, information and network theories, game theory, catastrophe theory and hierarchy theory. This appears once more as a rather "physicalist" way of thinking, to which a physical geographer like Fränzle is inclined. His opinion, however, that the complexity of ecosystems can only be expressed in mathematical models, has to be questioned because these ignore or omit the incalculable behaviour of living agents, leaving the modeller with incomplete or only partial knowledge of the system which he seeks to understand. As Schaffer (1985) put it, ecologists will never be able to write down the complete governing equations for any natural system.

Society's view of science, it is true, still consists of a trilogy of theories, hypotheses, and laws to explain observed facts and to allow prediction of events. But in ecology, this legacy of a reductionist, quantifying scientific problem approach has failed in the face of the huge complexity of the environment, which is much greater than the earth's climate system of which, as we know, only probable futures can be assessed. Environmental impact assessments (EIA) of projects, which are mandatory in many countries, require such predictions but they simply cannot be provided with the reliability taken for granted – an example of unqualified and irresponsible legislation.

Outlook: A Plea for Realism and Transdisciplinarity

O'Neill's (2001) cautious answer to his question of burying the ecosystem concept read: "Probably not". All attempts to basically question it, not to mention abolish it, have failed as yet, proving both the power of the concept and its utility. It appears that we are unable, even as scientists, to handle complexity (the dominant property of the biosphere or the natural environment) on the exclusive basis of clear scientific reasoning utilizing Kuhnian principles and logical theory, without resorting to heuristics. Thus, refraining from reliable predictions of future environmental states, we have to search for a general understanding of the complex dynamics of the environment in order to find approaches for suitable, practicable management strategies. For this purpose, the ecosystem concept has opened the right pathways, offering opportunities to approximate, contribute to, or even achieve basic societal goals such as sustainable development, environmental health, safety and security, biodiversity handling as well as general welfare. Accordingly, new scientific ecological disciplines such as conservation biology, ecological economics, ecological planning and management are relying on, and supported by ecosystem ecology in order to increase the controllability of an uncertain world (de Laplante 2005). In these more applied research fields, ecosystem ecology is becoming an influential, diversifying branch of ecological science, meeting with growing public attention and support. Furthermore, it transcends traditional disciplinary boundaries, bridging the gap separating it from social science and humanities (Cantlon 2002; Haber 2004). In a time of rapid social and environmental change, such transdisciplinarity – a visualisation already emphatically pursued by Eugene Odum–has become indispensable.

Orthodox scientific circles, however, even in ecology, still frown upon such "soft" science and its value-laden, issue-driven approaches. Although "hard" reductionist science based on quantitative mathematical models is proving unable to devise practicable means for solving complex environmental problems, its epistemological scruples should not be dismissed altogether. Ford and Ishii (2001), despite their strong support for "integrative concepts" in ecology, warned both of the conjectures encumbering them and of the lack of a recognized procedure for synthesis. Following Taylor and Haila (2001), such concepts are unstable and elusive and cannot be taken for granted, so we have to reconceptualise them from time to time. Even when the ecosystem concept has already proven indispensable for environmental understanding and management, we should be aware of such scruples (particularly of its theoretical weaknesses and "birth defects" mentioned above) and stick to understanding it as a "metaphysical construct" or metaphor *sensu* Schwarz (2003). Whenever we apply this construct to a real piece of nature whereby we delimit ourselves according to the purpose of ecological research, we have to be careful that the application is precisely defined, even if ecology still lacks terminological rigour (Breckling and Müller 1997). There will always be a temptation to slide from the metaphysical construct into tangible reality of nature. But it remains a helpful and fruitful image.

Ecology in general, and ecosystem ecology in particular, must no longer be confronted with the dichotomy between reductionism and holism – even less by insinuating that the latter is scientifically reprehensible. This dichotomy, an application of Hegelian dialectic, might have served a heuristic purpose, but diverts attention from the principal goal of ecological research: to explain life in its organization, its universality as well as its diversity of change, and all of them in their dependence upon, adaptation to, and limited capability of alteration of the non-living, physicochemical environment. The "third way" suggested by Schwarz (2003), mentioned above, can lead in the right direction, making use of narrative and "good metaphors" because to use metaphor well is to discern similarities." (Aristotle in Poetics 1459a: 6–7). In this way, we get back to, and revive the far-sighted and invaluable legacy of Alexander von Humboldt to science and art, "his sense of unity within the complexity of what we now call ecosystems" (Fränzle 2001, p. 63).

It is also appropriate to cite Georg Picht's (1979) definition of ecology as – "the science of the singularity of situations, founded on the generality of natural laws".[3] Picht's reasoning stems from an understanding of life (which he sometimes identified with "oikos") as a "contemplative conception of nature, being interpreted in a normative way" and insists upon this contemplative relationship, with nature being superior both to the scientific (ecological) and the economic-technical relationship with nature. The general tendency of science – in the eyes of the public supporting it – is oriented towards synthesis, analytical reductionism being only a – however important and indispensable – step in this direction. The concept of ecosystem is capable of synthesising, or integrating, its main properties as revealed through its analysis: dynamic connections, transience, scale-dependent views, irreversibility, short-term predictability and long-term unpredictability. The concept will therefore be an indispensable tool for the – hopefully sustainable – development of humankind in the twenty-first century.

References

Allen TFH, Hoekstra TW (1990) The confusion between scale-defined levels and conventional levels of organization in ecology. J Veg Sci 1:5–12

Allen TFH, Zellmer AJ, Wuennenberg CJ (2005) The loss of narrative. In: Cuddington K, Beisner BE (eds) Ecological paradigms lost. Routes of theory change. Elsevier Academic Press, Burlington, pp 333–370

Aristotle (1995) Poetics, ed. and transl. by S Halliwell. Harvard University Press, Cambridge Mass

Bergandi D (1995) "Reductionist holism": An oxymoron or a philosophical chimera of E. P. Odum's systems ecology? – Ludus Vitalis, Revista de filosofia de las ciencias de la vida. J Philos Life Sci 3(5):145–180, Mexico /Barcelona

Blandin P (2006) L'écosystème existe-t-il? Le tout et la partie en écologie. In: Gayon J, Martin T (eds) Le tout et la partie. CNRS Editions, Paris

Blandin P, Lamotte M (1988) Recherche d'une entité écologique correspondant à l'étude des paysages: la notion d'écocomplexe. Bulletin d'écologie 19:547–555

Breckling B, Müller F (1997) Der Ökosystembegriff aus heutiger Sicht - Grundstrukturen und Grundfunktionen von Ökosystemen. In: Fränzle O, Müller F, Schröder W (eds) Handbuch der Umweltwissenschaften. Ecomed Verlagsgesellschaft, Landsberg, chapter II-2.2

Brussard PF, Michael Reed J, Richard Tracy C (1998) Ecosystem management: what is it really? Landsc Urban Plann 40:9–20

Cantlon JE (2002) Ecological bridges revisited. Bull Ecol Soc Am 83:271–272

Carpenter RA (1995) A consensus among ecologists for ecosystem management. Bull Ecol Soc Am 76:161–162

Cherrett JM (1989) Key concepts: the result of a survey of our members' opinions. In: Cherrett JM (ed) Ecological concepts. The contribution of ecology to an understanding of the natural world. Blackwell, Oxford, pp 1–16

---

[3] die Lehre von der in der Allgemeinheit der Gesetze fundierten Einmaligkeit von Situationen (Picht 1979, p. 25).

Daily GC (1997) Nature's services: Societal dependence on natural ecosystems. Island Press, Washington, DC
De Laplante K (2005) Is ecosystem science a postmodern science? In: Cuddington K, Beisner BE (eds) Ecological paradigms lost. Routes of theory change. Elsevier Academic Press, Burlington, pp 397–416
Egler FE (1942) Vegetation as an object of study. Philos Sci 9:245–260
Ellenberg H (1973) Versuch einer Klassifikation der Ökosysteme nach funktionellen Gesichtspunkten. In: Ellenberg H (ed) Ökosystemforschung. Teil VII, Die Ökosysteme der Erde. Springer, Berlin, pp 235–265
Ellenberg H, Mayer R, Schauermann J (eds) (1986) Ökosystemforschung. Ergebnisse des Solling-Projektes 1966–1986. Ulmer, Stuttgart
Evans FC (1956) Ecosystem as the basic unit in ecology. Science 123:1127–1128
Evans GC (1976) A sack of uncut diamonds: the study of ecosystems and the future resources of mankind. J Ecol 64:1–39
Feibleman JK (1954) Theory of integrative levels. Br J Philos Sci 5:59–66
Ford ED, Ishii H (2001) The method of synthesis in ecology. Oikos 93:153–160
Fränzle O (1998) Grundlagen und Entwicklung der Ökosystemforschung. In: Fränzle O, Müller F, Schröder W (eds) Handbuch der Umweltwissenschaften, Part 3–2.1. Ecomed, Landsberg, pp 1–24
Fränzle O (2001) Alexander von Humboldt's holistic world view and modern inter- and transdisciplinary ecological research. Northeast Nat 1:57–90, Special Issue
Golley FB (1993) A history of the ecosystem concept in ecology. More than the sum of the parts. Yale University Press, New Haven
Haber W (2004) Landscape ecology as a bridge from ecosystems to human ecology. Ecol Res 19:99–106
Hölter F (ed) (2002) Scales, hierarchies and emergent properties in ecological models. Peter Lang, Berlin
Holling CS (1992, 1996) Cross-scale morphology, geometry, and dynamics of ecosystems. Ecol Monogr 62:447–502, and In: Samson, F.B. and F.L. Knopf 1996. Ecosystem management. Selected readings. Springer, New York, pp. 351–423
Jax K (2002) Die Einheiten der Ökologie. Peter Lang, Frankfurt/M
Johnson NC, Malk AJ, Sexton WJ, Szaro RC (1999) Ecological stewardship. A common reference for ecosystem management. Elsevier, New York
Likens GE (1992) The ecosystem approach: its use and abuse. Ecology Institute, Oldendorff/Luhe
Lindeman RL (1942) The trophic-dynamic aspect of ecology. Ecology 23:399–418
Malone CR (1995) Ecosystem management: Status of the federal initiative. Bull Ecol Soc Am 76:158–161
Mayr E (1997) This is biology. The science of the living world. Belknap Press of Harvard University Press, Cambridge
MEA (Millennium Ecosystem Assessment) (2003) Ecosystems and human well-being. Island Press, Washington, DC
MEA (Millennium Ecosystem Assessment) (2005) Ecosystems and human well-being: Synthesis. Island Press, Washington, DC
Mueller-Dombois D, Ellenberg H (2002) (Reprint 1974): Aims and methods of vegetation ecology. Blackburn Press, Caldwell
Novikoff AB (1945) The concept of integrative levels and biology. Science 101:209–215
Odum EP (1953) Fundamentals of ecology. First edition. 1959: Second edition. 1972: Third edition. Saunders, Philadelphia
O'Neill RV (2001) Is it time to bury the ecosystem concept? (With full military honors, of course!). Ecology 82:3275–3284
O'Neill RV, DeAngelis DL, Waide JB, Allen TFH (1986) A hierarchical concept of ecosystems. Princeton University Press, Princeton

Peterson GD (2005) Ecological management: Control, uncertainty, and understanding. In: Cuddington K, Beisner B (eds) Ecological paradigms lost. Routes of theory change. Elsevier Academic Press, Burlington, pp 371–395

Picht G (1979) Ist Humanökologie möglich? In: Eisenbart C, Eisenbart C (eds) Humanökologie und Frieden. Klett-Cotta, Stuttgart, pp 14–123

Pomeroy LR, Alberts JJ (eds) (1988) Concepts of ecosystem ecology. A comparative view. Springer, New York

Potthast T (2002) From "mental isolates" to "self-regulation" and back: justifying and discovering the nature of ecosystems. In: Schickore J, Steinle F (eds) Revisiting discovery and justification. Preprint 211. Max Planck Institute for the History of Science, Berlin, pp 129–142

Reiners WH (1986) Complementary models for ecosystems. Am Nat 127:59–73

Samson FB, Knopf FL (1996) Ecosystem management. Selected readings. Springer, New York

Schaffer WM (1985) Order and chaos in ecological systems. Ecology 66:93–106

Scheiner SM, Hudson AJ, van der Meulen MA (1993) An epistemology for ecology. Bull Ecol Soc of Am 74:17–21

Schultz AM (1967) The ecosystem as a conceptual tool in the management of natural resources. In: Ciriacy-Wantrup SV, Parsons JJ (eds) Natural resources, quality and quantity. The University of California Press, Berkeley, pp 139–161

Schwarz AE (2003) Wasserwüste, Mikrokosmos, Ökosystem. Rombach, Freiburg

Sheail J (2005) Tansley and British ecology: The formative years. Bull Br Ecol Soc 36:23–25

Sheifer IC (1996) Integrating the human dimension in ecosystem/ecoregion studies – a view from the ecosystem management national assessment effort. Bull Ecol Soc Am 77:177–180

Stein SM, Gelburd D (1998) Healthy ecosystems and sustainable economies: the federal interagency ecosystem management initiative. Landsc Urban Plann 40:73–80

Szaro RC, Sexton WT, Malone CM (1998) The emergence of ecosystem management as a tool for meeting people's needs and sustaining ecosystems. Landsc Urban Plann 40:1–7

Tansley AG (1922) Elements of plant biology. George Allen & Unwin, London

Tansley AG (1935) The use and abuse of vegetational concepts and terms. Ecology 16:284–307

Taylor P, Haila Y (2001) Situatedness and problematic boundaries: conceptualizing life's complex ecological context. Biol Philos 16:521–532

Trepl L (1988) Gibt es Ökosysteme? Landschaft Stadt 20:176–185

Vogt KA, Gordon JC, Wargo JP, Vogt DJ, Asbjornsen H, Palmiotto PA, Clark HJ, O'Hara JL, Keaton WS, Patel-Weynard T, Witten E (1996) Ecosystems. Balancing science with management. Springer, New York

Weil A (2005) Das Modell "Organismus" in der Ökologie: Möglichkeiten und Grenzen der Beschreibung synökologischer Einheiten, vol 11, Theorie in der Ökologie. Peter Lang, Frankfurt/M

Weinberg GM (1975) Introduction to general systems thinking. Wiley, New York

Wiegleb G (1996) Konzepte der Hierarchie-Theorie in der Ökologie. In: Mathes K, Breckling B, Ekschmitt K (eds) Systemtheorie in der Ökologie. Ecomed, Landsberg, pp 7–25

Wilcove DS and F B Samson (1987) Innovative wildlife management: listening to Leopold. Trans North Am Wild Nat Resour Conf 52: 321–329

Willis AJ (1997) Ecology of dunes, salt marsh, and shingle. Chapman & Hall, London and New York

Wilson EO (1996) In search of nature. Island Press, Washington, DC

Zeide B (1996) Is "The scientific basis" of ecosystem management indeed scientific? Bull Ecol Soc of Am 77:123–124

Part VI
Local Conditions of Early Ecology

Chapter 19
Early Ecology in the German-Speaking World Through WWII

Astrid Schwarz and Kurt Jax

The scientific practice and theory of ecology in the German-speaking world arose simultaneously yet independently of each other in different places and in relation to different subjects. The new disciplining perspective took in lakes and fish ponds as well as native forest, heath and mountain landscapes, though it also included the flora and fauna of tropical and arctic regions. "German-speaking world" refers here not so much to an area determined by its political or natural borders but rather by its linguistic boundaries. A lively exchange of publications, objects and individuals took place within this scientific world. Cities and regions belonging to different spheres of political influence were a part of this *Sprachraum*, which encompassed Zurich, Vienna, Prague, Budapest and Berlin, as well as Bohemia, Silesia and Prussia, the Rhineland and the Valais. Perfect examples of the commonplace exchanges that took place in what we call the "German-speaking world" of that time were the botanists Simon Schwendener and Gottlieb Haberlandt, who were decisive for the formation of physiological plant ecology (see below). Schwendener was born and educated in Switzerland and spent most of his working life in Germany (Tübingen and Berlin; prior to that in Basel, Switzerland); Haberlandt was born in Hungary, educated in Austria and worked for most of his life in Austria (Vienna and Graz), though at times also in Germany (Tübingen and Berlin). So if – for the sake of brevity – we speak of "German" ecology in this chapter, we mean this region as delimited by the common use of the German language as a means of communication.[1]

---

[1] Even today there is a joint Society of German-speaking ecologists, the Gesellschaft für Ökologie (GfÖ), founded in 1978, which includes scientists mainly from Germany, Austria, Switzerland and Liechtenstein.

A. Schwarz (✉)
Institute of Philosophy, Technische Universität Darmstadt, Schloss, 64283 Darmstadt, Germany
e-mail: schwarz@phil.tu-darmstadt.de

K. Jax (✉)
Department of Conservation Biology, Helmholtz Centre for Environmental Research (UFZ), Permoserst.15, 04318 Leipzig, Germany
e-mail: kurt.jax@ufz.de

Indeed up until the first 20–30 years of the twentieth century, German was still one of the dominant languages of scholarly research, including the natural sciences. The international symposium "Deutsch als Wissenschaftssprache im 20. Jahrhundert" (German as a language of scholarship in the twentieth century), held in January 2000 and organized by the Mainz Academy of Sciences and Literature, arrived at the conclusion that "[f]rom the mid-nineteenth century through to the 1920s German was a world language for the sciences; after this – from the 1930s due to the Nazi policy of expulsion and from the late 1950s due to international developments – it became a marginal language, like French and Russian".[2] Thus in the period up until the Second World War there were scientific journals containing for the most part articles written in German – including pieces by non-native speakers. However, one would also come across isolated articles written in English or French in these journals, indicating that they were widely recognized publication forums for scientists throughout the world. Among these, for example, is the *Biologisches Centralblatt* (founded in 1881), published since 1997 under the title *Theory in Biosciences*, *Engler's Botanische Jahrbücher* (founded in 1880) and the *Zoologische Anzeiger* (founded in 1878) but also more popularly oriented journals such as the *Kosmos* founded in 1877 by Ernst Haeckel to disseminate the Darwinian Weltanschauung. These journals were part of what was called the "second wave of formation"[3] of natural history journals and they were all important forums for an evolving German-language ecology. In addition to these journals, the many varied publications of the "scientific societies" were also used to disseminate ecological ideas and scientific programmes.[4]

Just a short time later, the first journals to specialize in ecological topics were founded. The spectrum of journals diversified rapidly, encompassing terrestrial and aquatic ecology and reflecting the growing consolidation of the new field of research. The founding of more journals at the start of the twentieth century contributed towards the increasing stabilization of the field and a clearer delineation of ecological research questions. The first specialist ecological journals included the *Forschungsberichte der Biologischen Station zu Plön*, which continued to exist from 1906 onwards under the title *Archiv für Hydrobiologie und Planktonkunde*; in

---

[2] "Pörksen 2001, p. 29."

[3] Andreas W. Daum speaks of a "zweite Gründungswelle"of natural history journals in 19th century, especially in the domain of scientific popularization (Daum 1998, p. 359).

[4] There are numerous examples of these in the 1880s and 1890s especially, but exactly how publication strategy related to the constitution of the community – particularly with regard to the relationship between "lay people", independent scholars and employed scientists – has not yet been systematically studied. Those societies active in the publishing sphere include the scientific societies (Naturforschende Gesellschaften) in Zurich, Lucerne and Lausanne as well as the Natural History Society of the Prussian Rhineland and Westphalia (Bonn). A more general perspective on the popularization of science is given for instance in Daum (1998), who deals in one chapter also with the "Naturvereine" in 18th century.

1920 it was re-named the *Archiv für Hydrobiologie*[5]; the first volume of the *Internationale Revue der gesamten Hydrobiologie und Hydrographie* was published in 1908, while from 1923 onwards the International Association of Theoretical and Applied Limnology (IVL) produced a *Zeitschrift der Internationalen Vereinigung für theoretische und angewandte Limnologie*; the Geobotanic Institute Rübel in Zurich also had its own journal in 1924, as did the Biological Institute Helgoland, with its *Helgoländer wissenschaftliche Meeresuntersuchungen*, from 1937 onwards.

Shaping the Field

Thus the first steps toward a scientific ecology in the German-speaking world can be dated back to the second half of the nineteenth century. They were taken by botanists and zoologists, microbiologists and physiologists, hydrologists, geographers and chemists. From about the 1870s onwards an increasing number of research programmes were devised and scientific networks created around people such as zoologist Karl Möbius (1825–1908) and physiologist Victor Hensen (1835–1924) at the University of Kiel, Anton Frič (1832–1913) at the Institute of Prague University, botanist Carl Schröter (1855–1939) at the University of Zurich, Friedrich Zschokke (1860–1936) at the Institute for Zoology in Basel, and within the study group around botanist Simon Schwendener (1829–1919) in Berlin. The formation of networks was further supported by the founding of specialist commissions, whether it occurred in scientific societies, in organizations, or through political interventions.

For example, in 1870 the Prussian government appointed a commission for the scientific exploration of the German oceans (Kommission zur wissenschaftlichen Erforschung der deutschen Meere), of which not only Hensen and Möbius were members but also plant geographer Adolf Engler, anatomist Carl von Kupffer, physicist Gustav Karsten and other scientists. A "limnological commission" was set up at a meeting of the *Schweizerische naturforschende Gesellschaft* (Swiss Scientific Society) in Frauenfeld in the year 1887, whose task was to develop, coordinate and organize specific research projects and to procure and archive as well as

---

[5]This step-by-step process of re-naming can be seen as a means of asserting a continuity in editorship: Volume 10 from 1915 was the last to be edited by Otto Zacharias on his own; Volume 11 (1917) begins with an obituary for Zacharias written by August Thienemann, the new co-editor; and both are named in Volume 12 from 1920, the point at which the journal changed its name to Archiv für Hydrobiologie. Not until Volume 13 (1922) does August Thienemann appear as the sole editor; this is also the year in which the International Association of Theoretical and Applied Limnology was founded in Kiel. August Thienemann was the co-initiator, along with Einar Naumann (the "junior partner"), of this society.

facilitate publications.[6] Wilhelm Halbfass (1856–1938) gave a talk at the 13th German Geographers Congress in 1901 in Breslau entitled "The scientific and economic significance of regional limnological institutes" (Die wissenschaftliche und wirtschaftliche Bedeutung limnologischer Landesanstalten). The talk ended with a "resolution", addressed to the Prussian government, calling for the establishment of regional limnological institutes. This call for action, its political content toned down, was given a forum in *Petermanns Mitteilungen* in 1902: it was now limited to an exhortation to gather data on all the lakes in Europe.[7]

Likewise dedicated to "naturwissenschaftliche Landesdurchforschung" (regional scientific research) were *comités* set up in Austria-Hungary (their headquarters in Vienna and Prague), which were concerned mainly with cartographical images but also with determining national resources, including not only geological formations but also "the animal world of our lakes, ponds and rivers". Zoologist Anton Frič, secretary of the Bohemian initiative that emerged from the Prague-based museum milieu,[8] lamented nonetheless that "those working out of love for science and out of patriotism have to rely on a meager reimbursement of their cash expenses" while "in Germany the enterprise has thousands at its disposal",[9] naming in this context the "zoological station at Ploen".[10] This *Biologische Station zu Plön*, established in 1892, was one of the nodal points of activity around which the new field of research became institutionalized. Other early initiatives included the founding of the

---

[6] The initiator was François Auguste Forel. While in 1890 the commission still comprised three official members, in 1892 it had grown to twice this size. In addition to Forel, its members were: Forests Inspector Coaz from Bern (from 1887), Prof. F. Zschokke from Basel, Dr. E. Sarasin and Prof. L. Duparc from Geneva, and grammar school teacher Prof. X. Arnet from Lucerne. The composition of the commission is representative of early ecological research in two respects: first, it is interdisciplinary and, second, its members come from both academic and non-academic milieus (Proceedings of the Swiss Scientific Society, 74th Annual Meeting in Freiburg, Annual Report 1890–1891, Freiburg: Gebr. Fragnière 1892, pp. 100–103, 112–115, 142).

[7] Halbfass' negotiations with the representatives of the German Geographers Congress and the re-shaping of the text to become a publishable, politically correct version are recounted in Ein vergessenes Kapitel aus der Seenforschung (A forgotten chapter in the history of lake research) (2005) by Sylvin Müller-Navarra, pp. 188ff.

[8] An insightful discussion about the role of German-based Natural History Museums is given in a comparative study by Carsten Kretschmann (2006) Räume öffnen sich (spaces open up).

[9] "Die aus Liebe zur Wissenschaft und aus Patriotismus arbeitenden Kräfte auf einen kargen Ersatz der Barauslagen angewiesen. [...] in Deutschland (stehen) dem Unternehmen Tausende zur Disposition." Interestingly enough, Frič continues here: "Repeated attempts to encourage the large landowners to support this endeavour for the common benefit have borne no fruit" (Wiederholte Versuche, den Grossgrundbesitz zur Förderung dieses gemeinnützigen Vorgehens aufzumuntern, führten zu keinem Resultat) (queryFrič and Václav Vávra 1894, p. 7). Apparently, however, there were exceptions, as Frič also reports that Freiherr von Dercsényi had a "solidly built nice little house" (festgebautes, nettes Häuschen) erected, which then served as a zoological research station.

[10] The research results from these activities were published in the *Archiv für die naturwissenschaftliche Landesdurchforschung von Böhmen*. The latter also contained, in 1894, Anton Frič's *Untersuchungen über die Fauna der Gewässer Böhmens*, which essentially referred to two artificial bodies of water, the *Unterpocernitz* and the *Gatterschlag Pond*. However, the study of Bohemian forest lakes and other artificial bodies of water had been initiated much earlier, in 1872 (Frič and Václav Vávra 1894, pp. 5–7).

Biologische und Fischerei-Versuchsstation Müggelsee (Biological and Fisheries Experimental Station at Müggelsee) in Berlin-Friedrichshagen (1893) and the Biological Station in Lunz, Austria in 1906: all these contributed towards the institutionalization of ecological research.

"Limnology" – An Attractor in the Emerging Field?

The coming together of ecologists in the International Association of Theoretical and Applied Limnology (IVL) was a further step in establishing aquatic ecology[11] as a scientific discipline as well as in the perceptions of the wider society. In 1922, the first congress took place in Kiel; it was conducted and also recorded in writing in German – "as is customary for international congresses – in the language in which they are held", writes Secretary General Friedrich Lenz (1889–1972) in the Foreword, which is also printed as a Preface and an Avant-Propos. The participants at the inaugural conference numbered 67 members and 15 guests from a total of 13 countries. In 1923, however, when the Proceedings were published, more than 350 "full members" from 26 countries were registered,[12] along with 31 institutions as "extraordinary members", as can be seen from the members directory in this first volume of the "Proceedings from the International Association for Theoretical and Applied Limnology". The themes addressed include the metabolic rate in open water and near the bank – the so-called pelagic and littoral zones – and the role of bacteria and funghi in these processes. The metabolic characterization, or circadian rhythm, of plankton and its origins are discussed along with the problems of the fisheries and of hydraulic engineering, water quality and the protection of lakes and rivers. The newly founded society could count on a certain level of support in the public perception when it came to the protection of aquatic sites in particular, backed up at an institutional level by the *Heimatschutz- and Naturschutzvereine* roughly translated, the local societies for the protection of the country and conservation of nature. However, local water supply and wastewater disposal companies were also affected by the problems attending water pollution and water usage – as, indeed, were all those responsible for the administration of water as a resource at the national governmental level. The production of drinking water and the use of hydropower to generate electricity had already long become a matter of national interest in the nineteenth century, which made them a potential source of conflict. Indeed disputes were rife between Heimatschutz, state authorities and industry. Plans to

---

[11] Aquatic ecology here describes above all the study of inland waters, although marine biologists were also involved in establishing the IVL (August Pütter and Ernst Hentschel as well as Karl Brandt from the Oceanographic Institute at Kiel attended the 1922 conference in Kiel). Marine biology had been established prior to limnology, which initially disassociated itself from the former for reasons that had more to do with competition for research funding and prestige; it also maintained that it had a more progressive research programme: "...la limnologie peut appliquer l'expérimentation là où l'océanographie en est le plus souvent réduit à la seule observation." (Forel 1896, p. 596).

[12] In his opening speech Thienemann speaks of another 187 individuals who had responded to the mailing of 1,000 copies of "Calls to join" – a response to a campaign initiated by Naumann and Thienemann which was regarded as highly positive (1923, p. 1).

build the barrage on the Rhine at Lauffenburg (1914) using dynamite and dam constructions were contested, but in the end the barrage did get built. On the other hand, in 1942 another disputed project, the so-called Rheinwald three-stage project on the posterior Rhine as well as the building of a power station were stopped after a long dispute (Tümmers 1999, pp. 38 f., pp. 88 f.). This ongoing process of industrialization along the rivers also left behind traces that could not only be seen but also smelt. In 1901 the issue of the Rhine being a "cesspool" was debated in the German Reichstag. The conflicts over the quality and production of drinking water quickly led to the establishment of state institutions. The Royal Experimentation and Testing Institute for Water Supply and Wastewater Removal (Königliche Versuchs- und Prüfunganstalt für Wasserversorgung und Abwasserbeseitigung) was founded in Berlin in 1901 (Kluge and Schramm 1986, pp. 84 ff.).

One of the co-founders of the IVL, August Thienemann (1882–1960), was quick to recognize this interrelationship between politics and science and between industry and the state in the sphere of ecological research; he sought to make use of it not only in conceptual terms but also as a tool for institution building. Thienemann wanted limnology to be seen as a "bridging science" (Brückenwissenschaft), for herein, according to him, lay "its great cultural significance for our times" (1935, p. 20). Thienemann borrows this term from a piece entitled Die Struktur der Ganzheiten (The Structure of Wholes) by philosopher Wilhelm Burkamp (1929), a text from which he frequently quotes (for instance 1933; 1935). What seems especially noteworthy to Thienemann is Burkamp's characterization of new sciences as problem-based, as "*methodological and factually structural wholes*" (methodische und sachlich strukturelle Ganzheiten", Burkamp quote, emphasis AS, KJ), as this supports his own formulation in which the unique character of limnology is seen to be grounded in "the [research] object *and* in the methods" (1935, p. 19; emphasis in original).

With this statement, Thienemann was effectively placing limnology explicitly in a mediating position between various natural science disciplines, predominantly hydrography, biology, geology and oceanology, as well as between basic and applied research. "Theory will always remain the foundation for practice!" he exclaims, adding as a caveat that nonetheless "the study of wastewater (an extreme milieu with regard to its chemistry), which was originally undertaken exclusively for practical reasons, has made it quite considerably easier for us to gain a theoretical understanding of the population in a chemically normal natural bodies of water" (1923, p. 3).[13] What is

---

[13] "…das Studium der Abwässer, eines in chemischer Hinsicht ganz extrem gestalteten Milieus, das ursprünglich doch ausschließlich unter praktischen Gesichtspunkten vorgenommen wurde, uns das theoretische Verständnis der Besiedelung chemisch-normaler natürlicher Gewässer ganz wesentlich erleichtert hat"; Thienemann's own research work was located initially in both the theoretical and the applied sphere. Later, however, and certainly after his definitive decision to work at the Kaiser-Whilhelm-Institut (KWI) Plön – the former *Biologische Station zu Plön* - rather than to succeed Bruno Hofer at the Royal Bavarian Biological Testing Station for Fisheries (Königliche Bayerische Biologische Versuchsanstalt für Fischerei) in Munich, his research can be described more as theoretical in the sense of conceptual work. Nonetheless he remained active in fisheries biology, in committees and organizations, and reference was frequently made to his work on biological studies of wastewater. In 1956 the Department of Agriculture and Horticulture at Humboldt University, Berlin conferred on him the degree of Dr. agrar. h.c.

especially remarkable about this statement is that, rather than supporting the familiar idea of knowledge transfer from theory to practice, or from the natural sciences to the engineering sciences, it also recognizes the reverse trajectory and acknowledges it as a source of cognitive insight and as a fruitful heuristic.

Thus Thienemann offers not only methodological arguments for the bridging function of limnology or justifies it by reference to a specific set of research objects; he also argues that the conceptual framework of this new science enables it to function as a bridge. He brings into play the link to concepts of natural philosophy, taking up once again a romantic conception of nature; at the same time, though, he also mentions modernity and the progressiveness and experimental character of limnology, which he claims makes it different from the "merely" descriptive sciences of natural history.

Thienemann returns again later to the term "bridging science" in order to characterize ecology, whose purpose, as the study of nature's household, is to connect "all the branches of the study of nature" ("alle Zweige der Naturkunde").[14] In the same year he also uses the metaphor *Grenzland Limnologie* (borderland limnology) – a place located between the "motherland" of biology and physiography. Much like the term "bridging science", the notion of a "borderland", or border zone, or even "border science" (Grenzwissenschaft) are part of a stabilizing terminology which serves to describe ecology to this day.[15]

In his attempts to describe ecology in conceptual terms, Thienemann refers repeatedly to limnology. This is the case, for example, with his use of the conceptual scheme of the three stages of ecology, which he apparently published for the first time in his 1942 article *Vom Wesen der Ökologie* (On the essence of ecology). This, too, is an attempt to give ecology a conceptual framework (given that it was a field still relatively untried in scientific practice), using limnology as a model example of an integrating science; this he could do because limnology was already institutionalized and possessed recognizable contours in the form of research programmes (Fig. 19.1).

---

[14] Thienemann 1942, p. 324. In his obituary, G.H. Schwabe stresses that Thienemann's legacy lay in seeing the core task of ecology as one of nurturing links – not only as a bridging science between scientific disciplines, but also between the natural sciences and humanities, between small and large interdisciplinarity, so to speak. Schwabe himself – treading to an extent in Thienemann's footsteps – situates himself in the tradition of a holistic world view and sets his sights on founding a meaning-making ecology, which he believes capable of preventing "modern civilization from drifting without an anchor into the incomparable". He continues: "[E]cology is a logically necessary connecting link between the natural sciences and the humanities and as such is inescapably at the mercy of the tensions and conflicts that are in keeping with its very essence" ([D]ie Ökologie [ist] ein logisch notwendiges Bindeglied zwischen Natur- und Geisteswissenschaften und als solches unvermeidlich den Spannungen und Konflikten ausgesetzt, die seinem Wesen entsprechen; 1961, p. 316).

[15] Thienemann 1927, p. 33. Historian Robert Kohler has developed the concept of "border zone" ecology where objects, concepts and individuals constantly travel back and forth between lab and field (Kohler 2002). The bridge metaphor, too, is taken up time and again as a means of characterizing ecology. The tradition and tenability of the semantics of bridging are reconstructed by Hans Werner Ingensiep and Thomas Potthast in "Brückenschläge – zur Sprache der Ökologie" (Building bridges: On the language of ecology) and "'Ökologie' als Brücke zwischen Wissen und Moral?" ('Ecology' as a bridge between knowledge and morality?) (Busch (ed.) 2007).

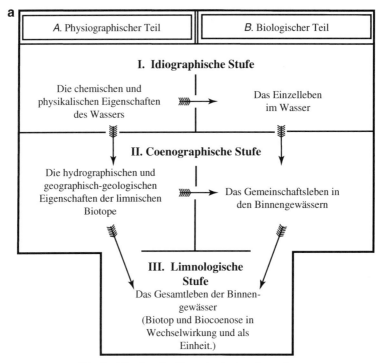

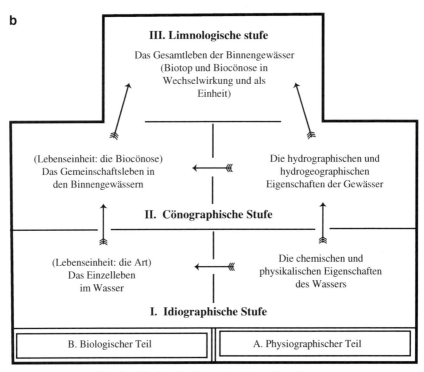

Die drei Stufen der limnologischen Forschung.

Fig. 19.1 (continued)

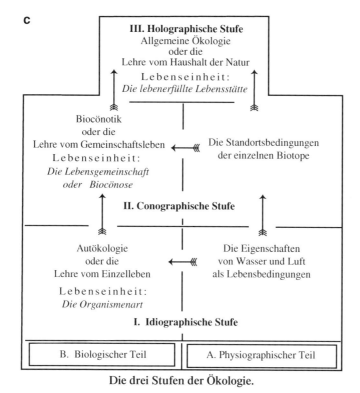

Die drei Stufen der Ökologie.

Fig. 19.1 (continued) (**a**) "The three stages of limnological research" (Thienemann 1925, p. 680) (**b**) "The three stages of limnological research" (Thienemann 1935, p. 18) are placed upside down and reversed sideways and are thus corrected, following the disciplinary mode of reading: In limnological research B goes ahead of A – the biological part takes precedence over the physiographic part (**c**) "The three stages of ecology" (Thienemann 1942, p. 325). In the text, autoecology is identified with the idiographic stage and synecology with the coenographic stage, although Thienemann agrees with Friederichs when he states: "Autoecology is a sub-concept of synecology", seeing in this a confirmation of his hierarchical three stages of integration (Thienemann 1942, p. 316)

Thienemann was careful to anchor the programmatic element of "his limnology" within a bourgeois educational canon, which included as a matter of course a Goethe-inspired natural philosophy and a set of Christian motifs.[16] Typically an encounter occurs here between cosmological and romantic – even sacred – elements. There is

---

[16] The emphasis placed on the personality of the researcher as the key ingredient which gives ecological knowledge that certain something to make it especially valuable and thus to set it apart from "just" making the scientific claim to objectivity was also a part of the bourgeois self-image: "Every scientific study that rises above the average has a markedly personal, that is to say, a subjective touch; and it is often this that makes such a study especially interesting. This is why I think we can only really fully appreciate the work of our scientific colleagues when we know them personally" (Jede wissenschaftliche Arbeit, die sich über das Durchschnittsniveau erhebt, hat eine starke persönliche, also subjektive Note, und diese ist es oft, die eine solche Arbeit besonders interessant macht. Deshalb meine ich, können wir die Arbeiten unserer Fachgenossen erst voll werten, wenn wir unsere Fachgenossen persönlich kennen; Thienemann 1923, p. 4 f.).

much talk, for example, of the "contemplation of the small with a view to the large" (Andacht zum Kleinen mit dem Blick aufs Große), that "the whole always exists prior to the parts" (das Ganze stets vor den Teilen da ist) and that "the parts are a world unto themselves, and yet intermesh with one another harmoniously" (die Teile jeweils eine Welt für sich sind, aber doch harmonisch ineinander greifen), while the lake is described as an "arena of life" (Bühne des Lebens) or as "a world in miniature" (eine Welt im Kleinen). Thienemann, for example, repeatedly establishes the connection to Goethe – this too occurring within a decades-old continuity that outlasts all political and epistemic breaks: In 1939 the "primary axiom of the holistic world view" (erste Axiom ganzheitlicher Weltauffassung; Meyer-Abich 1938 in a preface to Smut's Die holistische Welt) is "verified" by means of a quote from Goethe: "How everything weaves itself into the whole, one in the other works and lives" (Wie alles sich zum Ganzen webt, Eins in dem andern wirkt und webt; 1939, p. 12). This very sentence is also quoted in 1951, again with the reference to the "axiom of holism"; indeed the "intellectual world of Goethe also lives on in today's natural science, according to the authoritative judgment of modern scientific researchers"[17] (1951, p. 580). Finally, to name one last example, in 1954 we find this: "Our intellectual attitude towards nature must not be exhausted in the endeavour to identify its laws as the foundation of our material culture; instead, more than this, it must strive upwards to encompass a view of nature such as the one into which Goethe breathed life, who saw in living nature and in every one of its parts the whole, the great harmony in all disharmonious separate phenomena, and to whom it was a well-ordered whole, a cosmos"[18] (1954, p. 49).

This is the language of many, if not the majority, of the scientists in the formative ecological community. It is a style of writing distinguished by a remarkable degree of continuity and can still be found before, during and even after World War II. The titles of a few selected publications, again by August Thienemann, may serve to illustrate this: 1937 *Lebensgemeinschaft und Lebensraum* (Living community and living space), 1939 *Grundzüge einer allgemeinen Ökologie* (Principles of a universal ecology) accompanied by the motto "Gemeinschaft ist die Lebensform der Natur" (Community is nature's way of life),[19] 1944 *Der Mensch als Glied und Gestalter der Natur* (Man as part of and shaper of nature), 1951 *Vom Gebrauch und vom Mißbrauch der Gewässer in einem Kulturlande* (Of the use and misuse of freshwater in a land of culture), and 1954 *Wasser: das Blut der Erde* (Water: the blood of the earth).

---

[17] "[die] geistige Welt Goethe's [lebt] nach dem berufenen Urteil moderner Naturforscher auch fort in der heutigen Naturwissenschaft".

[18] "Unsere geistige Haltung zur Natur darf sich nicht erschöpfen in dem Bestreben, ihre Gesetze zu erkennen als Grundlage unserer materiellen Kultur; sie muß vielmehr darüber hinaus sich emporringen zu einer Naturanschauung, wie sie Goethe beseelte, der in der lebenden Natur und in jedem ihrer Teile stets das Ganze sah, die große Harmonie bei allen disharmonischen Einzelerscheinungen, dem sie ein wohlgeordnetes Ganzes, ein Kosmos war".

[19] Originally an article in the Archiv für Hydrobiologie (1939, 35, pp. 267–285), the Schweizerbart'sche Verlagsbuchhandlung publishing house also published a brochure bearing this title. Of particular interest here is the form of the text which, apart from the closing comments, is numbered in sections from 1–60, in similar fashion to a manifesto. Historian and philosopher of

It is not hard to recognize that certain more or less radical nation-state tendencies may also be implied here. For example, a passage from Thienemann's writing states "that the biological *foundation* for shaping and consolidating our German world view must be *the* science, which explores the great interrelations in living nature, namely: general ecology"[20] (1942, p. 326). And almost 20 years earlier limnologist Erich Wasmund (1902–1945) had written the following in an article on the scientific provinces in the conservative journal *Deutsche Rundschau* (later instrumentalized by the National Socialist regime): "The language is a part of the style of the innermost being of a people, and thus its specific scientific type becomes essential to it. People and soil create a material conditionality of their scientific types"[21] (1926, p. 245).

A further aspect of holistic conceptual figures becomes clear here, which serve in some sense to elide the political dimension of the concept of nature. Nature is treated as a moral category and is invested as such with an authority that cannot be betrayed. Above all, though, whole and harmonious nature becomes an icon of a way of thinking rooted in cultural pessimism which, during the Weimar period, became an integral part of bourgeois self-perception and was further fuelled, for example, by Oswald Spengler's book *The Decline of the West* and by the philosophical anthropology of Max Scheler, particularly his piece entitled *Die Stellung des Menschen im Kosmos* (The place of man in the cosmos, 1928).

In this society, humans are perceived above all as destroyers of nature; at best they assume the role of benevolent stewards, but nature can definitely not be a discursive political term. This position – one that is critical of civilization and often technophobic as a result – was also widespread in early ecology, where the more or less explicit commitment to a holistic image of nature plays a central, integrating role and is more or less a part of the philosophical core of the discipline. Yet a position such as this, sceptical

---

biology Thomas Potthast highlights the fact that this "little book" appeared "in greater detail" in 1941, entitled "Leben", with several National Socialist slogans added in, while the new editions published in 1956 and 1958 were cleansed linguistically of the National Socialist style "by [the deletion of] about 10 passages" (2003, 252). What this proves is, first, the historical continuity of the rhetoric of holism and its political malleability. Second, though, it also implies criticism of the "mandarin" of ecology, Thienemann, expressed even more clearly at another point in the text: "Currying favour with the powers that be, remaining silent on certain political practices, and biased criticism are by no means necessarily mutually exclusive" (Anbiederung, Schweigen zu bestimmten politischen Praktiken und partielle Kritik müssen sich […] keinesfalls ausschließen [sic!]; 2003, 238). The volume *Leben und Umwelt: Vom Gesamthaushalt der Natur* (Life and Environment: Of the total household of nature) was published in the series rowohlts deutsche enzyklopädie (rde), which bore the motto "Paperback XXth Century Knowledge" and was designed as such to achieve a broad public impact; the fact that it appeared in paperback form – quite a new medium for the German reading public in the 1950s – was intended to contribute to this.

[20]"daß für die Gestaltung und Festigung unserer deutschen Weltauffassung […] [die] biologische Grundlage […] die Wissenschaft sein muß, die die großen Zusammenhänge in der lebenden Natur ergründet: die allgemeine Ökologie."

[21]"Die Sprache ist aber Teil des Stils, Teil des Wesens eines Volkes, und so wird sein spezifischer Wissenschaftstypus ihm wesentlich. […] Volk und Boden schaffen eine Materialbedingtheit ihrer Wissenschaftstypen."

towards civilization, is by no means necessary to a science geared towards holism, either conceptually or politically.[22] Accordingly, the understanding of holism contained in the ecological body of knowledge gets played out in very varied ways: it may refer more to practical knowledge, as envisaged by the planning sciences and scientific nature conservation; or it may be elaborated in a more systems theoretical way, as with the ecosystem models of 1950s American new ecology; or again it may present itself as extremely technophilic, as with the large-scale experiment "Biosphere 2".

Normalizing the Field: The Role of Textbooks

The first tentative steps taken in the new field of research were taken not within academia alone but also, as mentioned above, beyond the confines of universities and scientific institutions[23]: at high schools, museums, forestry, agriculture and fish farming facilities, in the sphere of drinking water supply and that of medical hygiene, as well as in the context of nature research societies in which lay people and scientists engaged in lively exchange.[24] In all these places, issues are addressed which were later to be characterized as "ecological". The natural science curricula were not yet fixed either, so that quite a few ecologists who later became well-known began their research initially as lay people; they included, for example, Josias Braun-Blanquet and Wilhelm Halbfass, who went on to complete their scientific career at a university or as internationally renowned independent scholars.

It is Ernst Haeckel who was credited with giving all these already existing activities a label, even if he himself did not pursue any empirical work in the field which he named "ecology" (see Chap. 16). The major contributions of German-speaking researchers to the formation and maturing of early ecology came from other quarters, before and after the publication of Haeckel's "Oecologie" of 1866. The research projects on the Baltic Sea and the North Sea, conducted by zoologist Karl August Möbius, are a good example of this[25]; they are also a good example of the intertwining of theoretical and applied activities in ecology, of scientific ecology and ecotechnology.

---

[22] Joachim Radkau draws attention to the fact that in nature and environmental conservation circles during the 1930s and 1940s very varied positions regarding "nature" existed which are more adequately described through reference to a polycratic model than to a theory of totalitarianism (2003, p. 43).

[23] Geographer and historian H.-G. Schultz makes the claim for geography that it was done mainly outside scientific institutions in the first instance. In the case of ecology, this is true only for certain areas, most obviously for those inspired by natural history and with their main focus on the observation and description of isolated objects or events.

[24] Unfortunately, this paper can not focus in detail on the relationship between academia and learned societies, even if this is an important issue in early ecology (see, for instance, footnote 4). The article by Patrick Matagne in this volume provides an initial sketch of this interrelationship, whose significance has probably been underestimated.

[25] Möbius' Fauna of the Kieler Bucht, published in 1865, was a book that was seen by historians of biology, as early as the 1920s, as furnishing ecology with a modern research programme and methodology.

Möbius began his field studies with a rather traditional programme of observation, including classical zoological studies on a defined systematic group, but he also addressed certain conceptual issues. This work was published in a book entitled *Fauna der Kieler Bucht* in 1865.[26] A short time later, in 1869, he was commissioned by the Prussian government to study the oyster banks of Schleswig-Holstein; this was related to the crisis in oyster mussel cultivation, where initial evidence suggested overfishing had occurred. He developed a research programme, which eventually resulted in the manuscript *Die Auster und die Austernwirtschaft* (The oyster and oyster-culture) in 1877.[27] It was this publication in which the term "biocoenosis" was coined – not in the sense of a speculative blank concept but as a consequence of his research experience and as a means to structure theoretical work in ecology as well. Accordingly, in Chap. 10 – entitled *An oyster bank is a biocoenosis, or living community* (Eine Austernbank ist eine Biocönose oder Lebensgemeinde) Möbius writes:

> Every oyster bank is in some sense a community of living creatures, a selection of species and a sum of individuals which find in this very place all the conditions for their emergence and maintenance, that is, the right soil, sufficient food, the appropriate salt content and tolerable temperatures favourable for development.[28]

Thus the "oyster project" represented one of the first ecological programmes to involve actual practical research, simultaneously influencing so-called applied sciences such as fisheries and its political and economic environment. In addition, however, biocoenosis represented a novel concept and method, which was taken up by a wider public beyond academia, mainly in tertiary and secondary education. A couple of books were published which navigated the boundary between textbook and specialized work, between general and expert knowledge. Some of these authors knew Möbius personally and referred directly to his studies. The book *Dorfteich als Lebensgemeinschaft* (Village Pond as a Living Community), published in 1885 and written by Friedrich Junge, is in some sense an offshoot of Möbius's teaching activities in teacher training colleges; at the same time it was a book that was to play an important role in the dissemination and popularisation of a way of looking at research objects and one that certainly helped to prepare the field for the appearance of a new discipline called limnology. Thus *Dorfteich* is not merely a textbook or a schoolbook,[29] a text that might be regarded as a means of establishing certain disciplinary

---

[26]This first volume was followed in 1872 by a second, based likewise on studies – mainly observations – conducted by Möbius from 1860 onwards in Kiel Bay with his friend, patron and co-editor of the two volumes, Adolf Meyer-Forsteck. In 1869, at the age of 43, Möbius was appointed Professor of Zoology.

[27]An authorised English version of this work (translated by H.J. Rice) was published in 1883.

[28]"Jede Austernbank ist gewissermaßen eine Gemeinde lebender Wesen, eine Auswahl von Arten und eine Summe von Individuen, welche gerade an dieser Stelle alle Bedingungen für ihre Entstehung und Erhaltung finden, also den passenden Boden, hinreichende Nahrung, gehörigen Salzgehalt und erträgliche und entwicklungsgünstige Temperaturen." Möbius 2006 (1877), p. 75; see also Reise 1980, Jax 2002, pp. 32 ff.

[29]Junge himself notes explicitly that "Village Pond is definitely not intended as a book from which one can teach" (der Dorfteich absolut nicht ein Buch sein soll, aus dem man unterrichten kann, Junge 1885, p. VIII).

facts, theories, and methods by closing the field, creating a sense of "normality", reifying accredited knowledge and excluding open questions. Instead, with an eye to popularization the book additionally opens up a new field of scientific culture, insofar as Junge calls for a "deeper study of life in nature" (tieferes Studium des Lebens in der Natur; Junge 1885, p. IX), advocating that the village pond is an ideal object for illustrating what we can learn about the laws of nature "out there" in its manifold forms. In this respect Möbius's concept of the *Lebensgemeinschaft* appears to Junge to be especially salient because it refers to a delimited space which can be examined and which at the same time facilitates greater knowledge about the whole Earth as a biocoenosis[30]: "Now every last corner could be regarded as a world in itself, so that it later became possible, using such mirror images of the whole, to cast our gaze upon the Earth as the largest living community"[31] (Junge 1885, p. IX).

Other textbooks dealing mainly with ecological issues followed soon afterwards, but none of them offered this particular mix of educational, scientific, anthropogeographical and natural history narratives. Either they were written in a more popularizing style, or else they were oriented towards an explicitly specialized readership. The *Handbuch der Seenkunde. Allgemeine Limnologie* by François-Alphonse Forel (1841–1912) in 1901 might be regarded as one of the first textbooks of this latter style in the field of aquatic ecology. The rapidly growing number of textbooks that appeared over the subsequent three decades points to the disciplinary consolidation of ecology, particularly of aquatic ecology. This is what Friedrich Lenz emphasizes in his *Einführung in die Biologie der Süsswasserseen* (Introduction to the Biology of Freshwater Lakes) in 1928:

> Since hydrology now appears to have reached a certain end point in its initial development, the time would seem to have come to present its research results and research issues in a form which does justice to differing expectations, giving both the autodidact and the research biologist, the teacher and the pupil something to work with. […] This [material] presents a specific topic out of the overall field of freshwater research and yet is quite particularly suited to providing an introduction to the research issues in this field. The lake is virtually paradigmatic for the entire field of hydrobiological research.[32]

---

[30] The expression "the whole Earth as a biocoenosis" is all the more interesting because the concept of "biosphere" was introduced at about the same time. Geographer Eduard Suess used it first in his popular textbook *Das Antlitz der Erde* (The Face of the Earth) 1883–1909. Although Suess did not come up with a research programme (this was developed by Vernadsky only from 1926 onwards), the biosphere eventually succeeded in becoming the concept that described the whole Earth as an organism.

[31] "Nun konnte jeder kleine Winkel als eine Welt für sich betrachtet, und später von solchen Spiegelbildern des Ganzen aus ein Blick auf die Erde als größte Lebensgemeinschaft geworfen werden".

[32] "Da nunmehr die Hydrobiologie einen gewissen Abschluß ihrer ersten Entwicklung erreicht zu haben scheint, dürfte der Zeitpunkt gekommen sein, ihre Forschungsergebnisse und Problemstellungen in einer Form zur Darstellung zu bringen, die den verschiedenen Ansprüchen gerecht wird, die also sowohl dem Autodidakten wie dem forschenden Biologen, dem Lehrer wie dem Schüler etwas gibt. […] Dieser [der Stoff] stellt zwar ein spezielles Thema aus dem Gesamtgebiet der Süßwasserforschung dar, ist aber in ganz besonderem Maße geeignet zur Einführung in die Problemstellung dieses Gebietes. Der See ist geradezu das Paradigma für die ganze hydrobiologische Forschung." (Lenz 1928, p. III).

Although Lenz is still addressing a broader audience – the autodidact, the biologist doing research, the teacher and the student – his textbook is written in a completely different style from that of Junge's text. Historical reflections are restricted to a rather short introduction and concentrate mainly on mentioning the most important scholars in the field. "These kinds of references give students and experts the feeling that they are participants in a solid historical tradition. And yet the tradition conjured up by the textbook in which the scientists think they are participating, actually never existed" (Kuhn 1988, p. 149). This is how philosopher Thomas S. Kuhn assessed the role of textbooks, and the "Introduction to the Biology of Freshwater Lakes" seems to support this statement.

In addition to Lenz's *Biologie der Süsswasserseen* there was August Thienemann's *Limnologie. Eine Einführung in die biologischen Probleme der Süßwasserforschung* (Limnology: An Introduction to the Biological Problems of Freshwater Research) (1926), the first nominally entitled *Einführung in die Limnologie* (Introduction to Limnology) (1930) by Vincenz Brehm, but also, as early as 1909, *Das Leben des Süßwassers* (Freshwater Life) by Ernst Hentschel, followed in 1923 by the *Grundzüge der Hydrobiologie* (Principles of Hydrobiology), covering both marine and lacustrian research objects. A book with a more explicitly zoological focus was *Biologie der Wasserinsekten. Ein Lehr- und Nachschlagewerk über die wichtigsten Ergebnisse der Hydro-Entomologie* (Biology of water insects: A teaching and reference work for the most important results of hydro-entomology) (1934) by H.H. Karny. In the 1930s, elements of "space and nation" were increasingly present in the ecological literature too, good examples being *Ökologie als Wissenschaft von der Natur oder Biologische Raumforschung* (Ecology as a Science of Nature or Biological Research of Space) by Karl Friederichs, and also, certainly much less influential, *Der Süßwassersee* (The Freshwater Lake) by Fritz Steinecke in 1940.

Thus, in the mid-1920s the first step was taken towards the disciplinary consolidation of aquatic ecology, manifested in the successful establishment of journals, research laboratories, a body of researchers, a scientific society, and recognition by the scientific and political establishment.

At about the same time, plant ecology was also beginning to take shape more clearly. There were above all two books, which became what might be considered foundational works: The *Handbuch der Pflanzengeographie* (Handbook of plant geography, 1890) by Oscar Drude and *Pflanzengeographie auf physiologischer Grundlage* (Plant geography upon an ecological basis, 1898) by Andreas Schimper.[33] In the 1920s two other textbook followed which had decisive influence on plant ecology, namely *Pflanzensoziologie. Grundzüge der Vegetationskunde* (Plant sociology: Foundations of phytosociology, 1928) by Josias Braun-Blanquet, and *Einführung in die allgemeine Pflanzengeographie Deutschlands* (Introduction to the general plant geography of Germany, 1927) by Heinrich Walter. Also in animal

---

[33] Both books were widely received also beyond the German language area. According to Tobey (1981), for example, Drude's book was a major influence for the development of Frederic Clements' ideas of ecology.

ecology, the 1920s saw the publication of important textbooks, including especially Friedrich Dahl's *Grundlagen einer ökologischen Tiergeographie* (Principles of ecological animal geography) (1921) and Richard Hesse's seminal work *Tiergeographie auf ökologischer Grundlage* (Animal geography upon an ecological basis, 1924).

Ecological ideas in applied sciences such as agriculture, forestry and pest control were promoted especially by Karl Friederichs in two volumes entitled *Die Grundfragen und Gesetzmäßigkeiten der land- und forstwirtschaftlichen Zoologie* (The basic questions and regularities of agricultural and forest zoology, 1930). Despite its supplementary title – "especially of entomology" – these books were appreciated as a generic contribution to the dissemination and standardization of ecological methods and concepts. Billed as the "ecological part", the first volume deals mainly with theoretical and conceptual questions and offers a discussion of concepts and laws in ecology, of population biology and autecological factors, of the interrelations between organisms and their environment, with special consideration of soil life (edaphon). The second volume, the "economic part", introduces the major issues in forestry and agriculture such as pest control, epidemiology, domesticated animals, breeding and also commercial regulation. Thus the two books offer a synthesis of an ecotechnologically oriented field[34] and of scientific ecology. Friederichs writes: "Recently the entirety of the applied biological disciplines has been brought together by some in a 'technical biology'. This designation is intended to stress that we should be more open to the spirit of invention, technical thinking and technical methods than we have been to date"[35] (1930, Vol. 1, p. 15).

Research Programmes and Networks

This survey of the topology of early German-speaking ecology substantiates the hypothesis that a distinction was made according to the field of objects on land (or "in the air") and in the water, that is, between terrestrial and aquatic ecology. It was a distinction that simultaneously accepted physiographic conditions as a central issue for ecology. Furthermore, it confirms that the field of aquatic ecology was the better organized and institutionalized ecological community, a feature that was also

---

[34]In a general sense, ecotechnology mainly develops local theories and practices and is referred to as applied research, or "use-inspired basic research". Good examples of recent ecotechnologies are restoration ecology, landscape ecology, and industrial ecology. By contrast, ecoscience is characterized by the development of more general concepts and theories, such as the competitive exclusion principle, models of predator–prey relationships, as well as models in ecosystem theory.

[35]"Neuerdings wird von einigen die Gesamtheit der anwendenden biologischen Disziplinen als 'Technische Biologie' zusammengefasst. Durch diese Bezeichnung soll betont werden, daß mehr als bisher Erfindergeist, technisches Denken und technische Methoden bei uns Eingang finden sollten".

recognized at the time.[36] However, the early establishment of common ground in aquatic ecology is reflected not only in the founding of journals and the compiling and disseminating of reference works and textbooks. Terrestrial ecologists also joined together institutionally and epistemically much later. While botanists were able to close the gap relatively quickly, zoologists formed much later into fields such as animal or population ecology – in contrast to the situation, for instance, in the US or the UK. One might argue that research and teaching in the classic disciplines of zoology and botany were oriented towards other scientific ideals and programmes, which were too predominant to allow for autonomous ecological disciplinary activities. By contrast, in the relatively new and institutionally still unstructured field of limnetic systems, zoologists and botanists could situate their inquiries within the subject matter of ecology much more easily.

The different traditions of early German-speaking ecology highlight perfectly the main roots from which ecology originated. We find an important *physiological* or autecological research tradition, which around 1900 merged partially with what we might call the *classificatory* or natural history tradition to become what was then termed the "ecological biogeography of plants and animals". At the same time, we find a *holistic* or systemic tradition which, in different manifestations, extended beyond the biological aspects of nature to encompass a description of whole entities constituted both by biotic and abiotic elements of nature (Jax 1998).

These traditions were also taken up, albeit with differing emphases, in an epistemological approach that offers to characterize ecology according to three conceptual templates, or basic conceptions, namely "microcosm", "niche" and "energy". These allow for a systematic description of the often disparate theories and narratives in ecology, such as those concerning the concepts of population, accommodation or fitness, ideas about the transport of nutrients and organic matter, but also models designed to simulate, for instance, the productivity of a system, or predation between certain units. The notion of "microcosm" is based roughly on a romantic philosophy of nature and approximates to the physiognomic tradition, while "energy" comes close to the holistic tradition and is basically the most physical one: like energy in physics, organic matter in ecology becomes the basic dimension of properties and agency and enables the integration, for instance, of the living and non-living in a single system – an ecosystem. Finally, the basic conception of "niche" concentrates on concepts that were developed within the

---

[36]Thienemann notes retrospectively (1939, p. 13): "As the first of three sub-sciences, limnology has returned to its universal ecological objective and pursues it deliberately (after plant ecology had developed at least partly in the same direction)" ("Die Limnologie hat sich als erste der drei Teilwissenschaften – nachdem sich die Pflanzenökologie zum Teil wenigstens in der gleichen Richtung entwickelt hatte – auf ihre allgemeine ökologische Zielsetzung besonnen und verfolgt sie bewußt.") However, Thienemann certainly did not deny that plant sociology had a tradition stretching back longer than limnology. But in order to genuinely be doing ecology – to be "biocoenotics" – the terrestrial zoologists and botanists would have to join together more closely, as was the case, he claimed, in aquatic ecology (1925, p. 75 f.).

framework of evolutionary biology to characterize the relationship between individual organisms, species and their environments; as such, it is similar to the physiological tradition.[37] These three basic conceptions became decisive for the establishment and stabilization of ecology as a scientific discipline. According to Schwarz, ecology developed as a pluralistic endeavour from the beginning, while each of these basic conceptions ultimately guides more implicitly the building of hypotheses and concepts (see Chap. 8). Following philosopher Imre Lakatos, they are hidden in the "hard core" of a research programme, where these unverbalised ideas of nature irreducibly influence theory building. Another philosopher, Gernot Böhme, proposes from a phenomenological perspective that we talk of the character of nature in analogy to the way we talk about the character of people. Both concepts may be helpful in grasping the idea of the basic conceptions that are to characterize a particular field of ecological knowledge embracing practices, narratives and theories about living beings in both aquatic and terrestrial environments.[38] In the following, both conceptualizations are used complementarily to one another in order to chart the field of early ecology in the German-speaking world.

Ecological Water Affairs

From the eighteenth century onwards, geomorphological and hydrological descriptions of rivers, limited regions of the sea, and lakes emerged.[39] However, these bodies of water were merely perceived as "water deserts" – devoid of life – and it was not until about 1870 that they were widely acknowledged as an environment that played host to a plethora of macro- and microorganisms. A number of concepts were offered to describe these lakes, suddenly full of life and attractive objects to be investigated by zoologists, botanists, and microbiologists. One of the first concepts to circulate in the biological community was that of "the lake as a microcosm"; virtually simultaneously the lake was introduced as an "organism", an "island" and

---

[37] The relation between the pattern of the three "traditions" and the three "basic conceptions" is certainly not always equally close but rather involves a roughly drawn analogy and would require further investigation. However, the two analytical patterns make even more clear the divide between terrestrial and aquatic ecology in early German-speaking ecology. To mention just one important difference, the use of the word "organism" varies between the different traditions. While the "physiological tradition" is closely related to autecology and thus focuses on the single organism, the basic concept "energy" includes models and metaphorical conceptualizations of the organism which refer to the physiology of a lake or parts of the ocean.

[38] See Schwarz Chap. 8 this volume. An example focusing on aquatic microbiology is given in Psenner et al. 2008.

[39] A neat hydrological description of Lac Leman was provided as far back as 1779 by the famous Swiss naturalist Horace-Bénedict de Saussure (1740–1799), and of Lake Constance by David Hümlin in 1783. Rivers were also subject to geomorphological and hydrological measurements at the time (Schwarz 2003a, pp. 116 ff.).

a "geographic individual".[40] The Swiss naturalist François-Alphonse Forel wrote of the lake as a system, a conceptualization that was to become crucial not only in turning limnology into a scientific discipline (1886) but also in providing a first, if still somewhat hazy, blueprint for a systemic understanding of the lake. Forel depicted the lake as a laboratory model for large-scale phenomena also happening "in the immensity of the ocean" yet not accessible to scientific analysis on this scale, concluding: "It is much easier to study a lake than an ocean" (Forel 1896). Forel paved the way for consideration of the lake as an experimental system, looking at the lake as a system in which organisms interact, certain functions are processed, and an exchange and circular flow of materials occurs. The lake is constituted as a system in which abiotic and biotic elements are drawn closely together. In this, the "organic substance" is subject to constant "transformations and migrations", is "incarnated" in living organisms and released again. "The organic substance returns time and again to the great provision house, the lake, be it in the form of animal secretions such as carbon dioxide, urea or other products of the animal metabolic processes, or after the organisms have died as products of decay. [...] The organic substance released provides an inexhaustible, continually renewed supply from which animals and plants renew their structure [...]. This microcosm, to use the term introduced by Prof. S.A. Forbes, would be sufficient unto itself in the long term, even if it were to become completely isolated from all surrounding media"[41] (Forel 1901, pp. 237, 239).

At the end of his handbook Forel proposes a "programme for limnological investigations" in which the biological study of the fauna and flora of a lake is just one of nine headings, the last heading in a long list of issues dominated mainly by physico-chemical and geoscientific perspectives. The programme represents an approach, which focuses mainly on research concerning the cycles of elements in lakes, the functionality of organisms and organic substance in these dynamic cyclical processes, as well as the flow of energy through aquatic systems. Following the conceptual scheme mentioned above, these elements are representative of the template

---

[40] Forel 1901; Otto Zacharias, Skizze eines Spezial-Programms für Fischereiwissenschaftliche Forschungen (Sketch of a special programme for scientific fisheries research) in: Fischerei-Zeitung 7 (1904), pp. 112–115; Über die systematische Durchforschung der Binnengewässer und ihre Beziehung zu den Aufgaben der allgemeinen Wissenschaft vom Leben (On the systematic exploration of the inland waterways and their relationship to the tasks of the general study of life) in: Forschungsberichte der biololgischen Station Plön 12 (1905), pp. 1–39; Das Süßwasserplankton (Freshwater plankton). Leipzig: Teubner 1907; the classic paper The lake as a microcosm by American entomologist Forbes was published in 1887.

[41] "Die organische Substanz kehrt immer wieder in die große Vorratskammer, den See, zurück, sei es in Form von tierischen Sekretionen, wie Kohlensäure, Harnstoff und anderen Produkten der tierischen Verbrennungsvorgänge, oder nach dem Tode der Organismen als Produkte der Verwesung. [...] Die gelöste organische Substanz stellt einen unerschöpflichen, stets erneuerten Vorrat dar, aus dem Tiere und Pflanzen das Material zu erneuertem Aufbau entnehmen [...] Dieser Mikrokosmos, um den von Prof. S.A. Forbes eingeführten Ausdruck zu gebrauchen, würde sich selbst auf lange Zeit genügen können, auch wenn er gegen die umgebenden Medien völlig isoliert würde".

"energy", expressed in concepts relating to the productivity of lakes (Waldemar Ohle and many others), to theories describing lakes as circular causal systems, and to systems ecology in general inspired by cybernetics, in which the analysis of bioenergetic transfers became a central research question (see also Part 10); the latter programme was developed largely in the US, at least to begin with (mainly by George E. Hutchinson, Raymond Lindeman, Howard Odum & Eugene Odum). By contrast, the emphasis in the German-speaking ecological world was on the productivity approach which, in a certain sense, referred to the theoretical idea of an "outer physiology", looking at the physiology of the "organism lake" (or ocean), while at the same time being geared in a less theoretical way towards studies in farming and fisheries.

Forel's "limnological programme" had been spelled out for the first time in 1886 in a French publication and was taken up soon after that by a number of naturalists. These activities also had institutional backing, namely from the limnological commission of the Swiss naturalist society. What started out as a mere observation programme about the freezing of Swiss lakes which brought together laymen and scientists eventually culminated in systematic long-term observations at Lake Zurich and nearby Lake Walen, at Lake Constance and, in particular, at Lake Lucerne. The latter study was unique at the time, not only because of the length of time over which data were collected, but also because of the density of the intervals: this ecological long-term study – probably the first of its kind – began in 1896 and lasted over a period of roughly 5 years.[42] The number of samplings per year was stipulated at the start of the campaign, and a precise description of sites as well as detailed methods were provided. Despite the different disciplinary and institutional background of the naturalists who participated, an attempt was made to systematize and standardize sampling, methods and instruments. The results were published in a series of papers, starting in 1900 and ending in 1917.[43]

Again, for the first time in the short history of limnetic ecology, chemical, physical and biological data were related to one another, facilitating an integrative view of the interrelationships, say, between temperature, quantity of phytoplankton, organic substance and carbon stored in calcium carbonate. Ultimately, it was not only the

---

[42] The programme was described in detail by Xaver Arnet, a secondary school teacher in Lucerne, and was published in the communications of the natural history society of Lucerne in 1895. It also included a call to all members of the society to participate in the data collection, which again documents the institutional openness of the field ("Programm zur limnologischen Untersuchung des Vierwaldstätter Sees. Programm für den physikalisch-chemischen Teil". In: Mitteilungen der Naturforschenden Gesellschaft Luzern 1, pp. 1–16).

[43] To list just a few of these publications: Bachmann, Hans (1904). Das Phytoplankton des Süsswassers (Freshwater phytoplankton), in: Botanische Zeitung 62, 82–103; Burckhardt, Gottlieb (1900). Quantitative Studien über das Zooplankton des Vierwaldstättersees (Quantitative Studies on the zooplankton in the Vierwaldstätter lake), in: Mitteilungen der Naturforschenden Gesellschaft Luzern 3, pp. 129–411, 686–707, 414–434; Amberg, B. (1904). Limnologische Untersuchungen des Vierwaldstättersees (Limnological studies of the Vierwaldstätter lake), in: Mitteilungen der naturforschenden Gesellschaft Luzern 4, pp. 1–142; Nufer, W. (1905). Die Fische des Vierwaldstättersees und ihre Parasiten (The fish of the Vierwaldstätter lake and their parasites), in: Mitt. naturf. Ges. Luzern 5: 1–232.

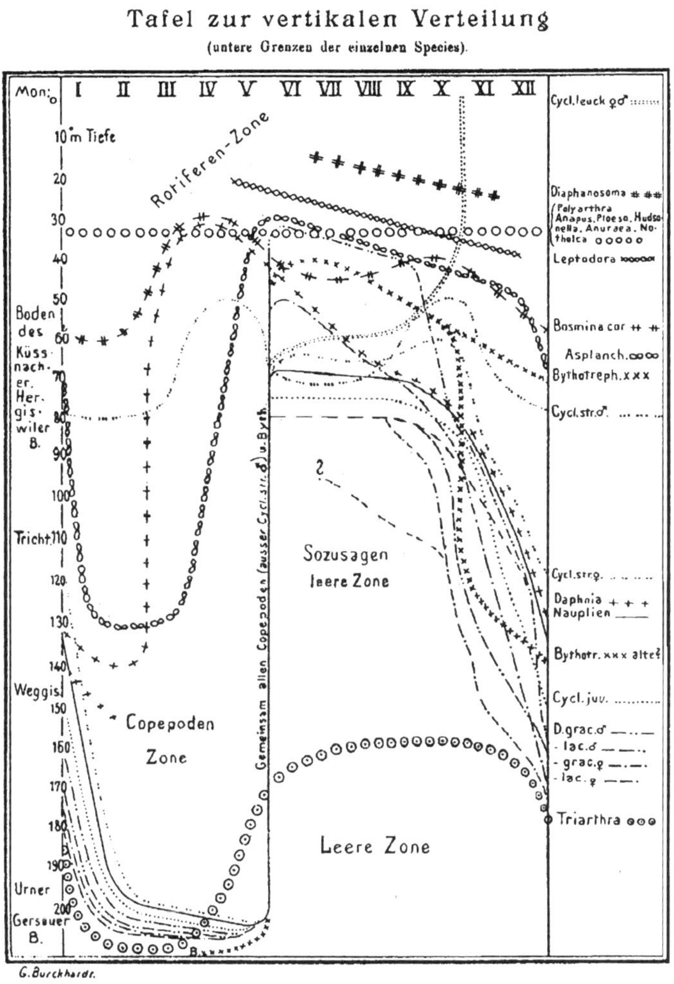

Fig. 19.2 Burckhardt uses the depth ordinate to depict the distribution of zooplankton over time. What makes the figure rather puzzling for contemporary, visual practice is the fact that he includes several sites in the same coordinate space without marking them (1900, p. 424)

research practice – the modality of acting in and on the lake – but also the style of representation that changed profoundly during the course of the project. Forel had already invented the "depth ordinate" (Tiefenordinate) by moving the abscissa to the top and depicting measured data downwards along the ordinate, thereby representing measuring in the lake into depth.[44] This type of diagram was also taken up in the Lake Lucerne studies and used in various representational modes while experimenting with time scales, multiple sampling sites and parameters (Fig. 19.2).

---

[44] A more detailed discussion on the impact of a visual language in aquatic limnology is given in Schwarz (2003b).

All these activities were considered to be rather avant-garde in limnological research, causing German plankton specialist Otto Zacharias to note: "Switzerland can lay claim to the fame of being the classical country not only of lakes but also of lake exploration"[45] (1888, p. 214), a statement he still stood by roughly 10 years later.

Zacharias was not only an admirer of Swiss aquatic research, and especially of Alphonse-François Forel, but was himself quite a well-known plankton specialist. He was interested in collecting plankton species, describing their manifold forms and their collectivization (Vergesellschaftung). More specifically, he took up Forel's main area of interest, namely to discover more about dead and living organic substance. However, for Zacharias this was just one among many other approaches. He was convinced that we need to discover, first of all, the many species not yet known to us, and also to learn much more about the species we do already know. Accordingly, a significant proportion of his publications is devoted to the description of plankton organisms and their behaviour in different lakes, seasons, and plankton communities, although he never delved too deeply into aspects of quantification.[46] In the thoroughly vitalized world of Zacharias, organisms were bound up within superordinate webs of connection. Zacharias's nature rests in a stable balance which prescribes certain functions to its organisms. There is no contradiction between this view and one that accepts the existence of a law of metabolism for the organic substance, a law which, he says, can be discovered in animal and plant communities, as the famous August Weismann had "astutely" established in his treatise on animal life in Lake Constance (1905, p. 31). However, in his description of the way this metabolism worked – which elements in it supposedly function in what way – Zacharias gets only a little further than the description given by Weismann about 30 years previously.[47]

---

[45]"Die Schweiz [...] darf den Ruhm für sich in Anspruch nehmen, das classische Land nicht bloß der Seen, sondern auch der Seendurchforschung zu sein".

[46]Zacharias was surely one of those nature researchers whom Lauterborn somewhat derogatorily referred to as "high climbers" – quoted thus by American limnologist Henry B. Ward in Science: "And what Lauterborn said five years ago is even truer today in the light of our more extended experience: 'For the question as to the distribution of organisms, the methods so cherished even up to the present day of fishing in the greatest possible number of lakes (which recalls, in many respects, the chase after summits on the part of our modern high climbers – Hochtouristen!), really have only limited claim to scientific value, since through them but a very incomplete picture of the faunal character of a water basin can be obtained." (1899, p. 499).

[47]Albeit this was a study which Weismann himself had presented neither as a research report nor to a scientific audience but rather as a "readily accessible talk" which he had "held in the main hall of the University of Freiburg in front of an audience consisting largely of ladies" (Weismann 1877, p. 3). The fact that nonetheless reference was repeatedly made to this talk given by an established scientific authority serves to illustrate the extent to which the status of aquatic ecology was still uncertain around the turn of the century. Another point of reference to do with the content of the talk may well have been that Weismann assumed the transmutation of species through the influence of external conditions and thus supported a Lamarckian rather than a Darwinian conception of evolution. Not until his later "genetic" studies did Weismann come to adopt a Darwinian position.

The descriptions given by Zacharias which proved to be relevant were, at best, those with which an otherwise indeterminate "whole" is replaced by the lake's "household of nature". This *Naturökonomie* (natural economy) can most certainly be measured and its productivity established in quantitative terms – an endeavour in which Zacharias was building on a successful research programme.

However, like many of his contemporaries in early ecology, Zacharias basically specialized in a scientific style geared towards natural history, that is, dedicated to collecting, organizing, putting in order and eventually establishing a physiognomic method. This method was common in geobotany, plant sociology and plant geography, including the description of aquatic plants; but it was also used to develop a typology of lakes and landscapes based mainly on either gestalts or typical animals and plants. To return to the conceptual pattern proposed earlier, all this was typical of the basic concept of "microcosm". The following quote may serve to demonstrate that Zacharias, and many other scientists as well, were devoted to a romantic vision of nature; in it, he vigorously opposes those positions in which nature is claimed to be a machine or something gruesome and threatening: "There is also a biological optimism in which nature, when viewed through its eyes, is most definitely not some eternally devouring, eternally ruminating monster, but is rather a goddess that conjures forth ever new inexhaustible life from death and decay, and whose reign challenges our admiration all the more as we become more intimately familiar with it through serious study"[48] (1907, p. 9).

Literary Genre and Data Representation

Up until the end of the 1920s, taxonomy and data collection were still a dominant part of most research activities in the aquatic realm. In a way, the review given by Kurt Lampert in *Das Leben der Binnengewässer* (The Life of the Inland Waterways),[49] a handbook in three editions expanded continually between 1899 and 1925, was a preliminary culmination of this first period of collection in aquatic ecology. Another project, even more encyclopaedic, was a handbook on plant ecology, *Die Lebensgeschichte der Blütenpflanzen in Mitteleuropa – spezielle Ökologie der Blütenpflanzen Deutschlands, Österreichs und der Schweiz* (The life history of flowering plants in Central Europe – special ecology of the flowering plants of Germany,

---

[48] "Es gibt auch einen biologischen Optimismus, mit dessen Augen angesehen die Natur durchaus kein ewig verschlingendes, ewig wiederkäuendes Ungeheuer ist, sondern vielmehr eine aus Tod und Verwesung immer neues unerschöpfliches Leben hervorzaubernde Göttin, deren Walten unsere Bewunderung um so mehr herausfordert, je genauer wir uns mit ihm durch ernste Studien bekannt machen".

[49] Das Leben der Binnengewässer was first published in 1899 by Kurt Lampert (Leipzig: Tauchnitz), the second edition in 1910 (by then it had grown from 591 to 856 pages), while the third edition of 1925 (892 pages) was edited by R. Lauterborn, V. Brehm and A. Willer. The recent project Die Süßwasserfauna von Mitteleuropa (Freshwater fauna of Central Europe), established by A. Brauer and encompassed in 21 volumes edited by J. Schwoerbel and P. Zwick between 1985 and 2000, also illustrates the fundamental importance of such encyclopaedic projects for ecological research.

Austria and Switzerland), edited by Carl Schröter, Oskar von Kirchner and E. Loew. The first volume was published in 1908 and the last in 1942. In animal ecology, again, classification from an ecological vantage point occurred rather late and was not initially manifested in the form of a handbook or series[50]; instead, this kind of knowledge appeared to be organized mainly in journals such as the *Zoologische Jahrbücher. Abteilung für Systematik, Ökologie und Geographie der Tiere* (Zoological Yearbooks. Department for Systematics, Ecology and Geography of Animals) (1926–1994), the *Zeitschrift für Ökologie und Morphologie der Tiere* (Journal of Ecology and Morphology of Animals) (1924–1967); around this time also, Richard Hesse published his monograph *Tiergeographie auf ökologischer Grundlage* (Animal Geography upon an Ecological Basis) (1924).

A similar type of classificatory work was the *Handbuch der biologischen Arbeitsmethoden* (Handbook of biological study methods). This – as its subtitle states – was a "comprehensive compendium of methods which embraces the entire scientific field of study and research". The handbook, edited by Emil Abderhalden, was begun in 1920 and covered a whole wide range of disciplines, from geology and palaeobiology to physiology and medicine; it also offered a detailed account of physical and chemical methods, instruments and materials. Einar Naumann, August Thienemann, Friedrich Lenz and others contributed to "the methods of freshwater biology" as part of the section entitled *Methoden der Erforschung der Leistungen des tierischen Organismus* (Methods for Researching the Functions of the Animal Organism, 1926).

Another basic literary genre were monographs on lakes and rivers. In some sense, these monographs fit very well with the ecological perspective regarding the individuality of its objects; the idea was to look at the lake as a geographical individual or as an organism that is born and then dies off. Both concepts were fairly common and were consolidated conceptually in the system of lake types, which encouraged aquatic ecologists to regard lakes – including their living and non-living parts – as evolving units. The three-volume *Le Léman: Monographie limnologique* by François-Alphonse Forel (1892–1904) was probably the most influential work here, functioning as a kind of blueprint. Of course, it was written in French, but it was well received nevertheless by the German-speaking community. This was almost certainly because Forel also published in German and, furthermore, was in close correspondence with his German-speaking colleagues. Other limnological monographs of a lake included the *Würmsee* (1901) and the *Ammersee* (1906) by Willi Ule and, to some extent, the *Vegetation des Bodensees* (Vegetation of Lake Constance) (1896, 1902) by Schröter and Kirchner. Insofar as the latter contains first a general scientific characterization of Lake Constance, however, the focus on botanical objects – plankton and macrophytes – is quite explicit. Running water was also the subject of monographic inquiries, most prominently the detailed description of the river Rhine by Robert Lauterborn, starting with several publications on "the geographical and biological structure of the river Rhine" between 1916 and 1918, based on and followed by studies of the Upper

---

[50]Die Ökologie der Tiere (Ecology of Animals) by Fritz Schwerdtfeger, published in three volumes between 1963 and 1975, might be regarded as a belated first series in animal ecology, comparable in some sense to Lampert's Binnengewässer.

Rhine and Lake Constance. In 1930 a first part of the monograph *Der Rhein: Naturgeschichte eines deutschen Stroms* (The Rhine: Natural history of a German river) was published, followed by a second and third part in 1934 and 1938.[51]

Classifying Lakes and Rivers – Relating Type and Process

Also part of the activities engaged in around classification was the development of a system of lake types which, to begin with, followed rather geographical and/or purely botanical or zoological criteria: *Coregonenseen und Zanderseen* (whitefish lakes and pikeperch lakes), *Dinobryonseen und Chlorophyceenseen* (dinobryon lakes and chlorophyceae lakes) are just some of the proposed types. Starting with more specific studies on the physiographic character of a lake, naturalists realized that the distribution and composition of organisms depends critically on the chemical and physical parameters of lake water. This was the start of a very influential programme, which sought to combine causal relations with physiognomic traits. The study of lake types absorbed a number of German and Scandinavian researchers and served as an overall conceptual framework in which much empirical research was embedded. While Thienemann and American limnologists Edward A. Birge (1851–1950) and Chancey Juday (1871–1944) conducted similar studies, they drew very different conclusions from the results. Both were interested in the relationship between physico-chemical conditions and the abundance of plants and animals in a lake. Birge and Juday concentrated mainly on the seasonal and diurnal dynamics of an individual lake, taking a physiological perspective on the lake as a system, whereas Thienemann tried to incorporate his findings into the overall scheme of a conceptual system of lakes. In his classificatory approach, he combined geographical zones, animal indicators and physical and chemical features – such as the thermocline or carbon concentration – and came up with a complex typology that was ultimately more confusing than illustrative, as the following comment rather unintentionally reveals: "The *Chironomus* lake had now acquired the designation 'Baltic lake', since it predominated in that area, whereas the *Tanytarsus* lake was called a 'subalpine lake'; neither was intended as a geographical term".[52]

---

[51] The story of this monograph is interesting in itself, but would go beyond the scope of this account; for more detail, see in RegioWasser e.V. (ed) (2009). 50 Jahre Rheinforschung. Lebensgang und Schaffen eines deutschen Naturforschers Robert Lauterborn (1869–1952). Freiburg: Lavori Verlag.

[52] "Der Chironomussee hatte nunmehr die Bezeichnung ‚baltischer See' erhalten, da er in diesem Gebiet vorherrschte, während der Tanytarsussee ‚subalpiner See' genannt wurde; beide sollten keine geographischen Begriffe sein". This comment was made by Friedrich Lenz, former assistant to Thienemann, at the IV. Hydrobiologische Konferenz der Baltischen Staaten in Leningrad, September 1933. The title of his talk was "Das Seetypenproblem und seine Bedeutung für die Limnologie" (The problem of lake types and its significance for limnology). At the time, the terminology of lake types had already shifted to the terminology of poly-/eu-, meso-, and oligotrophy. Even Thienemann himself had given up the geographical/zoological terminology. As early as 1921 he subsumes the Baltic type within the eutrophic type: "Vor allem im Flachland des Baltikums verbreitet, aber auch in den Alpen vertreten. Häufig in Nordamerika" (Common above all in the Baltic plains, but also found in the Alps. Frequently in North America); 1921, p. 345).

At about the same time, Swedish limnologist Einar Naumann (1891–1934) developed his "regionale Limnologie", the aim of which was likewise to conduct a "causal study of the distribution of types of water bodies in general with their specific world of organisms on Earth" (kausale Studium der Verbreitung der Gewässertypen im allgemeinen mit ihrer speziellen Organismenwelt auf der Erde; 1923, p. 75). From the very beginning Naumann concentrated on describing the productivity of lakes and, as early as 1918, he proposed an interesting typology that remains little known and indeed is not as interesting for the concepts proposed in it as for the method used. First and most notable, it was not based solely on empirical studies from different types of ponds at the fisheries experiment station at Aneboda; rather, Naumann drew his results from an experimental situation outside the laboratory. This approach was rather unusual at the time, as the following comment by Thienemann may illustrate: "The method to be applied in tackling our problem [establishing types of lake] is that of comparative observation in nature; experimental study of individual factors can – as indicated above all by Einar Naumann's studies – provide clarification and greater detail"[53] (1921, p. 344).

In his experimental system Naumann distinguished between a *Naturtypus* (natural type) and a *Kulturtypus* (cultural type), indicating mainly the depth of intervention: the natural type refers to ponds which "are neither given fertilizer nor used as feeding ponds", while the cultural type are ponds "which are either given fertilizer or serve as feeding ponds."[54] He proposes four types and comes to the conclusion that it is possible to consider "the production of phytoplankton, determined both quantitatively and qualitatively, as by far and away a highly accurate indicator of milieu" (die Produktion an Phytoplankton, quantitativ und qualitativ ermittelt, mit grossem Vorteil als einen sehr scharfen Milieu-Indikator verwerten [sic!], 1918b, p. II). He urgently cautions his readers, however, against using such an "algological crediting method" for purposes of fisheries biology as well – the link between the production of fish meat and that of phytoplankton, he says, is too complicated. In the following year he comes up with the concepts of eutrophic and oligtrophic lakes[55] and, two years later, comments: "When Kolkwitz and Marsson first analyzed systematically and in a modern way the effect of organic fertilizers on water (1908, 1909), the system of *saprobes* was established on this basis. Depending on the degree of pollution of the water, these latter were allotted to the zones of polysaprobes, mesosaprobes and oligosaprobes. [...] Now we might well wish to ask to what extent the system

---

[53]"Die Methode, die bei der Bearbeitung unseres Problems [der Aufstellung von Seetypen] anzuwenden ist, ist die vergleichende Beobachtung in der Natur; experimentelles Studium einzelner Faktoren kann, wie vor allem Einar Naumanns Untersuchungen zeigen, Klärung und Vertiefung bringen".

[54]Naumann 1918b, p. II. The quotes originate from a paper published in Swedish; only the summary is in German, and this again is contained "only in the publisher's offprints". Naumann used the biological laboratory in Aneboda for a number of field experiments and for what we today call mesocosm experiments, most likely beginning in 1916.

[55]Naumann points out that "eutroph" and "oligotroph" had been already used in 1907 by C.A. Weber in a study on swamps in Northern Germany (Steleanu 1989, p. 391).

proposed by myself is really very helpful. In order to settle this question, it is necessary to analyse the reciprocal relationships between these ecological systems – which, of course, may be used quite independently of each other. [...] (1) The physiological system proposed by me serves the purpose of a *pure analysis of the factors determining production* each for itself. (2) In contrast to this, the system of saprobes works with the *standard analysis* of water"[56] (1921, p. 19 f., emphasis in original). In conclusion, Naumann proposes a "pure analysis" of the "special factors" in order to investigate ecological systems scientifically using the "trophy standards", while the "standard analysis" indicating the saprobial index is appropriate for applied purposes. He expects that the "milieuspectra" of the different special factors, for instance the N and P household,[57] should then be useful for a "regional mapping of the various sub-spectra" and thus for an evaluation of the "overall biology of waterbodies" (1921, p. 20).

"Physiology" in Aquatic Ecology

The trophic system was quickly adopted in the 1920s, and the same applies for the "milieu-spectra": "Naumann's notion of 'milieu-spectra' has proven to be thoroughly stimulating and fruitful in limnology"[58] commented Thienemann in his review of the "nutrient cycle in water" (1927, p. 43).

Naumann used his method mainly to present the "lake" – however technically modified – as an ecological system which can and must be described in terms of physiological functions and with respect to its physico-chemical conditions; however, this same method was also used to represent the milieu standard (or milieu needs) of a single species. "Physiology" in aquatic ecology could thus refer to the physiology of a whole lake or to the physiology of a single organism (Fig. 19.3).

---

[56] "Als Kolkwitz und Marsson zuerst (1908, 1909) die Einwirkung von organischen Dungstoffen auf das Wasser systematisch in moderner Weise analysierten, wurde auf diesem Grund das System der *Saprobien* begründet. Je nach dem Verschmutzungsgrad des Wassers wurden dieselben in den Zonen der Poly-, Meso- und Oligosaprobien eingereiht. [...] Die Frage dürfte indessen gestellt werden können, inwieweit das von mir vorgeschlagene System wirklich weiter führt. Zur Erledigung dieser Frage ist eine Analyse der gegenseitigen Verhältnisse dieser ökologischen Systeme – die selbstverständlich ganz unabhängig voneinander gebraucht werden können – erforderlich. [...] 1. Das von mir vorgeschlagene physiologische System bezweckt eine *Reinanalyse der produktionsbestimmenden Faktoren* jeder für sich. 2. Das System der Saprobien arbeitet im Gegensatz hierzu mit dem *Durchschnittstandard* des Wassers".

[57] The special factors are gases, temperature, or mineral nutrients, such as calcium carbonate, phosphoric acid, nitrate, but also humin. Temperature, each gas or mineral nutrient has its own budget with a specific spectrum which is also classified in three realms: polytrophy, mesotrophy and oligotrophy (1921, p. 5 f.). Each lake can be evaluated on the basis of these spectra of the budget of each important chemical (or physical) factor.

[58] "Naumanns Gedanke der ‚Milieuspektren' hat sich in der Limnologie als überaus anregend und befruchtend erwiesen".

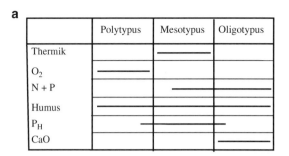

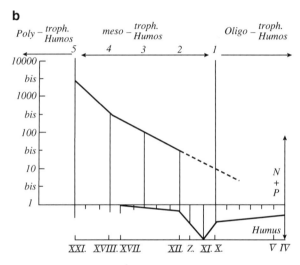

Fig. 19.3 (a) Depicted here are the "milieu needs" of *Holopedium gibberum*, a very common zooplankter which Thienemann had described in detail in the journal *Zoomorphology* in 1926 (Thienemann 1927, p. 43). (b) Hans Utermöhl offers a different, more precise representation of a diatom (*Cyclotella comta*), speaking of a "partial ecological spectrum" (ökologisches Teilspektrum) while commenting that the species need not necessarily occur at every line or level. "However, they may occur there", writes Utermöhl in his *Limnologische Phytoplankton-Studien* of 1925

This was also taken up in other parts of ecology. Zoologist Richard Hesse, in his book on animal ecology, talks about "the optimal conditions of a biotope" when the best possible number of species appears (1924, p. 18). The concept of "milieu standard" and, similarly, "partial ecological spectrum" also come close to today's concept of the environmental needs of a species or the distinction between physiological and ecological optimum.

For a number of limnologists, this kind of research did not go far enough in terms of physiological and morphological analysis of single organisms and ecological investigation of individual populations. Richard Woltereck was one of the scholars who deplored the fact that limnologists were interested only in "general questions of 'production biology', of regional limnology, of types of water bodies, 'milieu-spectra' etc. […] The reason for this phenomenon lies partly in the low esteem in which 'laboratory work' is held by those hydrobiologists who are unable to think physiologically, and partly also, one can assume, in the awkwardness of such work, which is nevertheless indispensable if hydrobiology is to achieve its objective as a science"[59] (1928, p. 543).

This statement is explicitly directed against Carl Wesenberg-Lund (1867–1955), and presumably also against August Thienemann. At the very least, this setting up of different fronts appears to be a polemical strategy as far as the implied concept of "milieu-spectra" is concerned, which – as shown above – undeniably reflects a physiological perspective, was applied to individual organisms and was also based on practical experimental work.

Woltereck was clearly after an experimental approach in his ecological research: He set up a laboratory at the *Biologische Station* in Lunz[60] and later a private laboratory in Seeon in which cladocerans served as laboratory organisms to investigate morphological, physiological and autecological questions. In addition, though, he was interested in field experiments: he exposed *Daphnia cucullata* in Lake Nemi (Italy) as an alien species (indigenous in Denmark), and he organized a survey of the biological repopulation of the same lake after it had been exhausted from water completely.[61] His in situ studies at Lake Lunz resulted in a model, the so-called *Nahrungs-Zehrungs-System* (N/Z-system, food-consumption system), to describe a three-part ecological system consisting of (1) several species of nannoplankton, (2) a population of *Daphnia*, and (3) "summer shoals of whiting" (Weißfische) (1928, p. 544). Before presenting the results of his study, he pointed out that the whole research design was merely a frame and that the main work had yet to be done; moreover, the empirical data he presented were indeed rather flimsy. Even so, he believed that the results of the study showed not only that the "ecological balance of the population numbers" (ökologische Gleichgewicht der Volkszahlen, 1928, p. 548) had been maintained, but that this situation of a system equilibrium represented law-like behaviour and could be described using the model $N : Z = K_Z : P_N$, or, in words: the number of daphnids is in proportion to the number of fish as the daily consumption of daphnids by a fish is in proportion to the daily reproduction of daphnids. Woltereck sought to pursue an ecology based on laws or law-like relationships in order to explain the behaviour of populations, the differences in lake types, and the harmony of a well-shaped system in general. However, his emphasis on explanation and laws in ecology was always accompanied by a strong

---

[59]"die allgemeinen Fragen der 'Produktionsbiologie', der 'regionalen Limnologie', der 'Gewässertypen', 'Milieuspektren' usw. [...] Die Ursache dieser Erscheinung liegt zum Teil in der Geringschätzung der 'Laboratoriumsarbeit' seitens solcher Hydrobiologen, die nicht physiologisch denken können, teilweise wohl auch in der Unbequemlichkeit solcher Arbeiten, deren Inangriffnahme gleichwohl unentbehrlich ist, wenn die Hydrobiologie ihr Ziel als Wissenschaft erreichen will".

[60]Despite being the station's first director, Woltereck never actually did any administrative work there, because he took up a position in Leipzig and later also in Ankara. Franz Ruttner was initially his deputy director, taking on the role of director of the station proper in 1908 (until 1957).

[61]Remarkably enough, this draining was done for the purpose of an archaeological project of national importance that was to recover the galleys said to have been built in ancient times by Caligula.

commitment to the existence of an "autonomous law" and an "inner indiosyncratic activity" of the organism – a view in which he essentially finds himself in agreement with Ludwig von Bertalanffy's biological holism (1940, p. 476). This belief in the autonomy of each individual might also be seen as substantiating his opposition to a type of explanation in biology requiring a supra-individual unity or superordinate power – such as Thienemann's cipher of the lake as "representing the unity of biotope and biocoenosis" (1925, p. 595).

The Physiological Programme in Terrestrial Ecology

The physiological (or autecological), approach in early ecology was, as far as terrestrial zoology and botany are concerned, basically a Darwinian, or at least adaptationist, one. This tradition has remained rather neglected within the history of ecology,[62] although it has been highly influential for the formation of ecology, both in Europe and in the United States. Taking up evolutionary theory, especially in the wake of Darwin's *Origin of Species* (1859), both botanists and zoologist strove to view organisms and their morphology in terms of their adaptedness to their environment. In its early stages this research remained a minor undercurrent of biological research, the mainstream of which was focused on morphology and physiology (as strictly separate fields) in the laboratory. The main type of research spawned by evolutionary theory during the second half of the nineteenth century was phylogenetic morphology, which focused on an analysis of questions relating to taxonomy and developmental biology. However, a group of younger scientists also found their way back from the lab to organisms in their natural habitat. As was the rule during the early years of terrestrial ecology, this occurred separately in zoology and botany and also with varying degrees of success and continuity.

Within botany the individuals who paved the way for an ecological perspective on the morphology and physiology of plants were Simon Schwendener (1829–1919) and Gottlieb Haberlandt (1854–1945). In opposition to a science that – in response to the speculative romantic biology of the early nineteenth century – focused exclusively on "exact", descriptive laboratory science and avoided speculation on the function of specific histological or morphological structures in relation to the lives of organisms, they founded what they called a "physiological anatomy". The most important hallmark of this research programme was the publication of Haberlandt's book *Physiologische Pflanzenanatomie* (Physiological plant anatomy) in 1884.[63] Although not motivated by what we today would call "ecological questions", the basic

---

[62] Although see Höxtermann 2001 and, especially, the elaborate study by Cittadino (1990).
[63] Haberlandt 1884. Five more (enlarged and revised) editions of the book were published before 1924. In 1919, an English translation (of the 4th edition) was published.

aim of this research programme was to identify the adaptive significance[64] of plant structures – involving the rigour of contemporary laboratory biology and even experimental approaches, but with the need to leave the laboratory and examine plants in their natural environments. Only in their natural environments, the researchers felt, could the potential physiological usefulness of morphological studies be elucidated (Schimper 1898, p. IV; Cittadino 1990). In order to see the potential differences in morphological adaptations to the environment more clearly (and supported by the growing colonial ambitions of Germany) many biologists took the opportunity to join expeditions headed for areas they perceived as extreme – or, at least from a European perspective, exotic – environments, especially the tropics, but also the deserts of Northern Africa. What began as a (comparative) analysis of the functional significance of plant structures – guided by an evolutionary and also mostly Darwinian perspective – gradually became a pillar for the ecology of plants in a broader sense. In 1881 one of Schwendener's first students in Berlin, Alexander Tschirsch (1856–1939), wrote his PhD thesis on the relationship between the vegetation zones described by August Grisebach (1814–1879) and the specific anatomical structures of the plants found in them,[65] particularly those of the stomata of the leaves. This was a first, important step in bringing together two hitherto independent research fields: physiology and plant geography.

The idea was taken up by other botanists (e.g. Georg Volkens, 1855–1917, another doctoral student of Schwendener) and culminated in Andreas Franz Wilhelm Schimper's book *Pflanzengeographie auf physiologischer Grundlage* (Plant geography upon a physiological basis), which was published in 1898. A.F.W. Schimper (1856–1901) was not directly related to the Schwendener circle, but was a student of de Bary in Strasbourg, who was opposed to the ideas of Schwendener and his collaborators. Only after his move to the University of Bonn in 1882 and the experiences acquired from his travels to the tropics did he become interested in ecological and biogeographical studies (Schenk 1901). His highly influential book was based in part on his own research but was for the most part a compilation of existing ecological knowledge relating to plant distribution and adaptation. Approaching the subject matter from an evolutionary and selectionist perspective (Cittadino 1990, p. 113), Schimper emphasized the need to examine the causes of the different species of flora as the new goal of plant geography (Schimper 1898, p. III) and argued that ecology had to stay in close contact with experimental plant physiology to serve this purpose (ibid., p. IV), furthering "*ökologische Pflanzengeographie*" (ecological plant geography; ibid.).[66] Schimper died only

---

[64] Although the scientists in this school were united in the aim of looking at the evolutionary significance of plant structures, some of the them (such as Schwendener himself, see Höxtermann 2001, p. 183) were opposed to Darwin's theory of selection and adhered to a more Lamarckian view of the mechanisms of evolution and adaptation.

[65] These vegetation zones were based on the physiognomy of plants; see below.

[66] Schimper made almost no reference to Eugenius Warming's *Plantesamfund*, which was published in 1895 (in Danish; German translation 1896, English translation 1909), whereas Warming credited the German researcher's work on physiological anatomy.

three years after the publication of his book – from a tropical disease contracted during one of his journeys.

The tradition of physiological (or ecological) plant geography was influential – not only in Germany but elsewhere too[67] – and was also well received by zoologists (e.g. Richard Hesse, see below). It did not, however, become the dominant tradition in the new plant ecology, which was influenced to a much greater extent by the classificatory research programme developed in the tradition of Humboldt, Grisebach and Drude and by the different schools of plant sociologists (see below). The most notable follower of a physiological strand of plant geography was Heinrich Walter,[68] who, however, remained an exception within German plant ecology.

The physiological programme, driven by an interest in validating Darwin's evolutionary theory, was also taken up by zoologists but, as in botany, this was beyond the mainstream zoology of the time. The person who took Haeckel's idea of ecology as "external physiology" most seriously and tried to develop it into a research programme was zoologist Carl Semper (1832–1893).[69] Semper, a morphologist at the University of Würzburg, was convinced that it was necessary to collect more empirical evidence for Darwin's theory, stating that "enough had been done in the way of philosophising by the Darwinists, and the task that now lay before us was to apply the test of exact investigations to the hypotheses we had produced in this way". (Semper 1881, p. v).[70] This comment was also a sideswipe at the speculative thinking of Haeckel. Thus Semper tried to combine the historical and comparative approach of evolutionary theory and morphology with the causal and even experimental research of the physiological biology of the time, both in embryology, and – to a lesser degree – with respect to ecology (Nyhart 1995, pp. 177ff.). As early as 1868 he postulated that it was necessary to investigate "the influence of temperature, light, heat, humidity, nutrition, etc. on the living animal" (die Einflüsse des Lichtes und der Wärme, des Feuchtigkeitsgrades, der Nahrung etc. auf die lebenden Thiere) and to find "*oecologische Gesetze*"(ecological laws) (Semper 1868, p. 229). The culmination of this area of his research came in a lecture series which he held in 1877 at the Lowell Institute near Boston and which was subsequently published in 1880 (German edition) and in 1881 (English edition). This book, entitled *Die natürlichen Existenzbedingungen der Thiere*[71] was the first book on animal ecology ever to be published. While Semper still used Haeckel's terminology in his publication of 1868, there was no mention of "ecology" or

---

[67] Thus, for example in the works of Frederic Clements and Henry Chandler Cowles; see Cittadino 1990, pp. 149ff. and Hagen 1988.

[68] Walter's students, especially Heinz Ellenberg and Wolfgang Haber, became highly influential in post-WWII ecology in Germany.

[69] Semper's first name is spelt "Carl" in his early publications and "Karl" in later ones.

[70] In the Geman version of 1880, the text reads thus: "...es sei von den Darwinisten doch schon genug philosophirt, und die Aufgabe träte nun in ihr Recht, die auf diesem Wege gewonnenen Hypothesen durch exacte Untersuchungen zu prüfen." (Semper 1880, Volume 1, p. v)

[71] *Animal life as affected by the natural conditions of existence* is the title of the American edition; the title of the British edition is *The conditions of natural existence as they affect animal life*.

indeed any other reference to Haeckel's wording in the later book.⁷² Making a similar distinction to that made by Haeckel, Semper divides physiology into a "Physiologie der Organe" (physiology of organs; i.e., physiology in the traditional sense) and a "Physiologie der Organismen" (physiology of organisms), characterizing the latter as "that branch of animal biology which regards the species of animals as actualities and investigates the reciprocal relations which adjust the balance between the existence of any species and the natural, external conditions of its existence, in the widest sense of the term."⁷³ His programme becomes explicit in the introduction to the book, where he writes:

> Da wir alle Theile des thierischen Körpers als echte Organe betrachten und die Gesamtsumme ihrer Leistungen die Lebensfähigkeit der Arten bestimmen sehen, so erkennen wir als Aufgabe des Zoologen zu untersuchen, wie die Lebensbedingungen auf die einzelnen Thiere und ihre Organe wirken müssen, um zurück schließen zu können auf die physiologischen Ursachen des Entstehens verschiedener Thierformen.⁷⁴

With Semper's emphasis on explaining the morphological adaptations of animals to their environment – in itself a means to furnish evidence of the causal mechanisms of evolution – it comes as no surprise that Semper's book is essentially (in modern terminology) autecological: it deals with the individual organism (or the species which it represents) and attempts to apply established laboratory methods to organisms in their natural environment. Although his book is – rightly – credited with being the first to point to food chains and trophic pyramids (concepts later formalized by Victor Shelford and Charles Elton), this was done only in passing, namely in dealing with food as one of the influences of the "inanimate" (sic!) environment. In fact, for Semper, ecology served as a tool to explain morphology and evolution. If we look at it differently, with the focus on ecology itself, we might justifiably say that Darwinian evolutionary theory served here as the structuring idea for ecological research.

Although influential, Semper was not able to form any kind of continuing tradition or school in German ecology, nor was he able to carry out his programme of a physiological animal geography himself. Animal ecology, as often acknowledged by

---

⁷² In a footnote to the ninth chapter, Semper (1880, Vol. 2, p. 268; 1881, p. 461) mentions Haeckel as a major proponent of an extremly dogmatic Darwinism. As documented in his lecture *Der Haeckelismus in der Zoologie* (Semper 1876), Semper had already developed an explicit distance towards Haeckel and his work, which he (and others) perceived as overly speculative and therefore unsound.

⁷³ Semper 1881, p. 33. The German version (Semper 1880,Vol. 1, p. 39;) reads: "jenen Theil der Biologie der Thiere [...], welcher die Species der Thiere als Wirklichkeit ansieht und die Beziehungen untersucht, welche zwischen der Existenz einer Art und ihren natürlichen äusseren Existenzbedingungen obwalten (wobei dieser letztere Ausdruck natürlich in seinem weitesten Sinne zu nehmen sein wird)."

⁷⁴ Semper 1880, p. 28. The English version reads: "For since we consider all the parts of the animal body as true organs, and see that the sum total of their functional activity determines the vital fitness of the species, we perceive that it is the task of the zoologist to enquire how the conditions of life must act upon individual animals and their organs, in order to be able to deduce our inferences as to the physiological causes of the origin of different animal forms." (Semper 1881, p. 23).

its protagonists themselves, lagged behind plant ecology[75] and, in particular, focused more on species interactions and communities in Germany. Following Semper's seminal book it took until the 1920s for comprehensive overviews of animal ecology to be produced in Germany. The tradition of physiological ecology and the merging of physiology and biogeography was taken up most succinctly by Richard Hesse (1868–1944). His book *Tiergeographie auf ökologischer Grundlage* (Animal Geography upon an Ecological Basis) (1924) stands in the tradition of Semper and of Schimper. As Hesse writes in his introduction, Schimper's book was a shining example for him, providing the basis on which he tried to model a zoological counterpart text (Hesse 1924, p. V). Like these two scientists Hesse was an explicit Darwinst and saw biogeography within the context of an adaptationist (and selectionist) programme[76]: "Ecological animal geography considers the animals in their dependence on the conditions of the area they live in, their 'adaptedness' to the conditions of their environment, regardless of the geographical location of their living area"[77]; and: "It is thus also one of the most important tasks of ecological animal geography to investigate the 'adaptations' of the animals to their environment."[78]

Like the physiological anatomist, he stresses the importance of experiments and the need to extend biogeography beyond a purely descriptive method: "Everything that has been called a process here is amenable to experimental confirmation and physiological analysis."[79]

Thus Hesse explicitly sees ecology in a physiological tradition: "These very examples demonstrate how ecology is but a continuation and a complement of physiological anatomy; the conditions of the environment are also included in the intellectual linking up of the individual processes."[80] Without neglecting the role of historical events in explaining biogeographical patterns, he expresses optimism that ecology will be able to explain many phenomena by physical and chemical

---

[75] E.g. Hesse 1924. p. 8: "Wir stehen zwar noch am Anfang einer experimentellen Ökologie; besonders sind die Zoologen noch weit hinter den Botanikern zurück." (We still are at the beginning of an experimental ecology; in particular the zoologists are still far behind the botanists.); Hesse 1927, p. 942: "Bis in die neueste Zeit wurden daher physiologische Anatomie und Ökologie vernachlässigt, wenigstens von den Zoologen." (Until very recently physiologogical anatomy and ecology were neglected, at least by the zoologists).

[76] See also Hartmann 1950 for Hesse's position on Darwinism.

[77] "Die ökologische Tiergeographie betrachtet die Tiere in ihrer Abhängigkeit von den Bedingungen ihres Lebensgebietes, in ihrem 'Angepaßtsein' an ihre Umwelt, ohne Rücksicht auf die geographische Lage ihres Lebengebietes" (Hesse 1924, p. 6).

[78] "So ist es auch eine der wichtigsten Aufgaben der ökologischen Tiergeographie die ‚Anpassungen' der Tiere an ihre Umwelt zu untersuchen." (ibid, p. 7).

[79] "Alles, was hier als Vorgang bezeichnet wurde, ist einer experimentellen Bestätigung und physiologischen Analyse zugänglich." (Hesse 1924, p. 8).

[80] "Gerade diese Beispiele zeigen, wie die Ökologie nur eine Fortführung und Ergänzung der physiologischen Anatomie ist; es werden die Bedingungen der Umwelt mit einbezogen in die gedankliche Verknüpfung der Einzelvorgänge." (Hesse 1927, p. 944).

laws: "In animal ecology we thus encounter many events, which we can disassemble into a sequence of processes. It is likely that it will be possible at some time to reduce these processes to physical and chemical laws."[81]

Hesse's work was well received beyond the German–speaking world, and an English translation of his 1924 book was published in 1937.[82]

The physiological and Darwinian approach never became a dominant tradition within German ecology during the first half of the twentieth century, becoming less important in comparison with the classificatory programme (i.e., plant sociology) in plant ecology, the biocoenotic programme in animal ecology, and also the system of lake types in aquatic ecology. A number of reasons may account for this decline in importance. First, as a major driving force behind the programme, Darwinism experienced what Julian Huxley called the "eclipse of Darwinism" (Bowler 1984), marked especially by a great scepticism towards natural selection as the major evolutionary mechanism. Many of the German ecologists who had been trained during the time when anti-Darwinist critique was at its height among biologists, i.e. between the 1880s and the 1930s, completely opposed the idea that the perceived "harmony of nature" was the result of "mere chance"[83] or at least ignored evolutionary issues. Especially in animal ecology, too, the importance of interactions of species within communities (biocoenosis in German ecology) – or, from a functional point of view, the roles of animal species within communities as well – seemed much more conspicuous and important than the animal's relations to abiotic factors, given the greater mobility of most animal species compared with plants. This latter tradition of German animal ecology is expressed most succinctly in the works of Karl August Möbius, August Thienemann, and Karl Friederichs as discussed above.

The Classificatory Programme of Terrestrial Plant Ecology: Plant Sociology

What we have called the classificatory programme here has its main roots in plant geography. While biogeography also became important for the formation of ecology in connection with the physiological approach, as shown above, another development

---

[81]"So begegnen uns in der Ökologie der Tiere vielerlei Geschehnisse, die wir in eine Reihe von Vorgängen auflösen können. Es wird voraussichtlich einmal gelingen, diese Vorgänge auf physikalisch-chemische Gesetzmäßigkeiten zurückzuführen." (Hesse 1927, p. 946).

[82]Mitman (1992, p. 81) comments on the translation of Hesse's book: "The volume was significant because it made available in English one of the first books, apart from Shelford´s Animal Communities in Temperate America, to offer an account of the worldwide distribution of animal life on a physiological as opposed to a historical basis".

[83]E.g. Friederichs 1927, p. 156: "Es gibt die Einheit der Natur. Hätten wir mit diesem Bewußtsein je im Banne der Darwinschen Theorie, die diese Einheit außer acht läßt, stehen können?" (The unity of nature exists. Being aware of this, would we ever have been under the spell of Darwinian theory, which disregards this unity?).

emphasized the distribution and environmental relations not of individual organisms or species but of whole *groups* of species. It thus followed a route later labelled by Schröter and Kirchner (1902) as "synecology", namely: "the science of the plants that live together, and at the same time the science of the plants that seek out analogous ecological conditions."[84]

The classificatory approach is rooted in Humboldtian plant geography, i.e. in the early nineteenth century. Alexander von Humboldt (1769–1859) was the first to describe systematically different recurring groups of plants. Humboldt's classification (Humboldt 1969), which he developed as a result of his travels to South America, was based on physiognomic criteria, i.e. on plant *form* and not plant taxonomy. The physiognomic approach was, however, for Humboldt not a purely scientific one, but also explicitly related to aesthetic and emotional dimensions (see Hard 1969; Kwa 2005). Humboldt's ideas proved to be very influential. They gave rise to several schools of plant geography which eventually, in a long process, developed into an *ecological* plant geography, as one of the major pillars of ecology.[85] In this process, Humboldt's plant geography was gradually "cleansed" of its aesthetic dimensions and enriched with concepts that today we call "ecological". This becomes especially evident with the development of the notion of plant formation, being the first major concept of a unit describing whole assemblages of organisms (here: plants). On the basis of Humboldt's earlier classifications, August Grisebach in 1838 defined the "formation" as an assemblage of plants composed of specific "physiographic character", i.e. based on plant forms.[86]

While Grisebach still referred to the aesthetic-emotional aspects of the physiognomic perspective, his main aim was already to interpret the plant forms as an expression of relations between vegetation and climate.[87] Later authors, especially Grisebach's student Oscar Drude, teaching in Dresden (1890) and Danish botanist Eugenius Warming (1895), completely abandoned Humboldt's aesthetic dimensions. Drude wrote: "It appeared necessary to me to remove the landscape-physiognomic aspects as far as appropriate from the characteristics of the vegetation formations and

---

[84] "Lehre von den Pflanzen, welche zusammen wohnen, und zugleich die Lehre von den Pflanzen, welche analoge ökologische Bedingungen aufsuchen". See also Chap. 14 this volume.

[85] See Trepl 1987 and Nicolson 1996 for an extended account of this transition.

[86] Grisebach 1838/1880, p. 2: "Ich möchte eine Gruppe von Pflanzen, die einen abgeschlossenen physiographischen Charakter trägt, wie eine Wiese, ein Wald usw., eine *pflanzengeographische Formation* nennen. Sie wird bald durch eine einzige gesellige Art, bald durch einen Complex von vorherrschenden Arten derselben Familie charakterisirt, bald zeigt sie ein Aggregat von Arten, die, mannigfaltig in ihrer Organisation, doch eine gemeinsame Eigenthümlichkeit haben, wie die Alpentriften, die nur aus perennirenden Kräutern bestehen." (I would term a group of terms which bears a definite physiognomic character, such as meadow, a forest etc., a *phytogeographic formation*. The latter may be characterized by a single social species, by a complex of dominant species belonging to one family, or finally, it may show an aggregate of species, which, though of various taxonomic character, have a common pecularity; thus the alpine meadow consists almost entirely of perennial herbs.) (translation as in Clements 1916, p. 116f.)

[87] See also Trepl 1987 pp. 103–113 and Du Rietz 1931.

instead bring in the biological element".⁸⁸ And he adds: "Forest, shrubberies, and meadows are different biological communities, which through their aggregation prepare the natural habitat for similar plants or those dependent on them; that they evoke a specific landscape impression is a highly enjoyable add-on, by which this direction of botany becomes dear to the nature lover and valuable to the descriptive geographer".⁸⁹

Bringing in the "biological element" here also meant that Drude brought in the specific *species* (in contrast to just life forms) as crucial components of plant communities, paving the way for the concept of plant association. Together with Andreas Schimper's (1898) systematic efforts to relate plant forms to abiotic conditions and Warming's (1895, 1896) emphasis on biological interactions between the organisms of a plant community, a genuinely *ecological* plant geography (and thus plant ecology) came into being and, with it, synecology in the modern sense.⁹⁰ Drude, Warming and Schimper have often even been considered as the very founding fathers of ecology as a discipline, (e.g. Worster 1985; Trepl 1987) because they merged different strains of biological and geographical research, providing a new perspective on the distribution of organisms, boosting what Allee et al. (1949) called a "self-conscious discipline" of ecology. In any case, the works of these three authors, all published in German, had a tremendous influence not only in the German-speaking countries but far beyond. Thus the young Frederic Clements was strongly influenced by the work of Drude (Tobey 1981), and Arthur Tansley also refers to the books by Warming and Schimper as major inspirations for his work.⁹¹

Ecological plant geography and the ensuing synecological plant ecology was a highly international enterprise from the beginning. Based on the seminal works of Drude, Warming and Schimper, however, and in line with the different emphases of their approaches, different schools of plant ecology developed very quickly.⁹² We will

---

⁸⁸Drude 1890, p. 23: "Es schien mir nämlich nötig, soweit als thunlich das landschaftlich-physiognomische aus den Merkmalen der Vegetationsformationen zu entfernen und dafür das biologische Element hineinzubringen".

⁸⁹"Wälder, Gebüsche und Wiesen sind verschiedene biologische Gemeinden, welche durch ihren Zusammenschluss ähnlich beanlagten oder auf sie angewiesenen Gewächsen die natürlichen Standorte bereiten; dass sie einen bestimmten landschaftlichen Eindruck hervorrufen, ist eine höchst angenehme Zugabe, durch welche diese Richtung der Botanik dem Naturfreunde lieb, dem beschreibenden Geographen wertvoll wird." (ibd.)

⁹⁰Schimper's work (especially Schimper 1898) is located at the boundary between the physiological approach and the synecological one.

⁹¹The original edition of Warming's book *Plantesamfund* was published in Danish in 1895. Most readers outside Denmark, however, used the German version of the book of 1896 (see e.g. Tansley 1947). A (strongly modified) English version of *Planetsamfund* appeared, somewhat tardily, in 1909. On Warming and his seminal book see Anker Chap. 23, this volume.

⁹²Overviews of the different schools of vegetation science can be found in Whittaker 1962; Mueller-Dombois & Ellenberg 1974; Shimwell 1971; Dierschke 1994; for the Russian tradition in particular and its politically hastened decline since the mid-1930s, see Weiner 1984, 1988. The early history of the discipline and its theory, in which the characterization of ecological units plays a major role, is described in Clements 1916; Rübel 1917, 1920; Gams 1918; and Du Rietz 1921.

describe here in particular those relating to the German-speaking world, which focused on an approach that came to be known as *plant sociology*.

The young science of synecological plant ecology (or vegetation ecology) saw heated debates early on, both regarding the definition and "nature" of the basic units of the discipline (in particular plant formation and plant association) and the appropriate methods to describe and classify these units and their dynamics. As a result of divergent opinions regarding the meaning of pivotal terms and the contents of the concepts connected with them, a report and accompanying proposal for the nomenclature of phytogeography (Flahault and Schröter 1910) was elaborated for the III. International Congress of Botany in Brussels. The aim of this proposal was to arrive at a uniform usage of the words "formation", "association" and others. This attempt was initially unsuccessful, however. In addition, a deeper split regarding the view of plant assemblages became apparent around the beginning of the twentieth century, which led to divided traditions of vegetation ecology in continental Europe and the Anglo-Saxon countries, a rift that can be perceived even today. On the one side there was the mostly descriptive and classificatory plant sociology in the tradition of Schröter, Sernander and later especially Braun-Blanquet, Du Rietz, and Tüxen, and on the other the more dynamically oriented vegetation ecology, i.e. oriented towards the notion of "development" and succession, as shaped by Clements, Tansley and Gleason, with a common theoretical basis but also with clear and specific differences.[93]

The main schools of European plant sociology here were the Central European "Zurich-Montpellier-school", founded by Carl Schröter and Charles Flahault, and the Scandinavian "Upsala school", founded by Rutger Sernander. These designations were not strict geographic ones: Austrian botanist Helmut Gams – although an exception – was considered part of the Upsala school, for example. Russia also had – especially up until the 1930s – a theoretically and empirically important tradition of research in plant sociology. The competing schools of plant sociology differed in many theoretical and methodological points, over which arguments were exchanged with great fervour; these disputes were, as Whittaker (1962, p. 27) put it, "waged with the special intensity of a civil war". Differences concerned things as the appropriate spatial dimension of their major ecological unit, namely the plant association, questions of quantitative vs. qualitative methods for its characterization, or the questions which kinds of species should be selected to characterize the association: *constants*, as species occurring in 90% or more of the samples, or *character species*, as species with a narrow ecological amplitude and thus restricted to particlar associations. A more fundamental difference concerned the question if associations only have to be "found" in nature or if are they a mere product of the ordering human mind. Is the association a concrete thing or a class concept? Du Rietz, for

---

[93] The British ecologists (as represented by the British Vegetation Committee, the predecessor of the British Ecological Society) included a whole successional sere into their definitions and classifications of plant formation, which most central European ecologists rejected as too full of hypotheses (see Flahault and Schröter 1910).

example insisted on the "reality" of the association: "The associations like the species are not produced in scientific treatments and textbooks. *They are species combinations existing in nature and delimited more or less sharply by nature itself.*"[94] In contrast, Braun-Blanquet in the same year (1921) emphasized: "There is, by and large, agreement that the association as well as the species is an abstraction, while we are faced with single association individuals and local stands in nature."[95] This question, about the "reality" of the plant association, evoked a flood of controversial publications, and Du Rietz called the issue in 1928 "one of the most important and burning main problems of modern plant sociology" (eine[s] der wichtigsten und aktuellsten Hauptprobleme der modernen Pflanzensoziologie; Du Rietz 1928, p. 20).[96]

Although there had always been publications in each native language (in particular French, Danish, Swedish, Norwegian and Russian), the main working language in continental European plant sociology was German. Plant sociology was an explicit research programme and became for a long time the dominant research tradition of plant ecology in the German-speaking world.

Josias Braun-Blanquet, a student of Flahault and Schröter and soon the leading figure in the Zurich-Montepellier school, developed the most detailed and most influential research programme for plant sociology, laid out in detail in his book *Pflanzensoziologie* (Plant Sociology) of 1928.[97] In the first pages of this book, Braun-Blanquet describes the status and overall aims of the field. First of all, "plant sociology" is considered to be synonymous with "vegetation science", which he, like many others (e.g. Du Rietz 1921) perceives as a distinct discipline, and not just as a sub-discipline of ecology or geography (Braun-Blanquet 1928, p. III). The object of the discipline is the "plant community as a social unit"[98] and its "clear but remote aim"[99] is "the characterisation and description of the social units, their causal explanation, the study of their development and distribution and their clear and systematic arrangement".[100] The research field is then characterized (Braun-Blanquet 1928, p. 1f.) by five "main problems", relating to (1) the organization and structure of plant societies, (2) their ecological relations (3) their (successional) development, (4) their spatial distribution and (5) their classification and systematics.

---

[94] Du Rietz 1921, p. 15; emphasis in original. "Die Assoziationen ebenso wie die Arten werden nicht in wissenschaftlichen Abhandlungen und Lehrbüchern fabriziert. *Sie sind in der Natur existierende, durch die Natur selbst mehr oder minder scharf und deutlich abgegrenzte Artenkombinationen*".

[95] Braun-Blanquet 1921, p. 311. "Man ist heute im grossen ganzen darüber einig, dass die Assoziation so gut wie die Art eine Abstraktion darstellt, während uns in der Natur einzelne Assoziationsindividuen oder Lokalbestände entgegentreten."

[96] See Jax 2002, pp. 110ff. for an more detailed analysis of this controversy.

[97] English translation 1932.

[98] "Pflanzengesellschaft als soziale Einheit"; ibid.

[99] "klares aber fernes Ziel"; ibid.

[100] "die Fassung und Beschreibung der Gesellschaftseinheiten, ihre kausale Erklärung, das Studium ihrer Entwicklung und Verbreitung und ihre übersichtliche systematische Anordnung"; ibid.

In a similar manner, Du Rietz, a student of Sernander and the dominant exponent of the Uppsala school, explained some years earlier what he considered to be the "*Endziel*" (ultimate aim) of plant sociology, namely:

> "eine[r] allseitige[n] Kenntnis von den in der Natur existierenden Pflanzengesellschaften, ihrem Aussehen, ihrer Zusammensetzung und ihrem inneren Bau, ihrer Entstehung und ihren Veränderungen, ihrer Verbreitung und Verteilung auf der Erde, ihren Lebensverhältnissen und ihrer Sukzession, nicht aber darin, daß man einzelne von diesen vielseitigen Forschungsaufgaben auf ein Piedestal über alle übrigen erhebt."[101]

In spite of many theoretical and methodological differences between the two schools of plant sociology, the overall research programme was thus very similar, and equally broad and ambitious. The direction and actual practice of research during the following decades, however, was much narrower, which already was anticipated by both Braun-Blanquet and Du Rietz. They both saw the *description and assessment of plant societies* as the first (although transient) task of plant sociology.[102] Description and classification was in fact the major work done within plant sociology in the German-speaking countries up until the end of WWII.[103] It was ecological in the sense of Drude's and Schimper's work, as it laid great stress on elucidating the relationships between plant communities and their (abiotic) environment. What was largely lacking, however, was an investigation of the species' relationships within the communities, as emphasized by Warming. The actual *classificatory* programme of terrestrial plant ecology in effect provided a huge amount of valuable empirical and spatially concrete data on the distribution of plant communities and their relation to factors such as climate and soil, as well as new methodologies. However, it did not produce much theoretical progress in the explication of plant communities and their dynamics.

Summary

To draw this overview to an end, let us briefly recapitulate some of the issues that have arisen in our account of German-speaking ecology. German-speaking ecology develops in two strands from the beginning – aquatic and terrestrial ecology. These evolve at different rates. This becomes apparent with respect to the institutional, cognitive and epistemic context. In aquatic ecology a process of institutionalization

---

[101] Du Rietz 1921, p. 248: "a general knowledge of the plant societies existing in nature, their appearance, their composition and their internal structure, their origin and their changes, their distribution and arrangement on the earth, their living conditions and their succession, but [its aim is] not to raise up any single one of these various research tasks on a pedestal above all the others".

This text appears towards the very end of Du Rietz's PhD thesis about the methodological foundation of modern plant sociology" (Zur methodologischen Grundlage der modernen Pflanzensoziologie).

[102] Du Rietz (ibid.), referring to his statement above, said that if one aspect were to gain some prominence, then it would have to be "the determination of the plant societies existing in nature and of their natural boundaries" (die Feststellung der in der Natur existierenden Pflanzengesellschaften und ihrer natürlichen Grenzen) – as a necessary prerequiste for all the other tasks.

[103] See e.g. Dierschke 1994, p. 20.

takes place between the 1890s and the 1920s: laboratories and field stations, positions, scientific societies and journals are all established. Terrestrial ecology – especially plant ecology and plant geography – also develops these structures, yet remains a rather disparate field, especially when it comes to animal ecology, until about the 1920s. We have attempted to take account of these differences by proposing two different conceptual patterns to accommodate the different structures and transformations in the respective fields of knowledge. Of course, we are aware that this is just an initial historical reconstruction, which still contains many unrelated fragments and cognititve fissures. Despite this, we are convinced that it is time to rearrange at least some of the narratives about German-speaking ecology while adjusting the vantage point on hitherto unseen ruptures and associations, most of them due to these differences between aquatic and terrestrial ecology. Taking them into consideration will hopefully facilitate the development of a novel and productive perspective on the constitution of ecological knowledge in the German-speaking world.

References

Allee WC, Emerson AE, Park O, Park T, Schmidt KP (1949) Principles of animal ecology. Saunders, Philadelphia
Bowler PJ (1984) Evolution. The history of an idea. University of California Press, Berkeley
Braun-Blanquet J (1921) Prinzipien einer Systematik der Pflanzengesellschaften auf floristischer Grundlage. Jahrbuch Sankt Gallener Naturwissenschaftlichen Ges 57:305–351
Braun-Blanquet J (1928) Pflanzensoziologie. Springer, Berlin
Brehm V (1930) Einführung in die Limnologie. Springer, Berlin
Burckhardt G (1900) Quantitative Studien über das Zooplankton des Vierwaldstättersees. Mitt Naturforschenden Ges Luzern 3:129–411, 686–707, 414–434
Burkamp W (1929) Die Struktur der Ganzheiten. Junker und Dünnhaupt, Berlin
Busch B (ed) (2007) Jetzt ist die Landschaft ein Katalog voller Wörter. Beiträge zur Sprache der Ökologie. Valerio 5, Die Heftreihe der Deutschen Akademie für Sprache und Dichtung. Göttingen, Wallstein
Cittadino E (1990) Nature as the laboratory. Darwinian plant ecology in the German Empire, 1880–1900. Cambridge University Press, Cambridge
Clements FE (1916) Plant succession. An analysis of the development of vegetation. Carnegie Institution of Washington, Washington, DC, Publication No. 242
Dahl F (1921) Grundlagen einer ökologischen Tiergeographie. Gustav Fischer, Jena
Daum AW (1998) Wissenschaftspopularisierung im 19. Jahrhundert. Bürgerliche Kultur, naturwissenschaftliche Bildung und die deutsche Öffentlichkeit, 1848–1914. Oldenbourg Verlag, München
Dierschke H (1994) Pflanzensoziologie. Ulmer, Stuttgart
Drude O (1890) Handbuch der Pflanzengeographie. Verlag von J. Engelhorn, Stuttgart
Du Rietz GE (1921) Zur methodischen Grundlage der modernen Pflanzensoziologie. Adolf Holzhausen, Wien
Du Rietz GE (1928) Kritik an pflanzensoziologischen Kritikern. Botaniska Notiser 1–30
Du Rietz GE (1931) Life-forms of terrestrial flowering plants. Acta Phytographica Suecica III:1–95
Flahault C, Schröter C (eds) (1910) Phytogeographische Nomenklatur. III. Internationaler Botanischer Kongress, Brüssel 1910. Zürcher & Furrer, Zürich
Forel F-A (1886) Programme d'études limnologiques pour les lacs subalpins. Arch Sci Phys Nat 3:548–550
Forel F-A (1892–1904) Le Léman. Monographie limnologique, vol 1–3. F. Rouge, Lausanne

Forel F-A (1901) Handbuch der Seenkunde. Allgemeine Limnologie. Engelhorn, Stuttgart

Forel FA (1896) La limnologie, branche de la géographie. Rep. Sixth Int. Geogr. Congress held in London 1895 593–602

Frič A, Václav V (1894) Untersuchungen über die Fauna der Gewässer Böhmens IV. Die Thierwelt des Unterpočernitzer und Gatterschlager Teiches. Archiv für die naturwissenschaftliche Landesdurchforschung von Böhmen 9, Prag

Friederichs K (1927) Grundsätzliches über die Lebenseinheiten höherer Ordnung und den ökologischen Einheitsfaktor. Naturwissenschaften 8:153–157, 182–186

Friederichs K (1930) Die Grundfragen und Gesetzmäßigkeiten der land- und forstwirtschaftlichen Zoologie (insbesondere der Entomologie), vol 1, 2. Verlagsbuchhandlung Paul Parey, Berlin

Gams H (1918) Prinzipienfragen der Vegetationsforschung. Ein Beitrag zur Begriffsklärung und Methodik der Biocoenologie. Vierteljahresschridft der Naturforschenden Gesellschaft Zürich 63:293–493

Grisebach A (1838) Über den Einfluß des Klimas auf die Begrenzung der natürlichen Floren. In: Grisebach A (ed) Gesammelte Abhandlungen und kleinere Schriften zur Pflanzengeographie. Verlag von Wilhelm Engelmann, Leipzig, pp 1–29

Haberlandt G (1884) Physiologische Pflanzenanatomie. Engelmann, Leipzig

Hagen JB (1988) Organism and environment. Frederic Clements's vision of a unified physiological ecology. In: Rainger R, Benson KR, Maienschein J (eds) The American development of biology. University of Pennsylvania Press, Philadelphia, pp 257–280

Hard G (1969) "Kosmos" und "Landschaft". Kosmologische und landschaftsphysiognomische Denkmotive bei Alexander von Humboldt und in der geographischen Humboldt-Auslegung des 20. Jahrhunderts. In: Pfeiffer H (ed) Alexander von Humboldt. Werk und Weltgeltung. Piper-Verlag, München, pp 133–177

Hartmann M (1950) Nachruf auf Richard Hesse. - Jahrbuch der Deutschen Akademie der Wissenschaften zu Berlin, 1946–1949, pp 160–170

Hentschel E (1909) Das Leben des Süßwassers. Eine gemeinverständliche Biologie. Ernst Reinhardt, München

Hentschel E (1923) Grundzüge der Hydrobiologie. Gustav Fischer, Jena

Hesse R (1924) Tiergeographie auf ökologischer Grundlage. Gustav Fischer, Jena

Hesse R (1927) Die Ökologie der Tiere, ihre Wege und Ziele. Naturwissenschaften 15:942–946

Höxtermann E (2001) Die Schwendener-Schule der Physiologischen Anatomie - ein "Grundpfeiler" der Pflanzenökologie. Verhandlungen zur Geschichte und Theorie der Biologie 7:165–189

Jax K (1998) Holocoen and ecosystem. On the origin and historical consequences of two concepts. J Hist Biol 31:113–142

Jax K (2002) Die Einheiten der Ökologie. Analyse, Methodenentwicklung und Anwendung in Ökologie und Naturschutz. Peter Lang, Frankfurt

Junge F (1885) Der Dorfteich als Lebensgemeinschaft. Lipsius & Tischer, Kiel

Karny HH (1934) Biologie der Wasserinsekten. Ein Lehr- und Nachschlagewerk über die wichtigsten Ergebnisse der Hydro-Entomologie. Fritz Walter, Wien

Kluge T, Schramm E (1986) Wassernöte. Sozial- und Umweltgeschichte des Trinkwassers. Alano-Verlag, Aachen

Kohler RE (2002) Landscapes and labscapes: Exploring the lab-field frontier in biology. The University of Chicago Press, Chicago

Kretschmann C (2006) Räume öffnen sich. Naturhistorische Museen im Deutschland des 19. Jahrhunderts. Akademie-Verlag, Berlin

Kuhn TS (1988) Die Struktur wissenschaftlicher Revolutionen. Suhrkamp, Frankfurt am Main

Kwa C (2005) Alexander von Humboldt's invention of the natural landscape. Eur Legacy 10:149–162

Lenz F (1928) Einführung in die Biologie der Süsswasserseen. Biologische Studienbücher, vol IX. Berlin, Julius Springer

Mitman G (1992) The state of nature. Ecology, community, and American social thought, 1900–1950. University of Chicago Press, Chicago

Möbius KA (1883) The oyster and oyster culture. Report of the comissioner for 1880. United States Comission of Fish and Fisheries. Government Printing Office, Washington, DC, pp 683–751

Möbius KA (2006) Zum Biozönose-Begriff. Die Auster und die Austernwirtschaft 1877 (2nd ed. by Thomas Potthast; 1st edition and comment by Günther Leps 1986). – Frankfurt am Main: Harri Deutsch

Mueller-Dombois D, Ellenberg H (1974) Aims and methods of vegetation ecology. Wiley, New York

Müller-Navarra S (2005) Ein vergessenes Kapitel der Seenforschung. Martin Meidenbauer Verlagsbuchhandlung, München

Naumann E (1918a) Försök angående vissa avfallsprodukters och gödselämnens inverkan på vattnets biologi. Särtryck Ur Skrifter, Utgivna Av Södra Sveriges Fiskeriförening 1917 (3–4):10–44

Naumann E (1918b) Undersökningar över fytoplanktonproduktionen i dammar vid aneboda 1917. Sartryck Ur Skrifter, Utgivna Av Södra Sveriges Fiskeriförening 1:62–75

Naumann E (1921) Einige Grundlinien der regionalen Limnologie. Lunds Univesitets Årsskrift NF 17:1–22

Nicolson M (1996) Humboldtian plant geography after Humboldt: the link to ecology. Br J Hist Sci 29:289–310

Nyhart LK (1995) Biology takes form. Animal morphology and the German universities, 1800 – 1900. University of Chicago Press, Chicago

Pörksen U (2001) Was spricht dafür das Deutsche als Naturwissenschaftssprache zu erhalten? Abhandlungen der Deutschen Akademie der Naturforscher Leopoldina NF 87:5–31

Potthast T (2003) Wissenschaftliche Ökologie und Naturschutz: Szenen einer Annäherung. In: Radkau J, Uekötter F (eds) Naturschutz und Nationalsozialismus. Campus, Frankfurt, pp 225–256

Psenner R, Alfreider A, Schwarz AE (2008) Aquatic microbial ecology: water desert, microcosm, ecosystem. What comes next? Int Rev Hydrobiol 93:606–623

Radkau J (2003) Naturschutz und Nationalsozialismus – wo ist das Problem? In: Radkau J, Uekötter F (eds) Naturschutz und Nationalsozialismus. Campus, Frankfurt, pp 41–55

RegioWasser eV (ed) (2009) 50 Jahre Rheinforschung. Lebensgang und Schaffen eines deutschen Naturforschers Robert Lauterborn (1869–1952). Lavori Verlag, Freiburg

Rübel E (1917) Anfänge und Ziele der Geobotanik. Vierteljahresschrift der Naturforschenden Gesellschaft Zürich 62:629–650

Rübel E (1920) Die Entwicklung der Pflanzensoziologie. Vierteljahresschrift der Naturforschenden Gesellschaft Zürich 65:573–604

Schenk H (1901) A.F. Wilhelm Schimper. Berichte der Deutschen Botanischen Gesellschaft 19:954–970

Schimper AFW (1898) Pflanzengeographie auf physiologischer Grundlage. Gustav Fischer, Jena

Schröter C, Kirchner O (1896) Vegetation des Bodensees. 1. Band. Stettner, Lindau

Schröter C, Kirchner O (1902) Die Vegetation des Bodensees, 2. Teil. - Lindau: Kommissionsverlag der Schriften des Vereins der Geschichte des Bodensees und seiner Umgebung von Joh. Stettner, Thom

Schwabe GH (1961) August Thienemann in memoriam. Oikos 12:310–316

Schwarz AE (2003a) Wasserwüste - Mikrokosmos - Ökosystem. Eine Geschichte der Eroberung des Wasserraumes. Rombach-Verlag, Freiburg

Schwarz AE (2003b) Die Ökologie des Sees. Diagramme als Theoriebilder. Bildwelten des Wissens Kunsthistorisches Jahrbuch für Bildkritik 1:64–74

Semper K (1868) Reisen im Archipel der Phillipinen. Zweiter Teil: Wissenschaftliche Resultate. Erster Band: Holothurien. Verlag von Wilhelm Engelmann, Leipzig

Semper K (1876) Der Haeckelismus in der Zoologie. W. Maukes Söhne, Hamburg

Semper K (1880) Die natürlichen Existenzbedingungen der Thiere. Brockhaus, Leipzig

Semper K (1881) Animal life as affected by the n-atural conditions of existence. D. Appleton & Co, New York

Shimwell DW (1971) The description and classification of vegetation. University of Washington Press, Seattle

Steinecke F (1940) Der Süßwassersee. Die Lebensgemeinschaften des nährstoffreichen Binnensees. Quelle und Meyer, Leipzig

Steleanu A (1989) Geschichte der Limnologie und ihrer Grundlagen. Haag und Herchen, Frankfurt am Main
Tansley AG (1947) The early history of modern plant ecology in Britain. J Ecol 35:130–137
Thienemann A (1921) Seetypen. Naturwissenschaften 9:343–346
Thienemann A (1923) Zwecke und Ziele der Internationalen Vereinigung für theoretische und angewandte Limnologie. Verhandlungen der Internationalen Vereinigung für theoretische und angewandte Limnologie 1:1–5
Thienemann A (1925) Der See als Lebenseinheit. Naturwissenschaften 13:489–600
Thienemann A (1926) Limnologie. Eine Einführung in die biologischen Probleme der Süßwasserforschung. - Breslau
Thienemann A (1927) Der Nahrungskreislauf im Wasser. 31. Jahresversammlung zu Kiel 1926. Zoologischer Anzeiger (Verhandlungen der Deutschen Zoologischen Gesellschaft 31) 2(Supplementband):29–79
Thienemann A (1933) Vom Wesen der Limnologie und ihrer Bedeutung für die Kultur der Gegenwart. Verhandlungen der Internationalen Vereinigung für theoretische und angewandte Limnologie 6:21–30
Thienemann A (1935) Die Bedeutung der Limnologie für die Kultur der Gegenwart. Schweizerbart'sche Verlagsbuchhandlung, Stuttgart
Thienemann A (1939) Grundzüge einer allgemeinen Ökologie. Schweizerbart'sche Verlagsbuchhandlung, Stuttgart
Thienemann A (1942) Vom Wesen der Ökologie. - Biologia Generalis 3/4 (special edition):312–331
Thienemann A (1951) Vom Gebrauch und vom Mißbrauch der Gewässer in einem Kulturlande. Arch Hydrobiol 45:557–583
Thienemann A (1954) Wasser - Das Blut der Erde. In: Uns ruft der Wald. Handbuch der Schutzgemeinschaft Deutscher Wald. Rheinhausen: Verlagsanstalt Rheinhausen, pp 45–49
Tobey RC (1981) Saving the prairies. The life cycles of the founding school of American plant ecology, 1895–1955. University of California Press, Berkeley
Trepl L (1987) Geschichte der Ökologie. Vom 17. Jahrhundert bis zur Gegenwart. Athenäum, Frankfurt am Main
Tümmers HJ (1999) Der Rhein. Ein europäischer Fluss und seine Geschichte. Beck, München
Ule W (1901) Der Würmsee (Starnbergersee) in Oberbayern, eine limnologische Studie. Leipzig
Ule W (1906) Studien am Ammersee in Oberbayern. Riedel, München
Utermöhl H (1925) Limnologische Phytoplanktonstudien: Die Besiedelung ostholsteinischer Seen mit Schwebpflanzen. - Archiv für Hydrobiologie, Suppl. 5
von Humboldt A (1969) In: Meyer-Abich A (ed) Ansichten der Natur. Reclam, Stuttgart
Walter H (1927) Einführung in die allgemeine Pflanzengeographie Deutschlands. Gustav Fischer, Jena
Ward HB (1899) The freshwater biological stations of the world. Science 9:497–507
Warming E (1895) Plantesamfund. Grundtræk af den økologiske plantegeografi S. Philipsen, Kjobenhavn
Warming E (1896) Lehrbuch der ökologischen Pflanzengeographie. Eine Einführung in die Kenntnis der Pflanzenvereine. Gebrüder Bornträger, Berlin
Warming E (1909) Oecology of plants: An introduction to the study of plant communities. Oxford University Press, Oxford
Wasmund E (1926) Wissenschaftsprovinzen. Deutsche Rundschau 52(12):243–253
Weiner DR (1984) Community ecology in Stalin's Russia: "Socialist and bourgeois" science. Isis 75:684–696
Weiner DR (1988) Ecology, conservation, and cultural revolution in Soviet Russia. Indiana University Press, Bloomington & Indianapolis
Weismann A (1877) Das Thierleben im Bodensee. Schriften des Vereins für Geschichte des Bodensees und seiner Umgebung 7:132–161
Whittaker RH (1962) Classification of natural communities. Bot Rev 28:1–239

Woltereck R (1928) Über die Spezifität des Lebensraumes, der Nahrung und der Körperformen bei pelagischen Cladoceren und über "Ökologische Gestalt-Systeme". Biologisches Zentralblatt 48:521–551

Woltereck R (1940) Ontologie des Lebendigen. Ferdinand Enke, Stuttgart

Worster D (1985) Nature's economy. A history of ecological ideas. Cambridge University Press, Cambridge

Zacharias O (1888) Vorschlag zur Gründung von zoologischen Stationen behufs Beobachtung der Süßwasser-Fauna. Zool Anz 11:18–27

Zacharias O (1904) Skizze eines Spezial-Programms für Fischereiwissenschaftliche Forschungen. Fischerei-Zeitung 7:112–115

Zacharias O (1905) Über die systematische Durchforschung der Binnengewässer und ihre Beziehung zu den Aufgaben der allgemeinen Wissenschaft vom Leben. Forschungsberichte aus der biologischen Station Plön 12:1–39

Zacharias O (1907) Das Süßwasserplankton. Teubner, Leipzig

Chapter 20
The History of Early British and US-American Ecology to 1950

Robert McIntosh

Introduction

Scientific ecology was anticipated by a long history of natural history observations from classical times, recently termed protoecology (Glacken 1967; Egerton 1976). Ecology (Oekologie) was coined in 1866 by Ernst Haeckel, a German biologist and advocate of Darwin (see Chap. 10). One of the earliest uses in English was by Patrick Geddes, a British botanist. In 1880, 20 years before the term ecology was in general use, Geddes offered a hierarchy of the sciences putting ecology under sociology rather than biology (Mairet 1957), anticipating later connections with sociology. Geddes taught the brothers Robert and William Smith who later joined with Arthur G. Tansley in furthering vegetation studies and plant ecology in Britain (Tansley 1911). In 1893 the president of the British Association for the Advancement of Science described "oecology" as a branch of biology coequal with morphology and physiology and "by far the most attractive" (McIntosh 1985).

Tansley was influential in establishing a British Vegetation Committee in 1904 and in producing a volume on *Types of British Vegetation* in 1911 and organizing a field trip, which marked the first meeting of the pioneer British and American plant ecologists with continental ecologists. The British Ecological Society and the *Journal of Ecology* were initiated in 1913 and Tansley gave the first presidential address to any ecological society in 1914.

In the United States formalization of ecology similarly lagged behind coinage of the term. Among its earliest proponents was Stephen Alfred Forbes. Following service in the Civil War Forbes undertook extensive studies in natural history including insects, fish and birds, initiating an Illinois State Laboratory of Natural History in 1877. Forbes produced many of the earliest and most insightful publications on ecology notably, in 1887, *The Lake as a Microcosm*, treating the lake as

R. McIntosh (✉)
Formerly Professor at the Department of Biological Sciences, University of Notre Dame, Notre Dame, Indiana, USA

producing an equilibrium, stressing its holistic nature; and in 1894 he recognized ecology as a science including "economic entomology", "the whole Darwinian doctrine" and agriculture (Croker 2001).

Early ecology in the United States flourished largely in the Midwest, in universities and state natural history agencies. In Wisconsin E.A. Birge initiated limnological studies of plankton in the 1890s as director of the State Natural History Division, continuing these and other aquatic studies with Chancey Juday for three decades at the University of Wisconsin. Plant ecology in the US, as in Britain, was prominent in early ecology, notably at the universities of Nebraska, Minnesota and Chicago. In 1893 the Madison Botanical Congress formally adopted the term ecology, dropping the earlier diphthong "oe". Prominent among early plant ecologists were F.E. Clements at the Universities of Nebraska and Minnesota and H.C. Cowles of the University of Chicago. Clements became the major systematizer for plant ecology and produced two of its seminal volumes (Clements 1905, 1916). Clements considered "dynamic ecology", emphasizing the succession of the plant community to a stable endpoint, the climax association, under the control of the climate. Clements regarded the community as an integrated organism or even a superorganism. (see Chap. 4) In 1916 he formulated his ideas as a "universal law" and organismic concepts constituted the major synthesis of early twentieth century ecology.

The concept of nature as an organism, or superorganism, grew out of the tradition of a divinely organized balance of nature and was metaphorically extended to the complex of organisms or community. S.A. Forbes wrote, "A group or association of animals or plants is like a single organism" Clements, and his co-author Victor Shelford, stated that the organismic concept in ecology "is a veritable magna carta for future progress" (Clements and Shelford 1939). F.S. Bodenheimer wrote "every modern textbook of ecology stresses the highly integrated supraorganismic structure of communities" but noted, "there is no scientific evidence to support it", an observation commonly ignored (McIntosh 1998).

Cowles, working on the sand dunes of the Lake Michigan shore, based his ideas, like Clements, on succession stemming from studies of vegetation. He differed from Clements in recognizing succession as a tortuous process not leading to a stable climax. His famous axiom was that succession is "a variable approaching a variable rather than a constant (...)", a problem frequently faced by ecologists.

Animal ecology followed on the heels of plant ecology in the US and Britain. In the US animal ecology was advanced by C.C. Adams and Victor Shelford, both having early association with S.A. Forbes in Illinois. Adams and Shelford contributed to the study of animal communities and provided important early volumes on animal ecology (Adams 1913; Shelford 1913). Adams (1935) was an early observer of the linkage of general ecology to human ecology. Both were involved in the formation of the Ecological Society of America in 1914, and Victor Shelford was its first president. Its journal, *Ecology*, was founded in 1920.

British animal ecology was stimulated by Charles Elton's (1927) primer, *Animal Ecology*, following his extended surveys (1921–1924) of Arctic animal communities. Elton elaborated key ideas of community organization, food chain (trophic structure) pyramid of numbers, niche and, like Charles Adams, described ecology as scientific natural history. Elton's long association with the Hudson's Bay Company and its fur data provided early insights into population dynamics of predator and prey. Elton established a research area, Wytham Woods, in Oxfordshire that became one of the most studied areas on earth (Cox 1979) and was later used as a site for one of the well known Inspector Morse mystery series on TV.

Populations and Mathematics

Ecology developed as a loose amalgam of marine biology, limnology and plant and animal ecology. Oddly, parasitology, an intrinsically ecological study, remained largely separate from ecology until the 1960s. Population ecology developed as a major element of ecology as the study of the numbers of individuals of a species, the changes in numbers and interactions among species, such as predation of birds on insects, a one-time concern of Benjamin Franklin's, and competition between species. Human population growth described by Thomas Malthus (1798) was a major stimulus to Charles Darwin's theory of evolution.

Population counts, or censuses, were a common aspect of early ecology, followed by the rise of statistics in ecology. In 1928 Aldo Leopold began his studies of game populations in the US that led to his pioneer volume on Game Management and subsequent career in ecology at the University of Wisconsin. Charles Elton, in Britain, formed a Bureau of Animal Populations and the *Journal of Animal Ecology* in 1932, and began studies leading to his important volume on animal populations, *Voles, Mice and Lemmings* (Elton 1942).

The 1920s saw the beginning of the "Golden Age" of theoretical mathematical population ecology. Raymond Pearl and L.J. Reed rediscovered the logistic equation describing population growth over time and advanced it as a "law of population growth" (McIntosh 1985; Kingsland 1985). The logistic curve describes the growth of a population over time as an S-shaped or sigmoid curve.[1] The crux of the logistic is that the rate of growth decreases as N increases and approaches K. The logistic equation was expanded to the relations of two species, predation and competition, independently by A.J. Lotka, a physicist, and Vito Volterra, a mathematician in the 1920s and later to n species. These ideas were pursued experimentally by a Russian zoologist, G.F. Gause, working in the US for a period of years. These and similar studies led to the formulation of "Gause's Law" or the "competitive exclusion principle" which was extensively pursued as ecological theory.

Mathematical population theory was expanded to n populations and was widely criticized and praised as contributing to ecology, but in 1949 a volume on the *Principles of Animal Ecology* (Allee et al. 1949) asserted that "theoretical population

---

[1] It is commonly represented as $dN/dt = rN(1-N/K)$ with r and K as constants, r being the maximum rate of dt K population increase in an unlimited environment, K the limiting population, N is the number of individuals, t is time.

ecology has not advanced to a great degree in terms of its impact on ecological thinking". Nevertheless, mathematical ecology flourished as applied to competition and predation. Here Robert May (1981) pointed out that the classical deterministic logistic equation could under specific circumstances produce random looking dynamics, not always a smooth sigmoid curve. In Australia, the ecologist Alexander J. Nicholson and the physicist Victor A. Bailey believed that animal populations were controlled by their density, or numbers, governed by competition for limited resources (Nicholson and Bailey 1935). In the 1930s experimental studies were used for further evidence for a balance or equilibrium of populations in nature. Disputes about population control and theories thereof persisted in postwar ecology theory.

Australian ecology had begun in the 1920s largely as studies of plants and animals with an economic basis. Ecology appeared in 1939 in a Conference of the Australian New Zealand Association for the Advancement of Science. In 1951 the Australian government agency, CSIRO, established a Section of Ecology, and an Ecological Society of Society of Australia was formed in 1960.

Creating Larger Entities: Communities and Ecosystems

Study of aggregations of plants or animals, commonly called communities or associations, was another aspect of protoecology and early ecology. Early British marine biologists, called "dredgers", studied marine bottom-dwelling organisms. Edward Forbes (1844) recognized "zones" with species peculiar to them. Well ahead of their time dredgers recorded location depths, species and numbers of individuals. A British biogeographer, H.C. Watson, advocated census of measured areas (1 square mile) to determine the number of species present. His sage advice was to increase the number of samples and decrease their extent. Fifty years later Roscoe Pound and Frederic E. Clements did so, introducing the "quadrat" of one square meter in larger numbers which was a major step in quantitative community ecology (McIntosh 1985). In the smaller area it was feasible to count and measure individuals. Plant ecologists had the advantage of stationary and more readily visible organisms. They came to recognize frequency as the number of samples in which a species occurred, density as the number of individuals per area, cover as the area of ground covered, and biomass (weight) as the measure of size. Studies of the effect of number, size and shape of quadrats on these quantitative measures and statistical analyses thereof occupied plant ecologists in subsequent decades (Greig-Smith 1957). In 1949 G. Cottam and J.T. Curtis provided an alternative to the quadrat by devising methods using distances from points in space.

Collection of data was followed by statistical analysis. In the US the efforts of European ecologists, Paul Jaccard and O. Arrhenius, to examine the relation of number of species and area were pursued by Henry Allen Gleason (1922). Gleason (1920) had also examined the distribution of individual plants and used statistics to test if they were dispersed at random but found they were most often distributed contagiously in patches or non-randomly. The perception of communities as mosaics or patchworks was advanced by William S. Cooper in the US and A.S. Watt in

Britain (McIntosh 1985). Watt (1947) coined the apt term "gap-phase" for the small-scale disturbances in communities.

Animal ecologists also pursued studies of animal communities, complicated by the mobility and difficulties of sampling animals. Birge and Juday pursued their studies of Midwestern lakes with sampling methods determining numbers of species and individuals per area or volume and relating these to environmental measures. They provided what came to be an all too familiar lament of ecologists, "The extension of our acquaintance with the lakes has been fatal to many interesting and at one time promising theories" (Birge and Juday 1922). S.A. Forbes used quantitative sampling methods and, in 1907, developed an early statistical index to show co-occurrences of species among stands and conducted a survey of birds in a cross section of the state of Illinois. Shelford studied insect communities on the dunes of Lake Michigan and compiled a volume on *Animal Communities in Temperate America* (Shelford 1913).

The British animal ecologist Charles Elton's early work was largely on animal communities and the interactions of populations in them. He considered the traditional analogy of the community as a clock but observed that the animal works occasionally moved to another clock, calling into question the traditional balance-of-nature concept and design metaphor of the clock (Elton 1930, pp. 16–17). He later participated in a review of ecological surveys of animal communities (Elton and Miller 1954) although much of his work was on populations.

In 1935 Arthur G. Tansley, in Britain, introduced a concept he termed "ecosystem" in the context of a debate about F.E. Clements' superorganism concept of community, then strongly advocated by South African ecologists and even Prime Minister Jan C. Smuts who was influenced by Clements' ideas. Tansley defined ecosystem as the "whole system" including "the organismal complex" (biotic) and the "whole complex of physical factors" (abiotic) called the environment. "Ecosystem" had several antecedents, such as S.A. Forbes' microcosm, and W.C. Allee's geobio-ecology in the English language literature, and others in European languages, suggesting that the time was ripe for it. It fitted in with holistic traditions in ecology. Tansley's ecosystem incorporated a "hierarchy of systems of the most various kinds and sizes" (Tansley 1935, p. 299). This generality persisted as a problem for ecosystems ecology. "Ecosystem" came to be the major term encompassing the complex of biotic community and physical environment and its later products, ecosystem ecology and systems ecology, constituted for many a revolution in ecology.

Tansley's term fell on well prepared ground, particularly in aquatic ecology. As the president of the Ecological Society of America, W.P. Taylor (1935), had written, "The emphasis placed by bio-ecology on organism and environment as a great unitary problem is an inspiring one". Among those inspired was a young limnologist, Raymond Lindeman, who developed his studies of Minnesota lakes into a famous article, "The Trophic-Dynamic Aspect of Ecology" (Lindeman 1942). Its departures from the traditions of limnology were evident in that it was initially rejected by the scientific journal *Ecology* on the basis of negative reviews by two distinguished limnologists which were finally outweighed by the intervention of a third (Cook 1977). Lindeman used Tansley's ecosystem concept, perhaps the first to do so, in his "trophic-dynamic aspect" of ecology emphasizing the transfer of energy in the ecosystem. Lindeman's contributions were to emphasize and quantify trophic function, establish a theoretical orientation in ecology and identify energy flow as a fundamental process in long-term processes of community change (Cook 1977). George E. Hutchinson, Lindeman's mentor and supporter of his manuscript wrote, in an addendum to the article, that Lindeman's approach "may even give some hint of an undiscovered type of mathematical treatment of biological communities" (Lindeman 1942), although the paper's critics were dubious about Lindeman's data and use of mathematics. Ecosystem ideas were the basis of Eugene Odum's influential ecology textbook of 1953 which contributed to the introduction of the ecosystem concept and its descendant, systems ecology, into ecology in the 1960s.

Another revolution in ecology was described by Barbour (1995) as appearing in the 1950s. This revolution was against the tradition of the ecological community as an organism or superorganism of integrated species developing to a stable climax which was widely accepted in Anglo-American ecology (McIntosh 1998). The revolution was instigated by belated recognition in the 1940s of three largely unrecognized publications from 1917 to 1939 formulating the "individualistic concept" of community. H.A. Gleason's (1939) concept was based on his idea that each species had individualistic properties, the environment varied in time and space, and species aggregated into communities with a large stochastic component according to vagaries of dispersal and the available environment. In 1947 Gleason's long ignored concept was resurrected by three prominent ecologists and in the 1950s it was supported by extended studies of plants and animals by J.T. Curtis and R.H. Whittaker, their students and associates. It became widely accepted by plant and animal ecologists (McIntosh 1975, 1995). Even widespread acceptance of Gleason's concept and the ideas of continuum and gradient advanced by Curtis and Whittaker did not entirely lay to rest the concept of equilibrium community and even organism which persisted in theoretical ecology.

A Search for Laws and Principles

Coincident with concepts of population and community was a search for laws or principles in early ecology. Both terms appeared with limited consensus on their applicability in ecology. Victor Shelford (1913) introduced a "law of toleration" which asserted that a species could grow in a range of environments with a minimum, maximum and optimum represented by a single humped curve of distribution. Such curves were the substance of niche theory. Hopkins (1920) formulated the "bioclimatic law" (Hopkins law) which recognized the familiar phenomena of biological events moving north with the spring in the northern hemisphere. The law stated that a biotic event lagged 4 days per degree of latitude in spring and early summer, 5° of longitude eastward and 400 ft of altitude and gave rise to the aspect of ecology, phenology, that followed such events. As ecology became more sophisticated new laws and principles appeared. Preston (1948) described the distribution of numbers of individuals of species distributed according to a "log-normal law".

Principles were more frequent and numerous. Gause (1936) offered an article on "The Principles of Ecology", Allee et al. (1949) indexed 25 principles and Odum's (1953) textbook in its 3rd edition published in 1971 recorded 30-odd principles. Consensus on ecological laws and principles has been difficult to achieve and the body of principles hoped for by Allee et al. (1949) did not readily emerge.

Nature as a Resource

Concern for nature and the conservation movement antedated ecology in Britain and America. In Britain several public societies and government agencies were formed in the 1860s to 1880s, and in 1894 a National Trust for the Preservation of Places of Historic Interest and Natural Beauty was formed. In the US the 1870s saw formation of a Fisheries Society and a US Fish Commission and the Audubon Society. In 1872 Yellowstone National Park was formed and in 1892 the Sierra Club was founded by the premier conservationist, John Muir, in time to defeat attacks on the young national parks.

Many early ecologists were involved in the problems of land and wildlife management and in conservation. A New Zealand ecologist, Leonard Cockayne (1918), wrote that agriculture was applied plant and animal ecology. Aldo Leopold graduated from forestry school, turned his attention to game management and ecology, and his influential book, *A Sand County Almanac*, published posthumously in 1949, became the classic statement of conservation in America. Paul Sears became a leading ecologist and exponent of conservation with his book, *Deserts on the March* (Sears 1935), and his subsequent founding of a conservation program at Yale University.

In Britain Elton and Tansley were on a committee recommending a national survey of soils and resources that was shelved due to wartime concerns, but during the dark days of WW II Tansley and other ecologists served on a committee on nature preserves. In 1949 the British Nature Conservancy was formed with Tansley as its first chairman. By 1959 there were 84 nature reserves comprising 56.000 ha. In the US Victor Shelford organized the Ecologists Union in 1946 that was transformed into The Nature Conservancy which developed into the largest conservation organization in the world.

Ecologists in Britain and America, well before the environmental crisis, recognized its imminence and William Vogt (1948) published an early warning of the crisis. Aldo Leopold (1949) extended ecology to ethics in his assertion, "That land is a community is the basic concept of ecology, but that land is to be loved and respected is an extension of ethics".

Although "human ecology" was considered as early as 1913, the British Ecological Society included it in its first meeting in 1914 and Clements (1935) quoted H.G. Wells saying "Economics is a branch of ecology" and the Prime Minister of South Africa saying "Ecology is for mankind", widespread acceptance of human ecology lagged. It was not until after WW II that ecology became widely recognized as an integral component of human culture.

References

Adams CC (1913) Guide to the study of animal ecology. Macmillan, New York
Adams CC (1935) The Relation of general ecology to human ecology. Ecology 16:316–335
Allee WC, Emerson AE, Park O, Park T, Schmidt KP (1949) Principles of animal ecology. Saunders, Philadelphia
Barbour M (1995) Ecological fragmentation in the fifties. In: Cronon W (ed) Uncommon ground: toward inventing nature. Norton, New York, pp 75–90
Birge EA, Juday C (1922) The inland lakes of Wisconsin, the plankton 1. Its quantity and composition. Wis Geol Nat Hist Surv Bull 64:1–222
Clements FE (1905) Research methods in ecology. University Publishing Co., Lincoln
Clements FE (1916) Plant succession: an analysis of the development of vegetation. Carnegie Institution of Washington Publ. 242, Washington DC, pp 1–512
Clements FE (1935) Experimental ecology in the public service. Ecology 16:342–363
Clements FE, Shelford VE (1939) Bio-ecology. Wiley, New York
Cockayne L (1918) The importance of ecology with regard to agriculture. N Z J Sci Tech 1:70–74
Cook RE (1977) Raymond Lindeman and the trophic-dynamic concept in ecology. Science 198:22–26
Cottam G, Curtis JT (1949) A method for making rapid surveys of woodlands by means of randomly selected trees. Ecology 30:101–104
Cox DL (1979) Charles Elton and the emergence of modern ecology. Ph.D. dissertation, Washington University, Washington, DC
Croker RA (2001) Stephen Forbes and the rise of American ecology. Smithsonian Institution Press, Washington, DC
Egerton FN (1976) Ecological studies and observations before 1900. In: Taylor BJ, White TJ (eds) Issues and ideas in America. University of Oklahoma Press, Norman, pp 311–351
Elton C (1927) Animal ecology. Sidgwick and Jackson, London
Elton C (1930) Animal ecology and evolution. Claredon Press, London
Elton C (1942) Voles, mice and lemmings: problems in population dynamics. Clarendon, Oxford
Elton CS, Miller RS (1954) The ecological survey of animal communities with a practical system of classifying habitats by structural characters. J Ecol 42:460–496
Forbes E (1844) On the light thrown on geology by submarine researches. New Philos J Edinb 36:318–327
Forbes SA (1883) The food relations of the Carabidae and Coccindellidae. Bull Ill State Lab Nat Hist 1:33–64
Forbes SA (1887) The lake as a microcosm. Bull Peoria Sci Assoc 111:77–87. (Reprinted Bull Nat Hist Surv 15:537–550, Nov 1925)
Gause GF (1936) The principles of biocoenology. Q Rev Biol 11:320–336
Glacken CJ (1967) Traces on the Rhodian Shore. University of California Press, Berkeley
Gleason HA (1920) Some applications of the quadrat method. Bull Torrey Bot Club 47:21–33
Gleason HA (1922) On the relation of species and area. Ecology 3:158–162
Gleason HA (1939) The individualistic concept of the plant association. Am Midl Nat 21:92–110
Greig-Smith P (1957) Quantitative plant ecology. Butterworths Scientific, London
Hopkins AD (1920) The bioclimatic law. J Wash Acad Sci 10:34–40
Kingsland SE (1985) Modeling nature. Episodes in the history of population ecology. University of Chicago Press, Chicago
Leopold AS (1949) Sand County Almanac. Oxford University Press, New York
Lindeman RL (1942) The trophic-dynamic aspect of ecology. Ecology 23:399–418
Mairet P (1957) Pioneer of sociology. The life and letters of Patrick Geddes. Humphries, London
Malthus TR (1798) An essay on the principles of population. Johnson, London

May RM (1981) The role of theory in ecology. Am Zool 21:903–910

McIntosh RP (1975) H.A. Gleason, "individualistic ecologist", 1882–1975: his contributions to ecological theory. Bull Torrey Bot Club 102:253–273

McIntosh RP (1985) The background of ecology: concept and theory. Cambridge University Press, Cambridge

McIntosh RP (1995) H.A. Gleason's 'Individualistic Concept' and theory of animal communities: a continuing controversy. Biol Rev 70:317–357

McIntosh RP (1998) The myth of community as organism. Perspect Biol Med 41:427–438

Nicholson AJ and Bailey VA (1935) The balance of animal populations. Proceedings of the Zool Soc Lond, 3: 551–598

Odum EP (1953) Fundamentals of ecology, 1st edn. Saunders, Philadelphia

Preston FW (1948) The commonness and rarity of species. Parts I and II. Ecology 43:185–218, 410–432

Sears PB (1935) Deserts on the March. University of Oklahoma, Norman

Shelford VE (1913) Animal communities in temperate America as illustrated in the Chicago region. Bulletin of the Geographical Society of Chicago, Chicago

Tansley AG (ed) (1911) Types of British vegetation. Cambridge University Press, Cambridge

Tansley AG (1935) The use and abuse of vegetational concepts and terms. Ecology 16:284–307

Taylor WP (1935) Significance of the biological community in ecological studies. Q Rev Biol 10:291–307

Vogt W (1948) Road to survival. William Sloane Association, New York

Watt AS (1947) Pattern and process in the plant community. J Ecol 35:1–22

Chapter 21
The French Tradition in Ecology: 1820–1950

Patrick Matagne

Conceptual history reveals that scientific ecology was built up around work conducted during the nineteenth century by authors who often came from what was, broadly speaking, a German cultural background. However, France also had a presence in this field, in particular through authors who worked to develop the study of botanical geography. Research in the history of ecology rarely explores the literature of the learned societies in the French provinces. One reason for this is that conventional historiography treats these groups and their work as social phenomena that, however interesting to historians of French society, are assumed *a priori* to be of little interest in terms of scientific content, except for local history and archaeology, where their contributions are recognized.

In fact, a cross-analysis looking at the histories of ecology and these learned societies together results in a contradiction of this conventional wisdom. The naturalists of the French provinces, who are usually relegated to the margins of science, can be seen instead to have helped to structure the scientific ecology.

Introduction

Since the 1960s, at a time when the general public was discovering the word "ecology" in a context of growing awareness of the insidious effects of certain human activities on the environment, historians of ecology in the United States and then in Europe demonstrated that the founding concepts of scientific ecology had appeared before the term was actually invented in 1866. They sketched out the conditions underpinning the development of ecology as a scientific discipline based on work conducted during the 19th century by authors who often came from what was, broadly speaking, a German cultural background (Acot 1998). The institutional period of ecology, characterized by congresses and university departments, began in 1900. During that period France stood out in particular due to the work of Augustin

---

P. Matagne (✉)
Université de Poitiers, I.U.F.M., 40 avenue du Recteur Pineau, F-86000 Poitiers, France
e-mail: patrick.matagne@univ-poitiers.fr

Pyramus de Candolle (1778–1841) – a Swiss man who worked in Montpellier – as well as Gaston Bonnier (1853–1922) and Charles Flahault (1852–1935), who developed the study of ecological botanical geography (or phytogeography).

An entire domain of the scientific literature of the 19th century has, however, remained unexplored: the impressive mass of publications of the learned societies. This is due to several factors. The provincial learned societies had been gradually relegated to the rank of associations of dilettantes who practiced science as amateurs. This raises the issue of the relevance of the distinction between amateurs and professionals. Unlike in the United States, where the main fields of science were professionalized during the first half of the twentieth century, in France this matter remained unsettled for much longer. The question remained in the nineteenth century of what determined the boundary between the amateur naturalist and the professional: was it the level and field of training, scientific style and method, institutional affiliation (learned society, museum, library, laboratory, university, etc.) or the origin of any remuneration? Was the professional distinguished from the amateur by their system of theoretical reference, their working environment, their social standing, their way of communicating or of disseminating science, or even by how close they were to Paris?

In practice, Alain Corbin observed that the capital "marked the successful culmination of careers". "At every level, Paris lived off the energy of the provinces, it took in its men and tended to become for the provinces the paramount center of culture, wealth and power, to the extent that the province evoked disgrace, remoteness from the center and a wasted life".[1] Another factor was a tendency in the second half of the 19th century to see in the naturalist literature of the learned societies only unoriginal work, anecdotal or obsolete knowledge, the result of outdated practices, leading to tiresome, pointless inventories. This was the view of the physiologist Claude Bernard (1813–1878), who, on the occasion of the Universal Exposition of 1867, which exposed the unsatisfactory state of affairs in France, wanted to bring the natural sciences into the laboratory and the university. After the debacle of the Franco-Prussian war of 1870, this desire to professionalize French science, which was spurred by a spirit of revenge, led to hastening the disqualification of amateurs. The learned societies were, as a consequence, relegated to the margins of science (compare the situation in Spain at this time in Chap. 22).

As a victim of this dominant discourse, conventional historiography took an interest in the contributions of these societies only in the fields of local history and archaeology. It was not until the 1970s that their contributions to natural science began to be evaluated without any *a priori* assumptions. It was in this spirit that a parallel analysis was made between the history of the learned societies and the history of ecology (Laissus 1976, pp. 41–68; Dupuis 1979, pp. 69–106; Bange 1988,

---

[1] "À tous les échelons, Paris se nourrit de la substance de la province, assimile ses hommes et tend à devenir pour elle le centre primordial de culture, de richesse, de puissance, au point que la province évoque la disgrâce, l'éloignement du centre et la moisissure de l'existence" (Alain Corbin 1992, pp. 793–794).

pp. 157–172). This revealed that the development of ecology did not merely take place within conventional scientific institutions and that its actors were far from all being professionals in the ordinary sense of the term. In addition, the day-to-day presence of the naturalists in the local areas (natural environment, botanical garden, laboratory, university, museum, library, biological station) and their membership in a network of societies helped lay the basis for the organization of the first schools of ecology in France, even before the well-known school of plant sociology in Zürich-Montpellier. Examples of this include the École d'Auvergne, the École de l'Ouest and the École méditerranéenne. These came together around the phytogeographic paradigm, but were divided by their different problematics.

The Naturalists in the Provincial Learned Societies

Out of the thousand provincial societies created between 1808 and 1914 (Fox, and-Weisz 1980), about 350 conducted naturalist activity as part of their programme (Fig. 21.1). The bulk of the societies' publications consisted of works on flora and fauna of local interest, as well as catalogues of plants and animals and monographs and reports on outings. Their botanical work covered the entire territory of France (Fig. 21.2). I conducted a survey on 28 societies (124 authors). This survey indicated that botany was predominant (with 70%) over zoology (30%) (Fig. 21.3).

An analysis of this literature reveals certain aspects of the social dynamics of the learned societies (Matagne 1997b, 1999a). Even if a work was signed by a single author, many occasional contributors were cited or thanked collectively. For example, the "Flore du Haut-Poitou" (1901) by Baptiste Souché (1846–1915), President of the Deux-Sèvres botanical society, brought together contributions by botanists from the society from 1889, as well as from their predecessors in the Deux-Sèvres statistics society from 1870 onwards. James Lloyd (1810–1896), a botanist of English origin from Nantes, cited 97 contributors in his "Flore de l'Ouest de la France" (1854). The pharmacist and director of the botanical gardens in Angers, Alexandre Boreau (1803–1875), included 72 contributors in his "Flore du Centre de la France" (1857).

Information and samples were exchanged in the field, and others were sent by post. A variety of species were circulated for the purpose of verification, or as exchanges or gifts. Large herbaria and collections of insects, rocks, stones and shells were offered, sold or bequeathed in whole or in part.

The societies' meetings gave rise to reports listing the species discovered. The societies encouraged and handled exchanges and published lists of available samples, opening what amounted to trading centers. They even founded groups devoted exclusively to this activity, such as the Pyrenees association for the exchange of plants, founded in 1890.

Consider the case of the Deux-Sèvres botanical society. Founded in November 1888, by the next year it already had 153 members, and over 350 members by the turn of the century. It became a regional botanical society, with 632 members on the eve of the First World War. It organized 1,254 outings between 1889 and 1914.

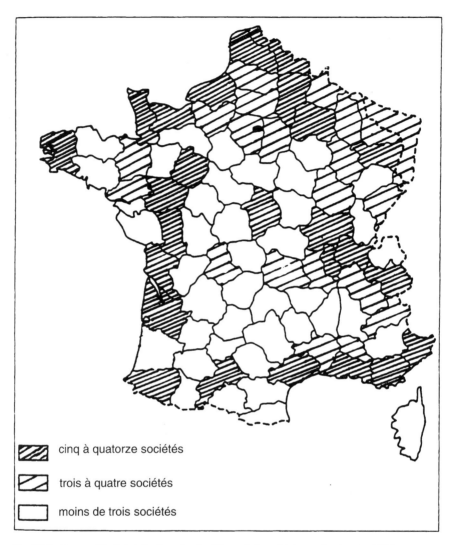

Fig. 21.1 Geographical distribution of scientific societies which practiced natural history

Societies		Authors	Botany	Zoology	Palaeontology	Entomology
Pluridisciplinary:	13	53*	30	13	0	0
Natural sciences:	8	39	24	14	1	0
Agriculture:	3	21	8	5	0	0
Geography:	2	5	4	0	0	1
Botany:	2	6	4	0	0	0
Total:	28	124	70	32	1	1

*some articles are written by several authors

Fig. 21.2 Number of authors of floristic books and catalogues (1800–1914)

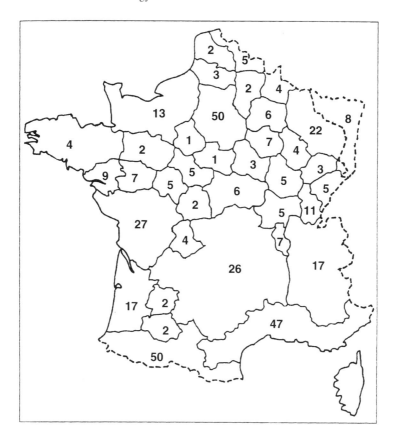

Fig. 21.3 Number of naturalist's publications of 28 learned societies in the "province"

In a single outing, 200 species were collected. In his field notes, Souché noted that on 17 May 1886 he had collected 160 samples of the same species. One month later, he filled his famous green botanic sample box (Fig. 21.4). In September of that same year, he prepared 52 samples of each of 12 species for the La Rochelle natural sciences society. The naturalists, whose curiosity was pushing them towards less accessible areas (the Alps, the Pyrenees), viewed their mission as collecting large quantities of rare species (Matagne 1988, 1997a) – an attitude that would be considered shocking in the twenty-first century! But at that time naturalists saw themselves as hunters of plants and animals.

The naturalists were long regarded as representatives of an outdated science, due to their passion for collection, classification and inventories. Their reports on their festive, sportive outings, events marked by their social character, replete with picnics, toasts, debates and anecdotal exchanges, could lead to confusion between the activities of these learned societies and those of other societies, such as singing or athletic groups. But promoting this social side had strategic value, as it attracted

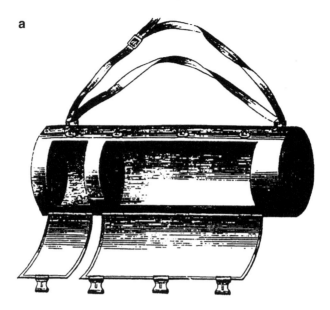

Fig. 21.4 (a–e): Basic tools of the botanising naturalist. (**a**) Herborization box, (**b**) and (**c**) tools for digging and breaking, (**d**) "échenilloir" et "sécateur", (**e**) knapsack (Figures taken from Bernard Verlot 1879. Le Guide du botaniste herborisant. - Paris: J.-B. Baillière et fils; Guillaume Capus 1883. Guide du naturaliste préparateur et du voyageur scientifique, ou Instructions pour la recherche, la préparation, le transport et la conservation des animaux, végétaux, minéraux, fossiles et organismes vivants. 2e édition, entièrement refondue par A.-T. de Rochebrune, avec une introduction par E. Perrier. Paris: J.-B. Baillière et fils)

people whose dues replenished the often meager funds of the groups. The newsletters also of course published work that reflected the interest of the learned societies in botanical geography.

The Practice of Botanical Geography

Nowadays, botanical geography is the study of the factors that influence the distribution of plants, the most important of which are geology, the climate, and the mechanism for the dispersion of the reproductive organs of the plant itself. Phytogeography thus includes the causal study of the distribution of species, the cartography of information showing the natural distribution of plants, and an index of the plant species and associations of a given region.

Baron Alexander von Humboldt (1769–1859), a German naturalist and explorer, is considered one of the founders of phytogeography. In 1805, he initiated an approach to landscapes that aimed at establishing phytogeographic classifications based on physiognomic analogies (Humboldt 1805). For example, the beech groves on the massif of the Grande Chartreuse (the Dauphiné area of the Alps) and the high-altitude beech groves in New Zealand and those in Croatia (below 1,700 m)

21 The French Tradition in Ecology: 1820–1950

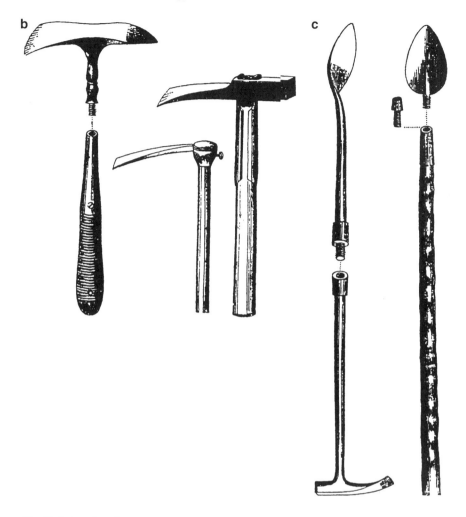

Fig. 21.4 (continued)

have a similar, characteristic physiognomy determined by the beech tree, even though the floristic composition of the plant community dominated by the tree varies with latitude and altitude.

The Swiss botanist Augustin Pyramus de Candolle originated the floristic tradition in phytogeography. The identification of plant communities is determined based on the most comprehensive comparison of inventories possible, which makes it possible to identify the species typical of associations that are always at least statistically measurable present in a specific environment. There, they could have only a minor impact on the physiognomy of the landscape. This is true for example of *Scilla lilio-hyacinthus* (Squill lily-hyacinth), which is characteristic of the association of the submountain beech groves with firs in Auvergne, which are found near Lake Pavin.

d

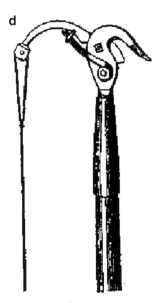

Échenilloir.

e

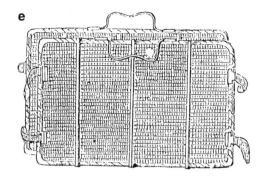

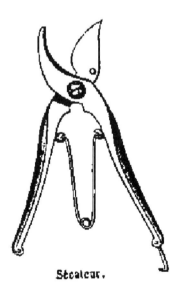

Sécateur.

Fig. 21.4 (continued)

The two traditions were developed by authors who attempted to develop a taxonomy that has the association as its basic unit (Matagne 1998). These traditions persisted through the century and formed part of the foundation of ecological botanical geography, the first treatise of which was published by the Danish botanist Eugenius Warming (1841–1924) in Danish (1895) and German (1896), and then in English (1909) (see for more details see Chap. 23). This was the era of the first generation of European and North American ecologists.

As early as 1810, the botanists from the Orléans Society for the Physical and Medical Sciences took an interest in the climatic and geological causes of the variation in flora between the Paris and Orléans regions. References to phytogeography became generalized in the 1820s, as was shown by the survey mentioned earlier which comprised 18 French départements (25% of the societies surveyed; Fig. 21.1).[2]

Phytogeographers studied the relations of plants with the type of terrain, temperature, altitude, and exposure in order to determine the laws governing their distribution. Sometimes the impact of human activity was considered. Starting in the 1850s, a period that saw the appearance of "Géographie botanique raisonnée" (de Candolle 1855) by Alphonse de Candolle (1806–1893) and "Etudes sur la géographie botanique de l'Europe" (Lecoq 1854) by Henri Lecoq (1802–1871), anyone who did not deal with these issues was marginalized by the Aveyron society of letters, sciences and the arts.

The survey showed that the new concepts in botanical geography, whether originating with Candolle or Humboldt, made sense out of discoveries that had theretofore caused confusion: the abnormal presence of *Pteris aquilina* (bracken fern) on limestone was cleared up if one invoked the nature of the bedrock, exposure and certain local atmospheric factors. Plants became environmental indicators: compiling a list of species was sufficient for the recognition of a limey soil. The *phytogeographic paradigm* served as the key to floristic transitions, the preferences of a given species, and the causal study of areas of distribution.

A.P. de Candolle drew the attention of naturalists and agronomists to potential applications (1809, pp. 335–373). Agricultural societies opened their newsletters to naturalists and chemists. Behind questions of yields, selections and attempts at acclimatization lay ecological issues concerning the relations between the plants and the soil (Dagognet 1973, pp. 51–52). References to Humboldt and to de Candolle, father or son, were rarely explicit. These could be identified rather by the use of various concepts, such as the "station", which was employed in different ways by Humboldt and by A.P. de Candolle (Drouin 1991, pp. 74f).

The botanists also undertook to identify and classify associations.[3] This research on associations changed profoundly botanical practice. In the field, the botanists, having identified species that were always associated with the same environment, attempted to determine what accompanied them: *the concept of association helped to identify regularities in nature*, to make and then verify predictions, and to identify

---

[2] Ain, Aisne, Ardennes, Aude, Aveyron, Cher, Doubs, Haute-Garonne, Gironde, Indre, Indre-et-Loire, Loire-Inférieure (actuelle Loire-Atlantique), Loiret, Maine-et-Loire, Saône-et-Loire, Deux Sèvres, Vendée, Vienne.

[3] The notion "association" was formulated for the first time by Humboldt in 1805.

anomalies and search for their causes. For example in 1912, Emile Château (1866–1952), a teacher and then primary school head, published an innovative work that has nevertheless remained relatively obscure, entitled "Les associations végétales", which was a precursor of plant sociology developped by the Zürich-Montpellier school in the 1920s (Château 1912, pp. 175–192). Others referred to "the concept of plant formation" (1838), formulated by the Göttingen botanist August Grisebach (1818–1879). In the tradition of Humboldt, a plant formation was understood as a group of plants with a definite physiognomy. Finally, some explored the physiognomic-floristic path laid out in 1863 by Anton Kerner von Marilaun (1831–1898). The Austrian botanist focused on floristic composition and on the physiognomy of plant groups, and began to free them from the environment, defining them in their own right, as in the case for species (von Marilaun Anton 1863). The consensus established around the phytogeographic paradigm showed the capacity of amateur naturalists to take on board rapidly the latest scientific contributions, originally humboldtian and candollian, incorporate them into phytogeographical practice and to adapt them to the local problems. Their attempts at a synthesis between the physiognomic and floristic traditions, and their search for a definition of an association that was considered sometimes as determined by the environment and at other times as a separate entity, all provided material for plant ecology during the 1920s and 1930s. This also anticipated the rupture that the Zürich-Montpellier school would make with phytogeography. All in all, the learned societies harbored pioneers in the ecological botanical geography of France. They were active in a number of fields where they had gained scientific credibility.

The Fields of Ecology

The natural environment is the terrain for botany par excellence. There are, however, other areas where naturalists exercise their skills and talents, geographical ones of course, but also institutional and social ones, such as gardens, museums, and biological stations.

During the nineteenth century, almost every large city created its own garden, under the pressure of the universities and learned societies, who wanted to be in charge of their administration.

Traditionally, the arrangement of plants was systematic (based on a classification system), so that one speaks of a "school of botany". The systems of Joseph Pitton de Tournefort (1656–1708), Carolus Linnaeus (1707–1778) and the methods of Antoine-Laurent de Jussieu (1748–1836) and A.P. de Candolle (1818–1821) succeeded one another or coexisted along corridors of labelled samples.[4] The arrangement of plant beds thus presented a summary of the geographic distribution of

---

[4]Joseph Pitton de Tournefort, Éléments de botanique ou Méthode pour connaître les plantes (1694); Carolus Linnaeus, Genera Plantarum (1737); Antoine-Laurent de Jussieu, Genera Plantarum (1789) and A.P. de Candolle, Reyni vegetalis Systema naturale (1818–1821).

plants. The societies promoted the use of these plant beds, as in Bordeaux (1847), and later in Béziers (1886) and Niort (1890). At the end of the century, some people wanted to create *ecological gardens* where the plants were arranged in their customary situation, in varying exposures Tarn (1882–1883). These gardens respected botanical affinities and highlighted the plant characteristics, importing the soil from each different region of origin Clermont-Ferrand (1893).

Alpine gardens, like the one started at the Pic du Midi observatory in 1878, were models. Edouard André, a landscape architect influenced by von Humboldt, argued that it was possible to reconstitute mountain associations in the plain. The French, following on the heels of the English, Swiss and Germans, attempted more or less successfully to reconstitute rock gardens (André 1894, p. 1228; Matagne 1992, 1999b, pp. 307–315). The various conceptions held by local scholars were reflected in the school garden (systematic), the ecological garden (relationship with the environment) and the geographic garden (by region of origin).

The passion for museums (traditional, revolutionary or commercial), often in association with libraries, reflected the rise of a new way of looking at the natural heritage bequeathed by the French Revolution. Here too the local societies were in demand, particularly since the means and skills needed to identify, classify and preserve the diverse collection of innumerable kinds of objects were lacking. In terms of natural history, the goal was comprehensiveness at the local level. Nîmes (1822), Nice (1828), Rouen (1834), Poitiers (1836), Le Havre (1845), Bayonne (1856), and Toulouse (1865), etc., all had their museum of natural history, created or supported by local societies.

The gardens, museums and libraries came to reflect the learned character of the societies that took responsibility for them. The last third of the nineteenth century then witnessed the boom in agronomic and botanical stations. Learned societies seized the opportunity to modernize by playing a role in these stations, with their modern laboratories (Matagne 1996).

Agronomic stations were established in France relatively late compared with Germany (Boulaine 1992). They had experimental fields and cattle pens as well as public laboratories. The technical methods used in chemistry, physics and physiology were harnessed to help agriculture. Farming societies linked to the stations promoted laboratory analysis and took part in developing agronomic maps. They opened their doors to naturalists by creating natural history sections.

Maritime stations were set up during this same period and opened their doors to amateurs. By the time of the First World War, the French coast had a dozen of them (Fischer 1997). One noteworthy case was the station at Arcachon, which was established in 1867 and for 30 years was owned and administered by the Arcachon scientific society. Innovative research programmes were launched in marine biology, oceanography, physiology and electrophysiology, and in the field of the ecology of the dune environment.

Also worthy of note is the station at Mauroc, established on 30 May 1912 by the University of Poitiers. For the next year the regional botanical society headed by B. Souché was extremely active meeting demand for the organization of outings and the creation of a botanical garden, a library and herbaria. In exchange, amateurs were invited into the laboratories.

Contrary to the general view, the learned societies thus had the desire and the capacity to take on board new scientific concepts, to apply them at various levels, and to familiar their members with laboratory practices. They were thus not pushed to the periphery of scientific inquiry. They even played an innovative role by taking charge of the promotion of ecological gardens and creating the first schools of ecology.

The First Schools of Ecology

The "École du Centre", the "École de l'Ouest" and the "École Méditerranéenne" reflected a diversity of approaches related to particular phytogeographic, sociological and local institutional features.

According to the phytosociologist Jules Pavillard (1868–1961), Auvergne was the "classic area of French phytogeography" (Pavillard 1926, p. 2). He is referring to the "Etude sur la géographie botanique de l'Europe et en particulier sur la végétation du Plateau central de la France" (1854) by Henri Lecoq, professor of natural history and director of the Clermont-Ferrand botanical garden. A number of well-known figures had spent time in Auvergne before Lecoq, including the botanists René-Louiche Desfontaines (1750–1833) and Adrien de Jussieu (1797–1853), the Pyrenees specialist Louis Ramond de Carbonnières (1755–1827), A.P. de Candolle, the explorer of India and the Himalayas Victor Jacquemont (1801–1832), the explorer of the Balearic Islands Jacques Cambessedès (1799–1863), the travelling botanist Auguste Prouvensal de Saint-Hilaire (1779–1853), the founder of modern geology Charles Lyell (1797–1875), the Montpellier professor Charles Flahault, the authors of Flores françaises Gaston Bonnier and Father Hippolyte Coste (1858–1924), and many others. A stopover in Auvergne seemed a must for any naturalist travelling around France or the world. Many amateurs published in the "Annales scientifiques", which had been founded by Lecoq and placed in the hands of the Académie at Clermont-Ferrand, which did not remain very active. Lecoq had difficulty establishing a following in his region before the creation of the natural history society in Auvergne (1894), which made use of his work.

From the 1820s, Lecoq, a follower of Humboldt, called on botanists to focus their work on plant associations, "these natural groups that form the subject matter of botanical geography".[5] He developed an original geological and paleontological approach. In a catalogue produced with his student Martial Lamotte (who died in 1883), he set out methodological guidelines for the readers of the "Annales scientifiques": "the geological study of such a varied soil made it much easier for us to resolve the previously unanswered question of the terrain's influence on plant stations".[6] He recommended breaking down the Massif Central into more limited areas based mainly on geology and secondarily on topography and plant physiology.

---

[5]"ces groupes naturels, dont l'étude constitue la géographie botanique" (Lecoq 1854, p. X).

[6]"l'étude géologique d'un sol aussi varié nous a donné de grandes facilités pour résoudre la question jusqu'ici indécise de l'influence des terrains sur les stations des plantes" (Lecoq and Lamotte 1847, p. IX).

Inspired by Humboldt's approach with regard to "the sociability of plants and their associations between similar individuals and between different species", Lecoq drew up a scale of dominance to evaluate the presence of one or another species, with a view that "the gathering of all the species of a single station constitutes a plant association".[7] This scale was used to distinguish dominant, essential, secondary and accidental species. Another student, S.E. Lassimonne, then drew on this for his study "principes de topographie botanique" (1892), which distinguished between isolated and associated species. Lecoq also developed an algebraic equation that described the relationship between the physico-chemical characteristics (atmospheric, edaphic) of a station. This attempt at establishing quantitative or mathematical relationships was a first in ecology.

A botanist from Aveyron, Joseph Revol, drew on Lecoq for his study of the flora of Southwestern France. So too did Charles Bruyant for his research into the evolution of plant groups in the lakes and peat bogs of the Mont-Dore area. Bruyant, a professor at the Clermont-Ferrand medical school, attempted to apply this methodology to zoogeography. Two works that dominated regional botany and phytogeography for a half-century were at the source of another school in the West of France "Flore de la Loire-Inférieure" (1844) and "Flore de l'Ouest" (1854), by James Lloyd. He had lived in France since the age of six, finished high school at Lorient, and then moved to Nantes and finally, from 1840, to Thouaré-sur-Loire (Loire-Inférieure, now Loire Atlantique).

In 1844, Lloyd argued that it was necessary for botanists to go beyond their administrative boundaries, and he urged them to study successions, striking examples of which were abundant along the Atlantic coastline. This abundance was due to a highly selective climate in the sea spray area. Lloyd noted that the strong winds and brisk, salty air determined the small-scale species, with fleshy leaves and thick, almost ligneous roots that dug deep into the soil in their search for fresh water. Lloyd's phytogeographic programme consisted of comparing the coastline flora with that of the interior so as to identify transitions:

> [...] if we start from the southern extremity of the coast, we find the salty marshes of Bourgneuf (area of Retz, Brittany marshland), the leading station for plants specific to salty silt[...]"; "[...]it is always useful to examine the points where the salt marshes arise, mixing with cropland, sand, marshy pastureland and fresh water streams. At Collet begins the maritime sand and the plants specific to it [...]"; "[...] before leaving the maritime area, [...] I would recommend that research botanists look at this part of the coast, which forms the transition between the maritime flora and the flora of the interior.[8]

---

[7]Lecoq 1854, p. 134.

[8]"[...] si nous partons de l'extrémité méridionale de la côte, nous trouvons les marais salants de Bourgneuf (pays de Retz, Marais Breton), première station des plantes propres aux vases salées[...]"; "[...] il est toujours utile d'examiner les points où les marais salants prennent naissance, en se confondant avec les terres cultivées, les sables, les pâtures marécageuses et les ruisseaux d'eau douce. Au Collet commencent les sables maritimes et les plantes qui lui sont particulières[...]", "[...]avant de quitter la région maritime, [...] je recommanderai aux recherches des botanistes cette partie de la côte qui forme la transition de la flore maritime à celle de l'intérieur" (Lloyd 1844, pp. 14–24).

Ten years later, Lloyd's Flore de l'Ouest covered eight départements. The boundaries were determined by the nature of the soil and climatic, physiognomic and physiologic factors. Lloyd had established his presence in the area and had numerous pupils in the learned societies of La Rochelle, Morbihan, Finistère, Deux-Sèvres and Loire-Inférieure, where the academic society officially introduced his research programme.

The son of a grocer, Emile Gadeceau (1845–1928), reviewed ecology's treatment of the causes of floristic and physiognomic successions in light of the concepts of American ecology. As a member and then president of the Loire-Inférieure academic society, he published studies influenced by Lloyd and Flahault. His most innovative work was entitled "Le Lac de Grand-Lieu. Monographie phytogéographique" (Gadeceau, Èmile 1909). The lake, located some 15 km southwest of Nantes, has been a nature reserve since 1980. There were three parts to Gadeceau's work: lake geography, aquatic plants, and biological ecology. He made use of the concepts of Warming and the Strasbourg botanist Wilhelm Schimper (1856–1901), who laid the physiological basis for plant ecology that Warming lacked (Schimper and Andreas Franz 1898).

The third part of Gadeceau's "Essai" introduced American notions of ecology for the first time. Starting from Lloyd's spatial view of succession, Gadeceau adopted the approach of the official botanist of the state of Minnesota, Conway MacMillan (1867–1929), who in 1897 studied plant distribution on the banks of a lake from a dynamic viewpoint. The point was to follow successions not only over space but also over time. Gadeceau also cited Henry Chandler Cowles (1869–1939), from the University of Chicago, who published a thesis on Lake Michigan (Cowles, 1899). The two authors were among the pioneers of the Chicago school of ecology (see Chap. 20).

Gadeceau read the Botanical Gazette and the Monde des Plantes published in Mans in the Sarthe region – a mine of bibliographical references – and was familiar with the latest thinking and applied it to the case of the Grand-Lieu Lake. His epistemological position was thus remarkable: an amateur botanist from a provincial learned society had introduced North American ecology into France before the First World War. The École de l'Ouest identified with Lloyd had of course created theoretical conditions that were propitious for integrating this new ecology, but this would still require moving from the conventional photographic vision of European phytogeography to America's cinematographic vision.

The first works on botanical geography in the Mediterranean region were carried out by Father Giraud Soulavie (1752–1813) (Soulavie 1780–1784), who evoked the value of territorial divisions based on climatic factors. Father Pourret (1754–1818) contributed material on the Eastern Pyrenees, and Marquis Gaston de Saporta (1823–1895) drew attention to the importance of the history of flora for an explanation of its current state.

The "Essai d'une géographie botanique du Tarn" published in 1862 was a precursor, even though it failed to gain a following (Gazel Larambergues 1862) as it set out an innovative programme that proposed to divide the département on the basis of topography and the nature of the soil rather than on the traditional basis of administrative limits (arrondissement, canton, commune).

In the neighboring département of Tarn-et-Garonne, Dominique Clos (1821–1908) laid the basis for the Mediterranean school of ecological phytogeography during a local congress in 1863. His programme was published in 1864 in Toulouse and in 1870 in Carcassonne (Clos 1864, 1870). Clos proposed seeking the limits of Mediterranean vegetation towards the west of Carcassonne and the south of Tarn by tracing the disappearance of certain species and sketching out their areas. The Carcassonne society of arts and sciences adopted this programme and identified the natural divisions within the département, which demarcated the Mediterranean phytogeographic region. Two contemporaries of Clos, the Hérault botanists Henri Loret and Auguste Barrandon, made use of this programme in their Flore de Montpellier (Loret and Barrandon 1876).

Clos, Loret and Barrandon thus were initiators of a Mediterranean school that adopted and adapted the phytogeographic paradigm based on Candolle. Their starting point was the principle that a département-based breakdown offered an easy and relevant framework of grids that botanists could then use to seek the causes of plant distribution and identify associations. They expected that when every département had its phytogeographic map, the puzzle would be complete, including for the country as a whole. Hector Leveillé (1863–1918), the head of "Le Monde des Plantes", supported Loret's quest.

The Béziers Society for the study of the natural sciences joined the programme, followed by the Aude Society for scientific studies. Publications came out up to 1909 that referred directly to Loret and Barrandon "Catalogue des plantes vasculaires du département de l'Ardèche" (Baichère 1891; Gautier 1898; Albert and Jahandiez 1908). Finally, a dozen authors out of the 50 who mattered in the region (Fig. 21.3) trained the botanists of a number of learned societies.

Nevertheless, Professor Charles Flahault took a dim view of this ecological work, which broke with the traditional practices of non-causal descriptive phytogeography (chorology). Even though he acknowledged that Mediterranean phytogeography was on the right path, he deplored the départemental confines that it set itself. He believed, for instance, that work on the flora of the limestone plateaus of the Causses region would be of greater interest than that in the départements of Aveyron and Lozère. He condemned those who studied "plants that have nothing in common other than being under the jurisdiction of the same civil servant".[9]

Flahault had in fact chosen the Mediterranean region as the grounds for an ambitious study. He wanted to make use of the local societies to spread his programme, which consisted of determining the limits of the natural areas of plants, without worrying about administrative boundaries. In other words, he wanted the learned societies to work on behalf of his programme, with a view to developing a cartographic summary at a regional and then national level.

---

[9]"des plantes qui n'ont en commun que d'être sous l'administration d'un même fonctionnaire" (Charles Flahault 1901, pp. 3–4).

Flahault's renown, his personality and his campaigning style had an influence well beyond the regional level. He took part in the debates and controversies raging among phytogeographers at the turn of the 19th and 20th centuries. It was urgent at that time to develop a consensus on phytogeographic nomenclature and on the definition of an association, in order to avoid a "Babelization" of the discipline (Bonnier 1900). Ultimately, it was the floristic line around which the Zürich-Montpellier school had organized that won the day in continental Europe. It defined an association in a way that ruptured with the approach of the phytogeographers: the association was to be identified by its floristic composition, and not by environmental factors.

Flahault made an appearance in 1888 at a conference in the Narbonne area of the French botanical society, where he met with people capable of spreading his ideas. In 1889, he helped set up the Aude scientific society, whose botanists officially declared that they were abandoning administrative limits. Twenty-five authors (out of 250 members) in the society published work along these lines. The already-mentioned rather modest society in Carcassonne went over to Flahault's approach in 1890–1892. The Languedoc geographical society, which was linked to the botanical institute founded by Flahault in Montpellier, made a powerful contribution to spreading his ideas.

Various journals ensured the promotion of Flahault's approach beyond the region, including the Revue scientifique du bourbonnais, Monde des Plantes, and the Bulletin de la société botanique de France. Distant followers of Flahault were active, such as Eugène Simon (1871–1967) in the Deux-Sèvres area (Matagne 1990). Finally, Flahault wrote an introduction to the scholarly but popular "Flore de France" by Father Coste (1901–1906), in which he promoted his programme (Flahault 1901). In the absence of centralized, standardized data, it is difficult to arrive at an assessment of the work of the school that preceded Flahault. A tenth to an eighth of French territory could have been covered. Flahault considered the results unusable. A review of publications from the period 1920–1930 indicates the need for another analysis.

The Flahault school, symbolically founded in 1888, brought together some 30 authors. In 1909, they had a cartography of a tenth of the country. Despite this, Flahault gave up due to a lack of support and of funds. He was unable sufficiently to mobilize the local societies, who resisted a form of instrumentalization and preferred to stick to tried and proven methods of exploration. Moreover, phytosociology came onto the plant ecology landscape scene at the start of the 20th century and resolved the problems in terminology and nomenclature in which phytogeography had become bogged down.

Other schools also undoubtedly existed, in particular around Toulouse and in the northeast of France.[10] The three that have been presented here illustrate the epistemological work of the Catalan ecologist Ramón Margalef. He notes that

---

[10] In Toulouse this was Lavialle before 1914, then Henri Gaussen; in the north-east, Alexandre Godron (1807–1880) and Paul Fliche (1836–1908) contributed to phytogeography, zoogeography and forestry.

ecology is marked by a sort of Genius loci, which would have the dynamics of schools linked to local landscapes: hence the phytosociology school, which develops concepts and methods adapted to the mosaic of Alpine and Mediterranean vegetation, and North American ecology, which is absorbed in the gradual transitions of large spaces, plant successions and pioneering species (Margalef 1968, p. 26). Lloyd and Gadeceau encountered this type of environment on a smaller scale along the French Atlantic coastline. An analysis of the environment in Auvergne shows the importance of geological factors. In the Mediterranean area, the ecological problem encountered the political problem of the administrative limits set by the French Revolution, which determined those of the small provincial "motherland" to which the local scholars were so attached.

Conclusion: The Unique Features of French Ecology

The destiny of the first schools of ecology was linked to the fate of the learned societies that organized and supported them. Many society members perished in the First World War. Greater labor mobility also hindered the development of local roots. In addition, devaluation ruined many learned societies, which often depended on investments. Those that re-formed during the 1920s did so on a different basis. The first schools of ecology thus disappeared with the learned societies before the First World War.

Nevertheless, it was not uncommon to see references to their authors in the 1920s. Pierre Allorge (1891–1944), professor of cryptogamie at the Museum of Paris, used the work of Lecoq in his phytosociology study "Associations végétales du Vexin français" (1922), in a chapter reviewing the accomplishments of ecology. There are also specific references to Lecoq in the work of Aimé Luquet, "Essai sur la géographie botanique de l'Auvergne", subtitled "Associations végétales du Massif des Monts Dores", which also cites the work of Lamotte and Bruyant, who followed in Lecoq's footsteps. Nevertheless, in this "Essai", dedicated to Braun-Blanquet and to Pavillard, it is the contributions of the phytosociological school that Luquet belonged to that are given greatest emphasis. Even so, there are still references to 19th century authors from Auvergne in "Recherches sur la géographie botanique du Massif Central", also by Luquet, in 1937.

Lloyd and Gadeceau are referenced in botanical and phytogeographical studies. Nevertheless, the introduction of concepts from American ecology, discovered before 1914 by Gadeceau, continued its course with other authors in the 1920s and 1930s, including Allorge, Emmanuel de Martonne (1873–1955) and Henri Gaussen (Acot and Drouin 1997).

Plant ecology in the 1920–1950 period came to be marked in France by the phytosociology propounded by Braun-Blanquet. It is also characterized by biocenotics and by the unique approach to human ecology that grew out of the geography of Vidal de la Blache (1845–1918).

Due to the domination of phytosociology, for the participants in the first international colloquium of the "Centre nationale de la recherche scientifique" (CNRS) on ecology, held in Paris from 20 to 25 February 1950, it was important to draw attention to the study of animal groupings (animal synecology), which was lagging behind. Even though phytosociology was thus officially not on the agenda, it was omnipresent. On the other hand, the theory of ecosystems, then all the rage in the US (Odum and Odum 1953), was missing. According to Acot Pascal (1994), a historian of ecology, this reflected a unique feature of French ecology in the 1920–1950 period.

Another unique feature of the French tradition in ecology had its roots in the 19th century. It was born and developed through the work of countless authors whose names conceptual history has not retained. They contributed to developing a French approach to ecology based on the phytogeographic paradigm organized around schools distinguished by their local dynamics.

At the beginning of the twentieth century, the institutionalization of ecology, the First World War, the rise of phytosociology, and the sudden emergence of American ecology led to the marginalization of these local schools of ecology. What remained of them after the War was a dispersed heritage and a few individual authors who were decontextualized by those who later cited their works.

Fig. 21.5 The map displays all the locations of early French ecology activities mentioned in the article (arranged by C. Haak)

References

Acot P, Drouin J-M (1997) L'introduction en France des idées de l'écologie scientifique américaine dans l'entre-deux guerres. Revue d'histoire des sciences 50:461–479

Acot P (1994) Le colloque international du CNRS sur l'écologie (Paris, 20–25 février 1950). In: Debru C et al (eds) Les sciences biologique et médicales en France, 1920–1950. CNRS éditions, Paris, pp 233–240

Acot P (ed) (1998) The European origins of scientific ecology. Gordon and Breach Publishers, Amsterdam

Albert A, Jahandiez E (1908) Catalogue des plantes vasculaires qui croissent naturellement dans le département du Var. Paul Kliensieck, Paris

André E (1894) Les Fleurs de pleine terre comprenant la description et la culture des fleurs annuelles, bisannuelles, vivaces et bulbeuses de pleine terre. Vilmorin-Andrieux et Cie, Paris, p 1228

Baichère A (1891) Contributions à la flore du bassin de l'Aude et des Corbières. Bulletin de la société d'études scientifiques de l'Aude, p 73

Bange C (1988) La contribution des ecclésiastiques au développement de la botanique dans la région lyonnaise au 19e siècle. 112e Congrès national des sociétés savantes, Lyon, 1987. Section histoire moderne et contemporaine, Paris, pp 157–172

Bonnier G (1900) Projet de nomenclature phytogéographique. Actes du 1e Congrès International de Botanique tenu à Paris à l'occasion de l'Exposition Universelle de 1900. E. Perrot, Lons-Le-Saunier, pp 427–449

Boulaine J (1992) Histoire de l'agronomie en France. Technique et Documentation, London, New York, Paris

Boreau A (1857) Flore du Centre de la France. Librairie Encyclopédique de Roret, Paris

de Candolle A-P (1809) Géographie agricole et botanique. In: Nouveau cours complet d'agriculture ou Dictionnaire raisonné et universel d'agriculture, vol 6. Déterville, Paris, pp 335–373

de Candolle A-P (1818) Reyni vegetalis Systema naturale. Treuttel & Würz, Paris

de Candolle A (1855) Géographie botanique raisonnée. Libraire de Victor Masson, Paris

Château Emile (1912) Les associations végétales. Bulletin de la Société d'Histoire Naturelle d'Autun, pp 175–192

Clos Dominique (1864) Coup d'oeil sur la végétation de la partie septentrionale du département de l'Aude. Mémoires de l'Académie des sciences, inscriptions et belles-lettres de Toulouse, pp 421–422

Clos Dominique (1870) Mémoires de la société des arts et des sciences de Carcassonne, p 377

Corbin A (1992) Paris-province. In: Pierre N (ed) Les lieux de mémoire 3. Les France. Traditions, vol 2. Gallimard, Paris, pp 793–794

Coste AH (1901–1906) Flore descriptive et illustrée de la France. Paul Klincksieck, Paris

Cowles HC (1899) The ecological relations of the vegetation on the sand dunes of lake Michigan. Chicago

Dagognet François (1973) Des révolutions vertes, histoire et principes de l'agronomie. Paris, pp 51–52

de Jussieu A-L (1789) Genera Plantarum. Herissant, Paris

Drouin J-M (1991) Réinventer la nature. L'écologie et son histoire. Desclée de Brouwer, Paris

Dupuis C (1979) Histoire naturelle et naturalisme dans la France de 1904, année de la fondation des naturaliste parisiens. In: Bulletin des naturalises parisiens (ed) Cahiers des naturalistes 35:69–106

Fischer J-L (1997) In: Ambrière M (ed) Stations maritimes, Dictionnaire du 19e siècle européen. PUF, Paris, pp 1128–1129

Flahault C (1901) La flore et la végétation de la France. In: Coste Abbé Hippolyte (ed) Flore descriptive et illustrée de la France. Paul Klincksieck, Paris, pp 3–4, 1901–1906

Fox R, Weiss G (1980) The organization of science and technology in France, 1808–1914. Cambridge-University Press, Maison des Sciences de l'Homme, Cambridge, Paris

Gadeceau È (1909) Le Lac de Grand-Lieu. Monographie phytogéographique. Nantes

Gautier G (1898) Catalogue raisonné de la flore des Pyrénées-Orientales société agricole, scientifique et littéraire des Pyrénées-Orientales

Gazel Larambergues Dissiton de (1862) Essai d'une géographie botanique du Tarn. Société littéraire et scientifique de Castres, pp 317–327; 403–414

Grisebach AHR (1838) Über den Einfluss des Climas auf die Begrenzung der Natürlichen Floren. Linnaea 12:159–200

Laissus Y (1976) Les sociétés savantes et l'avancement des sciences naturelles. Les musées d'histoire naturelle, 100ᵉ Congrès national des sociétés savantes, Paris, 1975, Section histoire moderne et contemporaine et histoire des sciences. Paris, pp 41–68

Lassimonne SE (1892) Principes de topographie botanique. Durond, Moulins

Linnaeus C (1737) Genera Plantarum. Conrad Wishoff, Leiden

Lecoq H, Lamotte M (1847) Catalogue raisonné des plantes vasculaires du Plateau central de la France. Annales scientifiques, industrielles et statistiques de l'Auvergne 9

Lecoq H (1854) Etude sur la géographie botanique de l'Europe et en particulier sur la végétation du Plateau central de la France.Paris pp X et p 134

Lloyd J (1844) Flore de la Loire-Inférieure. Prosper Sebire, Nantes

Lloyd J (1854) Flore de l'Ouest de la France.(5. ed. 1897) – Nantes: Èmile Gedeceau

Luquet A (1937) Recherches sur la géographie botanique du Massif Central. Revue de géographie alpine 26(2):467–470

Loret H and Barrandon A (1876) Flore de Montpellier ou analyse descriptive des plantes vasculaires de l'Hérault. Montpellier

Margalef R (1968) Perspectives in ecological theory. University of Chicago Press, Chicago

Matagne P (1988) Racines et extension d'une curiosité: la Société botanique des Deux-Sèvres, 1888–1915. In: Corbin Alain (ed) Mémoire de maîtrise d'histoire contemporaine, Tours

Matagne P (1990) De la taxinomie à la phytosociologie: Eugène Simon à la Société Botanique des Deux-Sèvres (1898–1915). Mémoire de DEA

Matagne P (1992) La tradition des jardins et la culture régionale: le cas des Deux-Sèvres de la fin du 18ᵉ siècle à la première guerre mondiale. Bulletin de la société botanique de France 139. Lettres Botaniques 1:5–13

Matagne P (1996) Les naturalistes au laboratoire. Bulletin d'histoire et d'épistémologie des sciences de la vie 3:30–41

Matagne P (1997a) La botanique dans le Centre-Ouest (1800 à 1915). Bulletin de la Société historique et scientifique des Deux-Sèvres, 3ᵉ série, Tome 5, 1ᵉ semestre: 125–247

Matagne P (1997b) Les mécanismes de diffusion de l'écologie en France, de la Révolution française à la première guerre mondiale. Editions du Septentrion, Villeneuve-d'Ascq

Matagne P (1998) The taxonomy and nomenclature of plant groups. In: Acot P (ed) The European Origins of Scientific Ecology, Editions des archives contemporaines, vol 2. Gorden & Breach, Amsterdam, pp 427–519

Matagne P (1999a) Aux origines de l'écologie. Les naturalistes en France de 1800 à 1914. Paris: Comité des travaux historiques et scientifiques – histoire des sciences et des techniques

Matagne P (1999b) Des jardins écoles aux jardins écologiques. In: Fischer Jean-Louis (ed) Le jardin entre science et représentation. Comité des travaux historiques et scientifiques, pp 307–315

MacMillan C (1897) Observations on the distribution of plants along shore at Lake of the Woods. Minnesota Botanical Studies, Minneapolis

Odum EP, Odum HT (1953) Fundamentals of ecology. W. B. Saunders, Philadelphia

Pavillard J (1926) Etudes phytosociologiques en Auvergne. Clermont-Ferrand

Revol J (1910) Catalogue des plantes vasculaires du département de l'Ardèche. Paul Klincksieck, Paris

Schimper AFW (1898). Pflanzengeographie auf physiologischer Grundlage. Jena

Souché B (1901) Flore du Haut-Poitou. Clouzeau, Niort

Tournefort de JP (1694) Éléments de botanique ou Méthode pour connaître les plantes. Paris

von Humboldt A (1805) Essai sur la géographie des plantes, accompagné d'un tableau physique des régions équinoxiales. Schoell, Paris

von Marilaun Anton Kerner (1863) Das Pflanzenleben der Donauländer. In: The background of plant ecology. The plants of the Danube Basin. (Reprint 1977). Arno Press, New York

Warming E (1895) Plantesamfund, Grundträk af den Ökologiske Pflanzengeografi. Kopenhagen

Warming E (1896) Lehrbuch der ökologischen Pflanzengeographie, Eine Einführung in die Kenntnis der Pflanzenvereine. Berlin

Warming E (1909) Oecology of plants. An introduction to the study of plant-communities (expanded translation). Clarendon Press, Oxford

Chapter 22
Early History of Ecology in Spain, 1868–1936

Santos Casado

Introduction

The early history of ecology in Spain provides an eloquent illustration of the contradictory relationships between the emerging science of ecology and the tradition of natural history that preceded it. For in Spain, as in almost every other western country, the nascent discipline of ecology was undoubtedly grounded in the pre-existing knowledge, practices and institutional framework of natural history. However, in the comparatively small and underdeveloped turn-of-the-century Spanish scientific community, those who strove to develop aquatic ecology or plant ecology as new, differentiated and respectable disciplines found little support among their colleagues and were seldom paid any attention by official authorities. Borrowing from ecological and biogeographical parlance, one could say that in such a small area the ecological species was not able to find its own niche, or that competitive pressure prevented it from establishing a viable population.

This, in a nutshell, is the story that is to be told in this paper. A more detailed, fine-grained account of the reception and early development of ecology in Spain can be found in the book "Los primeros pasos de la ecología en España" (Casado 1997), which covers roughly the same period to be examined here, from the 1868 democratic revolution to the onset of the Civil War in 1936.

Young Darwinians

A detailed account of the institutional structure of Spanish science during the late nineteenth and early twentieth centuries is far beyond the scope of this study. A brief summary of the knowledge available on this topic (Sánchez Ron 1999; López-Ocón Cabrera 2003) should be enough, however, to highlight some characteristics of the institutionalization process relevant for further discussion.

S. Casado (✉)
Universidad Autónoma de Madrid, Spain
e-mail: santos.casado@uam.es

The nineteenth century was one of decline for Spain, which lost most of its colonies in America, and became a peripheral power in Europe. The internal political situation was dominated by social and ideological conflicts, which were met with repressive policies on the part of the absolutist and conservative forces that, with few exceptions, ruled the country. In this somewhat unfavourable context, it was left to several "intermediate" generations of Spanish scientists (López Piñero 1992) to continue pursuing research and publishing activities, thus acting as a bridge until, in the last third of the century, the situation improved.

The 1868 democratic revolution brought ideological freedom, which favoured scientific modernization, including the spread of Darwinism. It was also the starting point for a period of institutionalization of Spanish science. Naturalists led the way, founding in 1871 the "Sociedad Española de Historia Natural" in Madrid (compare the situation in France in this period Chap. 20). In the following years, similar initiatives were taken in other scientific fields, such as geography, anthropology and medicine. The re-establishment of the monarchy in 1875 led to some ideological persecution, with scientists such as Laureano Calderón and Augusto González de Linares, for example, temporarily losing their chairs at the University of Santiago (Cacho Viu 1962); however, it also stabilised the political and social situation and thus, in the long term, benefited the institutional development of science, which had begun during the revolutionary period. Yet neither the "Sociedad Española de Historia Natural" nor other private associations could support research activities in any direct way. The only official natural history research centres that existed at that time were the "Comisión del Mapa Geológico de España" (Commission for the Geological Map of Spain), which was mainly committed to the exploration of mining resources, and the "Museo de Ciencias Naturales", both based in Madrid. The latter, founded in 1771, had experienced a profound crisis when, in 1845, it had become a mere dependency of the "Universidad Central". All the natural history professors at the university were simultaneously members of the museum, which had no other research staff (Barreiro 1992). With such limited human and material resources it should come as no surprise that, in comparison with other western European countries, basic knowledge about the natural history of Spanish territory contained some enormous gaps (Casado 1994). Many taxonomic groups were almost unknown, and what little data existed was somewhat scattered. Vast regions remained largely unexplored from the geological, botanical and zoological points of view.

In fact, many of the most valuable contributions to Spanish geology, botany and zoology were made during the late nineteenth century by foreign naturalists from Northern and Central Europe, such as Édouard de Verneuil, Moritz Willkomm and Franz Steindachner, who travelled to the Iberian peninsula, attracted by its extraordinarily rich and little known natural diversity in comparison. The founders of the "Sociedad Española de Historia Natural" were well aware of this shortcoming. Consequently, their inaugural manifesto was an invitation to collaborate on a wide taxonomic, geographical and descriptive work on "the natural products of the country"[1] to accomplish a body of information comparable to those already existing in other European countries (Sociedad Española de Historia Natural 1872).

---

[1] "las producciones naturales del país" (Sociedad Española de Historia Natural 1872, p. VI).

It soon became clear that producing a natural history of Spain would not be a task that took matter of years but would occupy decades of research, so this collective project shaped a long-lasting scientific tradition in which several generations of naturalists were trained and their approaches and goals defined. What role, if any, did ecological ideas have in this context? First, it must be noted that, in spite of their lagging behind in surveying the national territory, Spanish naturalists were not unreceptive to new trends in geology and biology, and particularly to Darwinism (Núñez 1977; Glick et al. 2001). While the historical connections between Darwinism and ecology have been interpreted in a variety of ways (Stauffer 1960; Acot 1983; Coleman 1986), there is no doubt that in the Spanish case a positive correlation can be found, at least on the theoretical level. Those young naturalists who led the way to the new Darwinian faith (Sala Catalá 1981) were also the first explicitly to demand the inclusion of ecological issues in the research agenda of Spanish botanists and zoologists. The following examples serve to substantiate this claim.

Ignacio Bolívar (1850–1944) was the most prominent Spanish specialist in the taxonomic study of Orthoptera and as such was a contributor to the collective project aimed at establishing a basic body of knowledge about Spanish nature. However, in his first important work, a fauna of Iberian Orthoptera published when he was 25 years old, he cautioned his colleagues not to forget that the ultimate purpose of natural history was to unveil such general problems as the interrelationships between organisms and the environment, a protoecological issue typical of naturalists influenced by Darwinian theories.

> Finalmente, no debo terminar sin recordar á los jóvenes entomólogos, á quienes principalmente van dirigidas estas notas, que si es importante enumerar y dar á conocer las especies, sobre todo en un país como el nuestro, en que tanto hay por conocer, á pesar de la incesante actividad y de los laboriosos esfuerzos de varios naturalistas que á su estudio desde largo tiempo se dedican, no lo es menos tratar de investigar las relaciones que entre sí y con los medios exteriores guardan las diferentes especies, para indagar y dar solucion en lo posible á los grandes problemas de la naturaleza, fin último á que tiende nuestra hermosa ciencia[2] (Bolívar 1876, p. 86).

Odón de Buen (1863–1945) was one of Bolívar's students and he readily applied the same point of view in his first scientific paper, a survey of the flora of the steppe-like central regions of Spain, published when he was barely 20 years old. Instead of providing a conventional floristic catalogue, Buen attempted to elaborate a general phytogeographical study, characterizing several plant groupings according to different topographic and climatic conditions. The motivation for such an approach was explained clearly as follows:

> Los estudios geografico-botánicos han adquirido verdadera preponderancia. Iniciados por Humboldt y secundados por Schouw, A. De Candolle, Wahlenberg y algunos otros, recibieron considerable impulso con la aparicion de la teoría de Darwin en el horizonte de las

---

[2] "Finally, I should not finish without reminding young entomologists, to whom these notes are addressed in the first instance, that even if it is important to enumerate and publicize the existence of all the species, especially in a country such as ours where so much is still unknown despite the restless activity and laborious efforts of several naturalists long devoted to its study, it is no less important to try to investigate the relationships that the different species have among themselves and with their external media, in order to search for and, as far as possible, to provide solutions to the great problems of nature, the ultimate goal towards which our beautiful science is directed".

ciencias naturales. Buscando hechos en apoyo de las tendencias evolucionistas primero, y alentados con el triunfo despues, diferentes sabios estudiaron las relaciones entre la planta, el suelo que habita y el clima en que vive[3] (de Buen 1883, p. 421).

Around that time, José Gogorza, also a student and colleague of Bolívar's, was one of the Spanish naturalists who journeyed to the famous Naples Stazione Zoologica (Fantini 2000), a mecca for converts to the new biology. There he chose to study the response of marine animals to freshwater conditions, a physiological and ecological problem closely connected to the concept of adaptation (Gogorza 1891).

Both Buen and Gogorza were members of the "Sociedad Linneana Matritense" (Linnean Society of Madrid). This botanical association, which brought together mainly young naturalists, had been founded in 1878 by Tomás Andrés y Tubilla (1859–1882). One of the initial purposes of the society was to complete an inventory of Spanish flora; however, Andrés y Tubilla soon proposed a new project aimed at studying the plant geography of the Iberian peninsula. He led this biogeographical turn by proposing a new botanical division of the Iberian peninsula based on the distribution of plant species within two families (Andrés y Tubilla and Lázaro e Ibiza 1882).

Andrés y Tubilla's biogeographical paper was published just after his death, when he was only 22 years old. We do not know if he would have developed his ecological interests further. What we do know is that other naturalists who remained active – such as Bolívar and Gogorza – did not. It seems that they lost their enthusiasm for biogeographical and ecological subjects and soon joined the same kind of taxonomic research programmes that their predecessors had undertaken, that is, a mere cataloguing of Iberian flora and fauna. Thus, the initial reception of ecological approaches among Spanish naturalists was very limited and lacked continuity. It corresponded almost exactly with the formative years of the generation of naturalists born in the 1850s. From about 1890 onwards they apparently lost what could be called their ecologically-minded scientific personality and their work was somehow diluted in the collective project aimed at the completion of a natural history of Spain comparable to those already available in other European countries.

Many years later, Ignacio Bolívar, the central figure of that group, was to describe the situation in retrospect using rather telling words.

> Nuestros naturalistas se han ocupado principalmente, casi exclusivamente, en el estudio de las especies que viven en la Península, por el que necesariamente se había de empezar, pues aparte de que en su tiempo otras modalidades de la ciencia eran desconocidas o poco estudiadas, se imponía con toda urgencia hacer, por decirlo así, el inventario de los seres que pueblan nuestro suelo, como base de todo estudio ulterior[4] (Bolívar 1922, p. 65).

---

[3] "Plant geography studies have become extremely important. First Humboldt and later Schouw, A. de Candolle, Wahlenberg and a number of others led the way. Later, these studies were given a significant boost with the appearance of Darwin's theory in the realm of the natural sciences. Searching initially for facts in support of evolutionary theory, and then stimulated by their success in doing so, several scholars studied the relationships between plants and the soil and climate in which they live".

[4] "Our naturalists have devoted themselves mainly – indeed almost exclusively – to the study of the species that inhabit the [Iberian] Peninsula. This was the principal task that needed to be carried out because, apart from the fact that other scientific modalities were unknown or little pursued at that time, it was utterly imperative to compile an inventory of the natural beings that live in our land, as a basis for further study".

Certainly ecology was among those "other scientific modalities" that were "unknown or little pursued" because of Spanish naturalists' almost exclusive dedication to the descriptive and taxonomic "study of the species", and so it remained for a long time. Nevertheless, in the period between 1910 and 1930 a few scientists ran the risk of placing themselves outside the mainstream of Spanish natural history and launched new scientific programmes positioned firmly in the field of ecology.

Celso Arévalo and Aquatic Ecology

In 1912 Celso Arévalo (1885–1944) founded the first centre devoted to the ecological study of continental waters in Spain in Valencia, called the "Laboratorio de Hidrobiología Española". Both on account of this institutional success as well as his research and popularizing works, Arévalo must be considered to be the one Spanish scientist who was responsible for establishing that form of aquatic ecology known as limnology or, as Arévalo preferred, hydrobiology (Casado 1997, pp. 155–263).

Having graduated in the Natural Sciences, Arévalo undertook further research training in 1905 at the "Estación de Biología Marítima" in Santander, which had been founded in 1886 and at that time was the only institution involved in marine biological studies in Spain (Fig. 22.1). Although the ecological approach did not figure in Santander, Arévalo adopted such an approach when he founded the "Laboratorio de Hidrobiología Española" a few years later. Arévalo took the institutional model of the marine laboratories and transferred it to continental waters, while simultaneously shifting towards a different, more ecological orientation in

Fig. 22.1 Celso Arévalo (right) and an unidentified researcher at the "Estación de Biología Marítima" in Santander, circa 1905 (Courtesy of María Teresa Arevalo, Madrid)

aquatic biology, in imitation of the emerging limnological schools of Central Europe and North America. In Arévalo's own words, it was a matter of establishing "la Hidrobiología" in Spain "as a Science created by the new orientations of Natural History"[5], replacing the "criterio taxonómico" with what he called the new "criterio biológico"[6]. The aim of the latter was to focus not on the species but on the "grupo biológico", defined by its association to one particular "medio" (medium) (Arévalo 1914a).

In addition to the laboratory in Santander, new marine laboratories were founded between 1905 and 1915 in several locations along the Spanish coast. Arévalo argued that similar institutions should exist to study rivers and lakes, and he attempted to attract the government's attention to his project by emphasizing potential applications with regard to the improvement of fishing and other economic activities. He also argued that other European countries had already applied the model of coastal laboratories to inland waters, building limnological research centres in different lakes and rivers. One such example was the pathbreaking Biological Station at Plön, in Germany (Overbeck 1989), cited by Arévalo as an outstanding example among many others in Europe and the United States (Arévalo 1914b).

Arévalo's professional and institutional standing was not unproblematic, however. He taught natural history in the public educational system, taking up a post at the secondary school in Valencia, the "Instituto General y Técnico de Valencia", in 1912 (Fig. 22.2). It seems that the proximity to the school of a coastal lagoon named "L'Albufera" prompted Arévalo to start his project in imitation of foreign limnological centres. Like those centres, Arévalo's laboratory was set up specifically to study a single system, L'Albufera. The laboratory itself, however, was located not on the shore of the lagoon but a few miles away, in the city of Valencia. It had no permanent staff, no boats and no other equipment that would have been needed to carry out proper limnological surveys. In fact, in its early days, the "Laboratorio de Hidrobiología" was simply a personal initiative on the part of Arévalo that had no support except for that of the "Instituto General y Técnico de Valencia" where he taught natural history. Arévalo had obtained permission from the head of the school to use a free space formerly used as a corridor, where he managed to install a few research tables with microscopes and other research equipment, as well as some aquariums for the observation of live organisms. Only after several years of work, in 1917, was it officially recognized by the government as a research centre. This was largely a matter of status, because even then the laboratory continued to operate with almost no funding from any local or national authority.

---

[5] "como una Ciencia creada por las nuevas orientaciones de la Historia Natural" (Arévalo 1914a).

[6] By "biological criterion" Arévalo simply meant that he wanted to study actual biological communities rather than artificially-defined taxonomic groups. Of course, such approach was hardly new in the 1910s, but it must be noted that these kinds of ecological studies were almost inexistent in Spain at that time.

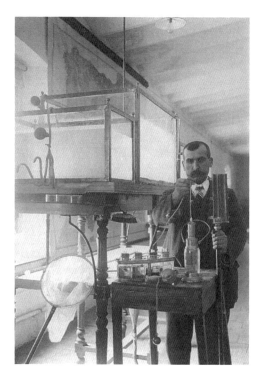

Fig. 22.2 An employee of the "Instituto General y Técnico de Valencia" shows part of the limnological equipment of the "Laboratorio de Hidrobiología", circa 1915 (Courtesy of María Teresa Arévalo, Madrid)

In spite of these limitations Arévalo managed to carry on doing remarkable research on the limnology of L'Albufera and nearby wetlands. However, soon enough he was forced to face a paradox. Having declared that it was necessary to change the "taxonomic criterion" dominant in Spanish natural history, he was still confronted with the very same problem – insufficient taxonomic knowledge – that had prompted a taxonomic-centered tradition among Spanish naturalists. The shortcoming was even worse for aquatic groups, which had been especially neglected. As a consequence, Arévalo was forced to begin his work by making taxonomic surveys of the aquatic organisms that were of interest to him, especially the planktonic groups. This contradictory situation prevented a deeper development of Arévalo's ecological research. However, the difference in his approach was clear. Instead of trying to specialize in one or a few groups for the whole Iberian peninsula, Arévalo concentrated on L'Albufera, studying as many groups as possible. This is how he became the first Spanish scientist to publish works on such important groups as Cladocera and Rotatoria (Arévalo 1916, 1917). In addition, he produced many short notes and gathered unpublished data on almost all the aquatic groups that inhabited L'Albufera, from algae to birds. Arévalo also investigated the spatial distribution of organisms in relation to the physicochemical conditions of the lagoon, and became especially interested in the temporal variation, both qualitative and quantitative, of planktonic communities. However, it seems that he did not attempt to study interspecific relationships.

Arévalo was a member of the "Sociedad Española de Historia Natural", where he talked to his colleagues about the new laboratory and its scientific goals (Arévalo 1914b), but his ecological projects did not enjoy much support from the scientific establishment in Madrid. In order to seek more local support, he founded a branch of the society in Valencia, becoming the leader of a small group of Valencian non-professional naturalists who provided some practical assistance for his project. Arévalo's main follower and collaborator was biologist Luis Pardo (1897–1957). Some foreign naturalists were temporarily affiliated with the laboratory, which proved useful for their specialized studies on aquatic groups. Such was the case with German malacologist Fritz Haas, German specialist on water mites Karl Viets, and Swiss ichtyologist Alfonso Gandolfi. Arévalo even managed to create a regular forum for his own publications by creating a journal for the Instituto, the "Anales del Instituto General y Técnico de Valencia". This contained a special series entitled "Trabajos del Laboratorio de Hidrobiología Española" which included 33 limnological works from 1916 to 1928. The "Trabajos" were reprinted separately for the purpose of scholarly exchange with foreign limnological journals and centres. It should be noted that with regard to limnology, most of Arévalo's scientific relationships were with foreign researchers and centres, especially in German-speaking countries. In 1921 he travelled to France, Switzerland and Germany to visit several limnological laboratories, including the one at Kastanienbaum, near the Swiss city of Lucerne, which was directed by Hans Bachmann. He also became a member of the "International Association of Theoretical and Applied Limnology" (best known as SIL) immediately after its founding in 1922, and attended the SIL congress in Rome in 1927.

In summary, Arévalo created his own institutional framework for his ecological project, taking an alternative route to the taxonomic approach favoured by the official centres devoted to natural history.

In 1919 Arévalo took up a new post to teach natural history at the "Instituto del Cardenal Cisneros" in Madrid. Subsequently, the "Museo Nacional de Ciencias Naturales" in Madrid created a new section of hydrobiology for Arévalo, who was appointed head of department. The "Laboratorio de Hidrobiología Española" in Valencia was incorporated as a branch of the museum. Did this mean that the museum was suddenly interested in Arévalo's ecological project? Clearly not – in fact, the museum's interest lay in transforming the laboratory in Valencia into a marine station. The museum had lost control over several coastal stations due to a previous institutional reorganization of those scientific centres controlled by the government, and it was in urgent need of alternative seashore locations if it was to carry on the taxonomic studies of marine groups that were the subject of several of its staff's zoologists. Nonetheless, this agreement gave Arévalo the chance to maintain some control over his laboratory in Valencia, which would have probably been lost otherwise; most importantly, it gave him the official support that he had sought for so long.

It was difficult for both the museum and Arévalo to safeguard their own respective interests. After some time the museum decided that Valencia was not the right place to establish a coastal station and lost all interest in the laboratory. In the meantime, Arévalo had succeeded in establishing an official post in Valencia for his collaborator Luis Pardo, who was appointed assistant researcher at the laboratory. For a few years Arévalo kept up his limnological research, focusing on the temporal variation of plankton and surveying various Spanish lakes, sometimes in collaboration with Pardo. It seems that this kind of research was not held in high regard by the museum's director, Ignacio Bolívar, or by other influential members of the centre, and soon the support given to Arévalo was reduced to a minimum. As mentioned in the previous section, Bolívar had long lost his youthful enthusiasm for protoecological subjects by this time. In 1928 Pardo moved to Madrid to attend to personal business and applied, without success, to continue his work as Arévalo's assistant at the museum in Madrid. In December 1931 Arévalo resigned his post at the museum. The laboratories of hydrobiology in Madrid and Valencia were finally closed down in 1932.

During his final years at the "Museo Nacional de Ciencias Naturales" Arévalo had sought alternative sources of support for his project. Instead of seeking help from scientific institutions controlled by naturalists, he addressed himself to government agencies dealing with natural resources of economic interest. He reinvented himself as a specialist in applied biology and drew attention to the state of neglect suffered by inland fisheries up to this point. This professional transformation was not Arévalo's preferred path – "[Y]ou know that I am not interested in technical aspects of fishing, but only in those that are purely scientific"[7] said Arévalo in a letter to Pardo in 1927. However, it proved to be the only effective way to institutionalize limnological studies in Spain. He also played an active role in various official commissions appointed to reform the legal regulation of fishing, insisting on the importance of biological studies as the only sound basis for improving the economic and social outputs of continental waters.

Management of inland fisheries was one of the tasks officially assigned to the corps of forestry engineers. As a distinct professional branch in the Spanish governmental administration, forestry engineers had their own official centre for applied research, the "Instituto Forestal de Investigaciones y Experiencias" (Institute of Applied Forestry Research). In 1931 this centre created a new "Section of Biology of Continental Waters". Thanks to the contacts made by Arévalo in previous years Pardo became a member of the technical staff of this new department.

Arévalo's own scientific career in aquatic ecology was virtually finished by 1931. But even after his death in 1944, Arévalo's legacy remained alive in his book "La vida en las aguas dulces", which, in spite of its popularizing character, gave a remarkably complete and updated view of the field and must be regarded as the first treatise on limnology published in Spanish (Arévalo 1929).

---

[7] "[Y]a sabes que yo no me intereso por los problemas tecnicos de la pesca y solamente por los puramente cientificos" (Casado 1997, p. 240).

An offshoot of Arévalo's project was Luis Pardo's affiliation with the "Instituto Forestal", where the latter continued studying Spanish continental waters, even if for the most part his focus was now on applied aspects, especially fishing activities. Apart from being Arévalo's collaborator, Pardo was significant because he was the first to attempt a comprehensive survey of Spain's rivers and lakes as natural resources. Even so, pure ecology was also represented at the "Instituto Forestal", as exemplified in the remarkable monograph on the river Manzanares written by forestry engineers Luis Vélaz de Medrano and Jesús Ugarte, which was the first ecological study of riverine ecosystems to be carried out in Spain (Vélaz de Medrano and Ugarte 1934). However, all opportunity to continue aquatic ecological research was eliminated by the outbreak of the Spanish Civil War in 1936. That is why, after the war, the young naturalist Ramon Margalef, unknown at that time, had to start his career in aquatic ecology from scratch. Ramon Margalef was to become one of the world's most influential ecologists in the 1960s and the 1970s – but that is a different story.

Emilio H. Del Villar and Plant Ecology

While Arévalo developed his project in the field of aquatic ecology, another Spanish naturalist chose plant ecology as his own scientific project. Emilio Huguet del Villar (1871–1951) is arguably the most influential of the scientists who undertook ecological research in Spain before 1936 (Casado 1997, pp. 265–352). He was the only one to gather around him a core group of researchers that might have become a genuine ecological school. He was also the most ambitious in defining his research programme, which was aimed at producing a complete, ecologically-based update of the plant geography of the Iberian peninsula. Finally, he was noteworthy for the originality of his theoretical elaborations, many of which he summarized in his book "Geobotánica" (del Villar 1929). But again, Villar's ecological project found little support among Spanish naturalists and he was not able to create a stable institutional structure for it.

Villar was a self-taught scientist without a university degree. This may well have been a key factor in the originality of his scientific thought. Villar's choice of plant ecology – a field almost completely ignored in Spain at that time – and the innovative way in which he developed it are characteristic of his overall personality. However, lacking any official academic qualification proved to be an additional obstacle in his somewhat eccentric career: having started out in journalism, he turned first to geography and later to botany and plant ecology, eventually ending up in the field of soil science (Martí 1984). From both an intellectual and institutional point of view, Villar is a prime example of the "scientific outsider", a characterization I have applied to early Spanish ecologists elsewhere (Casado 1997, p. 442).

Villar began his autodidactic training as a botanist and plant ecologist around 1912, but, lacking any professional links with scientific institutions, he earned his living writing geography books intended for a broad audience. In fact, the work that may be considered his first publication on the ecology of Iberian vegetation appeared in 1921 as part of one of these geographical books (del Villar 1921, pp. 176–192). Four years later Villar published a much more detailed report, entitled "Avance geobotánico sobre la pretendida estepa central de España" (Preliminary geobotanical study on the presumed central steppe of Spain) (del Villar 1925). For this paper, which was to be his official presentation as a fully-fledged ecologist, Villar cleverly chose a subject which demonstrated that modern ecological theories and methods could produce a radically different interpretation of vegetation. Adopting a dynamic approach which, by his own admission, was closely modelled on Frederic E. Clements' successional plant ecology, Villar showed that what had hitherto been known as the Spanish "central steppe" was in fact the result of a long process of anthropogenic deforestation and that the remnants of the original forests could be used to reconstruct the composition and structure of the climax vegetation.

During this period Villar had been engaged professionally as a plant ecologist, for the first and only time in his life, in his native Catalonia. In 1923 botanist Pius Font i Quer, who was serving as director of the "Museu de Ciències Naturals de Barcelona" at that time, had offered Villar a post in a new section of the museum devoted to plant geography. This museum was one of the scientific institutions that had been created during the first decades of the twentieth century as part of a broader cultural drive initiated by the Catalan nationalist movement. They represented an alternative to the central institutions based in Madrid, where Villar had found little support. Villar accepted this post but, wanting to establish his own research programme, he soon came into conflict with the museum and was dismissed in 1924. Villar had been a member of the Madrid-based "Sociedad Española de Historia Natural", where he published his first botanical works, since 1915, but his work attracted little attention. In 1919 he joined a new association, the "Sociedad Ibérica de Ciencias Naturales" (Iberian Society of Natural Sciences), based in Saragossa. A year later Villar founded a branch of this society in Madrid, clearly seeking an alternative scientific audience for his proposals and findings.

The quality and originality of Villar's work did not go completely unnoticed among Spanish botanists, especially the younger ones. The publication of his book "Geobotánica" (Villar 1929), the first treatise on plant ecology to be published in Spanish, further enhanced his prominent status by offering a concise and up-to-date summary of the field (Fig. 22.3). In addition, it proposed a large number of theoretical innovations, such as his classification of vegetation types and his concept of "sinecia", which Villar intended to introduce as a bias-free term for any unit of vegetation, equally distant from the sociologically-laden terms "community" and "association". These innovations endowed his work with the nationalist appeal of a genuinely Spanish contribution to vegetation science. Soon several university professors, graduate students and amateur botanists were following Villar's ideas and methodology. One of them was José Cuatrecasas (1903–1996), professor of Botany at the "Universidad Central" of Madrid, who combined his dedication as a plant taxonomist specializing in tropical flora with regional ecological studies of Iberian vegetation, which he carried out in close cooperation with Villar. This might have become the starting point for a Spanish school of plant ecology had Villar managed to pursue his

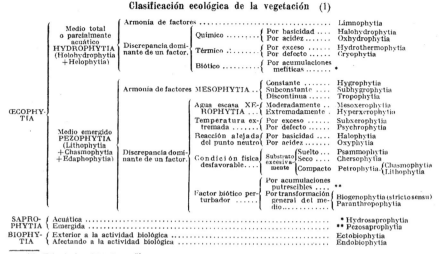

Fig. 22.3 Classification of vegetation types proposed by Emilio H. del Villar in his book "Geobotánica" (Villar 1929)

ecological interests further. Of course the term "school" is used here very loosely, for almost none of the conditions that have been proposed to define a research school (Morrell 1972; Geison 1981; Servos 1993) were met by the informal and somehow blurry group of naturalists clustered around a scientist without any institutional affiliation as Villar. In fact, the professional standing of Villar's followers was often much better than his own.

Such an anomalous situation may partially explain a new shift in Villar's scientific career towards soil science, which can be dated at about 1925. Villar started out with an interest in soils as one of the key factors in plant ecology. As soil science was almost unknown in Spain, he sought advice from leading specialists abroad and in 1924 attended the International Conference of Soil Science held in Rome. The International Association of Soil Science was created at the Rome conference and Villar was invited to organize the Spanish section. For Villar this was an excellent opportunity to ask for official support. Soil science, Villar argued, had obvious applications and potential economic benefits that the Spanish government could no longer afford to ignore. In 1925 an official commission was created to serve as the Spanish section of the International Association of Soil Science. Villar was appointed executive secretary and he even managed to include plant ecology – still his main interest – among the subjects covered by the new "Comisión de Edafología y Geobotánica" (Commission for Soil Science and Geobotany). But soil science was the commission's priority and Villar, who was the only researcher among its members, had to devote himself to these kinds of studies, largely abandoning his investigations in plant ecology. Once again, forestry engineers provided the institutional support that naturalists had been reluctant to give. From 1927 to 1932 Villar worked at the "Soils Section" at the "Instituto Forestal de Investigaciones y Experiencias". Villar is generally credited as being the founder of soil science in Spain, and he eventually published the first map of soils of the Iberian peninsula (Martí 1984).

Spanish botanists with an interest in ecology were "orphaned", as it were, by Villar's progressive withdrawal from ecology, and therefore turned their attention to Montpellier in France, where Josias Braun-Blanquet's phytosociological school had its headquarters at the "Station International de Géobotanique Méditerranéenne et Alpine", best known as Sigma (see also Chap. 21). Sigma had been founded by Braun-Blanquet in 1930 and was especially influential as an international training centre (Acot 1993). It was therefore an obvious choice for those wishing to specialize in vegetation research. In 1934 three young Spanish botanists attended courses at Sigma, and that same year an international excursion organized by Sigma and led by Braun-Blanquet visited Catalonia. José Cuatrecasas helped to organise these activities, which were intended to introduce the so-called Sigmatist school into Spain (Fig. 22.4). While Cuatrecasas himself was an adherent of Villar's concepts and methods, he understood that Sigma's structure could be used to fill the institutional gap that was preventing further development of plant ecology in Spain. There were profound differences, both in theory and practice (Nicolson 1989), between the taxonomic, static approach of Sigmatist phytosociology and the more holistic, dynamic approach of Clementsian plant ecology preferred by Villar. Phytosociology was concerned with floristic inventories and with the classification of vegetation types; Villar's plant ecology, by contrast, focused on the relationships between plant communities

Fig. 22.4 José Cuatrecasas (left), a local guide, and botanist Daniel Sans (right) at Sierra Magina (Jaén, Spain), where Cuatrecasas initiated his ecological investigations in the early 1920s (Courtesy of the Institut Botànic de Barcelona)

and environmental factors with the aim of producing a successional interpretation of vegetation. Nevertheless in the peculiar Spanish context of the 1930s, these influences should be considered as complementary rather than in opposition. Villar's works had aroused interest in plant ecology among Spanish botanists, and Sigma provided a methodological and institutional framework.

The outbreak of the Civil War in 1936 threw the scientific community into complete disarray and interrupted the ongoing acclimatization of plant ecology among Spanish naturalists. Both Villar and Cuatrecasas suffered the ideological persecution that followed the victory of General Franco, which forced many Spaniards into exile. Thanks to the international prestige he had achieved as a soil scientist, Villar was able to establish himself in Algeria and, later, in Morocco, where he continued studying soils until his death in 1951. Cuatrecasas settled in Colombia and later in the United States, where he continued his taxonomic studies as a specialist in tropical flora. He died in Washington DC in 1996.

Meanwhile, a new generation of botanists was in the process of filling the professional gap left in Spain by "émigrés" after the war. Initially, some of them superficially adopted Villar's methodology, but Sigmatist phytosociology was soon reintroduced by botanists such as Salvador Rivas Goday (Izco 1981) and Oriol de Bolòs (Pairolí 2001) – this time in an overwhelmingly dominant way, which has had a profound influence on vegetation research in Spain ever since.

Scientific Outsiders and Interrupted Projects

In some ways both Arévalo and Villar were "scientific outsiders" (Casado 1997, p. 442). They never held a university chair or a post in a pre-existing research department. They chose, instead, to create their own intellectual and professional opportunities, identifying previously unoccupied niches – that is to say, disciplines which had received little or no attention by Spanish naturalists – such as aquatic and plant ecology. They enjoyed more support – at least temporarily – in peripheral cities such as Valencia and Barcelona, which had more dynamic and open institutional settings, than in Madrid, where the central, traditional scientific institutions were based. Their strategy enjoyed mixed fortunes. On the one hand, they succeeded in establishing themselves as authorities in their fields, and created, albeit provisionally, an institutional basis for their projects, such as the "Laboratorio de Hidrobiología Española" and the "Comisión de Edafología y Geobotánica". On the other hand, their achievements proved to be built on sandy soil and were quickly washed away when unfavourable personal, social or political contingencies arose. The lack of support from the scientific establishment in the fields of natural history, botany and zoology, where their ecological projects should have been embedded, jeopardized the continuity of their pursuits, which they eventually abandoned.

Thus the early development of ecology in Spain was very limited and did not play a significant role in relation to international ecology. But it provides an interesting example of the complex processes and contradictory relationships involved in the development of new disciplines within the traditional framework of natural history.

In Spain, as in many other countries, naturalists were the natural substratum for the reception of new ecological approaches (Kohler 2002; Kingsland 2005). However, the fruitful synthesis envisioned in those early years by pioneer ecologists such as Charles C. Adams – "bringing together the best of old natural history and of the new laboratory biology" (Ilerbaig 2000, p. 459) – could not happen in Spain. Insufficient taxonomic knowledge certainly posed a problem for Arévalo and Villar, but this was not the real difficulty. The national and nationalist tradition that had shaped turn-of-the-century Spanish natural history into a taxonomy-oriented project proved to be a major obstacle for those early ecologists in terms of acquiring a recognized institutional and professional status. Even though Arévalo and Villar were primarily interested in basic ecological research, both of them were forced to emphasize the applied potential of their research fields in order to obtain alternative support from official institutions that were controlled by forestry engineers and devoted to fisheries, forestry and agriculture.

As far as I am aware, Arévalo and Villar had no personal or professional contact with one another, nor did they appear to be conscious of the connection between their scientific interests and professional careers. However, their common ecological approach can surely be linked to similarities in the historical development of their projects, including the way they related to the intellectual and social context within the turn-of-the-century Spanish scientific community.

Fig. 22.5 The map displays all the locations of early Spanish ecology activities mentioned in the article (arranged by C. Haak)

References

Acot P (1983) Darwin et l'écologie. Revue d'Histoire des Sciences 36:33–48
Acot P (1993) La phytosociologie de Zürich-Montpellier dans l'écologie française de l'entre-deux guerres. Bulletin d' Ecologie 24:52–56
Andrés y Tubilla T, Lázaro e Ibiza B (1882) Distribución geográfica de las columníferas de la Península Ibérica. In: Resumen de los trabajos verificados por la Sociedad Linneana Matritense durante el año 1881. Sociedad Linneana Matritense, Madrid, pp 25–33
Arévalo C (1914a) La Hidrobiología como Ciencia creada por las nuevas orientaciones de la Historia Natural. Ibérica 2:317–319
Arévalo C (1914b) El Laboratorio hidrobiológico del Instituto de Valencia. Boletín de la Real Sociedad Española de Historia Natural 14:338–348
Arévalo C (1916) Introducción al estudio de los Cladóceros del plankton de la Albufera de Valencia. Anales del Instituto General y Técnico de Valencia 1(1):1–67
Arévalo C (1917) Algunos rotíferos planktónicos de la Albufera de Valencia. Anales del Instituto General y Técnico de Valencia 2(8):1–50
Arévalo C (1929) La vida en las aguas dulces. Labor, Barcelona
Barreiro AJ (1992) El Museo Nacional de Ciencias Naturales (1771-1935). Doce Calles, Aranjuez
Bolívar I (1876) Sinópsis de los ortópteros de España y Portugal. Anales de la Sociedad Española de Historia Naural 5:79–130
Bolívar I (1922). Contestación. In: García Mercet R.Discurso leido en el acto de su recepción por el Señor D. Ricardo García Mercet. Madrid: Real Academia de Ciencias Exactas, Físicas y Naturales, pp. 49–70.
De Buen O (1883) Apuntes geográfico-botánicos sobre la zona central de la Península Ibérica. Anales de la Sociedad Española de Historia Natural 12:421–440
Cacho Viu V (1962) La Institución Libre de Enseñanza. Rialp, Madrid
Casado S (1994) La fundación de la Sociedad Española de Historia Natural y la dimensión nacionalista de la historia natural en España. Boletín de la Institución Libre de Enseñanza 19:45–64
Casado S (1997) Los primeros pasos de la ecología en España. Ministerio de Agricultura, Pesca y Alimentación, Madrid
Coleman W (1986) Evolution into ecology? The strategy of warming's ecological plant geography. J Hist Biol 19:181–196
Fantini B (2000) The history of the stazion Zoologica Anton Dohrn. An outline. In: Cariello L, Consiglio D (eds) Stazione Zoologica Anton Dohrn. Activity Report 1998/1999. ImPrint, Napoli, pp 71–107
Geison GL (1981) Scientific change, emerging specialties, and research schools. Hist Sci 10:20–40
Glick TF, Ruiz R, Puig-Samper MA (eds) (2001) The reception of Darwinism in the Iberian world. Kluwer Academic Publishers, Boston
Gogorza J (1891) Influencia del agua dulce en los animales marinos. Anales de la Sociedad Española de Historia Natural 20:221–271
Ilerbaig J (2000) Allied sciences and fundamental problems: C. C. Adams and the search for method in early American ecology. J Hist Biol 32:439–463
Izco J (1981) Prof. Salvador Rivas Goday. Lazaroa 3:5–23
Kingsland SE (2005) The evolution of American ecology. The John Hopkins University Press, Baltimore
Kohler RE (2002) Landscapes and labscapes: exploring the lab-field border in biology. University of Chicago Press, Chicago
López Piñero JM (1992) Introducción. In: López Piñero JM (ed) La ciencia en la España del siglo XIX. Marcial Pons, Madrid, pp 11–18
López-Ocón Cabrera L (2003) Breve historia de la ciencia española. Alianza, Madrid
Martí J (1984) Emilio Huguet del Villar (1871-1951). Cincuenta años de lucha por la ciencia. Universitat de Barcelona, Barcelona

Morrell JB (1972) The chemist breeders: the research schools of Liebig and Thomson. Ambix 19:1–46
Nicolson M (1989) National styles, divergent classifications: a comparative case study from the history of French and American plant ecology. Knowledge Soc 8:139–186
Núñez D (1977) El darwinismo en España. Castalia, Madrid
Overbeck J (1989) Plön history of limnology, foundation of SIL and development of a limnological institute. In: Lampert W, Rothhaupt KO (eds) Limnology in the federal republic of Germany. International Association for Applied and Theoretical Limnology, Plön, pp 61–65
Pairolí M (2001) Oriol de Bolòs. Una vida dedicada a la botànica. Fundació Catalana per a la Recerca, Barcelona
Sala Catalá J (1981) El evolucionismo en la práctica científica de los biólogos españoles del siglo XIX (1860-1907). Asclepio 33:81–125
Sánchez Ron JM (1999) Cincel, martillo y piedra. Historia de la ciencia en España (siglos XIX y XX). Taurus, Madrid
Servos JW (1993) Research schools and their histories. Osiris 8:3–15
Sociedad Española de Historia Natural (1872) Circular. Anales de la Sociedad Española de Historia Natural 1:5–7
Stauffer RC (1960) Ecology in the long manuscript version of Darwin's Origin of Species and Linnaeus' Oeconomy of Nature. Proc Am Philos Soc 104:235–241
Vélaz de Medrano L, Ugarte J (1934) Estudio monográfico del río Manzanares. Instituto Forestal de Investigaciones y Experiencias, Madrid
Del Villar EH (1921) El valor geográfico de España. Ensayo de Ecética. Sucesores de Rivadeneyra, Madrid
Del Villar EH (1925) Avance geobotánico sobre la pretendida estepa central de España. Ibérica 23:281–283, 297–302, 328–333, 344–350
Del Villar EH (1929) Geobotánica. Labor, Barcelona

Chapter 23
Plant Community, Plantesamfund

Peder Anker

The First Use of the Plant Community Concept

The plant community concept was first introduced by the Danish botanist Johannes Eugenius Bülow Warming (1841–1924) in his book "Plantesamfund" of 1895, where he suggested a general theory of explaining different geographical distributions of plants. The title "Plantesamfund" can be translated both as Plant societies and Plant communities, since the Danish word samfund means both "society" and "community" (or alternatively "Gesellschaft" and "Gemeinschaft" in German). To keep the broad meaning of the original title Warming chose the German title "Ökologischen Pflanzengeographie" (1896) and the English title "Oecology of Plants" (1909). The book addressed different factors limiting the geographical distribution of different plants. He used the concept of "community" or "Gemeinschaft" when describing smaller geographical distributions of plants, while "oecologie" or "Ökologie" had a broader geographical meaning corresponding to "society" or "Gesellschaft" as a whole. "Plantesamfund" was not translated into French, though Warming was inspired by the French botanical notion of "le commensal" (dinner partner) in his thinking about the plant community. "Plantesamfund" was also translated into Polish in 1900 and Russian in 1901.

Summary

The Danish botanist Warming coined the plant community concept in his book "Plantesamfund" in 1895. It has a neo-Lamarckian, morphological, and religiously informed understanding of plant geography. The community concept also drew its inspiration from the Danish political and social environment. Warming was a patri-

---

P. Anker (✉)
Gallatin School of Individualized Study and Environmental Studies Program at New York University, New York, USA
e-mail: pja7@nyu.edu

otic defender of the King's council's ambition to expand the Danish Empire and the exploitation of natural resources. The plant community concept provided a tool for management of nature that was inspired by the King's steering of human communities. Warming's morphologically informed research in Brazil and his geographical explorations of Greenland were also of key importance in the development of his plant community concept.

Main Phases of the History of the Concept

The plant community concept was first introduced in Danish. The English translation of 1909 is true to the original Danish definition:

> The term 'community' implies a diversity but at the same time a certain organized uniformity in the units. The units are the many individual plants that occur in every community, whether this be a beech-forest, a meadow, or a heath. Uniformity is established when certain atmospheric, terrestrial, and any of the other factors discussed in Section I [light, heat, humidity, air, nutrients, soil, water, etc.] are co-operating, and appears either because a certain, defined economy makes its impress on the community as a whole, or because a number of different growth-forms are combined to from a single aggregate which has a definite and constant guise. (Warming 1909, p. 91).

The plant community concept emerged from Warming's research in the Brazilian community Lagoa Santa in the early 1860s, research which was published in his book of the same name in 1892. His voyage to Greenland in 1884 was also important to the plant community concept, since it was during this expedition that he learned to appreciate the importance of having a managerial overview on a geographical landscape before analyzing its plants. Warming's patriotic political views and support of the King's geographical ambitions for enlarging the Danish Empire were also of significance for his plant community concept. It was assumed that only a stable and harmonious human community could be a true resource for the nation. Plant communities, by analogy, Warming argued, could only be a natural resource in so far as they lived in commensalism with the rest of nature. He would quote the French botanist Pierre Joseph van Beneden's (1809–1894) definition of commensalism – "Le commensal est simplement un compagnon de table" (The dinner partner is simply a companion at the table) – to evoke the sense of mutually benefiting way of living he thought both humans and plants were striving for (Warming 1909, p. 92). This commensalism was to describe a symbiotic relationship where different plants could live side by side at the same dinner table without harming each other's living conditions. He was particularly interested in cases where certain plants may benefit from living in co-relationship with other plants.

Warming was widely read and appreciated among Danish and Scandinavian ecologists (Prytz 1984; Söderqvist 1986). German plant geographers, such as Andreas Franz Wilhelm Schimper (1856–1901), were also inspired by Warming, and altogether three different versions of "Plantesamfund" appeared in the German language (Schimper 1898; Goodland 1975).

In Britain Arthur George Tansley (1871–1955) pursued a mechanist informed reading of Warming's plant community concept, while his rival Isaac Bayley

Balfour (1853–1922) pursued a morphological interpretation that was more in accordance with Warming's original ideas. The South African ecologist and Balfour-student John Phillips (1899–1987) would coin the phrase "biotic community" in reference to Warming and to the holistic philosophy of the South African statesman Jan Christian Smuts (1870–1955). (Smuts 1926; Phillips 1931).

Warming showed little interest in epistemology and philosophy, but regarded himself instead as a strictly empiricist. He had a religiously informed understanding of both the human and the plant community, believing that God's goodness and purpose was an acting force in nature and human communities alike.

Historical Background

The plant community concept Warming developed grew out of his patriotic as well as deeply religious views. He was raised on a farm in conservative rural Nørup west of Velje. This landscape in Denmark was dominated by the Randbøl heath where Warming would spend his youth nurturing a passion for nature. His father was a priest who died when he was only three years old, and his religious views were mixed with a lifelong longing for his father. His mother's family consisted of wealthy shopkeepers, and Warming would eventually inherit a fortune that enabled him to pursue his high-society botanical interests. This he did from the age of eighteen at the University of Copenhagen, where he would read for general exams in natural history and botany between 1859 and 1862 (Christiansen 1924–1926, pp. 617–665, 776–806).

This was a tumultuous political and social period in Danish history. The imperial ambitions of the King's cabinet caused much tension with respect to control of the rebellious duchies of Nord Schleswig, Holstein, and Lauenburg. A bitter war to defend the region between 1848 and 1851 did not settle the conflict, which re-emerged in another war between 1863 and 1864 in which Prussia took control of the duchies. In the coming years it became the King's council's official policy to expand the Danish Empire as well as trying to reunite Denmark with the lost land. Warming was on a lifelong crusade for the cause. It is telling that his biographer describes the reunion of Nord Schleswig (now Søderjylland) with Denmark in 1920 as "the most joyful event of his life" Christiansen (1924–1926, p. 780).

His religious beliefs were, like most of his fellow Danes, Lutheran protestant. Religion in Denmark was at the time a political matter. Though the throne provided citizens with religious freedom and the Church was declared independent of the State, the King's cabinet was in reality the head of Church. In this hierarchy the King's good will would secure the religious purpose and order of society, communities, and the use of natural resources. Warming was a religious patriot, which meant supporting the authority of the King, religious and social stability, and the Danish imperial ambitions. He saw the wisdom of the King's council in view of the larger purpose and goodness of all living things, a purpose which had its ultimate cause in the Creator. It was God who once started the evolution of the Creation, and the botanists could unveil His purpose in the successive development of the living things towards a gradually better world. It was then up to the King's council to wisely use botanical knowledge to guide the use of natural resources. According to his neo-Lamarckian views, plants adapting to their environment and God's purpose and goodness was behind these processes in nature. Though Warming in the late 1870s adopted the Darwinian principles of evolution, he could not agree with the view that this evolution was accidental or without a deeper aim (Coleman 1986). This purpose in biological evolution as well as human history he understood in view of the King's council's ambition of expanding the Danish Empire to secure the wealth of the Danish nation through exploitation of natural resources.

In 1863, the same year the King's council decided to pursue their geographical objective of trying to take back their lost duchies with military might, Warming left for Brazil. He was invited to serve as a secretary for the palaeontologist Peter Wilhelm Lund (1801–1880) who was working on an excavation site in the community of Lagoa Santa. According to Warming, it was a place of "light and joy and peace" Christiansen (1924–1926, pp. 624). It was also a place of loneliness. Lund was a rather asocial person whose only demand of Warming was to read and organize his correspondence. This forced Warming to stay in proximity of their house, and thus to limit his own research to the immediate surroundings. Over the next two and a half years, he consequently came to know the geographical location of almost every plant in the neighbourhood. Upon his return he would use three decades to describe the fourteen cases of species he collected into a book he eventually published as "Lagoa Santa" in 1892.

Warming was an admirer of Alexander von Humboldt (1769–1859), who in his books about plant geography relied on morphological methods as well as systematic botany (Nicolson 1983, pp. 12–73). His chief source of inspiration, however, was neo-Lamarckism and the idea that plants adapt to each other and their respective environments. Upon his return from Brazil he would plunge into morphological-organic studies of plants. His colleagues in Denmark initially found Warming arrogant and disagreed with his ideas about the importance of environmental factors in understanding the geographical distribution of plants. After extensive travelling at various universities in Germany, he eventually settled for a professorship in Stockholm where he would lecture and write about systematic botany. These lectures resulted in a series of textbooks, which were widely used in Scandinavian universities and beyond. His "Haandbok i den systematiske botanik" (Handbook in Systematic Botany) from 1879 and "Den almindelige botanik" (The common botany) from 1880 were both reprinted in several editions as late as 1891 and 1895 respectively. When he finally received a professorship at the University of Copenhagen in 1886 he would lecture regularly for medical and pharmaceutical students, lectures that were published as "Grundtræk af forelesninger over systematisk botanik" ("Outlines of lectures in systematic botany"), in 1896.

What brought Warming back to Copenhagen and the inner circle of Danish scholars was the desire of the King's council to map and explore the natural resources within the Danish Empire, such as those of Greenland, Faeroe Islands, and eventually Iceland. These explorations started in the winter of 1884 with a voyage to investigate the botany of Greenland. The sparse vegetation in this arctic landscape allowed Warming to achieve a swift and effective overview. The bare landscape gave him an opportunity to understand the geographical distribution of plants and see them in view of other plants and the entire habitat (Warming 1890). The possibility of seeing plant communities in relation to the ecological environment as a whole was an exciting turn in his research, and it became the methodology for organizing his Brazilian material published as "Lagoa Santa" in 1892. Botanical investigations into natural resources on contested Danish territory would occupy much of Warming's later work, such as in "Botany of the Færöes" (1901–1908) and in "Botany of Iceland" (1912–1918). The aim of these investigations was to establish Danish hegemony in the territories and open up for exploitation of plant communities. This correlation between botany and resource management was not accidental; Warming and his students sought to develop an ecological method suitable for Danish social control in a foreign region (Christiansen 1924–1926, pp. 799–800, 806–832).

Most of Warming's work prior to the Greenland expedition was morphological in content and systematic in outline. This research was widely respected among his fellow scholars, but it did not attract students' attention beyond preparing for exams. What created excitement among the young was Warming's introduction and development of the plant community and other ecological concepts. Throughout the 1890s, the plant community became a particularly central concept. It was to explain how plants could live and evolve together in "commensalism" without the dreadful struggle for existence described by Charles Darwin (1809–1882). Warming's religious neo-Lamarckian views implied that plant communities were in stable harmony slowly progressing towards higher development. This mirrored the concept of community and progress in the Danish society.

The original Danish edition of "Plantesamfund" from 1895 was a short book meant to provide the reader with a sense of overview. The book grew with each translation, however, since Warming continuously added more details to substantiate his claims. The key terms and concepts, however, hardly changed in the new versions of the book. The German and English translations were thus conceptually similar, while the number of examples and elaborations grew with each volume. Most non-Danish scholars learned about the plant community concept through the widely read German translation of "Plantesamfund" by Emil Knoblauch which was published as "Lehrbuch der ökologischen Pflanzengeographie" in 1896. In Britain, Tansley, for example, thought that Warming "opened [...] a new way of looking at the plant world" (Tansley 1924, p. 54) and he adapted the German translation as a textbook for his own botany classes at the University College, London. The Central Committee for the Survey of British Vegetation (with Tansley as Chairman) would dedicate their famous "Types of British Vegetation" from 1911 to Warming as "the father of modern plant ecology." It was the Edinburgh ecologist Balfour who arranged for the first English translation. Warming wrote a fully revised manuscript; he also upgraded the morphological content by leaning on the German versions. Balfour thus claimed in 1909 (with Warming as the authority and to Tansley's annoyance), that that University of Edinburgh was the center for plant community and ecological research.

Henry C. Cowles (1869–1939) at University of Chicago was much pleased with the morphological turn of ecological methodology, while Tansley wrote a long critical review of the English edition, in which he advised his readers to stick to the first German version (Cowles 1909; Tansley 1909). The tension among British ecologists with regard to how to read Warming soon evolved into a major debate between the mechanist inspired ecology of Tansley and the holistic "biotic community" reading of Warming by Balfour's student Phillips (Phillips 1931; Anker 2001). It was the ordering of plants by Warming according to geographical factors that intrigued Tansley. He was not against morphology as such when he first read Warming. The tension that developed between him and Balfour first became apparent when Tansley learned to appreciate genetic and biochemical research around 1901. Tansley now started to promote genetics and plant geography as the right ecological approach, while Balfour stuck to the morphological study of tracing the ancestral history of species as a methodological basis of ecology. Warming himself did not believe in the value of genetics, and the subsequent German editions of his book grew with a 600 page morphological enlargement in its final version of 1918. (Warming 1909, p. vi; Warming and Graebner 1918; Goodland 1975). Tansley would later renew his interest in Danish botany by studying the work of the Warming student Christen Raunkiaer (1860–1938), who over the years had elaborated on Warming's plant community concept in the direction of functional classification in what he called "life-form-systems" of plants. Tansley initiated a translation of Raunkiaer's collected papers to English, which appeared in 1934 (Raunkiaer 1934). Raunkiaer emphasized the importance of statistical methods in studying plant communities, something that caught Tansley's attention while working out his own "ecosystem" concept (Tansley 1935).

References

Anker P (2001) Imperial ecology: environmental order in the British Empire, 1895-1945. Harvard University Press, Cambridge
Christiansen C (1924) Den Danske botaniks historie. Hagerups Forlag, Kopenhagen
Coleman W (1986) Evolution into ecology? The strategy of Warming's ecological plant geography. J Hist Ecol 19:181–196
Cowles HC (1909) Ecology of plants. Bota Gaz 48:149–152, 465–466
Goodland RJ (1975) The tropical origin of ecology: Eugen Warming's Jubilee. Oikos 26:240–245
McIntosh RP (1985) The background of ecology. Cambridge University Press, Cambridge
Nicolson M (1983) The development of plant ecology 1790–1960. Ph.D. thesis, University of Edinburgh, Edinburgh
Phillips J (1931) The biotic community. J Ecol 19:1–24
Prytz S (1984) Warming: Botaniker og Rejsende. Bogan, Kopenhagen
Raunkiaer C (1934) The life forms of plants and statistical plant geography. Clarendon, Oxford
Rosenvinge LK, Warming E (eds) (1912–1932) The botany of Iceland, 4–5th edn. J. Frimodt, Kopenhagen
Rosenvinge LK, Warming E (eds) (1912–1932). The botany of Iceland Vols. 4–5. J. Frimodt, Kopenhagen
Schimper AFW (1898) Pflanzengeographie auf physiologischer Grundlage. Fischer, Jena

Smuts JC (1926) Holism and evolution. Macmillan, London
Söderqvist T (1986) The ecologists: from Merry naturalists to Saviours of the Nation. Almquist & Wiksell, Stockholm
Tansley AG (1909) Oecology of plants. New Phytol 8:218–227
Tansley AG (1911) Types of British vegetation. Cambridge University Press, Cambridge
Tansley AG (1924) Eug. Warming in memorian. Bot Tidsskr 39:45–56
Tansley AG (1935) The use and the abuse of vegetational concepts and terms. Ecology 16:284–307
Warming E (1879) Haandbok i den systematiske botanik. Philipsens Forlag, Kopenhagen
Warming E (1880) Den almindelige botanik. Philipsens Forlag, Kopenhagen
Warming E (1890) Botaniske exkursioner. Hovedbiblioteket, Kopenhagen
Warming E (1892) Lagoa Santa: Et bidrag til den biologiske Plantegeografi. Bianco Lunos Kgl. Hof-Bogtrykkeri, Kopenhagen
Warming E (1895a) Plantesamfund: Grundtræk af den økologiske plantegeografi. Philipsens Forlag, Kopenhagen
Warming E (1895b) A handbook of systematic botany. Swan Sonnenschin, London
Warming E (1896a) Lehrbuch der ökologischen Pflanzengeographie. Gebrüder Borntrager, Berlin
Warming E (1896b) Grundtræk af forelesninger over systematisk botanik. Det Nordiske Forlag, Kopenhagen
Warming E (1909) Oecology of plants: an introduction to the study of plant-communities. Oxford University Press, London
Warming E, Graebner P (1918) Eug. Warming's lehrbuch de ökologischen pflanzengeographie. Gebrüder Borntraeger, Berlin
Warming E (1901–1908) Botany of færöes, 1st–3rd edn. Det Nordiske Forlag, Kopenhagen

Chapter 24
Looking at Russian Ecology Through the Biosphere Theory

Georgy S. Levit

Introduction

The biosphere theory is crucial for all environmental sciences including scientific ecology. In Russia, the theory was from the very beginning a powerful factor affecting global and other holistic approaches in the life sciences. The theory was invented by Vladimir Ivanovich Vernadsky (1863–1945), who is regarded as one of the most famous Russian naturalists. In the history of Russian science he is referred to as a "savant" and influential thinker in rather different fields such as biogeochemistry, radiogeology, or crystallography, and also philosophy of science. In recent times Vernadsky is becoming appreciated also in the Western world. James Lovelock, author of the Gaia-theory, wrote: "We discovered him to be our most illustrious predecessor" (Lovelock 1986).

Therefore it is all the more astonishing that the origins of the theory and its perception by different scientific communities, such as ecologists, biochemists or geographers, are not very well known. This is even more true for the different national, language and institutional contexts. At present there are about 1,000 published works about Vernadsky, but his biosphere theory has not been adequately investigated, reconstructed and appreciated. One reason might be the ideological pressure and censorship in the former USSR. Another reason is the complexity and quantity of his scientific literature: Vernadsky produced about 200 articles an books in several languages directly connected to the biosphere theme and living matter.

Concerning the presence of Vernadsky in the history of science there is a rather paradoxical situation: In the Russian history of science Vernadsky is a super star of natural science and philosophy. His fame and influence certainly can be compared with that of Ernst Haeckel in the German speaking countries. This contrasts sharply to the position of Vernadsky in history of science in the West, where he is rather underevaluated or even ignored (Ghilarov 1995), which is also true for the community of Western biogeochemists that mostly do not know the founder of their own

G.S. Levit (✉)
History of Science & Technology Program, University of King's College, 6350 Coburg Rd., Halifax, NS, Canada, B3H 2A1
e-mail: Georgy.Levit@ukings.ns.ca

discipline. As the Vernadsky specialist Andrei Lapo claims: "To put the contemporary situation concerning Vernadsky's popularity on a world-wide scale in a graphic phrase, one can say that the dinner table has already been laid, but the guests are arriving late" (Lapo 2001). Yet Vernadsky's heritage is of crucial interest not only for geochemists, but also for ecologists, since "we should also credit Vladimir Vernadsky with the title of father of the global ecology [...]" (Grinevald 1996, p. 48).

The following study seeks to point at some of these shortcomings. The first part (Vernadsky and the Russian Science through the first half of the 20th century) gives some insights into the disciplinary and institutional influence of Vernadsky and the specific socio-political situation in Russia. In the second part (the essentials of the biosphere theory) the biosphere theory itself will be presented. The last part (Vernadsky's impact on ecology and global sciences) gives some examples illustrating the reception of the biosphere theory in the Russian- and non-Russian-speaking scientific communities.

Vernadsky and the Russian Science through the First Half of the Twentieth Century

Initial empirical impulse for the biosphere theory was the idea of interconnectedness and lawfulness of geological processes, accompanied by the idea of interplay between living and inert processes. The former is illustrated in Vernadsky's early works, e.g. by paragenesis, i.e. the regularities of mineral formation in an ore deposit. Already in the very early beginnings of developing the theory, Vernadsky founded a new scientific school detached from mineralogy and soil science. At that same time the American scientist Frank Wigglesworth Clarke (1847–1931) published comparable ideas in his "Data of Geochemistry" (Clarke 1908). However, in contrast to Clarke, Vernadsky paid a lot attention to the role played by living matter in the history of the earth's crust and the atmosphere. Already in 1909 Vernadsky delivered a paper to the "Meeting of the Russian Naturalists and Physicians" on the basic principles of a new science geochemistry (Aksenov 1994, p. 111).

At the same time Vernadsky was beginning to work in the field of radioactivity. In 1908 he took part in a conference sponsored by the "British Association for the Advancement of Science"[1], where he met John Joly, one of the pioneers of radioactivity research. Vernadsky was deeply impressed by the report of Joly and already in 1909 founded the first radiological laboratory in Russia. The combination of radioactivity studies with the idea of interconnectedness of living organisms ultimately (after the WWII) led to the occurrence of Russian radioecology.

---

[1] Vernadsky was a member of this Association since 1889.

Some Biographical Notes

Vladimir Ivanovich Vernadsky (1863–1945) originates from the centre of Russian cultural and academic life.[2] In 1881 Vernadsky enrolled at St. Petersburg University, where he was a student of such brilliant scientists as the chemists A. Butlerov (1828–1886) and D. Mendeleev (1834–1907), the botanist A. Beketov (1825–1902), the zoologist N. Wagner (1829–1907) and the physiologist I. Sechenov (1829–1905). The greatest influence on Vernadsky had however the soil scientist and mineralogist Vassilij Dokuchaev (1846–1903), the founder of modern soil science and genetic pedology. Dokuchaev founded a landscape science as a part of physical geography and created a concept of the natural climate related zones. Moreover, it is recognised now that he was the first to declare the necessity of a new synthetic science for studying "the genetic, eternal, lawful interconnections existing between the forces and bodies of inert and living nature" (Dokuchaev 1898). Thus Dukochaev can be also regarded as a fore-runner of the modern ecosystem approach in the life sciences.

Vernadsky completed his examinations for the degree of candidate of science in mineralogy and "geognosia", and in 1888 left St. Petersburg for Germany. Germany was regarded as the strongest academic land and German language was the language of international scientific publications. Vernadsky decided to study crystallography under the supervision of Paul Groth (1843–1927), who had a Chair in Mineralogy at the University of Munich.

In 1889 Vernadsky moved from Munich to Paris where he started to work simultaneously under the guidance of the chemist Henry Le Chatelier (1850–1936) and the mineralogist Ferdinand Fouqué (1828–1904). Le Chatelier helped Vernadsky to find his dissertation topic and, as Vernadsky later recognised, significantly influenced his scientific work. Working with the problem of crystalline polymorphism Le Chatelier indirectly contributed to Vernadsky's later space-time and biosphere theories. In 1890 Vernadsky returned to Russia and after completing two dissertations (master and Ph.D.) and in 1902 ultimately got a chair for mineralogy and crystallography at the Moscow University.

In 1910 Vernadsky visited Eduard Suess (1831–1914) in Vienna. Suess was the first scientist to use the term "biosphere" in the sense close to our modern usage, however without proposing a clear concept underlying the term. Already in 1911 this term appeared in the work of Vernadsky although without definition. The crucial step was made in 1912 as Vernadsky published a programmatic paper "On Gaseous Exchange of the Earth's Crust", where he emphasized that almost all of the earth's gases are of biogenic nature and involved in cyclical processes

---

[2] Vernadsky was born in the capital of the Russian Empire Sankt-Petersburg into the family of Ivan Vernadsky (1821–1884), a professor of economics and statistics in the Alexandrovsky Lyce, the elite college-like school, where, for example, the best known Russian poet Alexander Pushkin was educated.

(Vernadsky 1912). 60 years later the same observation delivered empirical basis of the so-called Gaia-hypothesis proposed by the English inventor James Lovelock (e.g. Lovelock 1972; Lovelock and Margulis 1974; Levit 2000). This was also the turning point in Vernadsky's orientation to biological phenomena. But in contrast to a purely biological approach, he began to think of life in terms of global geochemistry. After the October (Bolshevik) revolution (1917) Vernadsky moved to Kiev (nowadays the capital city of the Ukraine), where he was elected as the first President of the Ukrainian Academy of Science. In the same year he initiated *bio*geochemical scientific investigations. At the initial stages of this work Vernadsky formulated the basic tasks of the newly established science (Lapo and Smyslov 1989, p. 55): (1) to calculate a quantitative elementary composition of the different species; (2) to investigate the geochemical history of silicon, copper, zinc, lead, silver and some other elements; (3) to determine some other geochemical characteristics of living organisms like the average weight and water content as well as the percentage of carbon in the organisms.

In 1921 Vernadsky received a letter from the Rector of the Sorbonne University (Paris) with an invitation to teach a course on geochemistry and in 1922 Vernadsky arrived in Paris where he made crucial steps in the direction of the biosphere theory. Supported by the foundation of R. Rosenthal (a French "pears king" of Russian origin) Vernadsky laid the baselines for his "The Biosphere" (1926). In 1923 Vernadsky for the first time used the very term "biogeochemistry" (Mochalov 1982, p. 242).

Establishing and Widening the Concept in Russia

In March 1926 Vernadsky returned to Leningrad (St. Petersburg). Since he was well known for his anti-communist views, his decision to come back to Soviet Russia puzzled his biographers. Based on the archival materials of the Bakhmeteff Archive at the Columbia University, Bailes (1990) and Kolchinsky and Kozulina (1998) arrived at the conclusion that Vernadsky did make a considerable effort to remain in the West. Vernadsky was in no way a sympathizer of the Soviet authorities. In the time between two Russian revolutions (February and October 1917) he was involved in the anti-Bolsheviki resistance as a member of the Constitutional Democrats Party. However, he was unable to obtain permanent sufficient funding of his grandiose and expensive biogeochemical research program in the Western countries. Vernadsky decided to return to the USSR, realising that this was the only way to fullfill his scientific mission (e.g. Vernadsky 1998). Paradoxically the totalitarian state in Russia offered in this case more possibilities than institutions of the liberal world. Although after the October Revolution 1917 the land was plunged into chaos and economic disorder, it was at the same time the period of intensive innovations and foundations of new research institutes in the life sciences. New authorities were looking for a new scientifically based secular ideology. Life sciences and especially Darwinism were seen as basic for the Marxist worldview (Kolchinsky 2006, p. 273).

So Nikolaj Koltzov (1872–1940) was able to found in Moscow an Institute for Experimental Biology. In St. Petersburg Nikolaj Vavilov (1887–1943) initiated an Institute for Plant Studies known nowadays as the Nikolai Vavilov Institute of Plant Growing and Research. Another example is Alexei N. Sewertzoff (1866–1936), the founder of Darwinian evolutionary morphology, who in 1922 established a Department of Zoology at the University of Moscow, and seven years later (1930) a Laboratory of Evolutionary Morphology (as a structural unit of the Institute of Comparative Anatomy at Moscow University). In 1934, based on his Laboratory and including the Institute of Palaeontology, Sewertzoff founded the Institute of Evolutionary Morphology, which two years later was divided again into the Institute of Palaeontology and the Institute of Evolutionary Morphology, which was later reorganized into the A. N. Sewertzoff Institute of Ecology and Evolution. With more then 700 members, the institute is now one of the biggest ecological institutes worldwide.

Vernadsky initiated various institutions. For example, in 1926 he established a Commission for the History of Knowledge (1926–1932) which later, after some reorganisations, was transformed into the Institute of History of Natural Science and Technology (1946), still in existence. Two years later (1928) the official foundation of the Vernadsky's Biogeochemical Laboratory of the Academy of Science (BIOGEL) took place. After the series of reorganisations BIOGEL (1947) became a part of the V.I. Vernadsky Institute of Geochemistry and Analytical Chemistry with an emphasis on ecological-biogeochemical research, such as the ecological assessment of biogeocenoses. At the same time Vernadsky conducted his scientific research and in 1926 presented his views on the biosphere in the book of the same name ("Biosfera") published in Russian in Leningrad. Three years later the book was translated and published in French (Vernadsky 1929).

All these scientific developments took place in the cotext of political repressions and strengthening totalitarianism in the Soviet Russia. In 1928, with the end of the liberal economic policy (so-called New Economic Policy) and the emergence of forced of forced collectivization and industrialisation known as Stalin's Great Break, the repressive machine came into the a much more intensive phase.

It is astonishing that in spite of this accelerating terror machine Vernadsky enjoyed the liberty of traveling abroad. Thus Vernadsky spent the summer of 1929 in Germany and Czechoslovakia. In 1932–1933 he travelled in various countries including Germany, France, the UK, Poland and Czechoslovakia. In Münster he contributed a paper (1932) "Die Radioaktivität und die neuen Probleme der Geologie" (Radioactivity and new problems of geology) to the "Deutsche Bunsen-Gesellschaft für Physikalische Chemie" (German Bunsen-Society for Physical Chemistry). In England, Vernadsky communicated with Frederick Soddy (1877–1956) who founded a theory of isotopes. The study of the isotopic composition and radioactive elements in living matter became since that moment an important line of Vernadsky's research. In the 1930s Vernadsky also continued publishing his papers abroad (for example: Vernadsky 1930, 1934, 1935). It must be emphasized here that Vernadsky was not an absolute exception from the rule. As already mentioned, Sewertzoff published his major evolutionary book first in Germany

(Sewertzoff 1931) and only then in Russian, in the Soviet Union. The isolation of the soviet scientists in that period was not of an absolute nature. Table 1 presents, the publications of the Soviet scientists in Germany in only one, although central German biological publication. The table shows that even after the totalitarian regimes were established both in USSR and Germany there was no impenetrable borderline between the scientists of two countries. The publications of the Soviet scientists cease only after the beginning of the WWII.

In February, 1934 Sergej Oldenburg (1863–1934), the Permanent Secretary of the Academy of Science, died. He had been Vernadsky's closest friend and a strong supporter. His death symbolized also the end of the Petersburg period of the Academy of Science. Vernadsky's Biogeochemical Laboratory moved to Moscow together with the Academy. Approximately at that time Vernadsky came to the idea of writing a book where his holistic views on nature would be expressed both scientifically and philosophically. By 1936 Vernadsky understood that this was impossible to do in one work, and decided to separate this task into two books, one philosophical and one strictly scientific. Thus Vernadsky began to work on his main works "The Chemical Structure of the Earth's Biosphere and Its Environment", which was published only 20 years after his death, and laid foundations for modern biogeochemistry and global ecology (Vernadsky 1965). After the war with Germany broke out (22 June, 1941) Vernadsky was evacuated to the health resort Borovoje in Kazakhstan. The two years in Borovoje were highly productive. Vernadsky wrote the important third issue of series "Problems of Biogeochemistry" (1980) which he saw as his scientific will. He worked also on his main generalizing work "The Chemical Structure of the Earth's Biosphere and Its Environment" where his basic claims, expressed first in "The Biosphere" were revised and developed.

In Borovoje many outstanding scientists thus came to live and work together such as, for example, the founder of the nomogenesis theory (theory of directed evolution) and the theory of geographic zones Leo (Lew) S. Berg (1876–1950) (Levit and Hoßfeld 2005), the founder of the theory of biogeocenosis as an elementary unit of the biosphere, Vladimir Sukachev (1880–1967), and one of the architects of Russian evolutionary synthesis Ivan Schmalhausen (1884–1963). As is clearly documented they all experienced Vernadsky's influence and all of them performed a crucial role in the growth of Russian life sciences, including various branches of ecology.

The Essentials of Vernadsky's Biosphere Theory

In the theoretical system of Vernadsky, the concept of the biosphere is required by the new branch of science created by himself: *biogeochemistry*. Biogeochemistry studies the geological manifestations of life and considers biochemical processes in living organisms in relation to their impact on the geosphere (Vernadsky 1997, p. 156).

> The competence of biogeochemistry is defined, on the one hand, by the geological manifestations of life taking place under this aspect, and on the other, by the internal biochemical processes in the organisms the living population of our planet. In both cases

(for biogeochemistry is a part of geochemistry) one may identify as study objects not only chemical elements, i.e. the usual mixtures of isotopes, but also various isotopes of one and the same chemical element.[3]

Thus the specificity of biogeochemistry, in relation to classic geochemistry, includes its concentration on living matter as the major factor in biogenic migration of chemical elements. Neither living organisms by themselves nor their environment abstracted from them are, Vernadsky argued, the specific objects of biogeochemistry. A biogeochemist is interested, first of all, in studying the cyclic processes of the exchange of chemical elements between living organisms and their environment. The latter can only be described on the basis of a detailed study of interrelations of living and inert (non-living) matter in the space-time of Earth and throughout the history of Earth. How can the main subject of biogeochemical research be defined? Biogeochemistry never aims at the organismic level or at the environmental level alone. It concentrates, in Vernadsky's words, on the biologically controlled flow of atoms, which takes place in a specific geological domain.

In order to define this specific geological domain as the research field of the newly created science, biogeochemistry, Vernadsky introduced his interpretation of the term *biosphere*. The biosphere of the Earth appears as a geosphere occupied and organised by life and thus can be seen as a geological envelope.

Being a geological envelope, the biosphere can be also structured geologically (Vernadsky 1991, p. 120):

> The biosphere appears in biogeochemistry as a peculiar envelope of the Earth clearly distinct from the other envelopes of our planet. The biosphere consists of some concentric contiguous formations surrounding the whole Earth and called geospheres. The biosphere has possessed this perfectly definite structure for billions of years. This structure is tied up with the active participation of life, is conditioned by life to a significant degree and is primarily characterised by dynamically mobile, stable, geologically durable equilibria which, in distinction to the mechanical structures are quantitatively fluctuating within certain limits in relation to both space and time.

Under the various "geospheres" Vernadsky (1965, pp. 107–108) understands the troposphere, the hydrosphere, the land surface and the sphere of the subterranean life. However, Vernadsky's approach to the biosphere goes far beyond the purely stratigraphical statements. Examining living matter from the biogeochemical viewpoint, Vernadsky (1994b) arrived at the conclusion that the chemical compounds of the different species do not reflect that of their environment, but, on the contrary, living matter has determined the geochemical history of almost all the elements of the Earth's crust in the process of making the environment favourable to itself. Thus, living matter shapes the biosphere into a self-regulating system. The biosphere, being seen as a self-regulating system, embraces both the totality of living organisms (living matter) and their environment to the extent it is involved in the actual processes of life, that is, including the troposphere, the ocean, and the upper

---

[3] Vernadsky began the work on the "Scientific Thought as a planetary Phenomenon" (the book I quote here) in the late 1930s. The book was published in an uncensored form only in 1991. In 1997 appeared the English translation.

envelopes of the Earth crust, possibly down to the mantle. The structure of the biosphere is described as a dynamic equilibrium: "Not a single point of this system is fixed during the course of geological time. All points oscillate around a certain midpoint" (Vernadsky 1997, pp. 225–227).

A good example of such dynamic equilibrium is the troposphere. Vernadsky claimed that "all basic gases of the troposphere and of the higher gaseous envelopes N_2, O_2, CO_2, H_2S, CH_4, etc., are produced and quantitatively balanced by the total activity of living matter. Their sum total is quantitatively invariable over geological time [...]" (1965, p. 238). Thus, Vernadsky concludes, "life, i.e. living matter creates the troposphere and constantly maintains it in a specific dynamic equilibrium." It can be remarked here, that the first Gaian principle of atmospheric regulation (Lovelock and Margulis 1974) actually was derived by Vernadsky on the basis of his biogeochemical research 50 years before Lovelock (Levit and Krumbein 2000).

However the biosphere is, in Vernadsky's view, not only a self-regulating, but also an evolving system: "We can and must talk about the evolutionary process of the biosphere by itself" (Vernadsky 1991, p. 20).

Based on his experimental work, Vernadsky already in the beginning of the 1920s concluded that his concept of living matter would influence the evolutionary theory. Studying the natural history of the chemical elements, Vernadsky (1994a, pp. 66–68) arrived at the conclusion that living matter modifies the environment. Living matter, in its turn, is determined by the inert environments.

An important characteristic of the biosphere is its holistic nature, which is guarantied by the various biogeochemical functions. A biogeochemical function is a role which a taxon performs in the biospheric cycles. The major biogeochemical functions include, according to Vernadsky, five groups: (1) The gas-functions, regulating the gaseous structure of the atmosphere including submarine and subterranean environments. (2) The function of concentration, i.e. the ability of organisms to capture and concentrate the chemical elements of their environments. Concentration functions of the first kind describe accumulation by the organisms of elements composing all living organisms without exception; the functions of the second kind can be fulfilled only by the certain kinds of organisms, playing thus an unique role in the trophic chains, as for example molluscs concentrating heavy metals. (3) The oxidation-reduction functions; (4) the biochemical functions of organisms generating biogenic migrations of atoms connected with feeding, breathing, multiplication and destruction of organisms; (5) the biogeochemical functions of the mankind (Vernadsky 1965, p. 237).

Vernadsky has also shown how exactly different biogeochemical functions correspond to specific taxonomic groups. For example, the oxidising function is carried out by autotrophic bacteria and the function of destruction of organic compounds by chemoorganotrophic bacteria and fungi. Analysing these results Vernadsky came to the following three conclusions:

- All basic biogeochemical functions can be carried out by unicellular organisms;
- There is no species able carry out all these functions;
- In the course of geological time, different species may have replaced one another, but the biogeochemical functions must have been carried out.

Ultimately this means that life must have been occurred in the form of a biosphere-like from the very beginning and can exist only in the form of the biosphere, since various biogeochemical functions have to be fulfilled simultaneously: "The first occurrence of life in the biosphere could not be in form of separate organisms but only in the form of the sum total of organisms carrying out various geochemical functions. Biocoenoses necessarily had to occur from the very beginning" (Vernadsky 1994b, p. 459).

On these grounds Vernadsky negated the polyphyletic evolutionary model for the very first stages of biospheric evolution. In his late works, Vernadsky claimed that the biosphere as a self-regulating system has its clearly definable evolutionary "interests". A leading force of the evolution of the biosphere is living matter, which has its own process of evolution partially independent from the needs of adaptation. The biosphere as a whole behaves itself as if it had a peculiar evolutionary strategy: "We can and must talk about the evolutionary process of the biosphere by itself" (Vernadsky 1991, p. 20; 1997, p. 30). One of the basic methods of realisation of these "interests" is increasing the intensity and complexity of the biogenic migration of atoms.

Vernadsky's Impact on Ecology and Global Sciences

Vernadsky created a theoretical system that influenced the whole range of environmental sciences. The most evident is Vernadsky's influence in geochemistry and biogeochemistry. Vernadsky understood geochemistry as a natural history of terrestrial chemical elements. It is however important that his approach allowed scientific predictions about the ways of migration of the chemical elements including their compatibility in the various kinds of rocks. This was significant also for the applied geology, e.g. for the minerals search. One of the most known followers of Vernadsky in the field of geochemistry was his student Alexander Fersman (1883–1945), who gave the first regular course of general geochemistry as early as 1911 and became a founder of an influential scientific school. After Vernadsky's and Fersman's death the Russian school of geochemistry was led by Alexandr Vinogradov (1895–1975). Vinogradov was Vernadsky's deputy in the biogeochemical laboratory and after the WWII organised and headed the Vernadsky Institute of Geochemistry and Analytical Chemistry (1947–1975). He is also regarded as the major follower of Vernadsky in the field of biogeochemistry. In contrast to geochemists concentrating on migration of chemical elements and their composition in the rocks and minerals, biogeochemistry involves the study of cyclic biogenic migrations of chemical elements caused by activity of living organisms (Vinogradov 1953). Following Vernadsky Vinogradov proposed that the chemical composition of living organisms is a result of biological evolution, which proceeded in certain environments.

Along similar lines, based on Vernadsky's and Vinogradov's concept of biogeochemical provinces (see above), Victor Kovalsky (1899–1984) coined a concept of *geochemical ecology* (Kovalsky 1974). Kovalsky proceeded from the assumption that the biosphere is a biogeochemically heterogeneous body and that the chemical composition of elements is part of their adaptation to the environment. He showed that different species develop different strategies adapting to the specific environments. For example, *Pyrethrum parthenifolium* in the molybdenum reach soils of Armenia adapts by reducing molybdenum accumulation, while leguminous plants, on the contrary, increase molybdenum concentration.

An important line of Vernadsky's influence was his impact on the co-architect of the Modern Synthesis, the German-Russian biologist Nikolai Timoféév-Ressovsky (1900–1981).[4] In his own words, he was, first of all, interested in Vernadsky's biosphere theory. Supporting the idea of biogeocenoses as structural units of the biosphere, he wrote: "The biosphere in its entirety consists of more or less complex biotic and abiotic components, i.e. biogeocenoses. In other words, the biogeocenoses are the precise environments in which the evolutionary process of any group of living organisms takes place" (Timoféev-Ressovsky et al. 1975, p. 249). Timoféev-Ressovsky sought to create a new branch of evolutionary theory studying evolution of biogeocenoses (ecosytsems) and the biosphere. Specifically, following his forceful repatriation after WWII, Timoféev-Ressovsky founded a school of *radiation biogeocenology* and *radiation biogeochemistry* (Tjurjukanov and Fiodorov 1996, pp. 97–98). He defined biogeocenoses as "dynamic systems, which at the same time can be in a state of dynamic equilibrium over quite a long biological time period (in the course of many generations of living beings residing in this beogeocenosis)" (Timoféev-Ressovsky et al. 1975, p. 309). The biosphere is defined as the sum total of biogeocenoses. Timoféev-Ressovsky insisted that there is a significant difference to the term *ecosystem*, predominantly used in the Western world, because biogeocenosis comprises *all* abiotic factors and *all* biotic dependencies in a relatively isolated system occupying clearly detectable zones (e.g. a pine forest or a swamp). To study the input, output and cyclicity of these systems Timoféev-Ressovsky proposed using nuclear markers. Also, he pioneered the investigations into the impact of radioactivity on the living organisms and ecosystems and thus initiated studies in *radiation ecology*.

The influence of Vernadsky's biosphere theory and Timoféev-Ressovsky's vision of a new science studying evolution of the whole biogeocenoses influenced also the very recent studies in evolution of multi-species communities and *paleoecology*, as seen in the works of Vladimir V. Zherikhin (1945–2001), one of the central figures in the field. Methodologically Zherikhin proceeded from the in Russia unpopular organismic approach based on the ideas of Frederic E. Clements (1874–1945), and developed further by Stanislav Razumovsky (1929–1983) (Rautian 2003). Zherikhin claimed that biological communities can be analysed from the viewpoint of their "ontogeny" (endoecogenesis) and "phylogeny" (phyloecogensis). As a result Zherikhin – with co-authors – proposed a "biotocenogenetic" model of

---

[4] Timoféév-Ressovsky – who coined a somewhat awkward term "vernadskology" – wasn't a direct pupil of Vernadsky, although he met him two times in Berlin in the mid of 1920s (Timoféév-Ressovsky 1995).

feedback loops between taxa and biocoenoses. The direction of phylogenesis in this model will be, to an extent, directed by a biocenosis (biogeocenosis) so far the included taxa evolve co-adapting by specialization and thus their further specialisation is then "predestined" by the whole system. Vernadsky is important here, first of all, because of two points. First, the biosphere is the ultimate biogeocenosis and a self-regulating system directing evolution of the lower systems. Second, Zherikhin adapted Vernadsky's thesis on living matter as the major mover of geochemical cycles (Zherikhin 2003, p. 348).

The biosphere studies in Vernadsky's sense were never interrupted in the Russian speaking countries. In the most articulated contemporary version, represented – in my view – by Georgii Zavarzin (1997, 2003a, b), the biosphere theory claims that phylogenetically independent prokaryotes are basic for the running of biogeochemical cycles of the biosphere. This implies that (1) the Vernadskian approach excludes strict monophily at the very early stages of biospheric evolution, because life on Earth can exist only in the from of communities able to support closed biogeochemical cycles (there were functionally complete microbial communities even in the early Proterozoic); (2) evolution has an additive character ("new" plus "old" and not "new" instead of "old"); (3) the Biosphere functions as a well-balanced system of functionally complementary organisms, and the Darwinian laws work only on the lowest level of this system.

Although the above examples can be in no way seen as exhausting the subject, one can see that Vernadsky influenced nearly the whole range of global sciences in Russia.

However, his influence is of a disproportional nature. In Russia Vernadsky belongs to the pantheon of the most known and esteemed scientists. His name is to be found even in school textbooks. His biosphere-noosphere theory is included in the course "Basics of Natural Sciences" obligatory for all university students of all disciplines. Not surprisingly his theory is studied in depth by students of philosophy, evolutionary theory, ecology and, of course biogeochemistry. In the introductory chapter to a modern Russian university biogeochemistry textbook we read: "Theoretical foundations of biogeochemistry are composed of the theories of living matter and biosphere created by V.I. Vernadsky" (Dobrovolsky 1998)Vernadsky's ideas are also reflected in the very recent Russian governmental programs. One of the explicit mentions of Vernadsky's theories in the official documentation is found in the 1996 Presidential decree concerning the "Concept of Russia's Transition to Sustainable Development" (Oldfield and Schaw 2006).

By contrast, Vernadsky remains little known in the Western world. There are few contemporary researchers explicitly advocating Vernadsky's views. The English scientist James Lovelock and the American microbiologist Lynn Margulis (e.g. Lovelock and Margulis 1974; Lovelock 1986, 1996; Margulis 1996) number Vernadsky among their scientific "predecessors", but repeatedly stress that they came to their version of the biosphere theory (Gaia-hypothesis) independently of Vernadsky. In addition, the only book available for these English-speaking scientists is the recent translation of Vernadsky's early work "The Biosphere" (Vernadsky 1998). Vernadsky's major book "The chemical Structure of the Biosphere" remains untranslated in Western European languages. Vernadsky's follower, who clearly declared Vernadsky's theory as the basis of modern geophysiology, is the German geomicrobiologist Wolfgang E. Krumbein (Krumbein and Schellnhuber 1992; Krumbein and Lapo 1996). Vernadsky is mentioned in the recent introduction into geomicrobiology (Ehrlich 2002).

Yet the majority of introductory English-language books on biogeochemistry and global ecology do not even mention Vernadsky (Degens 1989; Schlesinger 1991, 2004; Libes 1992; Fenchel et al. 2000). The authors either somehow escape the question on the origins of the biosphere concept and biogeochemistry, or begin the story with the postwar developments. As Schlesinger in his most recent survey paper on the "Global Change Ecology" puts it: "I mark the beginning of global change science with the publication of 'The Biosphere' as a special issue of Scientific American in 1970" (Schlesinger 2006). To fully realize the absurdity of this situation one should imagine a book on evolutionary biology, which would begin with Ernst Mayr and William Provine's "Evolutionary Synthesis" (1980), with no references to Darwin or with a few cursory remarks about him. At the same time Vernadsky's ideas penetrated the theoretical landscape without being actually associated with his name. As the ecologist George E. Hutchinson (1903–1991) in the above mentioned book claimed: "It is essentially Vernadsky's concept of the biosphere, developed about 50 years after Suess wrote, that we accept today" (Hutchinson 1970). Yet in the situation of only fragmentary infiltration of Vernadsky's original concepts into the Western theoretical landscape, the proper estimation of his theory by the international scientific community seems to be a difficult task.

Summary

Vernadsky's most important contribution to modern science was his grandiose theory of the biosphere and living matter. The idea was developed through the first half of the twentieth century and influenced the whole range of natural sciences including the new field of ecology and evolutionary theory. Vernadsky understood geochemistry as a natural history of terrestrial chemical elements. The specificity of biogeochemistry, in relation to classic geochemistry, reflected his novel idea that living matter was the major factor in the migration of chemical elements in the biosphere. In Vernadsky's terms, the biosphere is both a geological stratum and a self-regulating system including both living organisms and their inert environments. Complemented by the concept of biogeocenosis coined (1940) by Vladimir Sukachev, the biosphere appeared to be a self-regulating system consisting of biogeocenoses as its elementary structural units, which in their turn represent self-regulating systems. Biogeocenoses comprise all abiotic factors and all biotic dependencies in a relatively isolated natural system occupying a clearly detectable zone, e.g., a pine forest or a swamp. This approach allowed scientific predictions about the ways of migration of the chemical elements in the biosphere and became crucially important for Russian ecology. A number of Russian scientists adopted and extended Vernadsky's new approach. These are, for instance, the concept of *geochemical ecology* (Victor Kovalsky), the whole school of *radiation ecology* and *radiation biogeochemistry* founded by Nikolai Timofeév-Ressovsky, important impacts on modern paleoecology (also Sukachev and Timoféev-Ressovsky's) and, more recently, the approaches of Vladimir V. Zherikhin, the founder of the so-called *naturalistic microbiology*. Alltogether, Vernadsky's theoretical system was one of the most crucial steps in shaping modern global sciences including ecology.

Acknowledgement My research on the history of life sciences was supported by the Deutsche Forschungsgemeinschaft (DFG) (Ho 2143/5-2). I am thankful to Astrid Schwarz and Christian Haak for helpful comments on an earlier version of this paper.

References

Aksenov G (1994) On the scientific solitude of Vernadsky. Probl Philos 6:74–87 [in Russian]
Bailes KE (1990) Science and Russian culture in an age of revolutions: V.I. Vernadsky and his scientific school, 1863–1945. Indiana University Press, Bloomington/Indianapolis
Barrow J, Tipler F (1986) The anthropic cosmoplogical principle. Claderon Press, Oxford
Clarke FW (1908) Data of geochemistry. Government Printing Office, Washington, DC
Dana JD (1852) Crustacea, Reprinted in 1972. Antiquariat Junk, Lochem
Degens ET (1989) Perspectives on biogeochemistry. Springer, Berlin
Dobrovolsky VV (1998) Basics of biogeochemistry. Vyschaja Schkola, Moscow [in Russian]
Dokuchaev VV (1898) The concept of zones in nature, 2nd edn., 1948. Moscow Geografgiz, [in Russ]
Ehrlich HL (2002) Geomicrobiology, 4th edn. Marcel Dekker, New York
Fenchel T, King GM, Blackburn TH (2000) Bacterial biogeochemistry: the ecophysiology of mineral cycling, 2nd edn. Academic, San Diego [u.a.], Reprinted
Fersman AE (1923) Khimtcheskije elementy zemli i kosmosa (Chemical Elements of the Earth and the Cosmos). Khimtekhizdat, Petrograd
Ghilarov AM (1995) Vernadsky's biosphere concept: An historical perspective. Q Rev Biol 70(2):193–203
Grinevald J (1996) Sketch for the History of the Idea of the Biosphere. In: Bunyard P (ed) Gaia in Action. Floris Books, Edinburgh, pp 115–135
Hutchinson GE (1970) The biosphere. Sci Am 223(3):45–53
Kolchinsky EI (1990) The evolution of the biosphere. Nauka, Leningrad [in Russ]
Kolchinsky E, Kozulina A (1998) The burden of choice: why did V.I. Vernadsky return to the Soviet Russia? Voprosy istorii estestvoznanija i tekhniki 3:3–25 [in Russian]
Kolchinsky EI (2006) Biology in Germany and Russia-USSR. Nestor-Istorija, St.-Petersburg [in Russ]
Kovalsky VV (1974) Geokhimitcheskaja ekologija [Geochemical ecology]. Nauka, Moscow
Krumbein WE, Schellnhuber H-J (1992) Geophysiology of mineral deposits a model for a biological driving force of global changes through Earth history. Terra Nova 4:351–362
Krumbein WE, Lapo A (1996) Vernadsky's biosphere as a basis of geophysiology. In: Bunyard P (ed) Gaia in action. Floris Books, Edinburgh, pp 115–135
Lapo AV, Smyslov AA (1989) Biogeochemistry: the foundations laid by V.I. Vernadsky. In: Yanschin AL (ed) Scientific and social significance of Vernadsky's creativity. Nauka, Moscow, pp 54–61 [in Russian]

Lapo AV (2001) V.I. Vernadsky (1863–1945), the founder of the biosphere concept. Int Microbiol 4:47–49

Levit GS, Krumbein WE (2000) The biosphere theory of V.I. Vernadsky and the Gaia theory of J. Lovelock: a comparative analysis of the two theories and two traditions. Zhurnal Obshchei Biologii (J Gen Biol) 61(2):133–144

Levit GS (2001) Biogeochemistry, biosphere, noosphere: the growth of the theoretical system of Vladimir Ivanovich Vernadsky (1863–1945), Series: "Studien zur Theorie der Biologie" (Edited by Olaf Breidbach & Michael Weingarten). VWB-Verlag, Berlin

Levit GS, Hoßfeld U (2005) Die Nomogenese: Eine Evolutionstheorie jenseits des Darwinismus und Lamarckismus. Verhandlungen zur Geschichte und Theorie der Biologie 11:367–388

Libes SM (1992) An introduction to marine biogeochemistry. Wiley, New York

Lovelock J (1972) Gaia as seen through the Atmosphere. Atmos Envir 6:579f

Lovelock J (1986) The biosphere. New Sci 1517:51

Lovelock J, Margulis L (1974) Atmospheric Homeostasis by and for the biosphere: the Gaia hypothesis. Tellus 26:2–10

Margulis L, Sagan D (1995) What is life? A Peter N. Nevraumont Book, New York

Margulis L (1996) James Lovelock's Gaia. In: P. Bunyard (ed) Gaia in action. Floris Books, Edinburgh, pp 54–65

Mochalov II (1982) Vladimir Ivanovich Vernadsky. Nauka, Moscow

Oldfield JD, Schaw DJB (2006) V.I. Vernadsky and the noosphere concept: Russian understandings of society-nature interaction. Geoforum 37(1):145–154

Por FD (1980) An ecological theory of animal progress – a revival of the philosophical role of zoology. Perspect Biol 23(3):389–399

Rautian AS (2003) O nachalakh teorii evoliutzii mnogovidovykh soobstchestv i ee avtore (On the beginnings of the theory of multi-species communities evolution – phylocenogenesis – and its autor). In: Lubarsky G (ed) Zherikhin V.V. Izbrannyje trudy. KMK Press, Moscow, pp 1–42

Schlesinger WH (1991) Biogeochemistry: an analysis of global change. Academic, San Diego [u.a.]

Schlesinger WH (ed) (2004) Treatise on geochemistry – Vol. 8: Biogeochemistry. Elsevier Pergamon, Amsterdam (u.a.)

Schlesinger WH (2006) Global change ecology. TREE 21(6):348–351

Sewertzoff A.N. (1931) Morphologische Gesetzmäßigkeiten der Evolution. Gustav Fischer Verlag: Jena

Sytnik K, Apanovich E, Stoiko S (1988) V.I. Vernadsky. Life and activity in the Ukraine. Naukova Dumka, Kiev [in Russian]

Teilhard de Chardin P (1961) The phenomenon of man. Harper & Row, New York/Evanston

Timofeév-Ressovsky NV (1995) Vospominanija (memoirs). Progress, Moscow

Timoféev-Ressovsky NW, Vononcov NN, Jablokov AN (1975) Kurzer Grundriss der Evolutionstheorie. Gustav Fischer Verlag, Jena

Tjurjukanov AN, Fiodorov VM (1996) N.V. Timoféev-Ressovsky: Biosfernyje razdumja. AEN, Moscow

Vernadsky VI (1902) O nauchnom mirovozzrenii (On the scientific worldview). Vorposy filosofii i psikhologii 1(65):1409–1465

Vernadsky VI (1903) Osnovy kristallografii (The Fundamentals of Crystallography). Izdatelstvo Moskovskogo Universiteta, Moscow

Vernadsky VI (1912) O gazovom obmene zemnoj kory (On gaseous exchange of the earth's crust). Izvestija Imp Akad Nauk Serija 6 6(2):141–162

Vernadsky VI (1924) La Géochemie. Alcan, Paris

Vernadsky VI (1926) Biosfera. NHTI, Leningrad

Vernadsky VI (1929) La Biosphère. Alcan: Paris

Vernadsky VI (1930) Geochemie in Ausgewählten Kapiteln. Autorisierte Übersetzung aus dem Russischen von Dr. E. Kordes. Akademische Verlagsgesellschaft, Leipzig

Vernadsky VI (1934) Le Problème du Temps dans la Science Contemporaine. Revue Générale des Sciences Pures et Appliquees 45(20):550–558

Vernadsky VI (1935) Le Problème du Temps dans la Science Contemporaine. Revue Générale des Sciences Pures et Appliquees 46(7):208–213, 47(10): 308–312

Vernadsky VI (1944) Problems of biogeochemistry. (Trans: George Vernadsky Ed and condensed: G E Hutchinson) Connecticut Academy of Arts and Sciences, New Haven [u.a.]

Vernadsky VI (1965) The chemical structure of the biosphere of the earth and of its environment. Nauka, Moscow [in Russian]

Vernadsky VI (1980) Problems of biogeochemistry. III, BIOGEL. Nauka, Moscow [in Russian]

Vernadsky VI (1988) Philosophical thoughts of naturalist. Nauka, Moscow, p 520 [in Russian]

Vernadsky VI (1991) Scientific thought as a planetary phenomenon. Nauka, Moscow [in Russian]

Vernadsky VI (1994a) Works on geochemistry. Nauka, Moscow [in Russian]

Vernadsky VI (1994b) Living matter and the biosphere. Nauka, Moscow [in Russian]

Vernadsky VI (1997) Scientific thought as a planetary phenomenon. Nongovernmental Ecological V.I. Vernadsky Foundation, Moscow

Vernadsky VI (1998) The biosphere. A Peter A. Nevraumont Book, New York

Vinogradov AP (1953) The elementary chemical compositions of marine organisms. Memoir Sears Foundation for Marine Research II. Yale University Press, New Haven

Vinogradov AP (1993) The geochemistry of isotopes and the problems of biogeochemistry: selected works. Nauka, Moscow [in Russian]

Zavarzin GA (1997) The rise of the biosphere. Microbiology/Microbiology 6(66):603–611

Zavarzin GA (2003a) Evolution of the geosphere-biosphere system. Priroda 1:27–35 [in Russian]

Zavarzin GA (2003b) Prirodovedcheskaja mikrobiologija (Naturalistic microbiology). Nauka, Moscow

Zherikhin VV (2003) Izbrannyje trudy (selected papers). KMK Press, Moscow [in Russian]

Part VII
Border Zones of Scientific Ecology and Other Fields

Chapter 25
Geography as Ecology

Gerhard Hard

Introduction: The Core Paradigm of Geography

The major theme and the core theory of classical geography, that is, of mainstream geography from the eighteenth to the twentieth century, can be summed up roughly as follows: the interaction and symbiosis between regional modes of life and entire cultures on the one hand and their concrete ecological milieu on the other. One could also express it using some rather misleading set phrases, namely, the man-nature, man-space or man-environment theme.[1] This concrete ecological or physio-biotic milieu could be taken to mean both the original milieu as well as that which had already been altered through history. This was the "Rittersche Wissenschaft" which, from the time of Carl Ritter (1779–1859) onwards and with regular and repeated reference to him, was to bring together a broad range of "raw material" and to refine that material until it had become a university subject suitable for research and teaching. Ritter's "Erdkunde" has had a major influence on many nations' geographies, not least thanks to his numerous students (and his students' students) scattered throughout the world.

Ritter and his followers, however, were merely formulating a programme that had already been an integral part of the common sense shared by educated people in the 18th and early 19th century. From now on, this – to put it in modern terms – "Kulturökologie" (cultural ecology) was no longer merely an almost ubiquitous theme in the "educated common sense" and popular science of the time; it also became the paradigm of a university discipline. As modern ecologists themselves are aware, such a constellation does not bring undiluted benefits.

In the English-language literature, this man-and-environment theme in geography is often simply called "the ecological tradition".[2] It is true that more recent

---

[1] On the theoretical concept of "Theoriekern" (theoretical core), synonymous with "Kerntheorie" (core theory) and "Kernparadigma" (core paradigm) etc., cf. Stegmüller 1973, 1979, 1980.

[2] Cf. e.g. Haggett 1965 (pp. 10 ff.) in the German edition of the book 1972 (p. 15) the term was translated literally using "ökologisch" and "Ökologie" and was explained in typically allencompassing terms as follows: "Geographie als die Erforschung der Beziehungen zwischen der Erde und dem Menschen" (Geography as the study of relationships between the earth and man).

G. Hard (✉)
Institut für Geographie, Universität Osnabrück, Seminarstraße 19 a/b, D- 49069 Osnabrück, Germany

English-language geography in particular contains more references to several (at least four or five) different "traditions", "schools", "major themes" or "focal points of interest" in geography's past and present, with the aforementioned "ecological" or "man-environment theme" as one amongst several others.[3] However, the history of the discipline shows that this "ecological" man-earth theme constituted – at least latently – the organising centre and actual legitimatory foundation for all important geographical research programmes from the 18th through to the 20th century.[4] It may have taken more of a background role due to the disciplinary breaks that occurred in the second half of the twentieth century, but it has by no means lost its influence.

From the very beginning, geographical research and literature consisted in regional or worldwide ("comparative") applications of this basic idea. Vast amounts of material from all manner of scientific and popular literature were dealt with in the light of the man-earth theme. The "ecological theme" determined not only the foundations and structure of regional geography but also the way in which geography as a whole constituted its objects. Whether consciously or not, even the Physical Geographers described the nature of the earth in such a way that their descriptions could still be linked to the man-earth theme.[5] Even today, many geographers and their scientific statesmen in particular (at least in the German-speaking world and when they are up on their soap-boxes) are convinced that the essential nature of "geography" and its inherent power make it a comprehensive ecology – indeed, make it the ecology and environmental science per se. But what do they mean by "ecology"? Where did this geographical ecology come from and what has become of it?

Geography as "Cultural Ecology"

Outside geographical circles, in cultural anthropology and ethnology, for example, the core paradigm of classical geography described above is characterised as "cultural ecology" or "the cultural ecological perspective" and, for some considerable time now, has been described roughly as follows: A comparative analysis, carried out between very different environments and cultural forms, of the relationships/interactions between the physico-biotic environment of human populations on the one hand, and their modes of behaviour, social organisation and culture on the other. "Culture" here was – and is – taken to mean not only the material culture

---

[3] In the German-language literature, Hard 1973 (pp. 79 ff.) and Bartels and Hard 1975 (p. 90) placed even greater emphasis on the diachronic and synchronic heterogeneity of geography, and described in detail each of the many research perspectives contained within this "diffuse discipline".

[4] Cf. Eisel 1980. A two-part anthology of geographical responses to the question of what geography is also provides impressive confirmation of this with regard to the self-perception of geographers from the 18th century up to the present day (Schultz 2003).

[5] For geomorphology see Böttcher 1979.

or the "technical culture" by means of which humans cope with their environment, but rather every aspect of society and culture that appears to be affected in any way by this process of "coping with the environment".[6] These and other descriptions of "cultural ecology" are a very accurate rendering of the aforementioned core paradigm of geography in the modern era, too; even the frequently quoted phrase introduced by the "founding father" of modern university geography – "the earth as mankind's dwelling and place of instruction"[7] – essentially had this cultural-ecological meaning and was consistently understood as such even after Ritter's death.

From time to time since then, even geography itself has explicitly defined its essence in terms of "ecology", as for example in American geography between about 1910 and 1930 (thereby establishing a clear link to the contemporary American plant ecology of the time). One example of this is the statement: "Geography is the science of human ecology", i.e. "the science of the mutual relations between man and his natural environment" (Barrows 1923, p. 3). The object of geography, however, was not to be seen as the "natural environment" as such, but rather its "habitat value" – in other words, the natural environment as an object of perception, judgment and utilisation by man. Thus, this "geography as human ecology" was also a version of the "Rittersche Wissenschaft".[8]

Geography as a Regional and Regionalising Cultural Ecology

Geographical attention was directed above all at the level of landscapes and regions. The whole logic of the programme described above was particularly well tailored to pre-industrial (especially peasant) life worlds, as long as they did not seem to have been significantly transformed (yet) by the global market, industrialisation and other kinds of modernisation and globalisation. Of course, themes such as industrialisation and urbanisation increasingly had to be addressed in geography, too, but it was now more difficult to establish a link between this modern world and the man-nature theme.

---

[6] For example, Steward 1955, pp. 40 f.; for further references, see Krewer and Eckensberger (1990). The term "cultural ecology" is used in a similar way in English-language geography as well (cf. e.g. Johnston 1988), while in German-speaking geography the term "human ecology" seems to have become established, cf. the recent anthology by Meusburger and Schwan (2003).

[7] "die Erde als Wohn- und Erziehungshaus des Menschengeschlechts" (Herder 1784–1791, 1966 edition; pp. 59–67).

[8] For more detail about this link to biological ecology from a history of science perspective, cf. Fuchs 1966, 1967. Around 1920, Chicago – an important location in the history of American geography – was not only a stronghold of plant ecology (see Chap. 20) it was also the stronghold of a variety of sociological urban research which thought largely in terms of ecological analogies. In this famous urban sociology of Chicago, however, "human ecology" (later to become "social ecology") usually meant something different from what it meant in the geography of the time, namely the study of the "social environment" of human populations. On the reception of this sociological "human" or "social" ecology in German urban research studies from the 1970s onwards see, amongst others, Hamm 1990.

The bulk of interest, moreover, was directed less towards individual behaviour than towards the relation between entire (groups of) life forms or entire cultures and their regionally differentiated "natural" or physico-biotic environments. Since the 19th century, therefore, geographical literature in every European language has provided an inexhaustible repertoire of original case studies and overviews extending across the entire globe.[9] One literary highlight in this field, indeed one highlight of classical geography as a whole, is the *géographie humaine* of Vidal de la Blache (1845–1918) and his school ("*l'école française de géographie*"), which remained dominant in French universities and schools until well into the mid-twentieth century, and which continues to define the general image of geography even today.[10]

In this geographical cultural ecology an extraordinary spatial perspective has emerged. Rather than looking at nature, natural conditions and culture as a whole, it was the *spatial* "natural plan" and the *spatial* "cultural plan" in particular (or even exclusively) that were placed in relation to one another and compared in terms of their congruity or lack of congruity, from the local through to the global scale. In other words, the focus was on comparison between the spatial patterns, boundary formations and "fault lines" in the natural (i.e. physical) landscape and those in the cultural landscape. This occurred, without exception, on the basis of a normative premise. On the culture and society side, only those elements were considered to be benevolent and permanent (or, to put in today's language of political ecology: "sustainable") (that correspond to natural, i.e. physical, areas) were felt to correspond to the basic features of a specific natural/physical environment. This applied to a range of elements, from the land use patterns of the cultural landscape, via state territories and borders, through to large economic and political areas. One might call this "normative natural determinism" or "normative geo-determinism". This "normative determinism of natural spaces" represents a continuation of both the physicotheology of the eighteenth century and the overtly and covertly physicotheological conceptual figures of Herder and Ritter.[11]

---

[9]Classic twentieth century texts in German-language geography include Waibel (1921), Troll (1931) and Bobek (1959). Only very few scholars attempted to apply the concept of life forms to highly industrialised countries, and little of significance emerged from this tradition.

[10]Cf. both critical and sympathetic descriptions by Claval (1964), Buttimer (1971) and Berdoulay (1981).

[11]On the geographical forms of physicotheology and the theory of Providentia, cf. e.g. Büttner (1975) and Hard (1988). In the geography of the nineteenth and twentieth centuries, this physicotheology of the regions of the earth turned into a largely implicit teleology of natural (i.e. physical) regions or landscapes: It is no longer God who speaks to us through His creation, but rather that concrete ecological nature (and later, the ecosystem) enables us – on its own authority, to a certain extent – to understand what is the right (or, nowadays, the sustainable) way of dealing with "planetary nature". This "eco-ethical" figure still seems to be fundamental to both modern ecological movements and the rhetoric of political ecology (see also Chap. 16 this volume).

The "physical regions" perspective described above and the associated search for the "right" (natural) physical regionalisation has been a leitmotif in geography for more than two centuries.[12] Among the late and dubious fruits of this ancient geographical obsession in the second half of the twentieth century are, for example, the equally extravagant and ultimately useless efforts to "divide" Germany into its ("natural spaces") "physical regions", or again, the regional systems and hierarchies devised by German landscape ecologists (especially Schmithüsen, Neef, Paffen and Leser). We might add to this the German tradition of landscape planning, geographically inspired from the 1930s onwards. This repeatedly attempted, albeit in vain, to "order" the economic landscape of a modern society on the basis of its "natural foundations". This involved identifying natural (i.e. physical) areas that were meant to function as a blueprint for wise usage patterns, "in order to harmonise the natural and the cultural plan".[13]

These regionalisations and their concepts of space generally have the explicit (and nearly always implicit) ambition to divide up the surface of the earth, in a universal and generally valid manner, into "natural units" or even into "ecosystems", in accordance with all its essential aspects and features. Those involved in this effort even believed that it was possible, so to speak, to produce "all-purpose regionalisations", the outcome of which was that the latter soon had no real use at all. This spatial projection of "integrated units of landscape" ("landschaftliche Wirkungsgefüge" – later to be called "ecosystems") was virtually never called into question. The constructionist logic and theory of regionalisation as a classification were never really accepted, either, not even after Bartels (1968, 1970) had introduced them into German geography.

Background and Original Form of the Cultural Ecological Man-Earth Theme

The cultural ecological paradigm of geography is older than university geography. Geography only became a university discipline to any significant extent (both in German-speaking countries and in other industrialised nations) in the latter part of the 19th century. The political intention everywhere was to create a discipline for national education, teacher training and school lessons. The paradigm itself, however, emerged during the 18th and early 19th century; around 1,800 it had almost become an integral part of the natural world view of educated Europeans, making an appearance in many literary genres, including philosophy, anthropology, (universal)

---

[12] For the overwhelming dominance of *this* theme of natural (and here this meant: the *right*) areas and boundaries, cf. amongst others Schultz (1997) , 2002, 2003).

[13] On the "principles and method" of "dividing Germany into natural (i.e. physical) areas", cf. amongst others Schmithüsen (1953).

history, political science, travel writing, research travel literature, popularised science and even in works of fiction.

It is possible to identify several points of origin for this paradigm. Among these are the traditional description of "land and people"; certain ancient philosophical notions, revived in the eighteenth century, concerning the human-nature relationship, especially the Hippocratic tradition[14] certain widespread notions from physicotheology about the wise and benevolent design of the earth, and finally, in German-speaking countries, Herder's philosophy of history, which could also be read as a natural philosophy and an anthropology.

In Carl Ritter's interpretation and that of the later geographers, Herder's philosophy of history already contained the entire core theory of geography and its complete basic programme. This was – to use the rich variety of phrases Herder himself used – to view "the earth as a arena of human history, a dwelling, an institution of instruction, as the nations' home and place of instruction"[15] and geography as studying the way in which "[t]he whole earth [...] plays for it [humanity] like the harmonic music of a stringed instrument, in which every sound is attempted or will be attempted".[16]

These are the kinds of premises that informed Carl Ritter's (physico-theologically inclined) geo-philosophy. It consisted in the idea of recognising the landscapes of the earth, both individually and in their entirety, as "the dwelling and place of instruction of the human race" (or "the race"). By means of "serious [geographical] science", the sculpture of the earth and the physiognomies of landscapes, as well as the natural plan of the earth, were to reveal the *nomos* and *telos* of the nature of the earth and of man, the "true destiny" of the earth's regions and its peoples – for the benefit of all peoples and for the whole of humanity. In other words, geography had found its mission.[17]

Thus, man and nature are teleologically "harmonised" with one another in the regions of the earth; the countries, cultural landscapes and cultures of the earth are entelechially "mature" outcomes of creative adaptations of regional life forms to concrete ecological nature, i.e. to their natural regions and landscapes. In this way, geography was able to make an important contribution to the idea and legitimation of the modern nation state (cf. e.g. Schultz 2002). The geographers endeavoured to

---

[14] On the history of these Hippocratic notions, from antiquity through to the 18th century, cf. amongst others Glacken (1967); on Herder's use of the Hippocratic tradition, cf. Hard (1988, pp. 183 ff).

[15] "die Erde als Schauplatz der Menschengeschichte, Wohnhaus, Bildungsstätte, Erziehungshaus und Bildungsplatz der Völker" (Herder 1784–1791, 1966 edition; pp. 59, 67).

[16] "die ganze Erde [...] ihr [der Menschheit] wie ein harmonisches Saitenspiel zutönet, in dem alle Töne versucht sind, oder werden versucht werden" (Herder 1784–1791, p. 201). On the interpretation of Herder's programme and his metaphorical thinking, cf. e.g. Hard (1988, pp. 189 ff).

[17] For some especially succinct, emphatic and brilliant formulations of this philosophy, cf. e.g. Ritter (1852, pp. 9 f.; 1862, pp. 1–23), especially pp. 14–16; for an interpretation and contextualisation of the key sections of text in Ritter, cf. Hard (1988, pp. 271 ff).

show that every nation state was underpinned by a natural plan and that every legitimate nation state was an "harmonic triad" consisting of nature (or a natural area), people (nation) and state. In places where this was obviously not the case, political correctives could be proposed.

This cultural ecology was *not* flatly geo-deterministic in and of itself. Cultures could not be "deduced" from nature, let alone determined by it. Instead, they were conceived of – at best – as happy man-nature balances, and this kind of optimum adaptation to nature was considered to be the *correct* form of emancipation from nature. This was a specific achievement, though, which could also entail failure, whether through lack of capability or immoderation. As Herder put it in his famous phrase: "climate, meaning the natural environment, the cosmic-earthly milieu, does not coerce, it inclines". He repeatedly expounds this phrase, borrowed from astrology: "astra non cogunt sed inclinant", in his work, e.g. "man may have been made the master of the earth, so that he might change the climate by art".[18]

But climates "which are all a living whole desire to be gently followed and improved, not [...] violently dominated for they avenge every sin done unto them and the silent breath of the climate has already obliterated, scattered and blown away many of these sinners, who had only robbed and devastated the nature of their country. By contrast, the historic opportunity for permanence belongs instead the silent growth that has acceded to the laws of nature".[19] However, a people that succeeded in achieving a cultural ecological balance and a harmonisation of man and nature was seen to fulfil the *general* destiny of humankind and history in a *unique* way, in its *concrete* region. This in turn gives rise to solutions of singular classicality. After all, to all who have eyes to see, cultural landscapes of individual beauty ("uniqueness and beauty"), diversity and harmony reveal the historical achievements of such peoples or nations.

This cultural ecological thinking clearly had normative, (eco-) ethical substance in its original form und was later also called "possibilistic".[20] In an attempt to adapt to the (misunderstood) causal thinking of the natural sciences, this thinking frequently slid into a coarse and often diffuse natural determinism in the nineteenth and twentieth centuries, albeit more outside geography than within it.

---

[18] "Das Klima zwinget nicht, sondern es neiget - zum Herrn der Erde gesetzt, dass er es [das Klima] durch Kunst ändere" (Herder 1784–1791, p. 187).

[19] "[Klimata,] die allenthalben ein lebendiges Ganzes sind [...] wollen sanft befolgt und gebessert, nicht [...] gewaltsam beherrscht sein [, denn sie] rächen [jeden] Frevel, den man [ihnen] antut [und] der Stille Hauch des Klimas [ist] verwehet und weggezehrt [... dem] stille[n] Gewächs, das sich den Gesetzen der Natur bequemte." (Herder 1784–1791, pp. 195 f.). This is the (conservative) set of concepts in which the idea of sustainability is rooted and makes sense; for more on this, cf. Körner and Eisel (2002).

[20] The "possibilistic" geographer (L. possibilis, "possible") always thought, at least implicitly, in terms of a range of (better or worse) possibilities for "adjustment" to concrete ecological nature. The French school therefore preferred to speak, for example, of a "vocation" than of a "détermination" of "régions naturelles".

Areas of Application

Ideal canvases that were available for projecting these ideas in the geography of the nineteenth and twentieth century (one could also say, for projecting this utopia) were the cultural landscapes of antiquity, then the rather more peripheral rural regions of the industrialised nations and, above all, "far-off countries". It was overseas especially where the countries and landscapes seemed to be just the way the paradigm taught people to see them: as the places of dwelling and instruction occupied by their cultures and as more or less harmonious adaptations to the concrete ecological nature that existed in each place. Here, as in antiquity and in the peripheral regions of Europe, industry, the global market and modern urbanism were still – whether in reality or only seemingly – absent, or else they could still be interpreted as being externally initiated transformations, or even "disruptions", of autochthonous harmonies and balances. Thus, geographers were not only writing their regional studies, but also, up until the late nineteenth century, their ethnographical works as well.

The idea that frequently guided geographers' interest in the nature and natural resources of far-off countries and national peripheries was that many of these landscapes, according to them, had not yet been fully recognized, valued or cultivated by their indigenous inhabitants; in other words, they still had to be led towards their true destiny. The landscape-oriented eye of the travelling and researching geographer could thus also be "an imperial eye". In the age of colonialism, geographers also served their societies by means of a theory that was already highly inappropriate to these modern societies themselves. The "inappropriateness" of the geographical paradigm lay in the fact that within its framework, society and culture as well as economics and politics had to be interpreted in principle as man-nature systems of adaptation, i.e. as adaptations and responses to physical landscapes.

Geography's cultural ecology often amounted, in fact, to a plea for progress, the global market and industry, whenever the zeitgeist suggested it.[21] For the most part, this was merely an opportunistic nod to the zeitgeist; however, it was also perfectly possible to link belief in progress to the classical paradigm – using the idea that the meaning and goal of human history as a man-nature conflict ultimately lay in emancipation from concrete ecological nature.[22] The implication of this idea, of course, was that progress would successively invalidate the geographical and ecological view of the world – "the (earth's) nature as man's opponent" – and would thereby render it irrelevant.

How was this geography able to survive exploratory travel research, the colonial period and even the blossoming of the scientific geo-disciplines? Not least by

---

[21] For German-language geography of the nineteenth and twentieth centuries and its move away from an apotheosis of progress (especially in the nineteenth century) towards a critique of civilisation (especially in the twentieth century), cf. Schultz (1997) etc.

[22] This "progressive" version of the geographical paradigm had already been expressed in succinct form by Kapp in 1845, drawing on Ritter and Hegel: the ultimate harmony, reconciliation and unity of man and nature lay, according to him, in a worldwide "spiritualization" and "liberation" of earthly nature by means of culture, technology and large-scale industry.

becoming a subject for instruction in schools, institutions of higher education and teacher training.

Geographical "Cultural Ecology" as a "Verstehende Naturwissenschaft" (Perception-Based or Hermeneutic Science)

In the context of its (human-nature) man-environment paradigm, classical geography consistently described nature in non-scientific terms. The language it used was largely of an everyday nature and included ordinary, colloquial forms of expression. Nature thus appeared and was read as "concrete ecological nature". This concept refers to a description of nature in which situations, events and objects appear more or less the way they are perceived as meaningful entities in the course of everyday life – and not, for example, as physical, chemical, (molecular) biological objects, situations or events.[23] In this, classical geography resembles both traditional and modern natural history, as well as large parts of modern ecology.

Geography directed its attention towards a physical-material world of objects, namely the "earth's regions". However, it did *not* view these objects – or indeed their physical assets – in strict scientific terms, but rather as "mankind's place of dwelling and instruction", i.e. as spaces which could be appropriated and made man's own for relatively simple, usually agrarian ways of life and practices. The physical regions and their ecology were described as the resources and outcomes of simple ways of life and a generally quasi-manual, physical use of nature. Essentially, they were described as intentional objects of human action and, especially, as the world of perceptions and effects of *homo agroregionalis*.

Geographical cultural ecology saw the physical-material world, then, as one that was structured by meanings related to social relevance; in other words, rather than being structured scientifically in any narrow sense, it was structured according to a hermeneutics of perception. Geography was rather like a hermeneutics (or a system of signs and symbols) of the landscape and of the surface of the earth; as paradoxical as it may sound, it was an"verstehende oder hermeneutische Naturwissenschaft" (understanding or hermeneutic natural science) (Eisel 1987).

This "perception-based natural science", however, was only inadequately aware of its own essential nature and the conditions of its existence. The "nature" of classical geography was so close in meaning to the "world of the natural attitude" (Husserl) and always contained so much common sense, that the question as to the particular object constitution of geography could barely be raised. Nonetheless, the geographers always had a certain awareness of the fact that their science of "man

---

[23] In contrast to this, "abstract nature" refers to a variety of nature that is described in physical, chemical and molecular biological terms; it is subject to experimental manipulation using scientific apparatus and tends to be expressed in mathematical formulae – in other words, it is the nature of the so-called exact natural sciences and of modern techno-science.

and the earth" played a special kind of role among the modern sciences. This was frequently manifested in strange over-valuations, familiar to us from modern ecology, as well: geographers liked to see geography – especially landscape science, or landscape ecology, and regional geography – as a superior synthesis of nature and man, "the physical" and "the social system". Geography, so it was said, could thus overcome the gap between the sciences on the one hand and the social sciences and humanities on the other.

"Landscape" in Geography's Cultural Ecology

In German-speaking countries in particular, the word "Landschaft" (landscape) became a key term from around 1,900 onwards, as it was the main figure through which geographers conceptualised what they saw and thought; it became the "the essential object of geography". "Landscape" could be used to refer both to the concrete ecological nature ("overall natural character") of a region, as well as to the material outcome of the historical "symbiosis" between this regional nature and the people who belonged to the relevant territory ("cultural landscape").[24] Above all, however, "landscape" stood for the idea that regional "nature" and "culture" form one single great, unified, objective and spatial whole, an "Wirkungsgefüge" (integrated unit) that is accessible via the "physiognomy of the landscape".

To the landscape geographer, this "landscape whole" consists of physical-material *and* intellectual (or social) components, that is, of objects that have physical-material *and* intellectual (or social) aspects.[25] "Landscape" thus became a cipher for the "entire geo-ecosystem" or "landscape ecosystem", consisting of the "lithosphere, hydrosphere, atmosphere, biosphere and anthroposphere" (numerous references, as in the collections by Storkebaum 1967 and Paffen 1973). While this "landscape" provided a kind of theoretical orientation for specific research endeavours, it was also more than this, namely a passe-partout and all-encompassing superstructure spanned across the actual disparateness of what was actually being researched in a field subject to increasingly rapid differentiation. One feature that can also be observed throughout this landscape geography is the reference to Alexander von Humboldt as the first great landscape geographer and the founder of landscape ecology[26] (see also Chap. 15).

It is well known by now where the concept of "landscape", especially in German geography, drew its enormous power to stimulate and legitimate certain ideas – namely,

---

[24] Today, the geography and "morphogenesis" of the cultural landscape has become a topic in its own right within historical geography, regional history and the preservation of historical monuments.

[25] On the philosophy that forms the background to this idea, see Hard 1970, reprinted 2002 (pp. 69 ff.).

[26] This reference did contain some measure of validity insofar as certain stylistic characteristics of "Humboldtian science" continue to exist to this day in modern geography, as well as in some parts of modern ecology.

from the semantics and, not least, the connotations of the word "Landschaft" itself. To put the matter succinctly and unequivocally: a community of researchers interpreted certain linguistic, particularly semantic structures as objective structure.[27]

"Landscape" is a concept that acquired its specific semantics in European discourses of art in the modern era. In German-speaking countries in particular (but not only here), "landscape" was associated with numerous connotations which were rich in meaning and thoroughly positive. There were powerful connotations of beauty, closeness to nature and naturalness, of unity, wholeness, diversity, universal interconnectedness and of harmony between man and nature. The word was associated with ideas of distinctive form, individuality and uniqueness, ideas of culture deeply rooted in the soil, of a positive tradition and a successful history. Moreover, "landscape" suggests ideas about bonds and obligations, about stability, balance and permanence or, in modern terms, "sustainability" – as well as about expectations of an holistic, immediate understanding of nature and culture. In the German language, anyone who speaks emphatically of "Landschaft" is soon inspired by these ideas; for many geographers, it is a straightforward matter of self-evident certainties. And this is how something which started out as an aesthetic figure of perception ended up being reinterpreted as an all-encompassing ecosystem. Indeed, given the associations mentioned above, it is easy to understand why "landscape" is capable of playing such a considerable role in conservative world views, in ecological movements and in political ecology.

Landscape Ecology, Geo-Ecology

Thus, first "landscape" and ultimately the "landscape ecosystem" ("geo-ecosystem" etc.) inherited the cultural ecological man-nature paradigm. A relationship (man-nature) and a noun ("landscape") were explicitly transformed into a "substance", a "totality", into "geographical reality" itself. Since the 1960s this "idea of landscape" (Schmithüsen) has been broadly criticised, in the spirit of a philosophical critique of language and of critical rationalism, as "a muddled holistic fantasy" (e.g. Bartels 1968, 1970; Hard 1970, 1973). Related critiques and corrections of the concept of "landscape" can also be found in the theoretical literature on ecology (e.g. Trepl 1987, 1996 etc.; Jax 2002).[28]

---

[27]For details, see Hard (1970) or, in summary form, Hard (2001).

[28]Jax (2002, pp. 213 ff.) proposes, on the basis of related epistemological and ontological considerations, that "Landschaft" – even in ecology itself – should no longer be conceived of as an ecological unity or as one (or even the) object of ecology; instead, it should be conceived of as a certain way of perspective based on space and distance. Thus, "Landschaft" comes to be seen as one or several spatial pattern(s), as the designation used for looking at organisms and societies of organisms in terms of their spatial patterns. The everyday linguistic use of "Landschaft" in particular suggests that the word should be used whenever spatial patterns on a medium-range, anthropocentric scale are the topic of consideration. This, Jax holds, approximates roughly to the meaning of "landscape" in the English term "landscape ecology".

Since then, almost the only place where "Landschaft" has appeared as a "geographical reality" and "holistic totality" is in the theories and terminologies of German landscape ecologists (or geo-ecologists), as, for example, in the form of terms such as "landschaftliches Ökosystem" and "Landschaftsökosystem" (landscape ecosystem), which is supposed to be the "Funktionseinheit" (functional unity) of the anthroposystem, biosystem and geosystem. Many geo-ecologists see the geo-(eco-)system in particular, i.e. the "Funktionseinheit von Klima-, Hydro-, Pedo- und Morphosystem" (functional unity of the climatic system, hydrosystem, pedosystem and morphosystem) as their object (cf. e.g. Leser 1991). Since the work of Schmithüsen, however, and especially in Leser, ideas expounded by German landscape ecologists have repeatedly sought to "ganzheitliche Erfassung" (capture in its entirety) the "realen Landschafts*ganzen*" (real *totality* of the landscape), the "Totalität des ganzen Landschaftsökosystems" (totality of the whole landscape ecosystem) (cf. e.g. Leser 2000). Linked to this is the normative idea that everything that happens or is planned in society and the natural sphere ought ultimately to be based on landscape ecological foundations and their spatial patterns (Leser 1991, 2000).

The term "Landschaftsökologie" was introduced into geography with increasing success by Carl Troll from 1938 onwards, initially for the purpose of formulating a multifaceted interpretation of aerial views of the earth, using the resources of physical geography and of ecological geo-botany in particular. In using the term "ecosystem", Troll was referring to Tansley and later also to Thienemann. The relationship to biological ecology remains a close one: vegetation, or biocoenosis, "*is at the center of the entire ecosystem*".[29] In the writings of "landscape ecologists" Schmithüsen, Neef and Leser, ecosystems are viewed – more obviously than in Troll – as spatial units with or without living beings; ecosystems, geosystems, geo-ecosystems etc. are conceived of as *both* functional unities *and* topographical unities ("sections of the earth").[30]

What the two similar versions or "schools" of landscape ecology around Neef and Leser have in common is that, alongside an overblown "theoretical" superstructure, there is a comparatively small-scale research practice whose methodology stems from neighbouring disciplines above all from applied soil science. On account of its methodology, this research practice was, or rather is – at least implicitly – geared towards assessments of potential agricultural capability and, as such, is focused on the abiotic sphere.[31] The significant objects of interest in this agrarian landscape perspective are linked to complex images of the "ecological landscape household" (Landschaftshaushalt, Standortregelkreis).[32]

---

[29] "steht im Zentrum des ganzen Ökosystems" (Troll 1966a, p. 38; emphasis in original).

[30] For a critique of this view in the ecological literature cf. Jax (2002) (pp. 43 f.) and Trepl (1988, 1996).

[31] This also explains Leser's curious separation of geo(eco)-systems from bio(eco)-systems, that is, of geo-ecology from bio-ecology.

[32] For a critique of Neef's "landscape ecology", cf. Hard 1973 (pp. 80 ff.); for a critique of Leser's landscape ecology, cf. Menting 1987, 2000, 2001; Lethmate 2000.

This is how a strange, special kind of geographical ecology emerged as the final form of the classical geographical paradigm.

Paradigm Lost

During the second half of the twentieth century a gradual change of paradigm could be observed – indeed, what it amounted to was the loss of a paradigm.[33] This event in the history of geography seems to me to have a significance that goes beyond the confines of geography itself: it indicates the likely future fate of those research programmes that are set up to address a "synthesis" or a "whole system" of "nature/environment and society".

One of the fundamental reinterpretations of the man-nature paradigm, or the man-space paradigm, consisted in geographers making "experienced" and "perceived space" the object of research instead of real space, and similarly the "perceived environment" instead of the "real environment" (for a summary of this research programme, cf. Hard 1990). Closer still to the old paradigm are the (few) modern "ethno-ecological" studies in geography, which explore the cognized environments of traditional societies, especially on the periphery of the inhabited planet and of world society (see e.g. Müller-Böker 1995); this work typically draws heavily on the cultural ecological orientation of English-language cultural anthropology.

Another influential reinterpretation of the classical paradigm consisted in focusing on the "spatial patterns" of human activities on the earth's surface. These "spatial patterns" were then described using geometrical, physical and cybernetic models, some of which were very complex, and were interpreted less and less as adaptations to natural environments and more and more as adaptations to social, economic, political and other contexts.

As the cultural, or human, ecological topic of the man-nature symbioses in space and history faded into the background, so too did the "unity of geography". Nowadays, almost all geographers – at least in practice but often also in terms of how they see themselves – are *either* social scientists *or* geo-scientists, i.e. natural scientists. Even if both groups are occupied with so-called environmental problems, they constitute completely different epistemic entities. The change, however, is often concealed by a traditional, nostalgic or utopian rhetoric which draws, for example, on the concepts of landscape used by landscape ecologists.

Thus, the split that runs through the modern scientific system now runs through geography as well. This is why, apart perhaps from a few "landscape ecologists" and "geo-ecologists", the geographers who in any kind of sense work ecologically

---

[33] The different stages and the logic of this paradigm change are described in Eisel (1980); a summary can be found, amongst others, in Hard (1982, 2002, pp. 203 ff.).

can barely be distinguished anymore nowadays from ecologists of a different provenance.[34]

However, there is no need to fear that the issue of the man-nature symbiosis or the landscape paradigm will be lost without a trace. They will live on, for example, in the curricula of geography and in other parts of the pedagogical domain (in part, for good didactic reasons); but they will also continue as a readily acceptable scheme for interpreting the world, as the folk science of common sense, educated people and political ecology. From time to time, inspired by the zeitgeist, they will doubtless also crop up again on the margins of established academic science.[35]

Nowadays, whenever geography attempts to revitalise the old themes, philosophy and ethics of the man-nature symbiosis using various terms such as "human ecology", it is interesting to note that even geographers no longer refer back to the geographical tradition and its classical texts (or even to landscape ecology). Instead, they absent-mindedly seek their future human ecological geography in the most disparate (e.g. poststructuralist and postmodernist) programmes and philosophies *outside* geography. In doing so, moreover, they are quite explicit about no longer seeking something that might be labelled geographical, but rather something interdisciplinary, transdisciplinary and multi-disciplinary.[36] This type of "dissémination" also seems to me to be a typical symptom of the historical end point of a singular scientific paradigm.

Paradigm Regained

Naturally there will always be activities in which natural and social scientific (or natural scientific and hermeneutical) issues and perspectives are linked to one another, indeed have to be linked to one another – for example, certain kinds of applied ecology and some professional practices that might be called "Handlungswissenschaften" (practical sciences).[37]

---

[34] This is not to say that ecologists with a geographical background cannot occasionally be recognized by certain specificities of their interests and research style that stem from their geographical socialisation; on this issue, cf. Hard (2003, p. 117 ff.).

[35] As occurred, for example, with the sociologists who, referring to "Rio" and "sustainable development", wanted to shake all prior sociological thought "to its foundations" in order "to provide a concise representation of the interactions between the social and the natural systemic context", between "the world society and the natural system of the planet", cf. Fischer-Kowalski and Erb (2003, p. 259). With this formulation, they too have arrived at the old geographical chimera, the supersystem "landscape" and the "landscape ecosystem" (Leser 2000, p. 108).

[36] Cf. in German geography, for example, the collection "Humanökologie" (Human Ecology) by Meusburger and Schwan (2003).

[37] On just such a "hybrid ecology" or "hybrid science", cf. Hard (1997), drawing on Trepl (1992), who of course did not use the fashionable word "hybrid". Concerning "ambigous disciplines" of this kind cf. Schwarz 2001 or also Potthast 2001. Some of these hybrids can be regarded as vital structural parallels to the hybrid geographical man-nature paradigm.

Almost everywhere today, the state of biocoenoses and ecosystems depends to a very large extent on what they mean, i.e., how they are perceived, interpreted and judged in political and popular ecology. Such folk ecologies are inseparably mixed in with scientific-ecological, everyday, political, moral and aesthetic considerations. One does not need to travel to the margins of the inhabited earth (e.g. to Nepalese peasants) in order to find such ethno-ecologies; they can also be found in the metropolitan centres, where very different groups and institutions produce, disseminate and practice such folk ecologies and ecological myths: lay people, professionals, administrations, civil society initiatives, social movements and so forth, along with their whole quasi- and pseudo-scientific supplier industry. Even to be able simply to assess the reality outside the laboratory and away from experimental sites, empirical – and above all applied – ecology will not be able to limit itself to ecological matters in the narrow sense. Instead, it will have to observe in its research practice:

1. the "real" ecological situation as it is perceived in the natural sciences;
2. how people perceive and interpret ethno-ecologically the ecological situation (and who sustains these ethno-ecologies);
3. the relationship between "real" ecology and symbolic ecologies;
4. how people behave and act on the basis of their symbolic ecology;
5. the impacts of people's behaviour and actions with regard to "real" ecology;
6. how the impacts of these often "unreal" ecologies are perceived, interpreted and legitimated

Only the first and fifth points can be answered in strictly ecological terms, i.e. scientifically; the others tend to be assigned more to the social sciences and similar fields in the contemporary scientific system – and for good reason. Classical geography, i.e. geographical cultural ecology, was occupied in principle with all six issues, even if this was often merely implicit. In this respect, the points formulated above are also a kind of modernising explication of classical geography.

References

Barrows HH (1923) Geography as human ecology. Ann Assoc Am Geogr 13:1–14
Bartels D, Hard G (1975) Lotsenbuch für das Studium der Geographie als Lehrfach. Verein zur Förderung Regionalwissenschaftliche Analysen e.V, Kiel
Bartels D (1968) Zur wissenschaftstheoretischen Grundlegung einer Geographie des Menschen. F. Steiner, Wiesbaden
Bartels D (ed) (1970) Wirtschafts- und Sozialgeographie. Kiepenheuer u. Witsch, Köln
Beck H (1973) Geographie. Europäische Entwicklung in Texten und Erläuterungen. Alber, Freiburg/München
Beck H (1979) Carl Ritter, Genius der Geographie: zu seinem Leben und Werk. Reimer, Berlin
Beck H (1981) Carl Ritter als Geograph. In: Lenz K (ed) Carl Ritter Geltung und Deutung. Reimer, Berlin, pp 13–24
Berdoulay V (1981) La formation de l'école francaise de géographie (1870-1914). Bibliothèque Nationale, Paris

Bobek H (1959) Die Hauptstufen Der Gesellschafts- Und Wirtschaftshaltung In Geographischer Sicht. Die Erde 90:259–298

Böttcher H (1979) Zwischen Naturbeschreibung und Ideologie. Versuch einer Rekonstruktion der Wissenschaftsgeschichte der deutschen Geomorphologie. Oldenburg: gesellschaft zur Förderung Regionalwissenschaftlicher Erkenntnisse e.V

Buttimer A (1971) Society and milieu in the French geographic tradition. Rand McNally, Chicago

Büttner M (1975) Regiert Gott die Welt? Vorsehung Gottes und Geographie. Studien zur Providentialehre bei Zwingli und Melanchthon. Calwer Verlag, Stuttgart

Claval P (1964) Essai sur l'évolution de la géographie humaine. Presses universitaires de Franche-Comté, Paris

Eisel U (1980) Die Entwicklung der Anthropogeographie von einer "Raumwissenschaft" zur Gesellschaftswissenschaft. Urbs et Regio 17:1–683

Eisel, Ulrich (1987) Landschaftskunde als "materialistische Theologie". In: bahrenberg, Gerhard, Jürgen Deiters, Mafred M, Fischer, Wolfgang Gaebe, Gerhard Hard and Günther Löffler (eds) Geographie des Menschen. Dietrich Bartels zum Gedenken. Universität Bremen, Bremen, pp 89–109

Eisel U (1997) Triumph des Lebens. In: Eisel U, Schultz H-D (eds) Geographisches Denken. Gesamthochschulbibliothek, Kassel, pp 39–160

Fischer-Kowalski M, Erb K-H (2003) Gesellschaftlicher Stoffwechsel im Raum. Auf der Suche nach einem sozialwissenschaftlichen Zugang zur biophysischen Realität. In: Meusburger P, Schwan T (eds) Humanökologie. Ansätze zur Überwindung der Natur-Kultur-Dichotomie. Erdkundliches Wissen 135. Stuttgart: steiner, pp 257–285

Fuchs G (1966) Der Wandel zum anthropogeographischen Denken in der amerikanischen Geographie: Strukturlinien der geographischen Wissenschaftstheorie; dargestellt an den vorliegenden wissenschaftlichen Veröffentlichungen 1900–1930. Dissertation, Philipps-Universität Marburg, Marburg

Fuchs G (1967) Das Konzept der Ökologie in der amerikanischen Geographie. Erdkunde 21(2):81–93

Glacken CJ (1967) Traces on the Rhodian shore. Nature and culture in Western thought from ancient times to the end of the eighteenth century. University of California Press, Berkeley/Los Angeles

Greverus I-M (1978) Kultur und Alltagswelt: Eine Einführung in die Fragen der Kulturanthropologie. Beck, München

Haggett P, Frey AE, Cliff AD (1965) Locational analysis in human geography. Wiley, London/New York

Haggett P (1965) Locational analysis in human geography. Edward Arnold, London

Hamm B (1990) Sozialökologie. In: Kruse L, Graumann C-F, Lautermann E-D (eds) Ökologische Psychologie: Ein Handbuch in Schlüsselbegriffen. Psychologie Verlags Union, München, pp 35–38

Hard G (1970) Die "Landschaft" der Sprache und die "Landschaft" der Geographen: Semantische und forschungslogische Studien zu einigen zentralen Denkfiguren in der deutschen geographischen Literatur. Dümmler, Bonn

Hard G (1973) Die Geographie. Eine wissenschaftstheoretische Einführung. De Gruyter, Berlin/New York

Hard G (1982) Ökologie/Landschaftsökologie/Geoökologie (sowie: Ökologische Probleme im Unterricht). In: Jander L, Schramke W, Wenzel H-J (eds) Metzler-Handbuch für den Geographieunterricht: Ein Leitfaden für Praxis und Ausbildung. Metzler, Stuttgart, pp 232–246

Hard G (1988) Selbstmord und Wetter Selbstmord und Gesellschaft. Studien zur Problemwahrnehmung in der Wissenschaft und zur Geschichte der Geographie. F. Steiner, Stuttgart

Hard G (1990) Humangeographie (bes. Wahrnehmungs- u. Verhaltensgeographie). In: Kruse L, Graumann C-F, Lautermann E-D (eds) Ökologische Psychologie: ein Handbuch in Schlüsselbegriffen. Psychologie Verlags Union, München, pp 57–65

Hard G (1997) Was ist Stadtökologie? Argumente für eine Erweiterung des Aufmerksamkeitshorizonts ökologischer Forschung. Erdkunde 51:100–113
Hard G (2001) Der Begriff Landschaft Mythos, Geschichte, Bedeutung. In: Konold W, Böcker R, Hampicke U (eds) Handbuch Naturschutz und Landschaftspflege. 6. Erg.-Lfg. 10/01. Landsberg: Ecomed, pp 1–15
Hard, G (2002) Landschaft und Raum. Aufsätze zur Theorie der Geographie. Bd. 1. Universitätsverlag Rasch, Göttingen
Hard G (2003) Dimensionen geographischen Denkens. Aufsätze zur Theorie der Geographie. Bd. 2. Universitätsverlag Rasch, Göttingen
Hellmann N, Post S (1985) Erarbeitung und Durchführung des Landschaftsplans nach Landschaftsgesetz Nordrhein-Westfalen. Diplomarbeit, Technische Universität Hannover, Hannover
Herder JG (1966) Ideen zur Philosophie der Geschichte der Menschheit. Wissenschaftliche Buchgesellschaft, Darmstadt
Jax K (2002) Die Einheiten der Ökologie: Analyse, Methodenentwicklung und Anwendung in Ökologie und Naturschutz. Lang, Frankfurt a. M
Johnston RJ (1988) On human geography. Basil Blackwell, Oxford
Johnsson RJ, Gregory D, Pratt G, Watts M (1988) The dictionary of human geography. Blackwell, Malden/Oxford/Carlton
Kapp E (1845) Philosophische oder vergleichende allgemeine Erdkunde als wissenschaftliche Darstellung der Erdverhältnisse und des Menschenlebens nach ihrem inneren Zusammenhang. George Westermann, Braunschweig
Kattenstedt H, Büttner M (1993) Grenz-Überschreitung: Wandlungen der Geisteshaltung, dargestellt am Beispielen aus Geographie und Wissenschaftshistorie, Theologie, Religions- und Erziehungswissenschaft, Philosophie, Musikwissenschaft und Liturgie. Festschrift zum 70. Geburtstag von Manfred Büttner. Brockmeyer Universitätsverlag, Bochum
Körner S, Eisel U (2002) Biologische Vielfalt und Nachhaltigkeit: Zwei zentrale Naturschutzideale. Geographische Revue 4(2):3–20
Krewer B, Eckensberger LH (1990) Die ökologische Perspektive in der Kulturanthropologie. In: Kruse L, Graumann C-F, Lautermann E-D (eds) Ökologische Psychologie: ein Handbuch in Schlüsselbegriffen. Psychologie Verlags Union, München, pp 49–56
Leser H (1997) Landschaftsökologie: Ansatz, Modelle, Methodik, Anwendung. Ulmer, Stuttgart
Leser H (1991) Ökologie wozu? Der graue Regenbogen oder Ökologie ohne Natur. Springer, Berlin
Leser H (2000) Geoökosysteme Ganzheiten oder Fragment? Gedanken zum Problem einer holistisch ansetzenden Landschaftsökologie. Klagenfurter Geographische Schriften 18:105–115
Leser H, Schaub DM (1995) Geoecosystems and landscape climate. The approach to biodiversity on landscape scale. Gaia 4:212–226
Lethmate J (2000) Ökologie gehört zur Erdkunde aber welche? Kritik geographiedidaktischer Ökologien. Die Erde 131(1):61–79
Menting G (1987) Analyse einer Theorie der Geographischen Ökosystemforschung. Geogr Z 75(5):209–227
Menting G (2000) Warten auf Godot. Die Erde 131:351–395
Menting G (2001) Geoökosystemforschung aufs Abstellgleis? Geographische Rundschau 52(6):34–40
Meusburger P, Schwan T (2003) (eds) Humanökologie. Ansätze zur Überwindung der Natur-Kultur-Dichotomie. Steiner, Stuttgart
Müller-Böker U (1995) Die Tharu in Chitawan: Kenntnis, Bewertung und Nutzung der natürlichen Umwelt im südlichen Nepal. Steiner, Stuttgart
Paffen K (ed) (1973) Das Wesen der Landschaft. Wissenschaftliche Buchgesellschaft, Darmstadt
Potthast T (2001) Gefährliche Ganzheitsbetrachtung oder geeinte Wissenschaft von Leben und Umwelt? Epistemisch-moralische Hybride in der deutschen Ökologie 1925–1955 Verhandlungen zur Geschichte und Theorie der Biologie 7:91–113

Pratt ML (1992) Imperial eyes: travel writing and transculturation. Routledge, London/New York
Ratzel F (1882) Anthropogeographie. Engelhorn, Stuttgart
Ratzel F (1897) Politische Geographie. Oldenburg, München
Ritter C (1852) Einleitung zur allgemeinen vergleichenden Geographie und Abhandlungen zur Begründung einer mehr wissenschaftlichen Behandlung der Erdkunde. Reimer, Berlin
Ritter C (1862) Allgemeine Erdkunde. Reimer, Berlin
Schmithüsen J (1953) Grundsätzliches und Methodisches. Einleitung zum Handbuch der naturräumlichen Gliederung Deutschlands. In: Emil M, Schmithüsen J (1953) (eds) Handbuch der naturräumlichen Gliederung Deutschlands. Bundesanstalt für Landeskunde und Raumforschung, Bad Godesberg, pp 1–44
Schmithüsen J (1964) Was ist eine Landschaft? Steiner, Wiesbaden
Schmithüsen J (1974) Landschaft und Vegetation. Gesammelte Aufsätze von 1934 bis (1971) Geographisches Institut der Universität des Saarlandes, Saarbrücken
Schultz H-D (1980) Die deutschsprachige Geographie von 1800 bis 1970. Ein Beitrag zur Geschichte ihrer Methodologie. Selbstverlag des Geographischen Instituts, Berlin
Schultz H-D (1981) Carl Ritter ein Gründer ohne Gründerleistung? In: Lenz K (ed) Carl Ritter Geltung und Deutung. Reimer, Berlin, pp 55–74
Schultz H-D (1997) Von der Apotheose des Fortschritts zur Zivilisationskritik. Das Mensch-Natur-Verhältnis in der klassischen Geographie. In: Eisel U, Schultz H-D (eds) Geographisches Denken. Gesamthochschulbibliothek, Kassel, pp 177–282
Schultz H-D (2002) "Jeder Raum hat sein Volk". In: Luig U, Schultz H-D (eds) Natur in der Moderne: Interdisziplinäre Annäherungen. Berliner Geographische Arbeiten, Berlin, pp 87–148
Schultz, H-D (2003) Geographie? Antworten vom 18. Jahrhundert bis zum Ersten Weltkrieg (Teil 1), Antworten von 1918 bis zur Gegenwart (Teil 2), vol 88, 89. Arbeitsberichte des Geographischen Instituts der HU Berlin, Berlin
Schwarz AE (2001) "Der See ist ein Mikrokosmos" oder die Disziplinierung des "uneindeutigen Dritten". Verhandlungen zur Geschichte und Theorie der Biologie 7:69–89
Stegmüller W (1973) Theorienstrukturen und Theoriendynamik. In: Stegmüller W (ed) Probleme und Resultate der Wissenschaftstheorie und analytischen Philosophie, vol. 2.2. Springer, Berlin
Stegmüller W (1979) Rationale Rekonstruktion von Wissenschaft und ihrem Wandel. Reclam, Stuttgart
Stegmüller W (1980) Neue Wege der Wissenschaftstheorie. Springer, Berlin/Heidelberg/New York
Steward JH (1955) Theory of culture change. The methodology of multilinear evolution. University of Illinois Press, Urbana
Storkebaum W (1967) Zum Gegenstand und zur Methode der Geographie. Wissenschaftliche Buchgesellschaft, Darmstadt
Trepl L (1987) Geschichte der Ökologie: Vom 17. Jh. bis zur Gegenwart. Athenäum, Frankfurt a. M
Trepl L (1988) Gibt es Ökosysteme? Landschaft und Stadt 20(4):176–185
Trepl, L (1992) Stadt-Natur: Ökologie, Hermeneutik und Politik. Bayerische Akademie der Wissenschaften (ed) Rundgespräche der Kommission für Ökologie. Pfeil, München, pp 53–58
Trepl L (1996) Die Landschaft und die Wissenschaft. In: Konold W (ed) Naturlandschaft Kulturlandschaft: Die Veränderung der Landschaften nach der Nutzbarmachung durch den Menschen. Ecomed, Landsberg, pp 13–26
Trepl L (1997) Ökologie als konservative Naturwissenschaft. Von der schönen Landschaft zum funktionierenden Ökosystem. In: Eisel U, Schultz H-D (eds) Geographisches Denken. Gesamthochschulbibliothek, Kassel, pp 467–492
Troll C (1931) Die geographischen Grundlagen der andinen Kulturen und des Inkareiches. Ibero-Amerikanisches Archiv 5:1–37
Troll C (1966a) Ökologische Landschaftsforschung und vergleichende Hochgebirgsforschung. F. Steiner, Wiesbaden
Troll C (1966b) Luftbildforschung und landeskundliche Forschung. F. Steiner, Wiesbaden
Waibel L (1921) Urwald, Veld, Wüste. Hirt, Breslau

Chapter 26
Border Zones of Ecology and the Applied Sciences

Yrjö Haila

Origins

Ecological thinking has evolved in close interaction with human subsistence practices. This is evident for two broad reasons. First, human subsistence depends ultimately on the utilization of ecological processes, whether through hunting and gathering, the breeding of domesticated animals and cultivation of plants, or through production based on other biological resources. Second, we humans ourselves are biological organisms and, like all other organisms, we lead our lives enmeshed in networks of ecological relationships.

Accordingly, all human societies have acquired knowledge of those elements and processes of nature on which their subsistence has depended. This practical knowledge predated the science of ecology by millennia. Historically, however, the process was not one of smooth transition from a variety of practical skills to a unified ecological science – far from it. Knowledge acquired by using various kinds of biological resources has given rise to a whole range of applied research traditions and disciplines such as agriculture, forestry, fisheries, and game and rangeland management. These are major applied sciences with which ecology has been in contact and contrast from its very inception. In fact, in the case of ecology the vague boundary between pure and applied science is perhaps at its vaguest.

If we are serious in claiming that the roots of ecological knowledge extend far back into the past, the question we have to address is this: what criteria can we use to distinguish ecology as a modern discipline from such traditional background knowledge? A brief practical answer to this question may be sufficient in this context. Ecology became a distinct discipline when ecologists – that is, people doing ecological research – adopted novel concepts and were able to put these concepts to work in their practical research. This created a "self-vindicating structure" that facilitated the stabilization of their science (Hacking 1992). However, a caveat is needed at this point. Hacking applied this idea to the laboratory sciences, whereas

Y. Haila (✉)
Department of Regional Studies, University of Tampere, 33014 Tampere, Finland
e-mail: yrjo.haila@uta.fi

a major part of ecological research is conducted in the field. This difference has created an additional challenge to the stabilization of ecological research (see Haila 1992, 1998; Kohler 2002); ironically, however, it has also strengthened the links between ecology and the applied disciplines: modern systematic use of ecological resources provides indispensable insights into ecological processes on a larger scale than controlled experiments usually allow. Ecologists have simply had to learn to use this experience.

Agendas

Fundamental and applied research are usually viewed in contrast to one another; a paradigmatic model of the contrast stems from the relationship between physical science and engineering. Thomas Gieryn (1983, 1999) has shown that the distinction was largely a result of rhetorical "boundary work", conducted in the mid-nineteenth century by John Tyndall and others, whose intention was to promote the social prestige of fundamental research (which they portrayed as theoretically demanding and disinterested) in comparison to engineering (which they depicted as pragmatic, shallow and profit-driven). Engineers commonly contributed to the boundary work on the other side of the divide by emphasizing the practical usefulness of their specialty.

A similar contrast has, at times, separated ecology from its corresponding applied fields, but the dynamics have their own peculiarities. For one thing, fundamental research in biology, let alone in the young sub-discipline of ecology, did not have such an established theoretical status as the physical sciences in the latter part of the nineteenth century. Aggressive boundary work on the part of ecologists would have been out of the question. Furthermore, ecologists at that time were still dependent to a considerable extent on many sorts of practical experience when it came to consolidating their science – much in the same way, in fact, as pioneering physicists had been in the seventeenth and eighteenth centuries.

A further distinction is that the practical disciplines that emerged out of the use of biological resources are themselves so heterogeneous that they would not easily fit a rhetorical caricature such as the "mechanics" ridiculed by Tyndall. Each of the applied fields that borders on ecology also has its own institutional history. Administrative bodies and governmental research institutes were established early on in the fields of agriculture, forestry, fisheries, and rangeland management in different countries according to their natural conditions and productive needs. At the core of such institutionalization, of course, lies the importance of biological resources for a country's economy. In other words, the political economy of the applied research traditions in ecology is an important factor conditioning the circumstances of their development (see the essay on agricultural research in Levins and Lewontin 1985).

There is one more anomaly in the profile of the applied sciences that border on ecology: the environmental sciences. The destruction of nature became a concern among pioneer ecologists quite early on in the nineteenth century, but those concerns acquired much greater force when environmental issues broke through into public consciousness during the last quarter of the twentieth century. Nowadays, environmental issues form a major part of the agenda of applied ecology. In fact, it would not be an exaggeration to claim that public recognition of ecology as an independent biological sub-discipline owes much to environmental concerns. This development has also influenced other traditional applied disciplines. It is evident that we need all the knowledge generated by the applied disciplines if we are to achieve the aim of sustainable management of biological resources.

In this chapter I aim to offer an overview of how the contacts between ecology and its corresponding applied disciplines have changed through time. Only the barest outline is possible; it goes without saying that each one of the practical fields deserves a chapter of its own. Furthermore, there is a shortage of background data. One particularly astonishing fact is that most histories of ecological science barely touch upon its rich connections to applied research. Obvious exceptions to this include Tobey (1981) on American plant ecology and prairie agriculture, and Golley (1993) on ecosystems ecology and environmental concerns; Donald Worster (1985), as a historian writing on the history of ecological thought, is sensitive to the social context throughout. Kohler (2002), in my view, gives a hint, albeit an indirect one, towards understanding this deficit. Ecologists building up an independent disciplinary identity have often striven to make their science respectable by emulating an ideal of basic research that is borrowed from the physical sciences via the laboratory, so that practical fields of endeavor were often relegated to the sidelines.

First, however, one more note on early origins is called for. Clarence Glacken (1967) describes the main stages in the early accumulation of insights into ecological processes in our western heritage. Some important summaries regarding knowledge of agriculture were written in Roman times and again during the late Middle Ages. In the latter period, "the infinitely better dissemination of knowledge that followed the invention of printing" (Glacken 1967, p. 317) greatly increased the possibility of capturing and comparing specific practical experience over increasingly long stretches of space and time. A broad comparative perspective was, of course, a necessary precondition for ecological regularities to be recognized as such at all. The eighteenth century, the "age of reason" in Europe, produced biological polymaths, such as Count Buffon in France and Carolus Linnaeus in Sweden, whose work provided broad inspiration for later ecological thinking.

Geographic explorations have provided essential material for ecological thinking. Throughout the early modern era, the European conquest and colonization of other continents led to contacts with a natural world which, in many instances, was strikingly different from the one familiar to Europeans at home. An important result of the conquest was the gradual realization during the eighteenth century that the inhabitable parts of Earth were already settled; this gave rise to "the idea of a closed space" (Glacken 1967, p. 623). Views concerning the future prospects of humanity diverged during the last few years of the eighteenth century into two mutually contradictory strands of thought: belief in historical progress on the one hand and doubt about progress on the other hand. These were famously represented by Marquis Condorcet and William Godwin, and Thomas Malthus, respectively, although both camps had their precursors, too. An analogous bifurcation has been reproduced in environmentalist discussions ever since. Ecological thought could not possibly have risen up independently of such contradictory assessments of the future prospects of humanity.

In addition, ideas about nature and ideas about society have long been in close interaction at the level of political philosophy. All human cultures have known that their existence depends on something that is "outside of" culture, namely, nature. As William Connolly (1993) notes, in modern societies this relationship has been increasingly articulated using the vocabulary of nature. This has taken two alternative forms. Some theorists have viewed nature as a set of regularities that can potentially be known and mastered by humans, while others have viewed nature as a meaningful order of which human beings are a part. These alternative views imply mutually contradictory presuppositions as to how human culture ought to deal with the rest of nature.

Overall, we have good grounds for concluding that views of living nature have developed in close interaction not only with productive practices but also with understandings of social order. The practical experience of previous generations has formed a backdrop against which the credibility of scientific views has been assessed as a preliminary test.

Use of Nature and Botanical Ecology

Metaphoric transfers of meaning have played an important part in the determination of ecological research objects (Worster 1985; Taylor 1988; Cuddington 2001; Jax 2002). Metaphors served as mediators of ideas and experience between the practical and metaphysical realms and scientific conceptualizations of living nature. The first research objects identified by the early ecologists were typically conceptualized using two complementary metaphors, organism and community; the latter was often further specified using the ancient metaphor of microcosm (Schwarz 2003). These concepts were used to characterize units consisting of a large number of different species. Early inspiration was provided by the results of geographic explorations in the early nineteenth century: pioneers such as Alphonse de Candolle and Alexander von Humboldt described geographic variation in vegetation using terms that paved the way for later community ecology. The early ecologists focused their studies on systems they believed to be pristine but their views on the nature of these systems were influenced by all the major productive practices based on biological resources.

Botanists predated zoologists by a few decades in commencing research that paved the way for modern ecology. This seems quite natural given that agriculture and forestry are the main productive practices in which humans are involved in managing multi-species communities, focusing mainly on plant resources. As I shall emphasize later on, some of the pioneers of applied community-level ecology were zoologists, but by and large botanists dominated the scene.

The significance of agriculture for early ecology is an under-researched topic. Ever since its inception agriculture has provided people with practical experience of how environmental conditions influence plant growth. The German organic chemist Justus von Liebig established a program of systematic research in agricultural chemistry and formulated his famous "law of the minimum" in 1840; according to Liebig's law, in Eugen Odum's succinct formulation, "an organism is no stronger than the weakest link in its ecological chain of requirements" (1959, p. 88). Agricultural experimental stations were established in the nineteenth century; the Rothamstead Experimental Station in England, established in 1843, is generally considered to be the oldest one. These stations began to produce data that were to be used later by ecologists. Botanists in particular remind us every now and then of the important role played by agricultural experimentation in early ecology. For instance, experimental data generated at Rothamstead were crucial for R.A. Fisher's statistical inventions, such as the analysis of variance, and long-term censuses of insect populations at Rothamstead have provided material for analyses of species-abundance distributions in samples of varying size (Williams 1964).

In North America, the subjection of the prairies to the plough stimulated research in plant ecology, but historians tend to disagree about the significance of this connection. For example, Donald Worster (1985) identifies a clear connection between the two, whereas Robert McIntosh (1985) hardly mentions the word agriculture at all in his historiography of ecology. The traumatic experience of the Dust Bowl in the 1930s triggered controversies among ecologists concerning the prospects of prairie agriculture. According to Worster's (1985) analysis, scholarly opinions bifurcated into two major strands, which largely correspond to the optimism vs. pessimism dichotomy that emerged around the turn of the nineteenth century concerning the prospects of human culture. The dominant figure in American plant ecology at that time, Frederick Clements, provided fuel for both views. On the one hand, his view of the regularity of vegetation succession could be used to bolster optimism concerning the ability of ecological science to provide tools for the management of the prairies; Clements was inspired by New Deal social engineering (see also Hagen 1992, p. 85). On the other hand, the idea of a climax community, which is in equilibrium with local environmental conditions, including the climate, does not leave much leeway for human initiative; prominent entomologists, in particular, adopted the latter position (Roger Smith, Paul Sears). The same basic controversy was reproduced in even more stark terms in the 1950s, when Walter Prescott Webb spoke out in favour of harmonious coexistence through livestock grazing, which imitates natural herbivory, and James Malin promoted strong intervention through mechanized agriculture.

A similar conflict was experienced in the Soviet Union in the 1930s, albeit in a much more brutal form. Ecologists demanded that care be taken in cultivating the vast plains of southern Russia and the Ukraine, whereas Stalin's ideologues condemned this as reactionary propaganda (Weiner 1988).

Forestry is another major form of human exploitation of ecological resources, but the deliberate growth of timber, or "silviculture", developed quite separately from the biological sciences in its early stages. George Perkins Marsh (1965) offers an overview of the art of silviculture in the mid-nineteenth century. The origins of this art go back to the second half of the eighteenth century, when German foresters developed methods of measuring the mass or volume of wood in forest plots (Lowood 1991).

The main interest of the early foresters in continental Europe, chiefly in France and in Germany, focused on the amount of wood in the forest, not on the forest as an ecological system. The idea of "sustainable forestry" originated at that time, but the sole indicator of sustainability was steady yield of timber from the forest.

Botanists in northern Europe began to study forest vegetation during the latter decades of the nineteenth century in order to identify ecological indicators of timber productivity. They wanted to establish scientific guidelines of management practices in different types of forest. One pioneer in this was the Finn A.K. Cajander, who was educated as a geobotanist and began his career by exploring the lowlands of the River Lena in eastern Siberia (at that time Finland was a Grand Dutchy of imperial Russia). Cajander developed a method of forest site type classification, based on the identification of a typical vegetation community found in different combinations of climate and soil (see Cajander 1949, which also includes a critical overview of the nineteenth century German tradition of forest classification; Cajander published his original magnum opus on forest site type classification, "Über Waldtypen", in 1909). Cajander's students applied his method in both Central Europe and North America.

Cajander describes the roots of his theory as follows: "The theory of forest (site) types can be traced back to the grouping of wooded lands, used of old in the forest-economy in Finland [...] and to the more exact analysis and definition of this grouping on the plant-topographic basis developed by Professor Cajander" (1949, p. 3). Norrlin was professor of botany at the University of Helsinki and published his first plant-geographic investigations in 1870 and 1871.

A.K. Cajander's scientific career in Finland demonstrates an interesting convergence of geographic explorations, ecological surveys and productive needs into a unified research tradition of forest ecology. A similar convergence could be documented in virtually every industrialized part of the world. An additional feature of Cajander's career is that he also became a successful politician, serving as prime minister in the late 1930s. He was elected into parliament on a "progressive party" ticket – an interesting analogy with New Deal progressivism in the US. Forest industries formed the backbone of the Finnish economy at that time, and forestry science, by implication, gained special prestige in the social realm.

In North America, the fate of the forest began to receive public attention toward the end of the nineteenth century. At that time it became clear that the extensive clearcuttings, which essentially amounted to timber mining, had badly damaged the forested landscapes in different parts of the continent (Williams 1989). This concern formed a major part of social debates at the time, in the guise of utilitarian resource management promulgated by so-called conservationists. I return to this point below.

Zoological Ecology: Focus on Populations

In early animal ecology, the varieties of inspiration gained from applied fields were just as multifaceted as in plant ecology. Major synthetic conceptualizations grew out of research in aquatic ecology in Central Europe. Charles Elton (1966), p. 30

pays homage to this tradition. Particularly important was the research of C.G.J. Petersen and his colleagues on marine life in Danish inland salt fjords. They made a number of methodological innovations in how to take quantitative samples of the most important components of the aquatic community. "The motive of this great survey was an economic interest in measuring the food resources available to fish" (Elton 1966, p. 31). Studies of oyster fisheries on the coast of Schleswig-Holstein by Karl August Möbius (1870s) were similarly motivated.

In the US Midwest, Stephen Forbes (1887) articulated a synthetic view of ecological communities in his article "The Lake as a Microcosm". Forbes and his followers, such as Victor Shelford, emphasized ideas of ecological community and actively sought to build a fruitful connection between basic ecology and limnology as an applied science (Golley 1993, p. 36).

By and large, however, botanists dominated research on ecological communities. Animal ecologists became more interested in the dynamics of single populations than in the composition of communities. In aquatic environments, the roles were often reversed, however (Schwarz 2003). Research on the dynamics of biological populations drew its original inspiration from human demography. D'Arcy Thompson (1992) tells the history of these connections in meticulous detail. Thomas Malthus, who published his "Essay on the Principle of Population" in 1798, was a pioneer in that he tried to detect systematic principles that determine human population growth. The Belgian demographer Pierre Francois Verhulst 1838 gave a more exact mathematical formulation to the theory of Malthus in the 1830s, drawing on population statistics available from various countries at the time and deriving the S-shaped "logistic curve" of population growth. Verhulst's work remained forgotten until the 1920s.

Human demography, together with epidemiology, medical entomology and chemical kinetics, inspired Alfred Lotka to write his great "Elements of Physical Biology" (1924/1956), which laid the foundations for the mathematical theory of population dynamics – even though his book was hardly read at all at the time. The co-founder of modern population dynamics, Vito Volterra, drew inspiration from the fluctuations in fish catches in the Mediterranean Sea during and after the First World War (Scudo 1971).

Fisheries have been an important productive practice all along. Sustainable yield was adopted from forestry into fisheries as a notional ideal in the late nineteenth century, but it was only in the mid-twentieth century that formal models were developed for optimal harvesting of renewable resources (Beverton-Holt model in fisheries, see Clark 1989). The models, however, assumed an unchanging environment and proved to be seriously deficient. It is only relatively recently that ecosystem dynamics have been included in models of sustainable fishery.

Kingsland (1985) demonstrated the close connections between population dynamic modelling and agricultural entomology in the inter-war years. Monitoring of forest insect pest populations has produced invaluable data sets for analyses of insect population dynamics (Clark et al. 1967). One particularly important case is the modelling of spruce budworm population dynamics in Canada by Holling, Clark and their colleagues. The models demonstrate "hysteresis" produced by the interaction between slow and fast variables in the same system (see Holling 1992).

Predator-prey, or parasitoid-prey modeling work has also progressed in close interaction with agricultural problems such as pest control. Classic experiments on the role of spatial heterogeneity for predator-prey interactions by Huffaker were made in greenhouse conditions, and C.S. Holling began his experimental work on predator behaviour – which resulted in the important distinction between "functional" and "numerical" response in predator populations to prey numbers – in a forest entomological setting (see e.g. Hassell 1978).

The Global Ecology of Humanity

As mentioned above, the ecological prospects of humankind came to the fore at the turn of the nineteenth century in the dispute between Malthus and his critics, albeit indirectly. The issue has lived on, however, and it is quite understandable that the knowledge produced by the new biological discipline of ecology should have been harnessed to address this problem. George Perkins Marsh, with his book "Man and Nature", originally published in 1864, became an unrivalled pioneer in assessing the effects of humanity on the global environment. He based his work on a rich collection of sources describing the detrimental effects of resource use in the mid-nineteenth century. While the book attracted considerable attention upon its publication and gave crucial inspiration to the late nineteenth century conservationists in North America, it apparently had little direct influence on research (for a history, see Turner et al. 1990). This is perhaps due to the synoptic nature of the book: it was a rich collection of experiences and observations from different fields of resource use rather than a starting point for a research programme. Marsh is better known among geographers than among biologists.

The historical background of Marsh's insights has created some controversy among historians lately. Marsh presented a path-breaking synoptic vision of the destructive influence of human culture on Earth, but his view was grounded in a broader popular conservation movement that was widespread in New England in the early and mid-nineteenth century (Judd 1997, 2004). New Englanders had the experience of being faced with a partially unfamiliar nature, of being all alone without external help, and they soon learned to feel respect for their new natural environment. It seems that similar experiences created moods favourable for conservation in other European colonies as well, including tropical islands (Grove 1995) and Australia and New Zealand (Wynn 2004). Although ruthless exploitation of riches and resources was commonly the driving motive of European colonization, once the colonists were left alone in the unfamiliar conditions of the new continents, their mood changed considerably.

A systematic research tradition in global ecology began to take shape in the early twentieth century on the basis of biogeochemistry, or the study of global material cycles and the role of living organisms in keeping these going. Alfred Lotka was one of the pioneers in this field. He included extensive analyses of the global cycles of water and the main chemical elements in his "Elements of Physical Biology".

He also established an energetic perspective in global ecology; "[t]he dynamics which we must develop is the dynamics of a system of energy transformers, or *engines*" (Lotka 1956, p. 325; emphasis in the original).

Lotka's interest in global ecology explicitly included an assessment of the human role in modifying global cycles. In Lotka's work, the boundary between "basic" and "applied" research is blurred throughout. Indeed, some of his summarizing statements amount to an early formulation of the goal of sustainable development, such as the following:

> The human species, considered in broad perspective, as a unit including its economic and industrial accessories, has swiftly and radically changed its character... (W)e are far removed from equilibrium – a fact which is of the highest practical significance, since it implies that a period of adjustment to equilibrium conditions lies before us, and he would be an extreme optimist who should expect that such adjustment can be reached without labor and travail (Lotka 1956, p. 279).

This quotation is taken from a chapter entitled "Moving equilibria"; in other words, Lotka's point of reference was not a stable "balance of nature" as the wording might suggest, but rather a dynamic, temporally shifting steady-state.

Another pioneer in the field of global biogeochemistry was the Russian V.I. Vernadsky. He included life in global ecology more explicitly than Lotka. Vernadsky brought the term "biosphere" into biological terminology through a French translation of his book "Le Biosphére", in 1929 (the Russian original came out in 1926) (see Chap. 24). Biosphere has become a key concept in the so-called Gaia theory, established by James Lovelock, Lynn Margulis and others in the 1970s (Lovelock 1979; Margulis and Sagan 1997). Gaia is a theory of how the evolution of life has created Earth in its current form; the role of humanity in this process is also explicitly considered.

Lotka's and Vernadsky's ideas paved the way for an energetic conception of ecosystems, which took shape in the US in the second half of the twentieth century through G.E. Hutchinson (see Golley 1993). Hutchinson is mainly known for his contributions in basic ecology, but among his intellectual heirs it was the brothers Eugene P. and Howard T. Odum in particular who developed the modern tradition of ecosystems research. Frank Golley's (1993) excellent history of this research tradition brings into focus its essentially applied nature. Most of the original funding for the empirical research carried out by the Odum brothers came from the Atomic Energy Commission, with the goal of finding out the ecological consequences of radioactive radiation on ecological systems. Golley writes: "By the 1960's active research programs on ecosystems were under way in the United States, supported mainly by the AEC" (1993, p. 74).

E.P. Odum's influential textbook "Fundamentals of Ecology" (1953, 2nd edn. 1959) gives testimony to the close relationship between ecosystems research and human global ecology. The book's third section is entitled "Applied ecology". It covers issues of both resource management and pollution as well as public health in a comprehensive manner. In his preface to the second edition, Odum expresses his optimism: "Fortunately, biologists and the public alike are beginning to realize that ecological research of the most basic nature is vital to the solution of mankind's environmental problems." Here, Odum draws no distinction between basic and applied ecology.

Inspired by ecosystems research, ecology became "big science" through the funding of the International Biological Program (IBP), conceived by the International Council of Scientific Unions in 1959 and launched officially in the first general assembly of the program in Paris in 1964. IBP was followed by the Man and Biosphere Program (MAB), launched by UNESCO in 1971. Golley (1993) describes the development of these programs. The aim of the IBP was to assess the total productivity of all ecosystems on Earth. Its guiding ethos was tied to human needs; for instance, the theme of a preparatory meeting in 1960 was "the biological basis of human welfare" (Golley 1993, p. 110).

The US Congress provided generous funding for the IBP. The issue was discussed at hearings held by the Subcommittee on Science, Research and Development of the US House of Representatives in the summer of 1967. Chunglin Kwa (1987) shows that broad environmental concerns were decisive for the positive funding decision. The ecologists defending the programme persuaded the committee to believe that it could provide scientific backing for solving problems posed by environmental pollution. At the core of the argument was the perception that nature is "a system that can be controlled and managed" (Kwa 1987, p. 425). In other words, the politicians involved came to perceive the IBP as a means of providing scientific guidelines for the management of the global ecological system.

The "Odumian" tradition of ecosystem research, however, has receded since the days of the IBP. There are two broad reasons for this. First, the concept was not analytically clear-cut. As Golley (1993) shows, it was mainly derived from loose analogical thinking, in addition to which the energetic background was outdated by being tied to equilibrium thermodynamics when, in fact, ecosystems are very far from constituting a thermodynamic equilibrium. It has been realized more recently that non-equilibrium thermodynamics offers conceptual tools for analyzing ecosystem energy flow (Morowitz 1968). Second, as many contemporaneous critics pointed out and as has become obvious since, the Odumian perspective could not adequately identify the relevant dynamic units. Furthermore, the approach to environmental problems as formulated in Odum's textbook was enumerative, consisting basically of only a listing of various kinds of problems, which had only vague links to the underlying theory. This was not even "synoptic" in the sense of the book by George Perkins Marsh a century earlier, but rather eclectic, and it offered few guidelines for systematic research.

Research on global change has come to dominate the ecosystem tradition more recently. Walker and Steffen (1996), for instance, give an overview of the approach applied to terrestrial ecosystems. While physiology is integrated into the research, there is a more precise focus on the dynamics of basic biochemical processes in organisms, and also of ecological interactions among individuals. In the last decade of the twentieth century, "sustainability science" has been advancing to become a new sub-discipline within the environmental sciences, after the term "sustainable development" was made popular by the Brundtland report "Our Common Future" in 1987. Systems modelers have a particularly prominent position within this field.

Ecology and Environmental Politics

The environmental awakening of the 1960s and the notion of environmental crisis gave rise to environmental movements whose concerns also entered ecology from the 1960s onwards. Rachel Carson's "Silent Spring" (1962) had a crucial influence on public consciousness of ecological issues, even though issues of human population growth and resource depletion had been acknowledged prior to this time – for instance in E.P. Odum's "Fundamentals of Ecology". Barry Commoner (1971) emphasized the ecological hazards brought about by the increasing application of new chemicals in production processes.

A "Neo-Malthusian" strand of environmental thinking originated in the late 1960s. According to this view, the growth of the human population is the main culprit for the environmental crisis. The single most famous example of neo-Malthusianism was the "ecocatastrophe" scenario depicted by Paul Ehrlich (1969). Ehrlich's scenario was derived from hypothetical competition between the two super-powers, the USA and the Soviet Union, about hegemony in the Third World by developing ever more efficient pesticides to be used in agriculture. The ecocatastrophe eventually followed from the leakage of new super-poisons into the oceans. Critics of Neo-Malthusianism such as Barry Commoner (1971) emphasized, instead, the significance of economic and political processes over sheer human population numbers for understanding the nature of the contemporary environmental problems. Ironically, the purely social character of Ehrlich's scenario lends actually support to such criticism.

Originally, the environmental awakening was greeted with enthusiasm by many academic ecologists, but it became clear quite soon that basic ecological theory had very little to offer in terms of understanding the problems (see, for instance, McIntosh 1985 and Boucher 1998). The main reason for this is that environmental problems are historical and contextual: it is very unclear whether ecology can provide any unified conceptual frame for assessing, let alone solving them (Haila and Levins 1992). As Andrew Jamison points out in his essay in this volume (Chap. 16), environmental movements produced elements for a new cognitive praxis, but this developed at some distance from academic ecology.

The controversy between utilitarian management ("conservationists") and total preservation is still very much alive, the latter represented by the "deep ecologists" – but deep ecology is essentially a philosophical stream of thought with only weak connections to ecological research. Latour (2004) presents an interesting critique of the arguments of deep ecology by showing that they lean on a stark and untenable nature-culture dualism.

Debates on the principles of sustainable management of natural resources carry further the legacy of the conservation-preservation controversy. Ecological economics is an important newcomer in economic thinking about the environment. The practical relevance of the field is, however, uncertain. Economics is still burdened by the heritage of Walrasian welfare economics and its outrageously unrealistic background assumptions about human economic behaviour (see Chap. 28).

Economic thinking still has a long way to go before it takes ecology seriously (Dyke 1988, 1997).

Conservation biology is a rapidly growing applied discipline that had its origin in academic ecology. Originally, in the nineteenth century, nature conservation was viewed as a cultural duty, inspired by the Romantic movement. Ecological science got involved in nature conservation through an increasing consciousness among ecologists about the threat of human-caused extinctions. This arose mainly through the devastation caused by big-game hunting, basically in all parts of the world. The "community" of colonial hunters provided an important germinating ground for this concern, their role was important also in organizing the first international meetings and establishing the first organizations dedicated to nature conservation in the early 1900s (Adams 2004).

Around the mid-twentieth century, nature conservation was generally accepted as a governmental responsibility; the founding of the International Union for the Protection of Nature in Fontainebleau, France, in 1948 was an important landmark in this regard (the organization developed later into the International Union for the Conservation of Nature; see Holdgate 1999) (see also Chap. 17). As a governmental duty, conservation required a systematic approach: a precise answer was needed to the question, what to preserve and why? – of all the possible species, habitats and sites that are around. It was not sufficient to argue for the uniqueness of specific sites or species, as was the case in the first half of the twentieth century (Adams 1996 presents a good analysis of this transition in Britain). In practice, this meant comprehensive inventories of endangered species, and eventually also habitats. The initiative for the compilation of Red Data Lists came from non-governmental organizations, World-Wide Fund for Nature being a particularly active proponent, and the lists acquired a systematic form in the 1960s. The first organisms covered were vertebrates and ("higher") plants in single countries, then, since the early 1990s, the lists have become international and cover increasingly also invertebrates (see Mace et al. 1998).

It was the perception of an imminent danger of a human-caused extinction wave that triggered academic ecologists to participate in the conservation movement and to establish the new discipline of conservation biology – a scientific society and a journal carrying this name were established in the mid-1980s. Michael Soulé (1985) famously characterized conservation biology as a "crisis discipline". In other words, the new discipline is driven by a strong commitment to produce knowledge that is useful for practical conservation policy and management. The concept of biodiversity arose into a central position in conservation biology in the 1980s. The launching of the concept was described by David Takacs (1996) who documented the deliberately political nature of the process.

The concept of biodiversity is used by conservation biologists to convey their message about the necessity of nature protection to decision-makers and the general public. However, paradoxically, as the concept is scientifically difficult to concretize, a sole emphasis on biodiversity also alienates the public. In fact, the adoption of the concept presents us with a shade of boundary work done by conservation biologists; in the words of Takacs: "By promoting and using the concept of biodiversity, biologists hope to preserve much of the biotic world, including the dynamic processes that shape that world, while simultaneously appropriating for themselves the authority to speak for it, to define and defend it" (1996, p. 99).

The all-encompassing nature of the term biodiversity was articulated by Edward Wilson as follows: "So, what is it? Biologists are inclined to agree that it is, in one sense, everything" (1977, p. 1). By such wording, Wilson clearly wanted to underscore the importance of biodiversity, but the effect may be different. If biodiversity, indeed, is "everything", it cannot offer practical guidelines for nature conservation: we need criteria for drawing critical distinctions, it is impossible to preserve "everything". Biodiversity is an important notion, and it has triggered important practical research but it has also been paralyzed by old controversies, particularly by a perceived dualism between humanity and the rest of nature (Haila 1999, 2004). With time, it may turn out that the notion contributes mainly to a deeper understanding of an ethical duty to cherish living nature, rather than to concrete conservation programs.

The establishment of governmental bodies responsible for environmental politics in all industrialized countries around the turn of the 1970s gave rise to "environmental science" that includes all the most variable natural scientific research perspectives on environmental problems (Weale 1992 documents the link between policy and science at the early stage). Specific research traditions within ecology, conservation biology being the most firmly established among them, have obtained a firm position among the environmental sciences. However, as environmental research is primarily focused on specific and often very concrete environmental problems, different background traditions combine together in specific research projects in a context-sensitive fashion. The research is genuinely transdisciplinary.

References

Adams WH (1996) Future nature. A vision for conservation. Earthscan, London
Adams WH (2004) Against extinction. The story of conservation. Earthscan, London
Boucher D (1998) Newtonian ecology and beyond. Sci Cult 7:493–517
Cajander AK (1949) Forest types and their significance. Acta Forestalia Fennic 56:1–71
Clark CW (1989) Bioeconomics. In: Roughgarden J, May RM, Levin SA (eds) Perspectives in ecological theory. Princeton University Press, Princeton, pp 275–286
Clark LR, Geier PW, Hughes RD, Morris RF (1967) The ecology of insect populations in theory and practice. Methuen & Co., London
Commoner B (1971) The closing circle. Alfred A. Knopf, New York
Connolly WE (1993) Voices from the whirlwind. In: Bennett J, William C (eds) In the nature of things. Language, politics, and the environment. The University of Minnesota Press, Minneapolis, pp 197–225
Cuddington K (2001) The "balance of nature" metaphor and equilibrium in population ecology. Biol Philos 16:463–479
Dyke C (1988) The evolutionary dynamics of complex systems. A study in biosocial complexity. Oxford University Press, Oxford
Dyke C (1997) The heuristics of ecological interactions. Adv Human Ecol 6:49–74

Ehrlich Paul R. 1969. Eco-Catastrophe! Ramparts (September): 24–28
Elton CS (1966) The pattern of animal communities. Methuen, London
Forbes SA (1887) The lake as a microcosm. Bull Illinois Nat Hist Surv 15:537–550
Gieryn TF (1983) Boundary-work and the demarcation of science from non-science: strains and interests in professional interests of scientists. Am Sociol Rev 48:781–795
Gieryn TF (1999) Cultural boundaries of science. Credibility on the line. The University of Chicago Press, Chicago
Glacken C (1967) Traces on the Rhodian shore. Nature and culture in western thought from ancient times to the end of the eighteenth century. University of California Press, Berkeley
Golley FB (1993) A history of the ecosystem concept in ecology. More than the sum of the parts. Yale University Press, New Haven
Grove RH (1995) Green imperialism. Colonial expansion, tropical island Edens and the origins of environmentalism. Cambridge University Press, Cambridge, pp 1600–1860
Hacking I (1992) The self-vindication of the laboratory sciences. In: Pickering A (ed) Science as practice and culture. The University of Chicago Press, Chicago, pp 29–64
Hagen JB (1992) An entangled bank. The origins of ecosystem ecology. Rutgers University Press, New Brunswick
Haila Y (1992) Measuring nature: quantitative data in field biology. In: Clarke AE, Fujimura JH (eds) The right tools for the job. At work in twentieth-century life sciences. Princeton University Press, Princeton, pp 233–253
Haila Y (1998) Political undercurrents of modern ecology. Sci Cult 7:465–491
Haila Y (1999) Biodiversity and the divide between culture and nature. Biodivers Conserv 8:165–181
Haila Y (2004) Making sense of the biodiversity crisis: a process perspective. In: Oksanen M, Pietarinen J (eds) Philosophy of biodiversity. Cambridge University Press, Cambridge, pp 54–82
Haila Y, Levins R (1992) Humanity and nature. Ecology, science and society. Pluto Press, London
Hassell MP (1978) Anthropod predator-prey systems. Princeton University Press, Princeton
Holdgate M (1999) The green web. A union for world conservation. Earthscan, London
Holling CS (1992) The role of forest insects in structuring the boreal landscape. In: Shugart HH, Leemans R, Bonan GB (eds) A systems analysis of the global Boreal forest. Cambridge University Press, Cambridge, pp 170–195
Jax K (2002) Die Einheiten der Ökologie. Analyse, Methodenentwicklung und Anwendung in Ökologie und Naturschutz. Peter Lang, Frankfurt a. M
Judd RW (1997) Common lands, common people: the origins of conservation in Northern New England. Harvard University Press, Cambridge
Judd RW (2004) George perkins marsh: the times and their man. Environ Hist 10:169–190
Kingsland S (1985) Modeling nature. Episodes in the history of population ecology. University of Chicago Press, Chicago
Kohler RE (2002) Landscapes and labscapes. Exploring the lab-field border in biology. The University of Chicago Press, Chicago
Kwa C (1987) Representations of nature mediating between ecology and science policy. Soc Stud Sci 17:413–442
Latour B (2004) Politics of nature. How to bring the sciences into democracy? Harvard University Press, Cambridge
Levins R, Lewontin R (1985) The dialectical biologist. Harvard University Press, Cambridge
Lotka AJ (1956) Elements of mathematical biology (original: elements of physical biology, 1924). Dover, New York
Lovelock JE (1979) Gaia: a new look at life on earth. Oxford University Press, Oxford
Lowood HE (1991) The calculating forester: quantification, cameral science, and the emergence of scientific forestry management in Germany. In: Frangsmyr T, Heilborn JL, Rider RE (eds) The quantifying spirit in the eighteenth century. University of California Press, Berkeley, pp 315–342

Mace GM, Balmford A, Ginsberg JR (eds) (1998) Conservation in a changing world. Cambridge University Press, Cambridge
Margulis L, Sagan D (1997) Slanted truths. Essays on gaia, symbiosis, and evolution. Springer, New York
Marsh GP (1965) In: Lowenthal D Man and nature. Or, physical geography as modified by human action. Belknap Press, Cambridge [original in 1864]
McIntosh RP (1985) The background of ecology. Concept and theory. Cambridge University Press, Cambridge
Morowitz H (1968) Energy flow in biology. Biological organization as a problem in thermal physics. Academic, New York
Odum EP 1959 (1953) Fundamentals of Ecology. Saunders, Philadelphia
Schwarz AE (2003) Wasserwüste - Mikrokosmos - Ökosystem. Eine Geschichte der "Eroberung" des Wasserraumes. Rombach, Freiburg
Scudo FM (1971) Vito Volterra and theoretical ecology. Theor Popul Biol 2:1–23
Soulé, ME (1985) What is conservation biology? BioScience 11:727–34
Takacs D (1996) The idea of biodiversity. Philosophies of paradise. The John Hopkins University Press, Baltimore
Taylor PJ (1988) Technocratic optimism, H. T. Odum, and the partial transformation of ecological metaphor after world war II. J Hist Biol 21:213–244
D'Arcy WT (1992) On growth and form. The complete revised edition. Dover, New York
Tobey RC (1981) Saving the prairies. The life cycle of the founding school of American plant ecology. University of California Press, Berkeley, pp 1895–1955
Turner BLII, Clark WC, Kates RW, Richards JF, Matthews JT, Meyer WB (eds) (1990) The earth as transformed by human action. Global and regional changes in the biosphere over the past 300 years. Cambridge University Press, Cambridge
Walker B, Steffen W (eds) (1996) Global change and terrestrial ecosystems. Cambridge University Press, Cambridge
Weale A (1992) The new politics of pollution. The Manchester University Press, Manchester
Weiner DR (1988) Models of nature: conservation, ecology, and cultural revolution. Indiana University Press, Bloomington
Williams CB (1964) Patterns in the balance of nature. Academic, London
Williams M (1989) Americans and their forests: a historical geography. Cambridge University Press, Cambridge
Worster D (1985) Nature's economy. A history of ecological ideas. Cambridge University Press, Cambridge
Wynn G (2004) On heroes, hero-worship, and the heroic in environmental history. Environ Hist 10:133–152

Chapter 27
Border Zones of Ecology and Systems Theory

Egon Becker and Broder Breckling

Introduction

> In any field of knowledge, the study of interactions leads logically to the concept of system organization.
>
> Khailov (1964)

For many years now, proponents of systems theory and advocates of ecology have been engaged in an intense exchange of ideas, principles, concepts, theories, models, and methods.[1] The dynamics of this exchange have given a boost to both fields. Individual pioneers (such as Eugene and Howard Odum) and innovative organisations (such as the Santa Fé Institute) have spurred on this reciprocal concept transfer. Ecology and systems theory thus form two research fields which are only partially separated and which display powerful internal dynamics and borders that are permeable from several sides. Both research areas are, however, riddled with controversy. Heterogeneous discourses have developed in both fields, each of these discourses possessing its own specific cognitive and social order, along with the corresponding theoretical concepts and scientific practices to match. While each discourse has its own history, the history of the relationship between the two remains unwritten.

Since the 1930s, the exchange between systems theory discourse and ecological discourse has been shaped by the paradoxical idea of a "living system". The idea is a paradox because – at least prima facie – "systems" are not living entities and "living beings" are not systems. Bertalanffy's (1932, 1968) proposition generalises the understanding of organisms as open systems. As a general conceptual reference point,

---

[1] This process will henceforth be referred to as "concept transfer".

E. Becker
Institute for Social-Ecological Research (ISOE) Frankfurt/Main (Germany)
e-mail: e.becker@em.uni-frankfurt.de

B. Breckling (✉)
University of Bremen, Germany (Center for Environmental Research and Sustainable Technology, UFT) and University Vechta, (Chair for Landscape Ecology)

the idea of a "living system" enables the distinctive ideas and conceptual frameworks of each discourse to be represented without requiring an explicit consensus.

The transfer process is closely linked with the critical question of whether the transformation of ecology based on the adoption of systemic concepts and methods is beneficial or harmful. The converse question, concerning the influence of ecological approaches on systems theory, is rarely asked. The rise of systems theory in ecology (Odum 1971, 1983) is the subject of considerable controversy. Some scientists welcomed the opportunity to change ecology from a traditional, descriptive discipline into a modern, explanatory science (Fränzle 1998). For others, however, the shift in ecology towards a systemic paradigm represents a move towards the mechanisation of living beings, linked to a technocratic turn in ecological research (Trepl 1987). Despite such criticism, recent decades have witnessed an expansion of systems thinking and formal, mathematical modelling – along with the development of computer applications – in science as well as in numerous other fields.

Criticisms and Controversies

In much of the current criticism, it remains unclear what exactly is being criticised. This is because, as previously mentioned, neither systems theory nor ecology comprise a coherent monolithic entity. In both areas, quite different theoretical perspectives and practical orientations coexist. Both fields continue to develop independently, responding constructively to criticism and exchanging concepts with one another. Through this process, many different kinds of connections have developed between the branches of the two fields. Despite this fact, all participants in the debate tend to operate with assumptions of homogeneity, and Ludwig von Bertalanffy's particular version of systems theory is widely considered to be systems theory as such.

Accordingly, there are many stories in circulation about the relationship between systems theory and ecology. These accounts are embellished with detail and vivid descriptions of controversy, enlivened by anecdotes, and interwoven with other stories. This enables developments in the discourse to be interpreted retrospectively as plausible. By and large, proponents of systems theory in ecology provide accounts of *progress* of how the transition from a metaphysical holism to a modern systems ecology occurred, how it has succeeded in overcoming the ideological controversy of the 1920s and 1930s, and how ecology developed from a descriptive discipline of natural history into a mathematical, model-based theoretical and empirical science. Critics of this viewpoint regard the situation from the opposite perspective and tell stories of *loss*: of how our understanding of life has become mechanized, how quantification has erased our understanding of the qualitative nature of living relationships, how organisms' individuality and idiosyncrasies have vanished from the heart of ecology, how systems thinking has led to methodological constraints, and how a mechanical model of the world and technocratic approaches have generally taken over.

Such accounts generally serve to reinforce one's own position in discursive struggles for power and conceptual delimitation, especially if they are put forward by pioneers. Detailed studies in the history of science have been able to clarify the role of such stories and to point out the myopic perspective both of the stories themselves and of criticisms of the same (McIntosh 1985). The following is a list of corrections referring to some of the most striking prejudices surrounding the critical discourse, particularly with regard to the implied suggestion of homogeneity in systems thinking:

1. There is no homogenous systems theory, only a heterogeneous systems discourse with a plethora of concepts and methods, background philosophies and practical applications. There is an enormous conceptual and methodological gulf between the systemic concepts used in cybernetics and communications engineering on the one hand and the sociological systems theory of Talcott Parsons or Niklas Luhmann on the other; as well as between the methods applied in operations research or mathematical game theory and those applied in ecosystem research.
2. The transfer of concepts did not occur unidirectionally from systems discourse to ecology; rather, ecological notions and concepts were imported into systems discourse early on.[2] The importance of population dynamics in Lotka's (1925) and Volterra's (1926) mathematical formulations for systems theory is often downplayed by the theory's founding fathers.[3]
3. Cybernetics is not merely a special case of General Systems Theory, nor has cybernetics ever developed fully within systems theory. The idea of circularity as a fundamental principle turned into the notion of "circular causality" in the broad theoretical outline of cybernetics. Cybernetics thus acquired its own discursive order, shaped by questions concerning regulation and information transfer. However, cybernetics is not identical with its technical applications in automation and regulation technology and in computer science, as is often implied by critics. Cybernetic concepts are also found in medicine, psychology, political science and cultural anthropology.
4. The classical cybernetics of Wiener and Ashby has not stagnated; rather, it has evolved into a "second-order cybernetics". The new cybernetics attempts to include the "observer" into the system; it explores the importance of "positive feedback" and concentrates on nonlinear aspects. This makes it possible to comprehend self-organization and emergence through the lens of cybernetics as well. Second-order cybernetics has had a strong influence on the more recent systems discourse and has helped to prevent its decline. In addition, numerous modelling approaches have emerged more recently. Examples of these are to be found in neuroinformatics, artificial life and biorobotics.
5. A systems ecology centered on the analysis of energy flows is not the only systems theoretical approach in ecology. Gregory Bateson, for example, taking his

---

[2] It would be highly instructive from an historical point of view to examine exactly what role the ecologist Evelyn Hutchinson played in the development of early cybernetics (Taylor 1988; Heims 1991; Schwarz and Schwoerbel 2001; Pias 2003). At any rate, cybernetics' central concept of "circular causality" was certainly influenced strongly by Hutchinson.

[3] Apart from the ecological import of Lotka's ideas, his book also anticipated Ludwig von Bertalanffy's development of General Systems Theory in the 1950s. Bertalanffy rather ungenerously downplayed the similarities between his method and Lotka's, but Lotka had clearly set down the basic procedure of systems analysis first (Kingsland 1985, p. 26.).

lead from early cybernetics, drafted a systemic concept of ecology in which he particularly emphasized the importance of information flows and communication networks in ecosystems (Harries-Jones 1995). However, this concept was not taken up in biological ecology; instead, it was the analysis of bioenergetic transfers (Odum 1971, 1983) that largely became dominant here. Something akin to a "semiotic shift" is currently occurring in cybernetics, a shift that has now extended as far as biology and has adopted the ideas put forward by Bateson. A new research field – biosemiotics – is emerging, in which communication and signification within and among organisms is central (Hoffmeyer 2008).

The heterogeneity of systems science discourse repeatedly gives rise to a fundamental dispute over the ontological and epistemological status of ecological systems. This dispute continues to be a source of controversy in the debate over the systems concept most appropriate for ecology, and it influences the concept transfer between ecology and systems science. The problems involved in this dispute are discussed in more detail in the following sections. Behind the various controversies lurks the question of the relationship between the formal representation of a "system" and specific ecological conditions.

Real World and Abstract Systems

Neither in systems discourse nor in ecology is there agreement regarding what exactly systems are, much less how they should be perceived and described. Accordingly, different opinions abound with regard to how the term "system" should be understood and used. In ecology, these issues continue to ignite controversy of the "realism versus nominalism" kind, in which concepts are hotly contested along with differing understandings of systems and the various ontological and epistemological implications involved. The main arguments resemble topoi from the famous controversy between nominalists and realists in medieval philosophy, which was waged vehemently between the eleventh and the fourteenth century, and which erupted again in the twentieth century, particularly within analytical philosophy.[4]

The original conflict was conducted over the issue of whether general concepts – *the universals* (the "Good", the "Divine", "Man" or "Animal") – exist as such (outside the mind), or whether universals are a property of things and thus merely a name. In classical Greek philosophy the universals were understood as an idea or principle (Plato), or again as a form or archetype (Aristotle). In later debates, the epistemological status of the universals was discussed using ethical or mathematical examples in particular. An issue for the medieval dispute were, on the one hand, observed forms and principles of thought and, on the other, questions about the mode of existence of thoughts themselves. As a scientific discipline, biology was

---

[4] In the philosophy of mathematics and quantum theory, this debate was played out in the contrast between "Platonist" and "constructivist" views (Stegmüller 1978; Khlentzos 2004).

dragged into the dispute as early as the Middle Ages: are genera and species real, or are they just a product of classification-oriented thought processes? The *realists* followed Plato. For him, universal concepts existed as ideas in their own realm, beyond the earthly world of things. According to Plato, they can be discovered by human thought. Aristotle, by contrast, held that universals existed only in things or individual beings because only the individual could be "real". The medieval *nominalists* radicalized Aristotle's notions and ultimately considered the universals as merely a name for spoken generalizations or philosophical abstractions. The universals do not represent a specific ontological status. Only individual things and living beings are real.

In ecology, the issue at hand is whether ecological systems really exist, or whether they are constructed conceptually by science. For "systems realists" ecological systems do exist as ontological entities, and therefore they cannot be defined freely or according to human preferences; for "systems nominalists" (who are nowadays usually called "constructivists") ecological systems are scientific constructions, such as logical classifications, which can be interpreted in terms of genus as well as of species (Trepl 1987, p. 140). It is possible to identify three different positions which gradually emerged in the course of ecological research practice. These are (1) image naturalism, (2) analytical realism and (3) constructivist realism.

Position 1: "*Image naturalism*" (Abbildrealismus) is an extreme variant of realism. While rarely advocated as an epistemological position, it is widespread in scientific practice. Ecological systems are understood as *given objects* in the real world, comprising a diversity of individual organisms and species, complex patterns of interactions, and entangled processes. Such systems can be delimited, identified and named in both spatial and temporal terms, and empirical research can generate increasing amounts of knowledge about them. In this sense, physicists speak of the "planetary system" while ecologists identify particular forests, frog ponds or ant colonies as "ecological systems". Empirical data collection, carried out as extensively as possible by means of impartial observation in the field or in the laboratory, along with inductive generalizations of such observation, are considered to be the foundation of scientific knowledge. If this knowledge is regarded as a more or less accurate image, or depiction, of nature as it is, then what follows from the ontological option for systems realism is an epistemological option for what we called *image naturalism*. This way of thinking stems from the "naturalistic myth of the given" (Hesse 2002). "System" is, in this case, merely a general name for any composite and internally structured part of reality. In this respect, naïve image naturalism contains a nominalist ontology. To put the point somewhat paradoxically, it is a "nominalist realism". Whenever the basic principles of this position are subjected to epistemological analysis, it quickly becomes apparent that "image naturalism" is in urgent need of corrective adjustment.

Position 2: Such revisions tend to occur especially in those areas where ecological questions are the subject of empirical investigation and mathematical modelling. This leads to a content-focused *analytical realism*: from an ontological point of view, it is a form of *realism* to the extent that it is based on the hypothesis that ecological systems exist in and of themselves, independently of any external observation or description. Such realism is *analytical* because it acknowledges that, in research, particular aspects of an ecological reality are always accentuated and distinguished analytically from the environment. It is also recognized that knowledge acquired as a result of the research process is then represented as a "system". Exactly which aspects are emphasized depends on the research interests and questions pursued by the scientists. This type of empirical-analytical understanding dominates scientific research in general and is not specific to systems analysis in ecology.

Position 3: The specific implications of the systems approach are given greater emphasis in cases where the understanding of the system is developed methodologically rather than on the basis of content. In this case, "system" is recognized as the hallmark of a particular method. In other words, an empirical context is interpreted according to its internal organization and its relationship to other objects in the world. According to this view, the object of perception does not exist independently of the act of perceiving it. Rather, a *systems object* must first be "constructed" according to specific thought patterns, concepts and methods. Only after an appropriate theoretical model for this object has been developed can the empirical phenomenon be reconstructed as a "system". *Systems methodology* reveals the operations undertaken to construct the system in a logical and transparent way. In this respect, the corresponding systems theory can be interpreted in constructivist terms. In this analytical-constructive view of systems the term "system", in terms of linguistic logic, is a so-called "abstractor" through which abstract objects can be constructed and named. Thus, what we have here is a kind of *nominalism*, insofar as "system" refers to a methodologically generated abstraction. This *constructivist* or *model-based nominalism* represents a counter position to both systems views discussed above, image naturalism as well as analytical realism. Sukachev (1964) formulated an explicit and conceptually pointed distinction between the idea of an ecosystem as a theoretical model and the notion of biogeocenosis as the empirically accessible object of observation. Weidemann and Koehler (2004) are currently working on this distinction within the field of ecological succession.

A veritable arsenal of abstracting methods has been developed in systems discourse. Image naturalism considers these methods to be instruments for depicting ecological matters; analytical realism sees them as a way of isolating and connecting particular aspects of a phenomenon. A more exact analysis reveals that all methods enable abstractions to be formulated. "Systems" can be constructed as *epistemic objects*, as things that humans can and want to know about, and as items which are described using appropriate language (Rheinberger 1997). A similar phenomenon of multiple interpretations occurs in everyday life. A thing of nature, such as an apple tree, can, in this sense, become an epistemic object. It can also be taken as an economic object (the apples could be sold as a commodity or be used as a raw material to make cider) or as a cultural-aesthetic entity (someone could write a poem about the apple tree). The tree's role in perception and analysis shifts depending on the context epistemic, producing or contemplating – in which it appears. Similarly, in ecology, an area of grass could be classified systematically as a particular vegetation association, it could be seen as a pasture system or again it could be designated as a habitat type. In specific cases, an area covered with grass may even be of interest only as a basis for defining the parameters of spatial resistance that are used to understand the dispersal dynamics of other organisms.

This example shows that the mere classification of an object (such as "grass") as a "system" brings into sharp relief the question of what mode of existence "systems" actually have. In *General Systems Theory* (GST), systems are generally viewed as *models* of real (material-energetic or communicative-symbolic) relations. Despite the constructivist epistemology that dominates GST, these models continue to be re-interpreted in everyday research in a realistic manner. This is because, according to prevailing scientific understandings, the models represent knowledge generated methodologically about a domain of ecological reality. In order to avoid the confusion caused by this situation, one should explicitly distinguish between the operations related to the *construction* of systems and the *interpretation* of those operations as models; this rarely happens, however. What we are left with are idealizing abstractions on the one hand and concretizing interpretations on the other.

If no ontological distinction is made between the real and the ideal world (for example, by interpreting the ideal world as a mere reflection of the real one – or vice versa), then the meaning of abstractions cannot be properly comprehended. In this respect, the understanding of model construction held by analytical realism is ontologically unsatisfactory. Within systems discourse, however, an analytical realism can certainly be justified epistemologically: According to this perception, systems are models of real phenomena limited by time and space. It is the *difference between model and reality*, the modelling relationship, which becomes the key epistemological question. Is it even possible to conceive of and define "reality" without models?

In order to avoid systems terminology, one can designate the "reality" to be modelled as a "spatio-temporal context of phenomena", as a "specific area of experience", as a "specific unit of investigation", or as a "part of reality". These formulations all contain elements of a realistic ontology, which holds that a thing-in-itself can be directly accessed as an object of empirical and theoretical insight. The critical question of whether the object of investigation is given naturally or whether it is constructed as an epistemic object within systems discourse is rarely posed. We maintain the position that modelling constitutes a specific class of "epistemic objects": An ecological context is observed and described *as if* it was a system; whether or not it exists ontologically as such need not be decided. This situation becomes clearer if one compares different ways of observing and describing the "same" object, such as, for example, in one case as "landscape" – in the other as ecosystem. This level of reflected understanding goes beyond a constructivist "nominalism". "System" is not merely a name for a methodologically generated abstraction; systems are models of knowledge about ecological issues.

This position can be called *model-based constructivism*. However, it can only be adopted seriously if systems are regarded as abstract objects in an ideal world. These objects may be of a logical mathematical kind, and can be implemented, at least in principle, in a computer program; graphic, metaphorical and conceptual models, as far as they represent complex networks of interaction, can also be considered as ideal objects in this sense.

A Genealogy of Systems Concepts

Conceptual ambiguity and confusion are always prevalent when one not only focuses on individual organisms, but also considers larger "syn-ecological units" (populations, animal or plant associations, communities) that are comprised of individual organisms. Do the complex relationships between individual organisms exhibit characteristics that the individuals themselves do not possess? And can these "syn-ecological units" themselves be understood as "living systems"? Taking the individual organism as a starting point, it follows, by analogy, that syn-ecological units would necessarily be seen as a kind of "super-organism". Taking the system as a starting point, however, syn-ecological units would have to be conceived of as emerging systemic levels, not of a super-organism but rather taking the form of a hierarchically organized system. The problems raised here feed into a *fundamental controversy* over whether organisms and syn-ecological entities can even be analysed adequately at all using the concepts and methods developed within systems theory discourse.

Numerous controversies exist over the systems concept best suited to ecology. These controversies appear to exist independently of the question of principle concerning the ontological and epistemological status of ecological systems. In general, the dispute has focused on the definition of the concept of ecosystem per se (Breckling and Müller 2003). This dispute can be comprehended particularly well if one examines the way in which ecological discourse and systems discourse are linked together by the concept of "living systems". This enables the developmental trajectories of different *systems concepts* to be traced and their genealogics mapped. What comes to light in the process are a number of notable conceptual differences between the various systems ecologies. In ecology, all the entities between which a relationship exists are called an *ecological system*. Such a system may also comprise the relationships between individual organisms and other entities in their environment. A spatio-temporally localized biocenosis, i.e. a community of organisms, together with its abiotic context, is called an *ecosystem*. This notion sometimes expresses a naïve view of systems ontology in terms of what we have called image naturalism; more frequently, however, it goes hand in hand with analytical realism. It might be possible to argue for a type of systems metaphysics using this approach, but no systems theory in the modern scientific sense can be constructed on this basis. It is only when the relationship between model and reality is reflected upon that an analytical understanding of the system can be achieved. All serious systems concepts are therefore analytical by nature. In the version of model-based constructivism for which we are arguing here, the difference between model and reality constitutes the starting point for systems theoretical reflection.

We neither wish nor are we able to present here a complete *genealogy* of modern systems thinking in its entirety. Even if one were to restrict oneself to biology and ecology alone, such a genealogy would be confusing enough on account of being closely linked with philosophical controversies and the advancement of biological knowledge. In natural scientific systems discourse of the late nineteenth and the twentieth century, a variety of distinct systems concepts and understandings were developed and set up against one another. Nevertheless, serious attempts to order, classify, assess and synthesize heterogeneous systems concepts were undertaken early on within this discourse (Klir 1972). Ordering attempts using classical pairs of opposites (such as realist/nominalist, concrete/abstract, real/ideal, material/formal) predominated. However, it was frequently unclear if a distinction was being made between different *systems concepts* or between different *types of systems*. A more fruitful approach appears to us to consist in arranging all the different variants of a general systems concept not in terms of pairs of opposites but rather in terms of the *basic distinction* that goes with each variant. While each basic distinction implies different specific limitations, it also implies distinct ontologies and epistemologies:

- One common conceptual strategy originates from the distinction between *system and environment*. A pattern of interaction is defined as a system and is thus separated from the surroundings in which it is embedded. This distinction can be traced more or less as far back as to Plato. The modern genealogy of this concept began in the 1930s when organisms were defined as thermodynamically open systems in a flow equilibrium. This was, notionally, a physical model, which Prigogine and others were later to elaborate in relation to nonlinear flows, introducing the systems concept of a "dissipative structure" in the context of thermodynamically irreversible processes (Glansdorff and Prigogine 1971; see also Prigogine and Stengers 1981). The "system" was originally characterized as a *black box* with an input, throughput and output of materials and energy. The relationship between input and output can then be described in functional terms on the basis of the system's transformation functions. This is known, therefore, as a *functional systems concept*.
- Classical cybernetics pursues a different conceptual strategy. In this case, a distinction is made between *regulation and perturbation*. According to the cybernetic view, systems are units of regulation which, on account of positive and negative feedback loops ("circular causality") as well as communication networks, are operationally closed. This enables them to react to external and internal disturbances. The distinction implied here is one between system components and system states. Classical cybernetics examines changes of states in systems with structurally stabilized components. This can be called an *operative* systems concept.
- Another conceptual strategy was elaborated within nascent systems theory after World War II. This strategy is based on the distinction between *element and*

relation. Systems are defined abstractly as sets of elements, above which sets of relations exist (Hall and Fagen 1956). Such an understanding involving the theory of sets is based on a *structural* concept of the system and introduces, in true Platonist style, the greatest conceivable abstraction: systems are defined as genuine mathematical objects, that is, as sets of related objects that can therefore, in principle, be isolated.

The different conceptual strategies give rise to genealogical lineages that sometimes run in parallel but repeatedly intersect. New concepts may originate from these points of intersection. One such concept is "autopoiesis" (Maturana and Varela 1992), in which "living systems" are defined as being materially and energetically open, but operationally closed. All the system's elements and the relations between these elements are generated by internal operations. In ecology one finds variations of all the various distinctions and conceptual strategies and therefore also a range of systems ecologies based on very different conceptual foundations. In the research context, these different systems ecologies can be drawn upon constitutively to various degrees by different sub-disciplines – providing another starting point for criticism. Just two examples will be mentioned here. In their standard textbook, Begon et al. (1986) generally avoided the concept of ecosystem, preferring the notion of relationships between individuals, populations and communities that can be interpreted both qualitatively and formally. In contrast to this, Odum (1983) places the concept of ecosystems at the center of ecology and focuses on energy flows as the key unifying measure.

With the exception of naïve *image naturalism*, one can detect a certain proximity between the different conceptual strategies identified here and *analytical realism* on the one hand and *constructivist nominalism* on the other. The *functional systems concept* is largely interpreted realistically in ecology. In contrast to this, sociological systems theory advocates a programme of de-ontologization using a radically constructivist epistemology by generalizing the system/environment distinction.[5] The *operative systems concept* developed in cybernetics balks at a realist interpretation because it abstracts from the material realization of the system. Second-order cybernetics attempts to establish a radical epistemological constructivism. The *structural systems concept* can be interpreted in either realist or constructivist terms. In a realist interpretation, however, "set" is not understood in the strict mathematical sense but rather as a diverse array of concrete objects. In a Platonically inspired constructivist interpretation, which we prefer here, systems are abstract objects in an ideal world.

---

[5]This approach attempts to define all existing concepts in terms of operations and to relate them strictly to "communication" as a point of both departure and of reference (Luhmann 1997). Whether or not this has led to a renewal of sociology's explanatory foundations, as claimed by its adherents, has been the subject of heated debate for a number of years (cf. Merz-Benz and Wagner 2000; Clam 2002).

Concept Transfers

The dispute between realist and constructivist positions was understood for a long time as a debate over the empirical content of General Systems Theory (GST).[6] The latter was viewed first as a general *theory* of very different systemic contexts as a general theory of physical, organic, psychological and social "systems". This is still claimed to be the case in systems philosophy, although it has been disproved in practice. Methodologically speaking, GST provides numerous ways of formalising networks of interaction (Hammond (2003); however, GST methods are simply not capable of capturing the structurally variable networks appropriate to describing the complexity and variety of "living systems". Structurally variable networks are used in the context of object-oriented modelling, for example, as a way of representing networks of interaction between individuals that are temporally, spatially and structurally variable (individual-based modelling) (deAngelis and Gross 1992; Breckling 2004).

Bertalanffy had originally conceived of *General Systems Theory* as a system of ordinary differential equations. However, although GST was applied in this form to organic, psychological and social phenomena, it achieved only modest success. Only by extending the theory to include systems based on non-linear, partial differential equations is it possible to formalize the emergence of organizational patterns and of emerging, self-organizing structures and differentiation processes. Turing raised this possibility in 1952 in the context of chemical morphogenesis, and the approach was taken further by Meinhard and Gierer (1972) and Meinhard (1982).

However, all of this does not necessarily lead to the oft-cited conclusion that GST has failed as a *theory* and that it can only be employed as a meta-theory or as a systems philosophy. What has failed, however, is GST's claim to universality and, along with this, its ambition to unify science. Thus far, no *mathematical* model has been found that is applicable to all empirical sciences – nor, presumably, will one ever be found. The endeavour to develop a general *empirical* theory for vastly different areas of reality has also failed. Systems science gave up on both of these attempts at generalisation long ago. In this respect, criticism has yielded success. Where these claims to generality continue to be attacked, criticism is directed either towards certain visionary philosophical exaggerations or towards an object in history; it has no relation to contemporary scientific practice. Nevertheless, one endeavor that remains is the pursuit of empirically oriented *systems research* in very different spheres of reality, as well as the search for adequate integrative concepts and mathematical models. During the course of its ongoing development, *General* Systems *Theory* has dissolved into a multiplicity of *specialised* systems sciences, and it continues to exist in the form of a heterogeneous discourse with pronounced ontological and epistemological contradictions.

---

[6] In 1950 Carl G. Hempel had already rejected the empirical relevance of GST and labeled it as "a branch of pure mathematics" (Hempel 1951, p. 314-15). In later works he then argued against simulation methods, against the functional explanation, against suggestions of isomorphism, and against the emergence thesis (Hempel 1965). Müller (1996 p. 245ff) provides a summary of the dispute over the empirical content and explanatory power of GST.

By contrast, however, another aspect of early systemic self-understanding proved to be highly successful in methodological terms: General Systems Theory is a field of *concept transfers* and possesses an arsenal of *ideal systems models*. The founding fathers – especially Ludwig von Bertalanffy and Anatol Rapaport – justified the possibility of such transfers by pointing to the existence of *isomorphisms* in the empirical sciences. They also recognized the methodological importance of systems theory in stimulating concept transfers and monitoring them methodologically. Metaphors and analogies play an important heuristic role in this process. Thus far, little research has been carried out with regard to exactly what is transferred from one field to another in a concept transfer – and what epistemological consequences can be expected to follow. Early General Systems Theory emphasized the formal structural similarity of laws and thus concluded that *principles and models* could be transferred.

Nevertheless, this sort of transfer is in no way innocuous because, along with the principles and models, theories and concepts are also transferred, and this is exactly how new *epistemic objects* are constituted. The phenomena of one field come to be regarded *as if* they structurally resembled those of another field. Moreover, a transfer of *principles* always denotes a similar transmission of *categories* from one realm of a being to another, for example from organisms to machines, or from ecosystems to cities – or vice versa. The transfer of *models* may involve a transmission of model structures as well as a transmission of modeling techniques. If concept transfer is differentiated in this way, different premises, problems and mistakes come to light – as, indeed, do interesting solutions. Similarly, though, *criticism* of concept transfers should also be differentiated – something that rarely happens. It makes a difference whether concept transfer is viewed from a realist or a constructivist perspective. In the one case, the existence of strong isomorphisms must be assumed, and in the other, concept transfer adds to the arsenal of logical-mathematical objects that can be used as models.

Systems as Objects in an Ideal World

It is sometimes questioned whether mathematics can be applied to the highly complex phenomena of biology, the psyche, society or history. Despite significant refinements, say the doubters, current mathematical tools are not yet up to the task of addressing the demands posed, say, by hydrodynamics, subatomic particle physics and cosmology (Castoriadis 1981, pp. 178f.). The special feature of biology and sociology is not only the complexity of their objects but, above all, the peculiarity and structural variability of the various relationships involved, as well as the specific character of their processes. Even very high levels of complexity can be handled mathematically, though. Moreover, mathematics is not simply an "instrument" of empirical science but rather, a theory of abstract objects and relationships. Despite the skepticism that exists regarding the performance of mathematics in its current stage of development, scarcely anyone concludes that hydrodynamics, subatomic particle physics or cosmology cannot be represented in mathematical terms. However, such conclusions are put forward time and time again in relation to biology and ecology. To us, it seems more productive to make use of advances in mathematics and of the many mathematical i.e. formal approaches that are already well-known for the benefit of ecology. This could help in the process of identifying and filtering out coherent elements and patterns from among the mass of details and data.[7]

Formalizations founded on set theory set in motion a powerful trend towards abstraction. It becomes possible to abstract *logical-mathematical systems objects*, such as partial differential equations, graphs and topological structures, from descriptions of ecological circumstances. More recent computer-based modelling techniques (cellular automata, genetic algorithms, etc.) are used to model the features of systems with a large number of elements and relations (and therefore high numerical complexity). Attempts are regularly made, especially in popular representations, to describe the properties of such logical-mathematical systems objects in words, using general principles. Hence, one reads: "Ecological Systems are complex, hierarchical, dynamic and adapted to an environment." This is how, by means of multiple concept transfer, new verbal languages emerge for describing ecological circumstances: mathematical objects are interpreted using concepts from a biological context (cells, genetics, hierarchy, adaptation, self-organization, etc.) while descriptions of the dynamic behavior of these mathematical objects in everyday language then be used to describe ecological patterns. Mathematics becomes a rich source of metaphors.

Nevertheless, it is misleading to view mathematics as an exact "language" in which "the characters of the book of nature are written," as Galileo Galilei conjectured almost 400 years ago. Pure mathematics is a system of signs and symbols with which formal implications in abstract networks of relations can be discovered and represented. It turns into a language only when it is interpreted semantically as a *model* of an area of reality. The newly described circumstances can then also be represented mathematically. Hence, metaphor can turn again into mathematics. The characteristics of a particular mathematical object which fulfills the criteria for self-organisation can be approximated in everyday language, and from these words a new, verbal language for describing ecological features can be derived. Such a language facilitates the emergence of new and unusual distinctions and descriptions. It then becomes possible to clarify something that was previously shrouded in mystery (in this case, processes of self-organization).

---

[7]In research practice, and especially in the analysis of ecological data, this is happening more and more. In landscape ecology, the methods of fractal geometry have maintained their hold (Turner and Gardner (1991)), while population ecology uses object-oriented modelling methods (Breckling 2004). In ecological research, techniques such as Petri-Nets (Gnauck 2001) are available. Techniques from the fields of machine learning and data mining are applicable in structure identification and analysis of large data sets (Dzeroski et al. 1994; Dzeriski 1995). Admittedly, these techniques did not originate in pure mathematics; they do, however, utilise formal descriptions.

Systems Theory as a Theory of Logical-Mathematical Objects

Between World War I and World War II, the ancient metaphysical scheme of part and whole was displaced in many fields of science as a result of the influence of mathematical set theory (Cantor 1895), the theory of logical classes of abstraction (Frege 1893), the theory of logical types (Whitehead and Russell 1910), as well as quantum physics. The formal scheme of *element/relation/system* took its place. Thus, a development that had begun with Newton's mathematical formulation of classical mechanics and the corresponding claim to universality reached its climax: the substitution of traditional substance metaphysics by an ontology of relations.

Although ecology was well prepared for this type of categorial change, it was not ready for the processes of logical-mathematical abstraction. It was certainly ripe for modern systems thinking beyond atomism and metaphysical holism: since the second half of the nineteenth century, ecology has been concerned with concrete "relations" as causal patterns (connections, interactions, exchanges) between organisms and their biotic as well as abiotic environment. In doing so, ecology has focused either on the related elements (for example, plants and animals) or on "patterns of relations" (for example, food chains, relationships of competition, forms of co-existence). Yet ecology's systems concept remained realistic and biological: "populations" and communities (biocoenoses) describe actually existing ecological circumstances; they represent the systemic character of the latter as supra-individual entities.

Only slowly and only through a process of concept transfer from General Systems Theory did ecology adopt the presumptions of logical and mathematical abstractions. In the 1950s (nearly 50 years after the foundational mathematical works of Cantor, Frege, Hilbert, Russell and Whitehead), the systems concept was strictly formalized. Using methods taken from set theory, Mesarovic (1972) attempted to summarize the different understandings of systems that were circulating in his day (open and closed systems, multilevel systems, control and decision-making systems, etc.) and to turn them into an axiomatic presentation. These understandings were strongly oriented toward cybernetic notions of systems. The most influential explanation of an abstract systems concept based on set theory was that put forward by Hall and Fagen (1956). Their definition was short and concise: "A system is a set of objects together with relationships between the objects and between their attributes." (p. 18)

In our view, this definition is the most significant conceptual innovation in the genealogy of systems discourse: even today, it is still capable of unravelling the tangle of errors found there. Once the systems concept had been explicated using set theory, this opened the way for a construction of logical-mathematical systems objects. It linked systems discourse to the developments and insights of modern logic, mathematics and meta-mathematics – including the fundamental questions these entail. The fact that such links can be fruitful has been acknowledged more in cybernetics than in organism-based systems theory.

Hall and Fagen defined systems in a completely abstract way as a set of distinguishable elements between which a set of distinguishable relations exists. Systems defined in this way do not constitute anything real; instead, they are classes of abstraction. The "elements" of an abstract system are not identical to the actual "individuals" in a real sphere of reality. A "system", as defined by set theory, contains no hares or foxes, no beeches or oaks – nor does it contain any people. At best, certain characteristics selected by the defining criterion might appear in the system – and even then only in abstract form. This is because in modern set theory "elements" of a system are no longer defined by real objects, but rather by the qualities of equivalence classes. Two classes qualify as "equivalent" if an unambiguous relation exists between them. In the set theory definition of a system, the "elements" that are defined by abstraction classes are only of interest insofar as they constitute a "system-like" context.[8] The circularity of the definition is avoided when the elements are determined by the classes of relations that exist between them. In other words, systems are related sets.

If one follows the path laid out by the abstract systems concept expounded by set theory, one leaves behind the actual reality of ecological circumstances and enters an ideal world of abstract objects. Systems exist in this world alone. The set theoretical explication and associated processes of abstraction make it possible to conceive of systems through the formal means of logic and mathematics (for example, through the logic of relations, through systems of interconnected differential or difference equations, through topological patterns, through graphs or networks). As a result, systems discourse gives rise to an ideal world of logical-mathematical system objects. This ideal world is undoubtedly a human creation and no divine revelation. However, once it has been created through the work of abstraction, the objects within it exist – just like Platonic forms – eternally. The only condition for the existence of these timeless objects is that they be free of inner contradiction and that there be a logical or quantitative link between the different objects. The ordering of systems objects in an organized network of logical connections is continuously threatened, however, by the fact that the absence of contradiction cannot easily be proven. Gödel (1931) used a study of number theory to provide formal proof of the fact that no sufficiently complex logical or mathematical system can simultaneously be complete and free of contradiction. This is because systems always contain statements, whose degree of truthfulness cannot be decided using the conceptual and technical resources of the system itself. A system can either be complete or free of contradiction, but not both. Thus, in the ideal world of abstract objects, ideas exist whose truth or falseness remains unknown.[9] Human activity creates eternal objects in the ideal world, and logical contradiction accounts for their mortality.

---

[8] The Lotka-Volterra equations from the late 1920s applied this principle – and Verhulst's (1838) growth equation from the mid-1800s anticipated such a formalization. Set theory provides a consistent basis for the possibility of mathematically representing specific connections between related entities.

[9] One can illustrate a related scenario in hydrobiology: Fish can be organized in two categories: those that are cannibalistic and eat their own kind if they happen to become accessible, and those that do not do this. And then there are species-specific trophic preferences. There may be a category of predatory fish that feeds only on those other fish that do not consume their own fry. What would a fish of that category do if one of its own offspring appeared in front of it? This setting is an ecological disguise of Russell's famous barber paradox. The poor fish cannot take up either option without becoming entangled in a net of contradictions – consuming its fry would violate the condition of not consuming it, and not consuming it would qualify the fry for consumption...

For ecology it is perfectly natural to be able to call concrete entities such as frog ponds or forests "systems" and to be able to describe their systemic characteristics verbally or graphically. Such descriptions may then be used to generate systems in the strict, abstract sense. This is because it is always possible in principle to find a set of objects and a set of relationships that, in the sense of intuitive set theory, form a "whole". However, applying the term "system" to the description of any biocoenotic context always brings with it the danger of reification: systems generated through abstractions are then confused with real frog ponds and forests, and statements about the qualities and dynamics of mathematical systems objects are misunderstood as statements about real circumstances; this then provokes the complaint that abstract systems are not concrete.

In order to reduce the confusion caused by reification and conceptual ambiguity, we propose *that use of the term "system" should be restricted to logical-mathematical objects and strictly distinguished from real circumstances*. A similar argument is made whenever a clear distinction is made in ecology between a model and the reality represented. If these objects conform to the systems definition given by set theory, they can be called *mathematical systems objects*. Systems theory would thus become a *theory of abstract logical-mathematical systems objects*. We doubt that it will ever be possible to develop a general theory for all potential systems objects. What would be the substance of a theory which encompasses not only systems of partial differential equations but also neural networks, cellular automata, fractals, chaos theory and all manner of other phenomena? Such a theory would presumably be just as broad and general as Hall and Fagen's (1956) set theoretical explanation of systems concepts – and correspondingly lacking in substance as well.

Modelling as an Interpretation of Logical-Mathematical Objects

As far as systems research and ecology are concerned, a theory of mathematical systems objects is relevant to practical research only if it can be connected with ecological issues of practical relevance. Thus, it is wise to treat the relationship between system and reality in the same way as that between mathematics and physics: mathematical objects can be used to generate a descriptive language for physical objects because their components (terms, operators, functions, etc.) can be given semantic meaning, in other words, they can be made into referents for the characteristics of physical objects. The same procedure is largely followed in ecological modelling.

An elegant procedure was developed for this situation in Einstein's theory of relativity and in quantum theory: reference to reality is secured through a mathematical description of *observables* and thus through mathematical expressions for measurable variables. Modern physics is thus a science of the measurable world. Similarly, it is possible to relate abstract systems objects to ecological circumstances. For example, the abstract variables in the Lotka-Volterra differential equation can be interpreted in the context of population ecology. Nevertheless, the question remains as to how the particular population-ecological variables can actually be observed.

It is common to use counting methods to determine population sizes empirically, although this says nothing about the interrelations that exist between different organisms. In predator–prey models, for example, these interrelations appear in the form of rates, which are derived from changes in population sizes over time.

For the purposes of ecology, models emerge in this way that represent complex structural hypotheses, and these can, in turn, be tested empirically. The abstract mathematical systems object is therefore interpretively connected with a concrete physical or ecological object. This object may be a technical artifact (a thermostat, a servo mechanism…) or a natural context (populations of hares and foxes in a landscape). In the same manner, cybernetics attempt to construct technical realizations of mathematical systems objects (for example, electrical circuits for Boolean algebra and for logical operations, analogue computers for mathematical functions, etc.). However, this does not mean that logical-mathematical systems objects are merely abstractions of technical objects. Certain systems objects can be technically implemented; others are substantiated biologically.

In systems research there is frequently no strict distinction between a logical-mathematical systems object and its practical representation by a physical or technical object. Thus, it often happens that physical objects (for example, a compound pendulum or a diffusion process) or technical objects (such as a thermostat) are used directly as a model for other types of objects, without the systems object having first been represented mathematically in an abstract form. This implicitly sets up a structural similarity between the physical object and the ecological context under examination, a supposition that can lead to familiar physicalistic errors. Certain epistemological interests may be subtly reinforced in this way and may shape the process of model construction. One example of this would be to plan and design landscapes or conservation areas in terms of matter and energy flow models in order to fulfill politically motivated environmental objectives. Such technocratic notions are subject to criticism in ecology, and rightly so.

References

Begon M, Harper JL, Townsend CR (1986) Ecology. Individuals, populations and communities. Blackwell, Sunderland

Breckling B (2004) Individuenbasierte Modellierung – Entwicklungshintergrund und Anwendung einer ökologischen Modellierungsstrategie. In: Fränzle O, Müller F, Schröder W (eds) Handbuch der Umweltwissenschaften. Grundlagen und Anwendungen der Ökosystemforschung, V–2.3. Ecomed, Landsberg am Lech, pp 2–25

Breckling B, Müller F (2003) Der Ökosystembegriff aus heutiger Sicht – Grundstrukturen und Funktionen von Ökosystemen. In: Fränzle O, Müller F, Schröder W (eds) Handbuch der Umweltwissenschaften. Grundlagen und Anwendungen der Ökosystemforschung, II–2.2. Ecomed, Landsberg am Lech, pp 1–21

Cantor G (1895) Beiträge zur Begründung der transfiniten Mengenlehre, Mathematische Annalen XLVL, pp 481–512

Castoriadis C (1981) Durchs Labyrinth. Seele, Vernunft, Gesellschaft. Europäische Verlagsgesellschaft, Frankfurt a.M

Clam J (2002) Was heißt, sich an Differenz statt an Identität orientieren? Zur De-ontologisierung in Philosophie und Sozialwissenschaft. Universitätsverlag Konstanz, Konstanz

DeAngelis DL, Gross LJ (eds) (1992) Individual-based models and approaches in ecology: populations, communities and ecosystems. Chapman & Hall, London

Dzeroski S, Dehaspe L, Ruck B, Walley W (1994) Classification of river water quality data using machine learning, Proceedings of the 5th international conference on the development and application of computer techniques to environmental studies, vol 1. Computational Mechanics Publications, Southhampton, pp 129–137

Dzeriski S (1995) Inductive logic programming and knowledge discovery in databases. In: Fayyad G, Piatetsky-Shapiro G, Smyth P, Uthurusamy R (eds) Advances in knowledge discovery and data mining. MIT Press, Cambridge, pp 118–152

Fränzle O (1998) Grundlagen und Entwicklung der Ökosystemforschung. In: Fränzle O, Müller F, Schröder W (eds) Handbuch der Umweltwissenschaften. Grundlagen und Anwendungen der Ökosystemforschung, II–2.1. Ecomed, Landsberg am Lech, pp 1–24

Frege G (1893) Grundgesetze der Arithmetik. Hermann Pohle Verlag, Jena

Glansdorff P, Prigogine I (1971) Thermodynamic theory of structure, stability and fluctuations. Wiley, New York

Gödel K (1931) Über formal unentscheidbare Sätze der Principia Mathematica und verwandter Systeme I. Monatshefte für Mathematik und Physik. Akademische Verlagsgesellschaft, Leipzig 38. pp 173–198

Gnauck A (2001) Kontinuierlich oder diskret? Zur Verwendung von Petrinetzen in der ökologischen Modellierung, Simulationstechnik. Gruner Druck GmbH, Erlangen, pp 453–458, ASIM 2001 Paderborn

Hall AD, Fagen RE (1956) Definition of system: general systems. In: Bertalanffy LV, Rappoport A (eds) Year book of the society for the Advancement of General Systems Theory, Vol I, Ann Arbor, Mich, pp 18–28

Hammond D (2003) The science of synthesis: exploring the social implications of general systems theory. University Press of Colorado, Boulder

Harries-Jones P (1995) A recursive vision: ecological understanding and Gregory Bateson. University of Toronto Press, Toronto

Heims SJ (1991) The cybernetics group. MIT Press, Cambridge

Hempel CG (1951) General system theory and the unity of science. Hum Biol 23:313–322

Hempel CG (1965) Philosophy of the natural science. Prentice-Hall, Englewood Cliffs

Hesse H (2002) Zur Konstitution naturwissenschaftlicher Gegenstände insbesondere in der Biologie. In: Lotz A, Gnädinger J (eds) Wie kommt die Ökologie zu ihren Gegenständen? Gegenstandskonstitution und Modellierung in den ökologischen Wissenschaften. Peter Lang, Frankfurt a.M, pp 117–127

Hoffmeyer J (2008) Biosemiotics: an examination into the signs of life and the life of signs. Scranton University Press, Scranton

Khailov KM (1964) The problem of systemic organization in theoretical biology. Gen Syst 9:151–157

Khlentzos D (2004) Naturalistic realism and the rntirealistic challenge. MIT Press, Cambridge

Kingsland SE (1985) Modelling nature: episodes in the history of population ecology. University of Chicago Press, Chicago

Klir GJ (ed) (1972) Trends in general systems theory. Wiley-Interscience, New York

Lotka AJ (1925) Elements of physical biology. Williams and Wilkins, Baltimore

Luhmann N (1997) Die Gesellschaft der Gesellschaft. Suhrkamp, Frankfurt a.M

Maturana HR, Varela FJ (1992) Tree of knowledge: biological roots of human understanding. Shambul Publications, Boston

McIntosh RP (1985) The background of ecology: concept and theory. Cambridge University Press, Cambridge

Meinhard H (1982) Models of biological pattern formation. Academic, London

Meinhard H, Gierer A (1972) A theory of biological pattern formation. Kybernetik 12:30–39

Merz-Benz P-U, Wagner G (eds) (2000) Die Logik der Systeme. Kritik der systemtheoretischen Soziologie Niklas Luhmanns. Universitätsverlag Konstanz, Konstanz

Mesarovic MD (1972) A mathematical theory of general systems. In: Klir GJ (ed) Trends in general systems theory. Wiley-Interscience, New York, pp 251–269
Müller K (1996) Allgemeine Systemtheorie. Geschichte, Methodologie und sozialwissenschaftliche Heuristik eines Wissenschaftsprogramms. Westdeutscher Verlag, Opladen
Odum EP (1971 (1953)) Fundamentals of ecology. Saunders, Philadelphia
Odum HT (1983) Systems ecology: an introduction. Wiley, New York
Pias C (ed) (2003) Cybernetics: the Macy conferences 1946–1953, vol 1, Transactions. Diaphanes, Zürich
Prigogine Ilya, Isabelle Stengers (1981) Dialog mit der Natur. Piper, München
Rheinberger H-J (1997) Toward a history of epistemic things: synthesizing proteins in the test tube. Stanford University Press, Stanford
Schwarz AE, Schwoerbel J (2001) George Evelyn Hutchinson. In: Jahn I, Schmitt M (eds) Klassiker der Biologie, Band 2. Beck, München, pp 215–232
Stegmüller W (1978) Das Universalien-Problem. Wissenschaftliche Buchgesellschaft, Darmstadt
Sukachev VN (1964) Basic concepts. In: Sukachev VN, Dylis NV (eds) Fundamentals of forest biogeocoenology. Oliver and Boyd, Edinburgh
Taylor PJ (1988) Technocratic optimism, H.T. Odum, and the partial transformation of ecological metaphor after world war II. J Hist Biol 21:213–244
Trepl L (1987) Geschichte der Ökologie. Vom 17. Jahrhundert bis zur Gegenwart. Athenäum, Frankfurt a.M
Turing A (1952) The chemical basis of morphogenesis. Philos Trans R Soc Lond B 237:37–72
Turner MG, Gardner RH (1991) Quantitative methods in landscape ecology. Springer, New York
Verhulst PF (1838) Notice sur la loi que la population suit dans son accroissement. Corr Math Phys 10(113):121
Volterra V (1926) Variazione e fluttuazioni del numero d'individui in specie animali conviventi. Mem Acad Lincei 6(2):31–113
von Bertalanffy L (1932) Theoretische Biologie, Band I: Allgemeine Theorie, Physikochemie, Aufbau und Entwicklung des Organismus. Gebrüder Borntraeger, Berlin
von Bertalanffy L (1968) General system theory: foundations, development, applications. George Braziller, New York
Weidemann D, Koehler GH (2004) Sukzession. In: Fränzle O, Müller F, Schröder W (eds) Handbuch der Umweltwissenschaften, Grundlagen und Anwendungen der Ökosystemforschung III, 2–1. Ecomed, Landsberg am Lech, pp 1–50
Whitehead AN, Russel B (1910) Principia Mathematica I. Cambridge University Press, Cambridge, UK

Chapter 28
Economy, Ecology and Sustainability

John M. Gowdy

Introduction

The umbrella covering the various pieces of economic theory is called *welfare economics*. It provides the basic framework for applying the tools of economics to policy including cost-benefit analysis of environmental policies and economic models of sustainability. Welfare economics provides the basic worldview of economists; giving answers to fundamental questions regarding the ultimate purpose of economic activity and the best policies to promote human well-being. For most of the twentieth century economic theory was dominated by a type of welfare economics called *Walrasian* economics, named after the 19th century political economist Léon Walras. The cornerstone of Walrasian economics is a theory of human preferences embodied in economic man or *Homo economicus*. Starting with the assumption that social welfare can be evaluated by summing the preferences of isolated, selfish individuals, Walras and his followers constructed a mathematical model of economic general equilibrium that defined the optimal allocation of scarce resources among alternative ends (Gowdy 2009).

Today, welfare economics is undergoing a revolution fundamentally changing the way economists see the world. Walrasian welfare economics is being replaced by a new experiment-based economics that recognizes the social and biological context of decision-making and the complexity of human behavior. The current sea change in economic theory offers a unique opportunity for ecologists and economists working together to move mainstream economic theories and policies toward a science-based approach to environmental policy and sustainability.

Central to the questions of sustainability and the interface between economics and ecology is the notion of value. In the Walrasian system, the question of how much a particular feature of the natural world is worth is simply how much people are willing to pay for it. Total social value is equal to the sum of human preferences as revealed by choices made under the behavioral assumptions embodied in *Homo economicus*.

J.M. Gowdy (✉)
Rittenhouse Professor of Humanities and Social Science, Department of Economics, Rensselear Polytecnic Institute, Troy 12180, New York, USA
e-mail: johngowdy@earthlink.net

Preferences are assumed to be stable, consistent, and independent of the preferences of others. The arena of choice is the market economy so that utility, or well-being, is equated with the consumption of (properly priced) market goods. If individuals rationally allocate their limited income to choose the goods that give them the greatest utility, then the sum of the value of market goods, including environmental goods, is a good indicator of social welfare. Sustainability in this framework means sustaining the capital stock necessary to produce economic goods and services.

Although the Walrasian framework has been discarded by leading mainstream theorists, including many recent Nobel Prize winners, it still dominates economic textbooks and underlies the day-to-day policy recommendations of most economists. Its assumptions lie behind cost-benefit analyses of environmental policies. It is also defended by many ecologists who are perhaps unaware of the core assumptions of the system and their implications for environmental valuation and policy. For these reasons it is useful to review the essential features of the Walrasian system.

The Walrasian System

The first building block of the Walrasian system is to establish that free exchange will lead to *Pareto efficiency* in a pure barter economy. Individuals with a predetermined amount of commodities are allowed to directly and freely trade valuable goods with each other and Pareto efficiency is achieved when no further trading can increase the well-being of one person without decreasing the well-being of another. The second building block of the Walrasian system is the demonstration that if market prices correctly reflect individual preferences a perfectly competitive market economy will lead to Pareto efficiency. That is, competition in free markets will exactly duplicate the result of direct negotiations and exchange in a barter economy. The third and final piece of the system is the recognition that the prices of market goods may be distorted for a variety of reasons. These reasons include the broad categories of externalities, market power, and public goods. In these cases governments have a legitimate role to play in correcting the failures of markets to establish the proper value (price) of things like environmental services.

To summarize the Walrasian system

1. Unfettered bartering by autonomous agents with stable preferences will lead to Pareto efficiency – a situation in which no further trading can make one person better off without making someone else worse off.
2. If prices correctly reflect consumer preferences then competitive markets are always Pareto efficient. Free markets will exactly duplicate the results of a direct barter system.
3. When necessary, enlightened government intervention can adjust market prices so that the best Pareto efficient outcome can be established.

These three building blocks provide the worldview of most economists. They are valid only if all the assumptions defining *Homo economicus* and perfect competition are met.

The Value of the Environment

The ultimate source of value in the Walrasian system is the preferences of Homo economicus whatever these preferences might be and however they are formed. Key implications for environmental analysis are: (1) the value of any environmental feature is the sum of the preferences of self-regarding, autonomous individuals, (2) all environmental features are treated as commodities on an equal footing with all other market goods, (3) environmental features are subject to trade and substitution with market goods, (4) if society is misusing a particular environmental feature this indicates solely that its price is incorrect and should be corrected by government action, and (5) since preferences are the ultimate source of value and that they are stable, the major task of environmental economists is to uncover the "true" preferences for environmental goods so that the proper price may be assigned and the correct Pareto efficient outcome may be achieved.

The Walrasian model is the basic framework for neoclassical models of sustainability (Stavins et al. 2002; Arrow et al. 2004). The only difference is that, instead of focusing on the optimal allocation of goods among individuals, the model is used to describe the optimal allocation of resources across generations to sustain per capita income. This optimal allocation of economic goods through time, including environmental features, is made by an average person or "representative agent" making self-regarding decisions at a particular point in time. Sustainability is cast as an ahistoric, static problem of resource allocation by an omniscient agent having prefect knowledge of future resource availability and future technology (see for example, the economy/climate model of Nordhaus 2001).

The Revolution in Welfare Economics

Recent theoretical and empirical advances within mainstream economics have undermined the model of self-regarding rational economic man, the lynchpin of Walrasian theory and the foundation for traditional environmental valuation and policy. First of all, the basis for making welfare comparisons using *Homo economicus* has been shown to be theoretically untenable. Identifying and correcting market failure requires a consistent framework for comparing alternative economic states and the Walrasian system does not provide such a framework (Gowdy 2004, 2005). To give one example, the so-called Boadway paradox shows that if the starting point is an economic state A and another state B is Pareto superior to this starting point A, it may also be possible that if state B is the starting point then state A is Pareto superior. This is called "preference cycling" and according to Boadway (1974, p. 926) this means that when comparing alternative policies, the one with the largest net gain is not necessarily superior. The upshot of the theoretical work in welfare economics is that no logically consistent comparisons of economic states can be made without interpersonal comparisons of utility. Useful summaries of this theoretical work are given by Bowles and Gintis (2000) and Chipman and Moore (1978).

The second nail in the coffin for the Walrasian system comes from empirical economics including behavioral economics and game theory. Research in game theory and experimental economics has demonstrated that the model of *Homo economicus* almost invariably yields false predictions (Gintis 2000). A widely-known example is the Ultimatum Game in which a proposer offers one of two players a sum of money and instructs that player to share it with the second player. The second player can either accept the offer or reject it in which case neither player gets anything. *Homo economicus* should accept any positive offer. For example, if the first player gets $100 and offers the second player $1, the second player should accept it because more is always preferred to less. Results from the ultimatum game, however, show that offers under 30% of the total are usually rejected because they are not "fair". The majority of proposers offer between 40 and 50% of the total (Nowak et al. 2000). These results have held up even when played with substantial amounts of real money—in one experiment an amount equal to 3 months wages (Fehr and Tougareva 1995). Game theory results show that, in a variety settings under a variety of assumptions, non-selfish motives are a better predictor of behavior than the purely selfish motives embodied in *Homo economicus*.

The theoretical structure of Walrasian economics has been shown to be internally inconsistent and its basic assumptions have been empirically demonstrated to be a poor guide to real human behavior. The demise of a system that dominated economics for most of the twentieth century presents a real opportunity for economists working with ecologists to formulate new science-based concepts of environmental valuation and sustainability (see also Chap. 26). Now is an opportune time for ecologists to bring new ideas about sustainability to the economic table.

New Directions for Environmental Valuation and Policy

Ecologists have made many important contributions to the sustainability debate, but three concepts are particularly important. First is the simple observation that humans are biological creatures that evolved within Earth's ecosystems. Humans have a variety of basic material, social, and biological needs and sustainability policies must, at a minimum, insure that the support systems for meeting these needs are maintained. Second is the importance of hierarchy in understanding complex systems. The human economy is part of an interrelated hierarchical system of social, ecological, and geological order. The current policy focus on one level of the hierarchy, the material economy, while ignoring the other levels, will eventually have negative consequences for sustaining the human species. The third idea is the recognition of the central role of energy flows in living systems. Ecologists have long emphasized the key role entropy plays in the evolution and organizational structure of ecosystems. Great strides have been made by ecological economists in applying these ideas to the analysis of human societies but there is much more to be done.

Basic Needs and Human Well-Being

A number of economists are returning to the roots of economics to once again define utility as well-being or happiness, not simply income or consumption. This forces the consideration of social and biological needs as well as culturally created material requirements. A striking result of happiness research is that, after a certain level, increasing incomes do not lead to greater happiness. For example, real per capita income in the US has increased sharply in recent decades but reported happiness has declined (Blanchflower and Oswald 2000). Similar results have been reported for Japan and Western Europe (Easterlin 1995). Studies of individual life histories also show a lack of correlation between increases in income and increases in happiness (Frey and Stutzer 2002). Again, the most widely-used indicator of sustainable well-being is per capita consumption yet the evidence overwhelmingly suggests that this is a poor measure.

What makes people happy? Behavioral experiments have identified key factors positively influencing well-being. These include health (especially self-reported health), close relationships and marriage, intelligence, education, and religion (Frey and Stutzer 2002). Age, gender and income also influence happiness, but not to the degree once thought. Security seems to be a key element in happiness implying that large welfare gains would come from increasing individual security through expanding health insurance, old age security, employment, and job security. Richer social relationships generally make people happier implying that welfare gains may be obtained from increased leisure time, and more public spending on social, environmental, and recreational infrastructure.

What are the implications of a policy focus on individual well-being for environmental sustainability? There is some evidence that when individuals are more secure financially (not necessary wealthier) they are more likely to care about the well-being of future generations and the well-being of the environment (Rangel 2003). Evidence suggests that focusing policy on direct measures of well-being, rather than income, would not only make people happier, it would also reduce consumption, place a higher value on environmental features, and place a greater emphasis on the well-being of future generations – in short, they would help put us on the path to a more sustainable society.

Focusing policy on basic needs rather than per capita consumption would have important positive implications for sustainability (Corning 2000). But even if sustainable welfare policies are based on scientifically measured individual "preferences" this leaves us with the problem that it may not insure the preservation of the life support systems of the planet. Examples abound of societies that apparently worked well in satisfying the preferences of their citizens but ended in ecological collapse (Tainter 1998). Identifying the factors contributing to human well-being is critical but we also need to understand the place of human society within the larger biophysical world. We need to expand the focus on individual well-being to include the well-being of human societies as ecological units.

The Sustainability of Hierarchical Systems

Economics focuses on the individual. But in the evolution of ecosystems individuals, even individual species, count for little. I have argued elsewhere that the value of the environment should be seen in hierarchical context (Gowdy 1997). Using biodiversity as an example and moving from higher to lower levels, biodiversity has direct market value that can be expressed in monetary units, it has social value which can be partially quantified using survey techniques, and it has ecosystem value which can described and modeled but not fully quantified. Focusing solely on market value ignores the importance of biodiversity to sustaining the human species. Leading biologists warn that the expansion of human economic activity has put us in, or on the brink of, the sixth major extinction episode in 600 million year history of multi-cellular animal life. Loosing a substantial portion of existing species cannot help but have negative consequences for the well-being of *Homo sapiens*.

Energy Flows and Sustainability

The analysis of sustainability by economists has been astonishingly ahistorical. Sustainability is typically examined in a framework devoid of institutional or evolutionary context. More promising is a growing body of work by ecologists and social scientists on the interrelationships between resources, complexity, and problem solving institutions (Allen et al. 2001; Ostrom 2009). A critical factor in the sustainability of diverse ecological systems, including human systems, is the availability and organization of energy flows.

Energy analysis has been a central concern of ecology for some time (Odum 1971). The critical role energy plays in economic systems has been addressed by some economists (Georgescu-Roegen 1971). Recent interdisciplinary by biologists and social scientists has focused on the different organizational traits that emerge from different patterns of energy use (Tainter et al. 2003). Among other things, the "energy return on investment" (EROI) (Hall et al. 1972) is a critical variable that fundamentally influences living systems. Systems with a high EROI look very different from systems with low energy rates of return. Common patterns have been identified in such diverse systems as fungus-farming ant colonies, beaver colonies and the Roman Empire (Tainter et al. 2003). In human systems, integrated political, religious and ideological institutions evolve under the influence of particular energy sources. The institutions and economic structure of the industrial North changed dramatically during the last 200 years under the overwhelming influence of inexpensive fossil fuel energy. The transition to a sustainable society will require overcoming the inertia of political and economic institutions that evolved in the context of cheap fossil fuel energy.

Conclusion

Economics and biology have a long history of the mutual exchange of ideas. The idea of natural selection came to both Darwin and Wallace from the writings of the political economist Thomas Malthus. Economics and biology share a common subject matter insofar as both fields deal with complex, hierarchical and evolving systems. After decades of being "Newtonian" and "Hermetic" to use E.O. Wilson's (1998) characterization, economics is in the process of transforming itself into an empirical science whose basic assumptions are consistent with known facts in psychology, biology and the physical sciences. The coming decades should be a fruitful time for ecologists and economists to work together to put sustainability policies on a sound scientific footing.

References

Allen T, Tainter JA, Pires C, Hoekstra T (2001) Dragnet ecology just the facts, Ma'am: the privilege of science in a postmodern World. BioScience 51:475–485

Arrow K, Dasgupta P, Goulder L, Daily G, Ehrlich P, Heal G, Levin S, Mäler K-G, Schneider S, Starrett D, Walker B (2004) Are we consuming too much? J Econ Perspect 18:147–172

Blanchflower D, Oswald D (2000) Well-being over time in Britain and the U.S.A., NBER Working Paper 7481. National Bureau of Economic Analysis, Cambridge

Boadway R (1974) The welfare foundations of cost-benefit analysis. Econ J 84:926–39

Bowles S, Gintis H (2000) Walrasian economics in retrospect. Q J Econ 115:1411–1439

Chipman J, Moore J (1978) The new welfare economics 1939-1974. Int Econ Rev 19:547–584

Corning P (2000) Biological adaptation in human societies: A basic needs approach. J Bioecon 2:41–86

Easterlin R (1995) Will raising the incomes of all increase the happiness of all? J Econ Behav Org 47:35–47

Fehr E, Tougareva E (1995) Do high stakes remove reciprocal fairness evidence from Russia. Manuscript, Department of Economics, University of Zürich, Zürich

Frey B, Stutzer A (2002) Happiness and economics: how the economy and institutions affect well-being. Princeton University Press, Princeton

Georgescu-Roegen N (1971) The entropy law and the economic process. Harvard University Press, Cambridge

Gintis H (2000) Beyond *Homo economicus*: evidence from experimental economics. Ecol Econ 35:311–322

Gowdy J (1997) The value of biodiversity: markets, society, and ecosystems. Land Econ 73:25–41

Gowdy J (2004) The revolution in welfare economics and its implication for environmental valuation. Land Econ 80:239–257

Gowdy J (2005) Toward a new welfare foundation for sustainability. Ecol Econ 53(2005):211–222

Gowdy J (2009) Microeconomics old and new: a student's guide. Stanford University Press, Stanford

Hall C, Cleveland C, Kaufman R (1972) Energy and resource quality: the ecology of the economic process. University of Colorado Press, Niwot

Nordhaus W (2001) Global warming economics. Science 294:1283–1284

Nowak M, Page K, Sigmund K (2000) Fairness versus reason in the ultimatum game. Science 289:1773–1775

Odum HT (1971) Environment, power, and society. Wiley-Interscience, New York

Ostrom E (2009) A general framework for analyzing sustainability of social-ecological systems. Science 325:419–422

Rangel A (2003) Forward and backward generational goods: why is social security good for the environment? Am Econ Rev 93(3):813–834

Stavins R, Wagner A, Wagner F (2002) Interpreting sustainability in economic terms: dynamic efficiency plus intergenerational equity. Resources for the Future Discussion Paper 02-29, Washington, DC

Tainter J (1988) The collapse of complex societies. Cambridge University Press, Cambridge

Tainter J, Allen T, Little A, Hoekstra T (2003) Resource transitions and energy gain: Contexts of organization. Conserv Ecol 7(3):4 [online at http://www.consecol.org/vol7/iss3/art4]

Wilson EO (1998) Consilience. Knopf, New York

Picture Credits

Chapter (10)

Fig. 10.1: Ernst Haeckel's ideas about the subdivisions of zoology. (a) From Haeckel 1866 (Vol. 1, p. 238), (b) from Haeckel 1902, p. 29). See text for explanation.

Chapter (19)

Fig. 19.1 (a): "The three stages of limnological research" (Die drei Stufen der limnologischen Forschung; Thienemann 1925, p. 680).

Fig. 19.1 (b): "The three stages of limnological research" ("Die drei Stufen der limnologischen Forschung", Thienemann 1935, p. 18).

Fig. 19.1 (c): "The three stages of ecology" (Die drei Stufen der Ökologie; Thienemann 1942, p. 325).

Fig. 19.2: Burckhardt uses the depth ordinate to depict the distribution of zooplankton over time. What makes the figure rather puzzling for contemporary, visual practice is the fact that he includes several sites in the same coordinate space without marking them (1900, p. 424).

Fig. 19.3 (a): Depicted here are the "milieu needs" of *Holopedium gibberum*, a very common zooplankter which Thienemann had described in detail in the journal *Zoomorphology* in 1926 (Thienemann 1927, p. 43).

Fig. 19.3 (b): Hans Utermöhl offers a different, more precise representation of a diatom (*Cyclotella comta*), speaking of a "partial ecological spectrum" (ökologisches Teilspektrum) while commenting that the species need not necessarily occur at every line or level. "However, they may occur there", writes Utermöhl in his *Limnologische Phytoplankton-Studien* of 1925.

Chapter (21)

Fig. 21.1: Geographical distribution of scientific societies which practiced natural history.

Fig. 21.2: Number of authors of floristic books and catalogues (1800–1914)

Fig. 21.3: Number of naturalist's publications of 28 learned societies in the "province".

Fig. 21.4 (a–e): Basic tools of the botanising naturalist. (a) Herborization box, (b) and (c) tools for digging and breaking, (d) "échenilloir" et "sécateur", (e) knapsack (Figures taken from Bernard Verlot 1879. Le Guide du botaniste herborisant. - Paris:

J.-B. Baillière et fils; Guillaume Capus 1883. Guide du naturaliste préparateur et du voyageur scientifique, ou Instructions pour la recherche, la préparation, le transport et la conservation des animaux, végétaux, minéraux, fossiles et organismes vivants. 2e édition, entièrement refondue par A.-T. de Rochebrune, avec une introduction par E. Perrier. Paris: J.-B. Baillière et fils).

Chapter (22)

Fig. 22.1: Celso Arévalo (right) and an unidentified researcher at the "Estación de Biología Marítima" in Santander, circa 1905 (Courtesy of María Teresa Arevalo, Madrid).

Fig. 22.2: An employee of the "Instituto General y Técnico de Valencia" shows part of the limnological equipment of the "Laboratorio de Hidrobiología", circa 1915 (Courtesy of María Teresa Arévalo, Madrid).

Fig. 22.3: Classification of vegetation types proposed by Emilio H. del Villar in his book "Geobotánica" (Villar 1929).

Fig. 22.4: José Cuatrecasas (left), a local guide, and botanist Daniel Sans (right) at Sierra Magina (Jaén, Spain), where Cuatrecasas initiated his ecological investigations in the early 1920s (Courtesy of the Institut Botànic de Barcelona).

Glossary

This glossary has been specifically selected for this volume. So-called main concepts are marked with an asterisk. Main concepts structure the problems and perspectives of a field of knowledge, in general they are rather undefined and open terms that attract and organize other concepts. They need not to be exclusively part of the ecological vocabulary.

A full catalogue of concepts is given on the website http://www.hoekweb.tu-darmstadt.de/. This catalogue is still work in progress but albeit gives an impression of the conceptual universe of ecology. It was put together by Kurt Jax and Astrid Schwarz going through the ecological literature of different scientific styles and languages. Especially textbooks, but also dictionaries of several types, were taken into account as well as articles dealing with particular concepts and their role in the scientific community. Subsequently the catalogue was circulated among a number of colleagues and expanded according to their comments and propositions.

Concepts	Begriffe	Concepts
* adaptation	Adaptation	adaptation
assemblage/assembly	Gefüge	assemblage
association	Assoziation	association
autecology	Autökologie	autécologie
* biocoenosis	Biozönose	biocénose
biodiversity	Biodiversität	biodiversité
biogeoc(o)enosis	Biogeozönose	biogéocénose
biome	Biom	biome
biosystem	Biosystem	biosystème
* biotope	Biotop	biotope
border zone	Grenzbereich	zone frontière
* boundary	Grenze	limite
* carrying capacity	Umweltkapazität	capacité de charge
* climate change	Klimawechsel	changement climatique
climax	Klimax	climax
coevolution	Koevolution	coévolution
* coexistence	Koexistenz	coexistence

(continued)

Concepts	Begriffe	Concepts
* community	Lebensgemeinschaft	communauté
* competition	Konkurrenz	compétition
competitive exclusion principle	Konkurrenz-ausschlussprinzip	principe d'exclusion compétitive
* complexity	Komplexität	complexité
* conservation	Naturschutz	Conservation/protection de la nature
* disturbance	Störung	perturbation
* diversity	Diversität	diversité
* ecosystem	Ökosystem	écosytème
ecotone	Ökoton/Saum-/Randbiotop	écotone
* energy	Energie	énergie
* environment	Umwelt, Umgebung	environnement
* equilibrium	Gleichgewicht	équilibre
food chain	Nahrungskette	chaîne alimentaire/ trophique
food pyramid	Nahrungspyramide	pyramide alimentaire
* food web	Nahrungsnetz	réseau trophique
habitat	Habitat	habitat
* heterogeneity	Heterogenität	hétérogénéité
holism-reductionism	Holismus – Reduktionismus	holisme – réductionisme
image of nature	Naturbild	représentation de la nature
individual	Individuum	individu
uncertainty	Unsicherheit	incertitude
* lake type	Seentypen	type de lac
* landscape	Landschaft	paysage, écocomplexe
law	Gesetz	loi
* life cycle	Lebenszyklus/Entwicklungszyklus	cycle biologique
medium	Medium	moyen
model	Leitbild	modèle
model (logistic, exponential, dynamic)	Modelle (logistisch, exponentiell, dynamisch etc.)	modèle (logistique, exponentiel, dynamique)
natural history	Naturgeschichte	histoire naturelle
* natural selection	natürliche Selektion	sélection naturelle
* niche	Nische	niche
* organism	Organimus	organisme
patchiness	Verteilungsmuster	Patchiness/mosaïque/ disposition en mosaïque
pattern	Muster	patron
political ecology	politische Ökologie	écologie politique
* population	Population	population
population ecology	Populationsökologie/ Demökologie	démécologie
* production	Produktion/ Produktivität	production
reductionism-holism	Reduktionismus – Holismus	réductionisme – holisme

(continued)

Concepts	Begriffe	Concepts
* reproduction	Reproduktion	reproduction
resilience	Resilienz	résilience
* resource	Ressource	ressource
restoration	Restauration	restauration
* scale	Maßstab	échelle
simulation	Simulation	simulation
* stability	Stabilität	stabilité
stochastic processes	stochastische Prozesse	processus stochastiques
* succession	Sukzession	succession
superorganism	Superorganismus	superorganisme
* sustainable development	Nachhaltigkeit/ nachhaltige Entwicklung	développement durable
symbiosis	Symbiose	symbiose
synecology	Synökologie	synécologie
* vegetation type	Vegetationstyp	type de végétation

Author Biography

Astrid Schwarz is senior researcher at the Technische Universität Darmstadt and the University of Basel. She was trained both in philosophy and biology. After finishing her diploma in experimental aquatic ecology (France, Germany) and her PhD in the history and philosophy of biology, she got a post-doc grant at the MSH Paris (Maison des Sciences de l'Homme). From Paris she came to Darmstadt and joined the Institute of Philosophy. Between October 2006 and September 2007 she was a fellow at the Center for Interdisciplinary Research (ZIF Bielefeld). Her field is the philosophy and cultural history of science and technology. Her main interest is the investigation of the status and power of concepts, objects and images in the process of generation, stabilization and demarcation of scientific/technoscientific knowledge. Aside from the edition project HOEK (http://www.hoekweb.tu-darmstadt.de/), she is running a project on "Visual Cultures of Ecological Research", a web-based information system for research and education (http://bildkulturen.online.uni-marburg.de/). Since August 2010 she is working on the "Genesis and Ontology of Technoscientific Objects" (ANR/DFG project), together with Bernadette Bensaude-Vincent and Alfred Nordmann. In the summer term 2010 she held a temporary professorship at the University of Applied Sciences in Darmstadt. All these projects and interests coalesce in the current book project on "Science at the Border – Permanent Tinkering with Unruly Conditions".

Kurt Jax is a senior scientist at the UFZ Helmholtz Centre for Environmental Research, Leipzig (Germany) and Professor for ecology at the Department of Ecology of the Technische Universität München, Munich. He was educated as a freshwater ecologist, working on the ecology of aquatic protozoa. He later moved into research on the conceptual foundations of ecology and conservation biology, which he now has followed since almost two decades. He did this first in connection with the development of a theoretical concept for ecosystem research in the German Waddensea (University of Oldenburg) and then moved more in more also into the philosophical dimensions of ecology, e.g. with a postdoctoral fellowship at the International Centre for Ethics in the Sciences and Humanities (University of Tübingen). Between 1999 and 2009, Kurt Jax has also been on the Editorial Advisory Board of the journal "Environmental Ethics". His habilitation thesis at the

TU München (2000) dealt with a theoretical and philosophical analysis of the concepts of ecological units (population, community, ecosystems). The special emphasis of his work is on the application of theoretical ecological concepts as tools for conservation biology and the adaptation of methods from the humanities (especially philosophy) to use them for interdisciplinary research in the environmental sciences. This is also one of the major interest as one of the editors of the long-term project HOEK. Major publications of the last years have dealt with the concepts referring to "ecological units", and the various uses of the function concept in the environmental sciences, and with the societal perceptions of nature and its practical and ethical implications (especially in the context of the research project BIOKONCHIL, on the evaluation of biodiversity in Southern Chile). Most recently he has published a monograph on the concept of "Ecosystem Functioning" (Cambridge University Press 2010).

Andrew Jamison has an undergraduate degree in history and science from Harvard University (1970) and a PhD from University of Gothenburg in theory of science (1983). He created and directed the graduate program in science and technology policy at the University of Lund from 1986 to 1995, and since 1996, has been professor of technology and society at the Department of Development and Planning at Aalborg University. He was coordinator of the EU-funded project, Public Participation and Environmental Science and Technology Policy Options (PESTO), from 1996 to 1999, and is currently coordinating a Program of Research on Opportunities and Challenges in Engineering Education in Denmark (PROCEED), from 2010 to 2013, funded by the Danish Strategic Research Council. He has published widely in the areas of environmental politics, social movements, and cultural history, most recently *The Making of Green Knowledge. Environmental Politics and Cultural Transformation* (Cambridge University Press, 2001) and, with Mikael Hård, *Hubris and Hybrids: A Cultural History of Technology and Science* (Routledge, 2005).

Egon Becker is a retired professor of Theory of Science and Sociology of Higher Education at the Goethe-University Frankfurt/Main, Germany and senior researcher at the Institute for Social-Ecological Research (ISOE). He studied mathematics and physics, sociology and philosophy, graduated in electrical engineering and physics, and has a doctoral degree in theoretical physics in the field of quantum theory of solid states from the Technical University Darmstadt, Germany. He was research fellow at Yale University (USA) and visiting professor in Brazil, Mexico and Sweden. Currently his fields of research are conceptual and methodological problems of social-ecological systems, complexity theory and philosophy of science, on which he has published extensively.

Patrick Matagne is trained both in biology and in history, he holds a PhD in the history of sciences. Matagne is Maître de conférences at Université de Poitiers (France) and also a member of the laboratoire ICOTEM-RURALITES EA 2252 (identity and knowledge of changing regions). His research is mainly focused on the history of practices in natural history and on the history of ecology. He is also involved in education programs on biodiversity as well as sustainable development. His recent publications are La naissance de l'écologie, Histoire des sciences (Ellipse 2009) and Histoire du concept d'écosystème (in: Ciência & Ambiente 39, 2009, p. 33–47).

Ludwig Trepl is born in 1946. He studied biology in Munich and Berlin. Between 1994 and 2011 he hold the Chair of Landscape Ecology at the Technische Universität München. After some years as a vegetation ecologist his main interests shifted to history and theory of ecology, in the last decade to history and theory of landscape as a cultural subject.

John M. Gowdy is Rittenhouse Professor of Humanities and Social Science, Department of Economics, Rensselaer Polytechnic Institute in Troy, New York. He is past president of the U.S. Society for Ecological Economics and is current (2010-2012) President the International Society for Ecological Economics. His current research interests include climate change, biodiversity valuation, behavioral economics, and evolutionary economics. He has been a Fulbright scholar at the Economic University of Vienna, Leverhulme Professor at Leeds University and a visiting scholar at the Autonomous University Barcelona, the University of Zurich, the Free University of Amsterdam, the University of Queensland and Tokushima University Japan. He is the author or co-author of over 160 articles and 10 books. His most recent books are M*icroeconomic Theory Old and New: A Students Guide*, Stanford University Press 2010*, Paradise for Sale: A Parable of Nature*, co-authored with Carl McDaniel, University of California Press 2000, and *Frontiers in Ecological Economic Theory and Application*, Edward Elgar Press 2008; co-edited with Jon Erickson.

Broder Breckling studied at the University of Bremen (Biology). In his dissertation he investigated the potential of individual-based modeling to analyse ecological processes and interactions. He worked at the University of Kiel in the Ecosystems Research Centre, and at the International Institute of Applied Systems Analysis (IIASA, Laxenburg, Austria). After obtaining the *venia legendi* in Bremen (Centre for Environmental Research and Sustainable Technology, UFT), he continues research in Bremen and at the University of Vechta focusing on ecological theory, computer simulation of ecological processes, and environmental impact of genetically modified plants in agricultural systems. He was involved in and co-ordinated various research projects. Together with Fred Jopp and Hauke Reuter he edited a textbook on ecological modelling (appearing early 2011).

Wolfgang Haber is Professor Emeritus in landscape ecology at Technische Universität München and holds an honorary doctor degree in agricultural sciences from the University of Hohenheim as well as an Einstein professorship of the Chinese Academy of Sciences. He has published about 435 titles, including 4 books, and participated in 12 editorships. Haber studied biology, chemistry, and geography at the Universities of Münster, Munich, Basel (Switzerland), and Hohenheim. He earned a M.Sc. in biology, specializing in orchid growing, and a PHD in soil microbial ecology and theoretical ecology. From 1958-1966 he served as a junior scientist, lecturer and also curator at the Westphalian Museum of Natural History in Münster, later he was responsible for research and administrative tasks in nature conservation and landscape management in Westphalia. In the following he became the director of the newly founded Institute of Landscape Ecology at the Technische Universität München and in same time professor of landscape ecology. Since then he did teaching and research in general ecology and landscape ecology, vegetation and ecosystem science, conservation biology, agroecology, and land use history. In 1970 he co-founded the Gesellschaft für Ökologie (GfÖ) (Ecological Society of the German-speaking countries), from 1980-1989 he served as President of the society. He was member of numerous academic institutions such as the Advisory Committees for Nature Conservation of the State of Bavaria Government and the German Federal Government, the German Federal Council of Environmental Advisors, and the German Council of Land Use Management. From 1990-1996 he was elected President of the International Association for Ecology (INTECOL), the umbrella organization of ecological societies all over the world. Since 1996 he acts as an ecological advisor and senior ecologist (also in Switzerland and China), as environmental expert, and is still active as author and supervisor of PhD students.

Annette Voigt is postdoctoral researcher in the working group "Landscape and Urban Ecology", Paris Lodron University, Salzburg, Austria. She received her diploma (Dipl. Ing.) from the Technische Universität Berlin, Germany, where she studied landscape planning and landscape architecture. After finishing her diploma she worked as research associate at the chair of Landscape Ecology, Technische Universität München (Munich) from 2000–2010, and received her doctor's degree for her dissertation "Theories of synecological units: a contribution to the explanation of the ambiguity of the ecosystem concept" (published book: Die Konstruktion der Natur. Ökologische Theorien und politische Philosophien der Vergesellschaftung. Steiner, Stuttgart, 2009). Her main field of interest is the philosophy, history, and cultural background of ecological sciences, mainly community ecology and ecosystem ecology, landscape planning and nature conservation, as well as theories of man and nature relationship. Her current scientific focus is on concepts of urban nature and challenges of nature management in urban regions.

Donato Bergandi is Maître de conférences (Associate Professor) in philosophy of ecology at the Muséum National d'Histoire Naturelle, Paris. His interests concern the philosophy of biology and the philosophy of ecology (methodology in particular), environmental ethics, environmental history, and the social, economic, environmental and ethical aspects of sustainable development. He has published, among others, papers in Ludus Vitalis, Kybernetes, Acta Biotheoretica, Revue d'histoire des sciences, and he is the scientific editor of the forthcoming book: Ecology, Evolution, Ethics: The Virtuous Epistemic Circle. He participated in setting up the code of ethics in the IUCN Biosphere Ethics Initiative.

Georgy S. Levit is an Assistant Professor at the University of King's College (Halifax, Canada) and a research fellow at Jena University (Jena, Germany). Levit's current research concerns the history of alternative (non-Darwinian) evolutionary theories as well as the (pre-)history of evolutionary developmental biology. He has published on a wide range of topics, including the history of evolutionary theory, German science under Hitler, global issues, and the history of Russian science. There is a common thread, however, running through these diverse topics, the interaction of science with society, religion, and philosophy. His recent publications include: Levit G.S., Hossfeld U. (2009) From Molecules to the Biosphere: Nikolai V. Timoféeff-Ressovsky's (1900-1981) Research Program within a totalitarian landscapes. Theory in Biosciences, 128:237-248, and Levit G.S., Simunek M., Hoßfeld U. (2008) Psychoontogeny and Psychophylogeny: The Selectionist Turn of Bernhard Rensch (1900-1990) through the Prism of Panpsychistic Identism. Theory in Biosciences, 127:297-322.

Gerhard Hard is Professor Emeritus for Physical Geography at the University of Osnabrück (1977-1999). In 1962 he earned his PhD with a study in the field of vegetation geography and seven years later habilitated with a book on the "'Landschaft' der Sprache und die 'Landschaft' der Geographen". In 1971 he became Professor of Geography and Didactics at the Pedagogical University of the Rhineland, Bonn. In his semantic analysis of landscape concepts he argues that the concept "Landschaft" as used in German geography embodies esthetic notions and refers finally to an aesthetic perception of a spatial land configuration. This statement was highly disputed in German geography – and still is. Hard rejected the concept "Landschaft" as being a scientific object, but instead argued for a conception of geography as a science of signs and symbols for instance in the book "Spuren und Spurenleser" published in 1995. This same conception was also applied to landscape ecology and architecture as well as open space planning and nature protection. With his critical research program Hard is perceived as a pioneer for a new perspective on geography, in research as well as in teaching. In 2007 Gerhard Hard was awarded a honorary doctor's degree by the University of Jena (Faculty of Chemistry and Geosciences).

Santos Casado is Associate Professor in the Department of Ecology at the Universidad Autónoma de Madrid (Spain). His research to date has focused on the history of natural history and environmental sciences in Spain and on the intellectual and cultural history of Spanish conservation and environmental movements. His books on these topics include "Los primeros pasos de la ecología en España" (Madrid: Ministerio de Agricultura, 1997), "La ciencia en el campo" (Tres Cantos: Nivola, 2001), and "La escritura de la naturaleza" (Madrid: Caja Madrid, 2001). His latest book, "Naturaleza patria" (Madrid: Marcial Pons, 2010), examines the way that scientific visions of Spanish nature were intermingled with discourses of national identity and political reform in turn-of-the-century Spain.

Robert McIntosh is Professor Emeritus at the University of Notre Dame, Indiana. He started his career as a plant ecologist and, as one of the first students of John T. Curtis, he received his PhD from the University of Wisconsin-Madison in 1950. Together with Curtis, he developed the "vegetational continuum concept" which resurrected Henry A. Gleason's "individualistic concept" of the plant community and substantiated it by a large body of empirical data. The studies established a well-founded and subsequently well-received alternative to Frederic Clements' organismic concept still dominant at the time. After teaching at colleges in Vermont and in Poughkeepsie (NY, Vassar College), McIntosh became professor at the University of Notre Dame, where he worked from 1958 through 1987, until his retirement. In addition to his theoretical and empirical work in ecology, Robert McIntosh early on became also interested in the history of ecology, on which he published already from the 1950s on. Especially well-known is his book "The Background of Ecology" (Cambridge University Press 1985) which as one of the first books at all provided a broad overview on the history of the discipline. Until today he is one of the most respected scholars of this field. From 1970 to 2002 McIntosh also was editor of the journal "American Midland Naturalist". Among the honors he received is the AIBS award (American Institute of Biological Sciences) for contributions to biology in1998.

Patrick Blandin is Professor Emeritus at the Muséum National d'Histoire Naturelle (Paris). He was Director of the laboratoire d'Ecologie Générale (1988-1998), Deputy Director of the Institut d'Ecologie et de Gestion de la Biodiversité (1995-1998), Director of the Grande Galerie de l'Evolution (1994-2002), Director of the laboratoire d'Entomologie (2000-2002). From 2003, he is a member of the Department "Hommes-Natures-Sociétés". Previously, as an Assistant Professor (Ecole Normale Supérieure, 1967-1973; Université Paris 6, 1973-1988), he studied spiders' ecology in the Lamto savannah (Ivory Coast) for his thesis (1969-1981). In 1983, he received an award from the Société Zoologique de France for his work on spiders. During the period 1975-1985, he organized a field station for forest ecology near the Fontainebleau forest. He directed interdisciplinary programs on suburban forests (1980-1983) and biodiversity of woodlots in openfield landscapes (1992-1994) near Paris. For the Environment ministry, he wrote a synthesis on bioindicators, for which he received an award from the Académie des Sciences (1987). Since 1968, he also studied taxonomy and biogeography issues on South and Central America butterflies, and he works on diversity issues, in the field, in Peru, since 2005. He published a monograph on the genus *Morpho* in 2007, and he received an award from the Société entomologique de France in 2008 for his researches on butterflies. Dr. Patrick Blandin was involved in Fontainebleau Forest conservation problems during the 90s and he chaired the Scientific Committee of the Fontainebleau MAB Reserve (2003-2006). He created and chaired the French National Committee of the International Union for Conservation of Nature (1992-1999). Presently, he is Co-chair of the IUCN Biosphere Ethics Initative. He has published two books on historical and epistemological aspects of conservation of nature and biodiversity. For the more recent, *Biodiversité, l'avenir du vivant* (2010), he has received a Great Award from the Académie Française.

Peder Anker is associate professor at the Gallatin School of Individualized Study and the Environmental Studies Program at New York University. His works include "Imperial Ecology: Environmental Order in the British Empire, 1895-1945" (Harvard University Press 2001) and "From Bauhaus to Eco-House: A History of Ecological Design" (Louisiana State University Press 2010). See www.pederanker.net <http://www.pederanker.net>.

Peter Taylor is a Professor at the University of Massachusetts Boston where he teaches in and directs undergraduate and graduate programs on critical thinking, reflective practice, and science-in-society (www.faculty.umb.edu/pjt). His publications focus on the complexity of environmental and health sciences in their social context, including Unruly Complexity: Ecology, Interpretation, Engagement (U. Chicago Press, 2005). This project on complexity and change had its beginnings in environmental and social activism in Australia which led to studies and research in ecology and agriculture. He moved to the United States to undertake doctoral studies in ecology, with a minor focus in what is now called science and technology studies (PhD Harvard University, 1985). Subsequently, he has combined scientific investigations with interpretive inquiries from the different disciplines that make up STS, his goal being to make STS perspectives relevant to life and environmental students and scientists.

Gerhard Wiegleb is a full Professor of General Ecology at Brandenburg University of Technology, Cottbus (Germany). He was educated as a freshwater ecologist at the University of Göttingen studying distribution and indicator value of macrophytes in water bodies of Northern Germany. Later he moved into the field of ecological restoration of streams, mires, and heavily disturbed landscape such as open cast strip mines and military training areas. This work culminated in the coedition (together with S. Zerbe) of the textbook „Renaturierung von Ökosystemen in Mitteleuropa (Spektrum Verlag, Berlin 2008). Since 1986 he has been teaching history of science, philosophy of science, and recently bioethics at the Masters and PhD level. Recent research concentrates on social, legal, economic and ethical aspects of biodiversity protection. Major publications deal with the ethical justification of the evaluation of biodiversity, the integration of biodiversity aspects into decision-making in landscape management, and the function of legal instruments such as the environmental liability legislation of the EU for biodiversity protection.

Yrjö Haila studied ecological zoology, with philosophy as his main secondary subject, at the University of Helsinki, and defended his PhD thesis on the ecology of land birds in the Åland Archipelago in 1983. In his later research, he specialized on ecological changes in environments intensively modified by humans, such as cities and commercially managed forests. He has been the professor of environmental policy at the University of Tampere since 1995. His main research interests have centered on the nature–society interface, from several complementary perspectives. He has published *Humanity and Nature* (with Richard Levins; Pluto Press, 1992), and *How Nature Speaks. The Dynamics of the Human Ecological Condition* (co-edited with Chuck Dyke ; Duke University Press, 2006), as well as several books in Finnish.

Author Index

A

Abbot, D., 100
Acot, P., 145, 146, 287, 303, 304
Adams, C.C., 278
Adams, W.H., 380
Agazzi, E., 33
Aksenov, G., 334
Albert, A., 301
Alberts, J.J., 188, 216
Allaby, M., 6, 166
Allee, W.C., 149, 153, 267, 279, 283
Allen, D.E., 101, 108, 110, 410
Allen, T.F.A., 101, 104, 105, 110, 124, 220, 410
Andersen, D.C., 59, 91
André, E., 297
Andrés y Tubilla, T., 310
Andrewartha, H.G., 103, 105, 167
Anker, P., 3, 53, 55, 99, 330
Arévalo, C., 312, 314
Arneson, R.J., 66
Arrhenius, O., 102, 280
Arrow, K., 407
Ayala, F.J., 31, 47, 48

B

Baichère, A ., 301
Bailes, K.E., 336
Baker, G., 19
Baker, W.L., 209
Bange, C., 288
Barbour, M., 282
Barreiro, A.J., 312
Barrows, H.H., 353
Bartels, D., 352, 355, 361
Bartha, S., 110
Beatty, J., 123
Beckermann, A., 31
Beck, H., 53, 67
Begon, M., 97, 102
Bentley, A.F.35, 42, 212
Berdoulay, V., 354
Bergandi, D., 33, 35, 37
Berg, G., 23
Bertalanffy, L. v., 49, 53, 72, 134, 184, 186, 260, 386, 387, 395, 396
Bews, J.W., 53
Birch, L.C., 103, 105, 107, 167
Birge, E.A., 255, 278
Blanchflower, D., 409
Blandin, P., 47, 205–213, 215, 217, 221
Blondel, J., 209
Boadway, R., 407
Bobek, H., 354
Bock, G.R., 47
Bödeker, H.E., 23
Böhme, G., 21, 137, 139, 248
Böhme, H., 137
Bolívar, I., 309, 311
Bonnier, G., 288, 298, 302
Bookchin, M., 196, 199
Boreau, A., 289
Bormann, F.H., 102, 107, 188
Botkin, D.B., 54, 185
Böttcher, H., 352
Boucher, D., 379
Boulaine, J., 297
Bowles, S., 407
Box, E.O., 103
Braun-Blanquet, J., 54, 58, 108, 166, 242, 268, 269, 303, 319
Bray, J.R., 101, 102, 108
Breckling, B., 224, 385–401
Breger, H., 132
Brehm, V., 245, 253
Bremekamp, C.E.B., 100
Brenner, A., 69
Browne, J., 99, 167

Brown, J.H., 77
Brussard, P.F., 217
Bueno, G., 47
Burckhardt, G., 250, 251
Burgess, R.L., 171
Busch, B., 237
Busse, D., 22, 26
Buttimer, A., 354

C
Cacho Viu, V., 308
Cajander, A.K., 374
Callicott, J.B., 69
Calow, P., 166
Campbell, D.T., 33
Candolle de, A.-P., 100, 288, 293, 295, 296, 298, 309, 310, 372
Canguilhem, G., 19, 20, 24
Cantlon, J.E., 224
Capra, F., 45
Carnap, R., 118, 121
Carpenter, R.A., 105, 107, 110, 157, 158, 217
Carpenter, S., 105, 107, 110, 157, 158, 217
Carrier, M., 89, 123, 124, 126
Carson, R., 198, 199, 379
Cartwright, N., 121, 122
Casado, S., 307–321
Cassirer, E., 51
Castoriadis, C., 396
Château, E., 296
Cherrett, J.M., 97, 101, 215
Chipman, J., 407
Christiansen, C., 327–329
Cittadino, E., 101, 162, 163, 260–262
Clam, J., 394
Clark, C.W., 375
Clarke, F.W., 375
Clark, H.J., 375
Clark, L.R., 375
Clark, W.C., 375
Claval, P., 354
Clements, F.E., xii, xiv, 16, 36–38, 40, 54, 55, 57, 58, 68, 101, 106, 108, 110, 119, 134, 153, 162, 164, 168, 175, 177, 189, 219, 245, 262, 266, 267, 277, 278, 281, 283, 319, 342, 373
Clos, D., 301
Cloud, P.E. Jr., 175
Cockayne, L., 283
Coleman, W., 309, 318
Commoner, B., 198–200, 379
Connell, J.H., 102
Connolly, W.E., 372

Connor, E.F., 102, 108
Cook, R.E., 281, 282
Cooper, G.J., 4, 120, 121, 125, 126, 280
Corbin, A., 288
Costanza, R., 70, 74
Coste, A.H., 298, 302
Cottam, G., 280
Cowles, H.C., xiii, 146, 153, 161, 162, 167, 176, 177, 262, 278, 300, 330
Cox, D.L., 279
Cramer, J., 198, 202
Crick, F., 52
Croker, R.A., 278
Crowcroft, P., 173
Cuddington, K., 372
Curtis, J.T., 72, 101, 102, 108, 280, 282
Cushman, J.H., 105, 107

D
Dagognet, F., 295
Dahl, F., 145, 156–158, 246
Daily, G.C., 243
D'Arcy, W.T., 375
d'Arge, R., 70, 74
Daum, A.W., 232
Davidson, M., 184
Dawkins, R., 50
DeAngelis, D.L., 89, 395
Debatin, B., 26
De Buen, O., 309, 310
Degens, E.T., 344
de Groot, R.S., 74
De Laplante, K., 215, 221, 224
Delcourt, P.A., 104
Del Villar, E.H., 316, 317
Desmond, A., 67
Detering, K., 73
Dewey, J., 35, 212
De Wit, C., 106
Diamond, J.M., 108
di Castri, F., 207, 208, 210, 211
Dice, L.R., 109
Dickson, D., 201
Diederich, W., 127
Dierschke, H., 267, 270
Dilthey, W., 48, 49
Disko, R., 70
Dobrovolsky, V.V., 343
Dobzhansky, T., 47
Dodson, S., 97, 103, 104, 109
Dokuchaev, V.V., 335
Drengson, A., 45
Driesch, H., 49

Drouin, J.-M., 295, 303
Drude, O., 100, 245, 267, 270
Drury, W.H., 63
Dubos, R., 200, 201
Duhem, P., 37, 48
Dupuis, C., 288
Du Rietz, G.E., 151, 177, 266, 267, 269, 270
Dutt, C., 25
Dzeriski, S., 397
Dzeroski, S., 397

E
Easterlin, R., 409
Eckebrecht, B., 71
Egerton, F.N., 99, 100, 145, 277
Eggebrecht, H.-H., 23
Eggleton, F., 4
Egler, F.E., 72, 102, 219
Ehrlich, H.L., 344
Ehrlich, P., 198, 200
Ehrlich, P.R., 198, 200
Eisel, U., 45, 47, 54, 55, 66, 67, 352, 357, 359, 363
El-Hani, C.N., 33
Ellenberg, H., 188, 220, 221, 267
Elster, H.-J., 172
Elton, C.S., 36, 101, 103, 130, 156, 165, 171, 278, 279, 281, 374, 375
Emerson, A.E., 149, 161, 267, 279, 283
Engelberg, J., 188
Engels, F., 3, 67
Erwin, T.L., 207
Eser, U., 69, 70
Evans, F.C., 219, 220

F
Fagen, R.E., 189, 394, 398, 400
Fantini, B., 310
Farber, S., 74
Farber, S.C., 74
Fehr, E., 408
Feibleman, J.K., 33, 38, 39, 219
Fenchel, T., 108, 344
Ferguson, B.K., 70
Fersman, A.E., 341
Fischedick, K.S., 171
Fischer, J.-L., 297
Fischer-Kowalski, M., 364
Fisher, R.A., 101, 102
Flahault, C., xi, xii, xiv, 4, 146, 165, 268, 269, 288, 298, 300–302
Forbes, E., 175, 280

Forbes, S.A., 36, 37, 40, 54, 118, 129, 131, 249, 277, 278, 281, 375
Ford, E.D., 223, 224
Forel, F.-A., 131, 234, 235, 244, 249, 251, 252, 254
Forman, R.T.T., 104
Foucault, M., 47, 155
Fox, R., 289
Fränzle, O., 220–224, 386
Frazier, J.G., 4
Frey, A.E., 351
Frey, B., 409
Friederichs, K., 49, 54, 63, 158, 166, 185, 186, 239, 245, 246, 265
Fuchs, G., 353

G
Gams, H., 54, 151, 177, 185, 267, 268
Gaston, K.J., 105
Gates, D.M., 102, 103, 110
Gauch, H.G., 108
Gause, G.F., 103, 130, 279, 283
Gautier, G., 301
Geison, G.L., 318
Georgescu-Roegen, N., 143
Ghilarov, A.M., 210, 213, 333
Giere, R., 124
Gieryn, T.F., 370
Gintis, H., 407, 408
Glacken, C.J., 277, 356, 371
Gleason, H.A., 16, 36–38, 40, 54, 60, 61, 101, 102, 107, 108, 110, 133, 164, 185, 268, 280, 282
Glick, T.F., 309
Gloning, T., 14
Gnädinger, J., 63
Gnauck, A., 106, 108, 397
Godwin, H., 173
Gogorza, J., 310
Goldstein, K., 54
Golley, F.B., 47, 55, 99, 101, 107, 108, 185–189, 206, 219, 220, 371, 375, 377, 378
Goodland, R.J., 101, 326, 330
Goodman, D., 55, 110, 206–208
Gordon, J.C., 188, 220
Gotelli, N.J., 106, 108
Gowdy, J., 405, 407, 410
Grasso, M., 74
Graves, G.R., 106, 108
Green, R.H., 103
Greiffenhagen, M., 66, 67
Greig-Smith, P., 280
Grime, J.P., 98, 103, 107

Grimm, V., 4
Grinevald, J., 334
Grisebach, A.H.R., 49, 100, 146, 163, 261, 262, 266, 296
Grove, R.H., 376
Gumbrecht, H.U., 23

H
Haas, H.-D., 167, 168
Haberlandt, G., 163, 231, 260
Habermas, J., 64
Haber, W., 215, 224, 262
Hacking, I., 122, 369
Haeckel, E., 36, 101, 145–147, 149–152, 155, 156, 158, 163, 165, 167, 232, 242, 262, 263, 277, 333
Haefner, J.W., 21
Hagen, J.B., 53, 55, 88, 101, 162, 164, 185, 187–189, 262, 373
Haggett, P., 351
Hagner, M., 135
Haila, Y., 5, 87, 89, 90, 92, 93, 120, 224, 369, 370, 379, 381
Hairston, N., 107
Haldane, J.S., 49, 53
Hall, A.D., 189, 394, 398, 400
Hall, C.A.S., 189, 410
Hamm, B., 353
Hampe, M., 124
Hannon, B., 74
Hard, G., 70, 266, 351, 352, 354, 356, 360–364
Harper, J.L., 101–103, 105, 107, 109, 110, 177
Harries-Jones, P., 388
Harrington, A., 54
Harroy, J.P., 205
Hartmann, M., 117, 264
Harvey, P.H., 51, 106
Haskell, B.D., 70
Hassell, M.P., 376
Hauhs, M., 109, 188
Hedgpeth, J.W., 172
Heims, S.J., 187, 387
Hempel, C.G., 32, 395
Hentschel, E., 235, 245
Herder, J.G., 54, 353, 354, 356, 357
Hesse, H., 389
Hesse, M., 26
Hesse, R., 162–164, 167, 176, 246, 254, 258, 262, 264, 265
Higashi, M., 188
Hoekstra, T.W., 102–105, 220
Hoffmeyer, J., 388

Höhler, S., 3
Holdgate, M., 380
Holling, C.S., 216, 375, 376
Hölter, F., 220
Hopkins, A.D., 282
Horder, T.J., 20
Höxtermann, E., 260, 261
Hubbell, S.P., 72, 101, 103, 107, 108
Huber-Fröhli, J., 167, 168
Huffaker, C.B., 102, 376
Humboldt, A.v., 49, 100, 147, 163, 224, 236, 262, 266, 292, 295, 296, 298, 299, 309, 310, 328, 360, 372
Hurlbert, S.H., 108
Hutchinson, G.E., 39, 101, 102, 130, 177, 186, 187, 210, 250, 282, 344, 377, 387

I
Ilerbaig, J., 321
Illich, I., 201
Illies, J., 167
Ishii, H., 223, 224
Izco, J., 320

J
Jahn, I., 99, 100, 157, 158
Jamison, A., 195, 199, 203, 379
Jansen, A.J., 99
Jax, K., 3, 4, 11, 14, 21–23, 37, 49, 54, 59, 102, 104, 149, 153, 155, 161, 164, 166, 171, 175, 185, 190, 219, 220, 231, 243, 247, 269, 361, 362, 373
Johnson, N.C., 218
Jones, C.G., 110, 178, 189
Jones, R., 31, 33
Jørgensen, S.E., 51, 72, 73, 106, 108, 110, 188
Juday, C., 255, 278, 281
Judd, R.W., 376
Juhász-Nagy, P., 102
Junge, F., 243–245
Jussieu, A.-L.de., 296, 298

K
Kapp, E., 358
Karny, H.H., 245
Kay, J.J., 188
Keller, D.R., 47, 101
Khailov, K.M., 385
Khlentzos, D., 388
Kieffer, J.J., 54, 55, 186
Kiester, R.A., 120

King, G.M., 344
Kingsland, S., 88, 90, 375
Kingsland, S.E., 101, 130, 177, 279, 321, 387
Kirchhoff, T., 47, 54, 67
Kirchner, O., 176, 254, 266
Kitchell, J.F., 105, 107, 110
Klink, H.-J., 178
Klir, G.J., 393
Klötzli, F., 104, 105
Kluge, T., 3, 236
Knobloch, C., 25, 26
Köchy, K., 49
Koehler, G.H., 390
Kohler, R.E., 22, 155, 162, 237, 321, 370, 371
Köhler, W., 54, 183
Kolasa, J., 102, 110
Kolchinsky, E.I., 336
Körner, S., 66, 70, 357
Koselleck, R., 21, 25
Kovalsky, V.V., 341, 345
Krebs, C.J., 97, 102, 103, 110, 167
Kretschmann, C., 234
Krewer, B., 353
Krumbein, W.E., 340, 344
Kühnl, R., 66
Kuhn, T.S., 47, 48, 66, 98, 223, 245
Küppers, G., 99
Kurt, J., 3, 121, 149, 155, 161, 171, 175, 231, 253
Kwa, C., 188, 199, 266, 378

L
Laissus, Y., 288
Lakatos, I., 21, 98, 119, 122, 126–128, 139, 248
Lamotte, M., 186, 217, 298, 303
Lange, H., 109, 188
Langthaler, R., 45
Lapo, A.V., 333, 336, 344
Latour, B., 5, 136, 137, 379
Lawton, J.H., 178, 189
Lázaro e Ibiza, B., 310
Lecoq, H., 295, 298, 299, 303
Lecordier, C., 210
Legendre, P., 102
Leigh, E.G., 206
Lenoble, F., 54
Lenz, F., 235, 244, 245, 254, 255
Leopold, A.S., 69, 279, 283
Lepenies, W., 136
Leser, H., 167, 168, 355, 362, 364
Leslie, P.H., 102
Lethmate, J., 362
Levin, S.A., 209

Levins, R., 33, 46, 71, 88, 92, 93, 110, 370, 379
Levit, G.S., 333, 336, 338, 340
Lewontin, R., 33, 93, 110, 370
Lewontin, R.C., 46, 71, 91
Libes, S.M., 344
Likens, G., 102, 105, 107, 167
Likens, G.E., 167, 188, 216
Lilienfeld, P.v., 134, 184
Limburg, K., 74
Lindeman, R.L., 36, 38, 73, 101, 102, 105, 107, 186, 216, 220, 250, 281, 282
Linnaeus, C., 99, 100, 107, 198, 296, 371
Little, P., 92
Lloyd, J., 289, 299, 300, 303
Löbner, S., 15
Loeb, J., 50
Lomborg, B., 196
Looijen, R.C., 4, 47, 64
López-Ocón Cabrera, L., 307
López Piñero, J.M., 308
Loret, H., 301
Löther, R., 99, 100
Lotka, A.J., 100, 102, 124, 187, 279, 375–377, 387, 399, 400
Loucks, O.L., 206, 209
Lovejoy, A.O., 22
Lovelock, J., 333, 336, 340, 343, 377
Lovelock, J.E., 71, 377
Lovins, A., 202
Lowe, P.D., 101
Lowood, H.E., 373
Lübbe, H., 19, 20
Luckin, B., 3
Luhmann, N., 387, 394
Luquet, A., 303
Lussenhop, J., 101

M
MacArthur, R.H., 101–105, 107, 168
Mace, G.M., 380
Macfadyen, A., 206
MacMahon, J.A., 59, 91
Mägdefrau, K., 99
Maienschein, J., 20
Mairet, P., 277
Major, J, .167
Malone, C.R., 217
Malthus, T.R., 99, 200, 279, 371, 375, 376, 411
Margalef, R., x, 72, 73, 102, 108, 167, 188, 206, 302, 303, 316
Margulis, L., 336, 340, 343, 377
Marquard, O., 136–138
Marsh, G.P., 198, 373, 376, 378

Matagne, P., 146, 242, 287–304
Maturana, H.R., 285, 394
Mayer, H., 5, 23
Mayr, E., 4, 31, 51, 52, 64, 100, 218, 344
McCoy, E.D., 4, 102, 110, 120, 121, 125, 126
McIntosh, R.P., 21, 40, 53–55, 72, 73, 88, 97, 120, 161, 166, 168, 175, 185, 277–283, 379, 387
McLaughlin, P., 51, 65, 74
McMillan, C., xiii, 167
McNeely, J.F., 207
Medin, D.L., 15
Meinhard, H., 395
Menting, G., 12, 251, 258, 362
Merz-Benz, P.-U., 394
Mesarovic, M.D., 398
Meusburger, P., 353, 364
Meyer-Abich, A., 49, 54, 240
Meyer-Abich, K.M., 69
Miller, R.S., 91, 281
Mitchell, S., 125, 126
Mitman, G., 265
Mittelstraß, J., 51, 52, 137
Möbius, K.A., 101, 134, 152, 233, 242–244, 265, 375
Mocek, R., 49
Mochalov, I.I., 336
Mooney, H.A., 105, 107
Moore, J., 67, 407
Morowitz, H., 378
Morrell, J.B., 318
Mosimann, T., 167, 168, 362, 364
Motterlini, M., 120
Mueller-Dombois, D., 220, 267
Müller-Böker, U., 363
Müller, F., 73, 106, 109, 127, 188, 201, 224
Müller, E. 23–25
Müller, K., 54, 72, 73, 184, 190, 392, 395
Müller-Navarra, S., 234
Müller, P., 167

N
Naeem, S., 70, 74
Naess, A., 45, 195, 198
Nagel, E., 33, 47, 51, 64
Naumann, E., 172, 233, 235, 254, 256, 257
Naveh, Z., 212
Needham, J., 53
Nelson, G., 100
Nichols, G.E., 165
Nicholson, A.J., 280
Nicolson, M., 266, 319, 328
Niehr, T., 22, 26
Nielsen, S.N., 188

Nordenskiöld, E., 67
Nordhaus, W., 407
Nordmann, A., vii, 137
Norton, B.G., 70, 74, 282
Novikoff, A.B., 102, 104, 219
Nowak, M., 408
Núñez, D., 309
Nyhart, L.K., 262

O
Odum, E.P., 38–40, 72–74, 101–105, 167, 177, 187–189, 206, 216–218, 220, 250, 282, 283, 304, 373, 377–379, 388
Odum, H.T., 91, 93, 102, 103, 107, 108, 110, 187, 188, 198, 199, 217, 250, 304, 377, 385, 386, 394, 410
O'Hara, J.L., 188, 220, 283
Oldfield, J.D., 343
O'Neill, R.V., 102, 105, 188, 216, 218, 223
Oppenheim, P., 32, 48
Ostrom, E., 410
Oswald, D., 241, 409
Overbeck, J., 312

P
Pace, M.L., 188
Paesler, R., 167, 172
Paffen, K., 355, 360
Paine, R.T., 102, 206, 209
Pairolí, M., 320
Palmer, M.W., 102
Palmiotto, P.A., 188, 218
Park, O., 149, 161, 177, 267, 279, 281, 283
Park, T., 149, 161, 177, 267, 279, 281, 283
Patten, B.C., 72, 73, 188
Pavillard, J., 298, 303
Petersen, C.G.J., 101, 107, 110, 375
Peterson, G.D., 215
Peters, R.H., 4, 5, 97, 98, 101, 107, 110, 120, 135, 215
Petit, G., 205
Peus, F., 54, 55, 59, 72, 185
Phillips, J., 36–38, 40, 53, 54, 185, 189, 327, 330
Pias, C., 187, 387
Picht, G., 225
Pickett, S.T.A., 5, 57, 90, 102, 107, 110, 209
Pielou, E.C., 107–109
Pimm, S.L., 5
Pires, C., 91, 101–105, 108, 110, 124, 220
Pomata, G., 13
Pomeroy, L.R., 188, 216
Popper, K.R., 21, 33, 122, 127

Pörksen, U., 24, 232
Porter, W.P., 102
Portmann, A., 49
Potthast, T., 69, 70, 221, 237, 241, 364
Preston, F.W., 102, 282
Prigogine, I., 185, 393
Prytz, S., 326
Psenner, R., 248

Q
Querner, H., 100
Quine, V.O.W., 34, 48

R
Rabinbach, A., 131
Radkau, J., 242
Ramensky, L.G., 16, 54, 185
Rangel, A., 409
Rapport, D.J., 70, 98, 108
Raskin, R.G., 70, 74
Raunkiaer, C., 103, 330
Rautian, A.S., 342
Real, L.A., 7
Recknagel, F., 189
Regier, H.A., 98, 108
Reichholf, J.H., 57, 71
Reiners, W.H., 216
Reise, K., 101, 243
Rheinberger, H.-J., 135, 390
Ricklefs, R.E., 91, 167, 209, 210
Rifkin, J., 202, 203
Ritter, C., 23, 351, 353, 354, 356, 358
Rodhe, W., 172
Rosenberg, A., 51
Roughgarden J., 120
Rowe, J.S., 105, 168
Rübel, E., 233, 267
Ruse, M., 31, 32, 47, 124, 125

S
Saarinen, E., 47, 72
Sagan, D., 377
Sagoff, M., 4
Salisbury, E., 110, 171
Salt, G.W., 39
Salthe, S.N., 33
Samson, F.B., 217, 218
Sánchez Ron, J.M., 307
Schaffer, W.M., 223
Scheiner, S.M., 219
Schellnhuber, H.-J., 344
Scherzinger, W., 57, 66

Schimper, A.F.W., 100, 153, 162–164, 245, 261, 264, 267, 270, 300, 326
Schimpf, D.J., 59, 91
Schmidt, K.P., 149, 161, 267, 279, 281, 283
Schmithüsen, J., 355, 361, 362
Schoenichen, W., 69
Scholz, B.F., 25
Schramm, E., 236
Schröter, C., 4, 165, 176, 233, 254, 266, 268, 269
Schultz, A.M., 219, 221, 222, 242
Schultz, H.-D., 352, 355, 356, 358
Schwabe, G.H., 73, 237
Schwan, T., 353, 364
Schwarz, A.E, 3, 11, 19, 36, 45, 54, 66, 72, 117, 145, 149, 155, 172, 175, 183, 184, 222, 224, 231, 364, 372, 375, 387
Schweitzer, B., 123, 125
Schwenke, W., 177
Schwerdtfeger, F., 72, 177, 254
Schwoerbel, J., 253, 387
Scudo, F., 101, 375
Sears, P.B., 283, 373
Semper, K., 145, 153, 162, 262–264
Servos, J.W., 318
Sewertzoff, A.N., 337
Sheail, J., 171, 219
Sheifer, I.C., 217
Shelford, V.E., 54, 100, 104, 105, 107, 156, 162, 164, 165, 168, 176, 185, 189, 263, 265, 277, 278, 281–283, 375
Shimwell, D.W., 267
Shrader-Frechette, K., 4, 110, 120, 121, 126
Sieferle, R.-P., 3
Simberloff, D., 72, 88, 98, 101, 108–110, 167
Skellam, J.G., 102
Slobodkin, L.B., 102
Smart, J.J.C., 123
Smith, D., 188
Smith, F.E., 206
Smuts, J.C., 37, 49, 53, 54, 281, 327
Sober, E., 71
Söderqvist, T., 199, 326
Souché, B., 289, 291, 297
Soulé, M, 380
Sousa, W.P., 209
Southwood, T.R.E., 105
Starr, T.B., 91, 188
Star, S.L., 5
Stauffer, R.C., 100, 151, 309
Stavins, R., 407
Stearns, S.C., 107, 110
Steffen, W., 378
Stegmüller, W., 127, 351, 388
Steinecke, F., 245

Stein, S.M., 217
Steleanu, A., 172, 256
Stengers, I., 5, 393
Steward, J.H., 241, 353
Stöcker, G., 13
Stöckler, M., 45
Storkebaum, W., 360
Streit, B., 168
Strong, D., 88, 108
Sukachev, V.N., 54, 58, 338, 344, 390
Sutton, P., 70, 74
Szaro, R.C., 217
Szasz, A., 202

T

Tainter, J.A., 409, 410
Takacs, D., 155, 380
Tansley, A.G., 36–38, 40, 72, 101, 102, 108, 165, 171, 173, 177, 185, 186, 189, 206, 215, 216, 219–221, 267, 268, 277, 281, 283, 326, 329, 330, 332
Taylor, P.J., 40, 87–94, 120, 187–189, 224, 372, 387
Taylor, W.P., 168, 281
Teichert, D., 25
Ter Braak, C.J.F., 108
Thienemann, A., 16, 49, 54–57, 69, 166, 172, 186, 233, 235–237, 239–241, 245, 247, 254–260, 265, 362
Tilman, D., 103
Timoféev-Ressovsky, N.V., 342, 345
Timoféev-Ressovsky, N.W., 342, 345
Tisdell, C.A., 210
Tjurjukanov, A.N., 342
Tobey, R.C., 55, 185, 190, 245, 267, 371
Todd, N.J., 201
Toepfer, G., 24
Tönnies, F., 67
Toulmin, S., 98
Tournefort de, J.P., 99, 296
Townsend, C.R., 103
Trepl, L., 5, 45–76, 98, 100, 110, 129, 130, 133, 152, 156, 163, 176, 178, 185, 221, 266, 267, 361, 362, 364, 386, 389
Trier, J., 14
Trojan, P., 72
Troll, C., 354, 362
Troll, W., 49
Tümmers, H.J., 236
Turing, A., 395
Turner, B.L.I.I., 376
Turner, M.G., 110, 397

U

Uexküll, J.v., 54, 157
Ugarte, J., 316
Ulanowicz, R.E., 72, 108, 110, 188
Ule, W., 254
Utermöhl, H., 258

V

van den Belt, M., 74
van der Klaauw, C.J., 157, 177
van der Meulen, M.A., 219
Varela, F.J., 185, 394
Vareschi, E., 109
Vávra, V., 234
Vélaz de Medrano, L., 316
Verhulst, P.F., 375, 399
Vernadsky, V.I., 187, 244, 333–345, 377
Vinogradov, A.P., 341
Vogt, D.J., 188, 220
Vogt, K.A., 188, 220
Vogt, W., 283
Voigt, A., 45–76, 129, 183–190
Volterra, V., 100–102, 107, 187, 279, 375, 387
von Ehrenfels, C., 54
von Marilaun, A.K., 100, 296
Vononcov, N.N., 342
Vorzimmer, P., 100

W

Wahrig-Schmidt, B., 135
Waide, J.B., 188, 216
Walker, B., 378
Walter, H., 110, 145, 168, 245, 262
Ward, B., 200, 201
Ward, H.B., 252
Wargo, J.P., 188, 220
Warming, E., 101–104, 107, 133, 146, 153, 163, 164, 167, 261, 266, 267, 295, 300, 325, 326, 329, 330
Wasmund, E., 241
Waters, C.K., 125
Watt, A.S., 102, 280, 281
Watts, M., 92
Weale, A., 381
Weber, M., 55, 125
Weidemann, D., 390
Weil, A., 65, 69, 190, 222
Weinberg, G.M., 219
Weiner, D.R., 267, 373
Weismann, A., 252
Wengeler, M., 22

Wertheimer, M., 54
Whitehead, A.N., 35, 37, 54, 122, 398
White, P.S., 90, 102, 107, 209
Whittaker, R.H., 72, 101, 108, 267, 268, 282
Wiegleb, G., 97–110, 220
Wiemken, V., 118
Wiens, J.A., 208, 209
Wilcove, D.S., 218
Wilke, S., 136
Wilson, D.S., 71, 98
Wilson, E.O., 102, 168, 207, 216, 219, 222, 381, 411
Wissel, C., 44, 107, 109
Wolf, K.L., 49
Woltereck, R., 134, 166, 258, 259
Woodger, J.H., 33
Worster, D., 55, 100, 101, 107, 185, 187, 196–198, 267, 371–373

Wuennenberg, C.J., 215
Wuketits, F.M., 100
Wynn, G., 376

Y

Yoda, K., 102
Younès, T., 207, 208, 210, 211

Z

Zacharias, O., 131, 233, 249, 252, 253
Zauke, G.P., 101, 107, 109
Zavarzin, G.A., 343
Zeide, B., 218
Zellmer, A.J., 215
Zherikhin, V.V., 342, 343, 345

Subject Index

A

Adaptation, 35, 56–58, 61, 68, 75, 127, 131, 133, 150, 210, 212, 224, 260, 261, 263, 264, 310, 341, 342, 356–358, 363, 397
Aesthetic/aesthetically, 63, 138, 163, 205, 266, 361, 365, 390
Agriculture, 68, 100, 109, 201, 242, 246, 278, 283, 290, 297, 321, 369–373, 379
Analogy, 37, 46, 63, 75, 132, 133, 190, 219, 248, 281, 326, 374, 392
Applied sciences, 109, 246, 369–381
Aquatic ecology, 104, 132, 134, 165, 166, 172, 176, 232, 235, 244–248, 252, 253, 257–260, 265, 270, 281, 307, 311–316, 374
Assemblage/assembly, xi, 36, 89, 108, 118, 119, 166, 176, 206–208, 211, 212, 266, 268, 378
Association, xi, xii, 13, 15, 16, 37, 38, 40, 54, 55, 59, 60, 73, 88, 89, 94, 129, 172, 185, 196, 205, 220, 221, 267–269, 271, 278, 279, 288, 292, 293, 295–299, 301, 302, 308, 310, 312, 317, 335, 361, 391, 392
Autecology, xi, 59, 104, 163, 175–177, 248

B

Balance of nature, 21, 99, 100, 205–209, 212, 278, 281, 377
Basic conception, 118–122, 126, 128–136, 138, 139, 247, 248
Bioc(o)enosis, 13, 15, 55, 56, 123, 134, 243, 244, 260, 265, 343, 362, 392
Biodiversity, 24, 105–107, 109, 205–213, 216, 218, 224, 380, 381, 410
Biogeocenoses, 337, 342, 344
Biological conservation, 7, 12, 109, 218, 224, 380, 381
Biological law, 123
Biological sciences, 123, 155, 373
Biome, 37, 104, 105
Border zone, 5, 22, 147, 165, 237, 369–381, 385–401
Botanical garden, 289, 297, 298
Boundary, 5, 60, 62, 162, 219, 220, 243, 267, 288, 354, 369, 377
Boundary work, 155, 370, 380

C

Causal explanation, 63–65, 222, 269
Causality, 101, 187, 222, 387, 393
Circular causality, 187, 387, 393
Climate change, 24, 196
Climax, x, xiii, xiv, 37, 55, 57, 58, 61, 62, 68, 69, 71, 72, 106, 130, 206, 208, 209, 278, 282, 317, 373, 398
Coevolution, 106, 212
Coexistence, 98, 211, 373
Collectivization, 252, 337
Community, xii, xiv, 4, 11, 13, 15, 16, 21, 37–39, 46, 52–63, 65–76, 88–90, 92, 97, 103–108, 118, 119, 125, 126, 131, 133, 134, 138, 146, 157, 163–167, 177, 178, 183, 185–187, 189, 198, 202, 206–212, 220, 232, 243, 244, 248, 252, 254, 264, 265, 267, 270, 278, 280–283, 314, 317, 325–327, 329, 330, 342, 343, 361, 372–375, 380, 392
Community, climax, 57, 58, 62, 68, 72, 373
Community concept, 15, 16, 59, 106
Community, ecological, 24, 67, 88, 104, 106–108, 129, 133, 167, 177, 188, 201, 216, 240, 246, 282, 375
Community, organic, 46, 53, 66, 67, 69, 70, 75, 76, 187, 189

Community, plant, 40, 100, 107, 131, 133, 164–166, 176, 221, 252, 267, 269, 270, 278, 293, 319, 325–330
Community, scientific, 168, 176, 178, 222, 307, 320, 321, 333, 334, 344
Competition, 13, 36, 57, 67, 71, 76, 88, 91, 102, 103, 106, 108, 120, 122, 128, 130, 131, 133, 158, 187, 235, 279, 280, 379, 398, 406
Competitive exclusion principle, 90, 130, 246, 279
Complexity, 5, 13, 26, 32, 39, 52, 87–94, 157, 185, 206, 221, 223, 224, 333, 341, 395–397, 405, 410
Concept, 11, 19, 31, 46, 87, 97, 118, 147, 149, 156, 161, 175, 183, 195, 206, 215, 237, 278, 287, 310, 325, 335, 351, 369, 385, 408
Conception, x, 25, 26, 47, 49, 71, 122, 125, 128–132, 135, 136, 139, 189, 197, 207, 252, 297, 377
Conception, basic, 118–122, 126, 128–136, 138, 139, 247, 248
Conception of nature, 135–137, 139, 198, 225, 237, 241, 278
Conception, triadic, 130, 137–139, 198
Concept of community, 15, 16, 59, 106
Concept of niche, 6, 21, 106, 129–133, 135, 136, 222, 247
Conceptual cluster, 11, 13–16, 22, 24, 27
Conceptual framework, 5, 16, 125, 127, 136, 199, 237, 255, 386
Conceptual work, 87, 92, 236
Conservation, 57, 68–71, 106, 109, 125, 130, 150, 151, 190, 205–207, 211, 213, 221, 235, 242, 283, 376, 379–381, 401
Conservation, biological, 7, 12, 109, 218, 224, 380, 381
Conservative, 54, 66–68, 75, 133, 134, 203, 241, 308, 327, 357, 361
Control, 24, 76, 87, 91, 93, 131, 132, 137, 138, 188, 190, 206, 246, 278, 280, 314, 327, 329, 376, 398
Cultural, x, xii, 4, 7, 23, 25, 47, 66, 70, 100, 178, 200, 213, 236, 241, 287, 317, 335, 352, 354–358, 360, 361, 363, 380, 387, 390
Cultural ecology, 351–355, 357–361, 365
Cultural type, 256
Cybernetic(s), 108, 183–185, 187, 188, 198, 199, 206, 250, 363, 387, 388, 393, 394, 398, 401
Cycle, 13, 56, 74, 75, 102, 105, 123, 132, 187, 188, 249, 257, 340, 343, 376, 377

D

Darwinism, 50, 57, 66, 99, 263–265, 308, 309, 336
Deep ecology, 45, 198, 379
Descriptive, 23, 66, 88, 101, 128, 237, 260, 264, 267, 268, 301, 308, 311, 386, 400
Discipline, x, 4, 5, 7, 22–24, 46, 87, 117, 129, 132, 135, 149, 150, 153, 161, 162, 168, 172, 175, 183, 184, 188, 201, 241, 243, 247, 254, 267–269, 302, 307, 320, 333, 343, 351, 352, 355, 362, 364, 369, 370, 380, 386
Discipline(s), applied, 109, 370, 371, 380
Discipline(s), biological, 123, 168, 246, 376
Discipline(s), ecological, 7, 22, 100, 128, 215, 224, 267, 307
Discipline(s), scientific, 22, 24, 35, 38, 40, 41, 117, 118, 121, 124, 126, 129, 136, 184, 205, 235–237, 248, 249, 287, 288
Disturbance, 70, 90, 92, 106, 107, 209, 211, 212, 281, 393
Diversity, biological, 206, 207, 217
Diversity, ecological, 209, 212
Diversity, genetic, 207, 210–212
Diversity, local, 209
Diversity, species, 39, 106, 206, 207, 209–212
Diversity-stability hypothesis, 107, 110, 207, 208
Divine, 21, 100, 388, 399
Dynamic equilibrium, 190, 340, 342
Dynamics, v, x, xii–iv, 11, 16, 19, 20, 38, 47, 53, 76, 87, 90–92, 100, 102, 105–107, 109, 117–139, 164, 186, 189, 190, 197, 209, 211–212, 217, 223, 225, 249, 255, 268, 270, 278–280, 289, 300, 303, 304, 317, 319, 320, 340, 342, 370, 375, 377

E

Ecological economics, 203, 224
Ecological function, 13, 15, 178, 190, 207
Ecological knowledge, 5, 7, 21, 22, 93–94, 99, 109, 110, 117–139, 223, 239, 248, 261, 271, 369
Ecological object, 15, 68, 186, 247, 361, 401
Ecological structure, 88, 91
Ecological unit, xi, 11, 13–16, 37, 66, 104, 187, 189, 209, 211, 267, 268
Ecology, 3, 11, 19, 31, 45, 87, 97, 117, 149, 155, 161, 183, 195, 205, 215, 231, 277, 287, 307, 329, 333, 351, 369, 385, 405
Ecology, cultural, 351–355, 357–361, 365
Ecology, deep, 45, 198, 379

Subject Index 439

Ecology, ecosystem, 36–40, 104, 107, 167, 177, 178, 189, 206, 215, 216, 218, 221, 224
Ecology, evolutionary, 106, 107, 178
Ecology, geographical, 105, 352, 363
Ecology, global, 334, 338, 344, 376–378
Ecology, human, 103, 197–201, 278, 283, 303, 353, 364
Ecology, individualistic, 98
Ecology, political, 45, 92, 168, 354, 361, 364
Ecology, population, 100, 101, 103, 104, 107, 175, 177, 247, 279, 397, 400
Ecology, scientific, ix, 55, 121, 126, 147, 233, 242, 246, 277, 287, 333, 365
Ecology, social, 353
Ecology, theoretical, 6, 109, 131, 221–223, 282
Ecology, urban, 104
Economics, 24, 198, 283, 335, 358, 379, 405, 407–411
Economics, ecological, 203, 224
Economics, welfare, 379, 405, 407–408
Economy of nature, 150, 153
Ecosystem, 5, 11, 21, 36, 51, 88, 102, 129, 164, 183, 197, 206, 215, 242, 280, 304, 316, 330, 335, 354, 371, 387, 408
Ecosystem dynamics, 189, 375
Ecosystem function, 74, 105, 167, 190, 216
Ecosystem health, 70, 71
Ecosystem management, 5, 215–218
Ecosystem services, 74, 215–218
Ecosystem theory, 38, 72–74, 76, 102, 135, 183, 185–190, 246
Ecotechnology, 242, 246
Emergentism, 5, 31–41
Energy, 24, 38–40, 51, 56, 73, 74, 88, 93, 102, 106, 118, 122, 128, 130–136, 165, 167, 178, 186–190, 198, 201, 202, 216, 222, 247–250, 281, 282, 288, 377, 378, 387, 393, 394, 401, 408, 410
Environment, 6, 20, 36–40, 50, 58, 59, 61, 73, 118, 119, 122, 130, 132–134, 150, 157, 158, 162–168, 175, 183, 185–187, 195–197, 215, 217, 223, 224, 243, 248, 260, 261, 263, 264, 270, 279, 281, 282, 287–289, 293, 295–297, 303, 309, 325, 328, 329, 339–342, 352–354, 357, 363, 375, 376, 379, 390, 392–394, 397, 398, 407, 409, 410
Environmental crisis, 3, 188, 203, 205, 206, 283, 379
Environmentalist, 99, 196, 200, 202, 203, 221, 371
Environmental movement, 70, 147, 168, 188, 195–203, 379

Environmental politics, 195, 196, 201, 203, 379–381
Environmental protection, 12, 99, 106, 195, 200
Environmental sciences, 4, 55, 77, 88, 93, 97, 147, 333, 352, 370, 378, 381
Epistemic break, 23, 240
Equilibrium, 21, 36, 69, 73, 90, 187, 190, 205, 206, 208–211, 259, 278, 280, 282, 340, 342, 373, 377, 378, 393, 405
Essentialism, viii, 20, 31, 40, 45, 48, 50, 52, 54–56, 117, 124, 162, 164, 168, 176, 185, 221–223, 234, 241, 260, 263, 299, 334, 338–341, 344, 352, 353, 355, 359, 360, 371, 374, 377, 379, 406
Ethics, 190, 283, 364
Ethology, 147, 155–158
Evolution, xiii, xiv, 35, 50, 93, 100, 110, 119, 128, 151, 197, 206, 210–213, 252, 261, 263, 279, 327, 328, 341–343, 377, 408, 410
Evolution theory, 64, 100, 156, 162, 260, 262, 263, 310, 340, 342–344
Experience, 21, 25, 26, 47, 130, 138, 139, 202, 243, 252, 261, 370, 372, 373, 376, 391
Experience, practical, 370–372
Experience, scientific, 137, 139
Experiment, 89–91, 100, 123, 127, 128, 138, 162, 197, 201, 202, 207, 242, 256, 259, 264, 370, 376, 405, 408, 409
Experimental system, 135, 249, 256
Experimentation, 198, 201, 202, 373
Explanation, causal, 63–65, 222, 269
Explanation, functional, 51, 395
Explanation, teleological, 51, 64

F
Field science, 7, 22, 33, 288, 398
Fisheries, 172, 235, 236, 243, 249, 250, 256, 315, 321, 369, 370, 375
Food chain, 100, 130, 263, 278, 398
Food web, 102, 106
Forces of nature, 132
Forestry, 68, 242, 246, 283, 302, 315, 318, 321, 369, 370, 372–375
Functional explanation, 51, 395
Functional group, 56, 103
Functionalist, 133, 208
Functional relation, 36, 73, 185, 206
Functional systems concept, 393, 394
Function, ecological, 13, 15, 178, 190, 207
Function, goal, 106
Function, statistical, 39

G

Gaia hypothesis, 71, 336, 343
General systems theory (GST), 72, 73, 183, 184, 387, 391, 395, 396, 398
Geography, vii, 46, 54, 100, 147, 149, 150, 163, 164, 167, 168, 176, 185, 197, 198, 210, 215, 242, 245, 246, 253, 261–267, 269, 271, 287, 288, 290, 292–296, 298, 300, 308, 310, 316, 317, 325, 328, 330, 335, 351–365
Geology, 145, 175, 197, 236, 254, 292, 298, 308, 309, 337, 341
Gestalt, 48, 49, 54, 134, 163, 183, 222, 253
Global, 3, 31, 117, 164, 172, 206–208, 211, 212, 218, 333, 334, 336, 341–345, 353, 354, 358, 376–378
Global ecology, 334, 338, 344, 376–378
Global process, 3, 209, 211, 212
Goal function, 106
Green knowledge, 203

H

Habitat, 6, 37, 55, 56, 58, 90, 97, 104, 106, 107, 109, 130, 157, 176, 177, 260, 267, 329, 353, 380, 391
Heterogeneity, 87–94, 102, 208, 209, 211, 352, 376, 388
Historiography of concepts, 20, 21, 25, 26
History of concepts, 8, 11, 19–27
Holism, 5, 31–34, 36–40, 45–66, 68–76, 190, 221, 222, 224, 240–242, 260, 386, 398
Holism-reductionism, 31, 36, 40, 45–76, 190
Holistic approach, 38, 40, 54, 61, 65, 69, 72, 208, 221, 333
Holistic science, 39, 40
Holistic theory, 56, 75
Homogeneity, 21, 104, 386
Household, 145, 147, 150, 164, 199, 237, 257, 362
Household of nature, 145, 166, 241, 253

I

Identity, 37, 162, 196–198, 203, 371
Idiographic, 157, 239
Image of nature, 56, 57, 68, 71, 75, 132, 134, 241
Individual, 15, 33, 37, 40, 46, 49, 50, 53, 55–61, 64–71, 73–76, 88, 90, 98, 102, 104, 105, 107, 118, 119, 123, 125, 127, 129–131, 133, 134, 138, 151, 157, 163–166, 168, 175–177, 184, 185, 187, 189, 197, 210–212, 235, 237, 243, 248, 249, 254–256, 258–260, 263, 264, 266, 269, 279–282, 299, 304, 326, 354, 357, 378, 385, 389, 392, 394, 395, 398, 399, 405–407, 409, 410
Individualistic, 37, 38, 40, 53, 59–62, 71, 76, 102, 107, 163, 282
Individualistic ecology, 98
Individualistic theory, 59
Intervention, 34, 233, 256, 281, 373, 406

J

Judg(e)ment, 64, 65, 69, 126, 240, 353

K

Knowledge, ecological, 5, 7, 21, 22, 93–94, 99, 109, 110, 117–139, 223, 239, 248, 261, 271, 369
Knowledge, green, 203
Knowledge, partial, 120, 121, 128, 223
Knowledge production, 4, 5, 8, 22, 93–94, 136, 202, 203, 376, 380
Knowledge transfer, 237

L

Laboratory, 5, 7, 22, 135, 155, 156, 162, 163, 202, 245, 249, 256, 259–261, 263, 271, 288, 289, 297, 298, 311, 312, 314, 315, 321, 334, 337, 338, 341, 365, 369, 371, 389
Laboratory biology, 156, 162, 261, 321
Laboratory organism, 259
Laboratory science, 7, 22, 135, 163, 260, 369
Lake type, 100, 254, 255, 259, 265
Landscape, x, xiii, xiv, 24, 41, 46, 54, 70, 89, 104–106, 138, 178, 198, 209, 211, 212, 217, 219, 221, 231, 246, 253, 266, 267, 292, 293, 297, 302, 303, 326, 327, 329, 335, 344, 353–364, 374, 391, 397, 401
Landschaft, 54, 360–362
Law, v, 7, 32–36, 39, 41, 62, 63, 65, 71, 73, 97, 109, 121, 123–125, 131, 136, 138, 158, 183, 205, 219, 222, 223, 225, 240, 244, 246, 252, 259, 260, 262, 265, 278, 279, 282–283, 295, 343, 357, 373, 396
Law, biological, 123
Law, physical, 33, 73, 123, 125
Liberal, 66, 76, 133, 134, 150, 221, 336, 337
Liberalism, 45, 66, 67, 134
Limits, 3, 8, 20, 25, 26, 32, 33, 36, 38, 90, 92, 123, 128, 132, 136, 176, 300–303, 328, 339, 365

Subject Index 441

Limnology, 101, 104, 161, 172, 176, 235–243, 245, 247, 249, 251, 255, 257, 258, 279, 281, 311, 313–315, 375

M

Machine, 53, 71, 75, 76, 134, 135, 189, 190, 253, 337, 396, 397
Man and nature, 91, 213, 351, 353, 356–358, 361, 363, 364, 376
Mathematical model, 88, 89, 109, 199, 223, 224, 386, 389, 395, 405
Mathematical object, 394, 396–401
Metaphor, 4, 22–27, 36, 41, 166, 209, 222, 224, 237, 281, 372, 396, 397
Microcosm, 36, 45, 118, 122, 128–132, 135, 136, 138, 165, 222, 247–249, 253, 277, 281, 372, 375
Model, 7, 26, 41, 49, 53, 70, 75, 76, 88–90, 92, 93, 97, 102, 106, 108, 118, 121, 123–126, 130, 133, 184, 188, 190, 198, 206, 207, 237, 242, 246–249, 259, 264, 297, 311, 312, 341–343, 363, 370, 375, 385, 386, 390–393, 396, 397, 400, 401, 405, 407, 408
Model(ing), computer, 189, 397
Model, conceptual, 26, 392
Model, ecological, 3, 129, 133, 400
Model, explanatory, 26
Model(ing), mathematical, 88, 89, 109, 199, 223, 224, 386, 389, 395, 405
Model, three world, 21
Moral, 205, 241, 365
Museum, 234, 242, 288, 289, 296, 297, 303, 308, 314, 315, 317

N

Narrative, 118, 124, 128–130, 132, 133, 135–137, 224, 244, 247, 248, 271
Natural history, 20, 88, 98–100, 123, 136, 147, 155, 156, 158, 161, 162, 164, 165, 197, 198, 232, 237, 242, 244, 247, 250, 253, 255–279, 290, 297, 298, 307–314, 320, 321, 327, 340, 341, 344, 359, 386
Natural selection, 35, 36, 130, 208, 265, 411
Nature, balance of, 21, 99, 100, 205–209, 212, 278, 281, 377
Nature, concept of, 135–137, 139, 198, 225, 237, 241, 278
Nature conservation, 57, 68–71, 106, 109, 130, 190, 221, 242, 380, 381
Nature, forces of, 132
Nature, history of, 136, 277
Nature, household of, 145, 166, 241, 253
Nature, image of, 56, 57, 68, 71, 75, 132, 134, 241
Nature protection, 69, 380
Nature reserve, 106, 283, 300
Nature, unit of, 38
Niche, ix, 6, 11, 14, 15, 21, 24, 102, 106, 122, 128–136, 165, 222, 247, 278, 282, 307, 320
Niche, concept of, 6, 21, 106, 129–133, 135, 136, 222, 247
Niche theory, 106, 282
Nomothetic, 123, 157
Normative, x, 4, 62, 63, 109, 124, 128, 132, 225, 354, 357, 362

O

Object, ix, x, xiii, 5, 6, 15, 21–24, 26, 31, 33, 35, 39, 46, 48, 49, 51–54, 59, 62, 64, 70, 73, 75, 76, 118, 123, 126, 135, 147, 164, 166, 176, 178, 185, 187, 189, 195, 218, 219, 221, 222, 231, 236, 237, 242–246, 248, 254, 269, 297, 339, 352, 353, 359, 360, 362, 363, 372, 389–392, 394–397, 401
Object, ecological, 15, 68, 186, 247, 361, 401
Objectivity, 126, 136, 239
Object, mathematical, 394, 396–401
Object, physical, 189, 400, 401
Object, real, 186, 190, 221, 399
Observation, v, 47, 89, 90, 99, 100, 104, 106, 120, 138, 207, 242, 243, 250, 256, 277, 278, 312, 336, 376, 389, 390, 408
Oceanography, 149, 161, 172, 176, 297
Ontological distinction, 391
Ontological status, 41, 389
Ontology, 31, 32, 34, 38–40, 136, 389, 391, 392, 398
Organicism, 31–41, 46, 47, 53, 71, 73, 75, 134, 189–190
Organism, 6, 11, 13, 15, 21, 34, 36–40, 46, 49–59, 61–66, 68–76, 90, 91, 100, 103, 106, 118, 129, 131, 132, 134, 138, 145–147, 149–153, 156–158, 162, 164, 166–168, 175–177, 183–190, 207, 219–222, 244, 246, 248–250, 252, 254–260, 263, 266, 267, 278, 280–282, 312, 313, 334, 336, 338–344, 361, 369, 372, 373, 376, 378, 380, 385, 386, 388, 389, 391–393, 396, 398, 401
Organismic, 46, 55, 98, 184, 187, 278, 339, 342

Organism, individual, 13, 15, 46, 49, 50, 56–59, 61, 64, 65, 68, 69, 71, 73, 74, 76, 90, 131, 134, 164, 168, 175–177, 185, 187, 189, 248, 259, 263, 266, 389, 392
Organism, living, 49, 50, 56, 59, 64, 249, 334, 336, 338–342, 344, 376
Organism, societal, 134

P

Paradigm, 23, 31, 38–41, 47, 66–68, 98, 110, 133, 188, 205–213, 218, 244, 289, 295, 296, 301, 304, 351–353, 355, 356, 358, 359, 361, 363–365, 386
Partial knowledge, 120, 121, 128, 223
Pattern, 12, 35, 47, 58, 60, 62, 63, 88–91, 121, 122, 128, 129, 132, 133, 156, 161, 168, 208, 211, 212, 217, 248, 253, 264, 271, 354, 355, 361–363, 389, 390, 393, 395, 397–399, 410
Pattern(s) and process(es), 89, 91, 156
Phenomenological, 37, 39, 119, 121, 248
Philosophy of biology, 123
Philosophy of nature, 118, 119, 124, 130, 139, 197, 247
Philosophy of science, 8, 26, 98, 117–119, 121–124, 126, 139, 333
Physical law, 33, 73, 123, 125
Physical object, 189, 400, 401
Physiognomic, 49, 108, 139, 147, 163–165, 247, 253, 255, 266, 292, 296, 300
Physiognomy, xiii, 163, 222, 261, 293, 296, 360
Physiographic, 165, 167, 239, 246, 255, 266
Physiological, 104, 147, 152, 162–165, 167, 176, 184, 231, 247, 248, 255, 257–265, 267, 300, 310
Physiology, 5, 46, 72, 109, 132, 147, 149–152, 155, 156, 161–165, 177, 222, 248, 250, 254, 257–264, 277, 297, 298, 378
Phytogeography/phytogeographical, vii, viii, 99, 105, 268, 288, 292, 293, 295, 296, 298–303, 309
Phytosociology/phytosociological, 108, 166, 172, 221, 245, 302–304, 319, 320
Planning, 71, 109, 198, 217, 224
Planning, landscape, 221, 355
Planning sciences, 242
Plant association, 13, 37, 38, 40, 221, 267–269, 298, 299, 392
Plant community, 40, 100, 107, 131, 133, 164–166, 176, 221, 252, 267, 269, 270, 278, 293, 320, 325–330
Plant ecology, 37, 88, 103, 104, 108, 130, 131, 146, 158, 168, 172, 175, 176, 231, 245, 247, 253, 262, 264–271, 277, 278, 296, 300, 302, 303, 307, 316–320, 329, 353, 371, 373, 374
Plant geography, xi, 100, 145, 163, 164, 167, 176, 245, 253, 261, 262, 265–267, 271, 310, 316, 317, 325, 328, 330
Plant society/societies, 130, 164, 177, 269, 270, 325
Plant sociology, x, xi, xiv, 60, 165, 166, 245, 247, 253, 265–270, 289, 296
Pluralism, 21, 97, 120–122, 136–137
Pluralism, scientific, 97, 120, 122
Pluralism, theoretical, 122
Plurality, 44, 118–126
Plurality in ecology, 121–126
Political ecology, 45, 92, 168, 354, 361, 364
Populariz(s)ation, 24, 107, 198, 232, 243, 244
Population, 11, 13–16, 38, 39, 50, 72, 88–90, 97, 99–101, 103–107, 124, 125, 164, 175, 177, 178, 186, 187, 189, 197, 200, 201, 210–212, 216, 220, 236, 246, 247, 258, 259, 279–282, 307, 338, 352, 353, 373–376, 379, 387, 392, 394, 397, 398, 400, 401
Population ecology, 100, 101, 103, 104, 107, 175, 177, 247, 279, 397, 400
Positivism, 33, 54
Practice, v, vii, ix, x, xi, xiv, 3, 5, 6, 19, 22, 38, 76, 92–94, 119, 122, 125, 126, 132, 136, 149, 164, 171, 183, 202, 203, 215, 221, 236, 237, 241, 246, 248, 251, 270, 288, 292–296, 298, 301, 307, 319, 359–365, 369, 372, 374, 380, 395
Practice, ecological, 6, 7, 38, 71, 164, 198, 203, 364, 369, 389
Practice, productive, 372, 375
Practice, research, 71, 92, 94, 251, 270, 362, 365, 389, 397
Practice, scientific, 24, 27, 87, 93, 94, 155, 231, 237, 248, 288, 385, 389, 395
Process, ix, x, xiii, xiv, 3, 6, 12–15, 19, 25, 31–35, 41, 50, 51, 56–58, 65, 67, 68, 73–76, 87–89, 91–94, 98, 101, 102, 106, 107, 117, 119, 125, 128, 131, 137, 147, 156, 157, 162, 163, 167, 168, 186, 189, 190, 200, 203, 205, 207, 209–212, 216, 219, 220, 223, 233, 235, 236, 249, 255–257, 264–266, 278, 282, 307, 317, 320, 328, 334, 335, 338, 339, 341, 353, 369, 377–381, 385, 386, 389, 390, 392, 393, 395–399, 401, 411

Subject Index 443

Process, ecological, 88, 90, 92, 207, 270, 369–371
Process, evolutionary, 123, 124, 133, 210, 213, 340–342
Process, intersecting, 91–92, 94
Production, 92–94, 118, 200, 201, 235, 236, 256, 257, 369, 379
Production biology, 100, 258
Production, biomass, 74, 190
Production, knowledge, 4, 5, 8, 22, 93–94, 136, 202, 203, 376, 380
Production, species, 209
Progress, xiv, 3, 20, 57, 66, 75, 90, 98, 101, 102, 108, 109, 133, 223, 270, 278, 329, 358, 371, 386
Psychology, 15, 46, 88, 92, 98, 158, 183, 185, 187, 387, 411

R
Reductionism, 5, 31–41, 45–76, 134, 189, 190, 221, 224, 225
Regularity/regularities, v, 91, 123–125, 246, 295, 334, 371–373
Representation(s), 19, 23, 48, 49, 52, 88, 123, 125, 129, 133, 136, 139, 149, 152, 251, 253–255, 258, 364, 388, 397, 401
Reproduction, 36, 150, 151, 156, 259
Research programme, 21, 45, 48, 97, 98, 101, 105, 107–108, 110, 120, 122, 126–129, 132, 133, 135, 139, 146, 147, 152, 153, 162, 188, 216, 233, 235, 237, 242–244, 246–248, 253, 260–262, 269, 270, 297, 300, 310, 316, 317, 352, 363, 376
Resource, 66, 68, 69, 88, 90, 94, 104, 106, 117, 118, 130, 133, 197, 198, 200, 206, 207, 210–213, 216, 217, 234, 235, 280, 283, 308, 315, 316, 326–329, 358, 359, 362, 369–377, 379, 399, 405, 407, 410
Restoration, 106, 109, 246
Romantic, 128, 130, 134–136, 138, 197, 237, 239, 247, 253, 260, 380

S
Scale, ix, 90–92, 104–105, 208, 210, 211, 217–220, 225, 249, 251, 299, 333, 354, 361
Scale, large-scale, 90, 104, 105, 178, 188, 197, 242, 249, 358, 370
Scale, small-scale, 58, 72, 105, 281, 299, 303, 362
Scale, spatial, 90, 92, 97, 104, 211–213, 219, 220
Scale, temporal, 92, 104, 211, 220
Scientific community, 168, 176, 178, 222, 307, 320, 321, 333, 334, 344
Scientific discipline, 22, 24, 35, 38, 40, 41, 117, 118, 121, 124, 126, 129, 136, 184, 205, 235–237, 248, 249, 287, 388
Scientific ecology, v, 55, 121, 126, 147, 233, 242, 246, 277, 287, 333, 365
Scientific field, 8, 132, 165, 175, 208, 254, 308
Scientific institution, 201, 242, 289, 315–317, 320
Scientific practice, 24, 27, 87, 93, 94, 155, 231, 237, 385, 389, 395
Scientific revolution, 98, 100
Scientific society/societies, 147, 171–173, 232–234, 245, 271, 290, 297, 302, 380
Self-conscious, 93, 97, 100, 147, 149, 171, 176, 267
Self-organization, 185, 212, 387, 397
Semantics, 6, 14–15, 22, 24, 31, 32, 34, 119, 124, 220, 237, 361, 400
Simulation, 7, 98, 189, 209, 395
Social ecology, 353
Social sciences, 66, 92, 103, 167, 187, 198, 224, 360, 365
Society/societies, ecological, 101, 129, 171, 277
Society/societies, learned, 242, 287–292, 296–298, 300, 301, 303
Society/societies, scientific, 147, 171–173, 232–234, 245, 271, 290, 297, 302, 380
Sociology, 46, 98, 165, 166, 185, 198, 277, 396
Sociology, plant, xi, xiv, 60, 165, 166, 245, 247, 253, 265–270, 289, 296
Sociology, urban, 353
Species assemblage, 36, 176
Species, co-occurrence(s) of, 281
Species, endangered, 207, 380
Species, number(s) of, 57, 209–211, 258, 279–282, 372
Species, pioneering, 303
Species production, 209
Species, rare, 211, 291
Stability, 55, 61, 69, 73, 89, 102, 105–107, 110, 119, 122, 125, 130, 135, 206–208, 210–212, 327, 361
Statistical function, 39
Structure, 6, 8, 11–16, 23, 32, 35, 38, 41, 46, 47, 49, 50, 52, 67, 72, 74, 75, 87–89, 91–94, 97–110, 119, 122, 123, 125, 130, 134, 151, 166, 167, 176, 185, 202, 206, 209, 211, 221, 236, 243, 249, 254, 260, 261, 269–271, 278, 287, 307, 316, 317, 319, 339, 340, 352, 361, 369, 393, 396, 397, 408, 410

Structure and function, 88, 167
Structure, conceptual, 46, 47, 69, 75
Structure, self-organizing, 395
Substance, 104, 123, 126, 132, 187, 249, 250, 252, 282, 357, 361, 398, 400
Substance, organic, 123, 132, 249, 250, 252
Succession, x, xii–iv, 55–58, 61–63, 68, 69, 73, 88, 90, 100, 102, 103, 106, 107, 133, 186–188, 190, 215, 220, 268, 270, 278, 299, 300, 303, 373, 390
Superorganism, 46, 56, 58, 59, 63, 69, 71, 73–76, 119, 131, 133, 185, 221, 278, 281, 282
Surrounding, 36, 91, 131, 145, 147, 150, 151, 165, 249, 328, 339, 387, 393
Sustainable development, 196, 203, 215, 217, 218, 224, 343, 364, 377, 378
Symbiosis, 351, 360, 364
Synecology, vii, 104, 165, 166, 175–177, 239, 266, 267, 304
Systems concept, 73, 101, 184, 187, 189–190, 388, 392–394, 398–400
Systems concept, functional, 393, 394
Systems concept, operational, 393, 394
Systems theory, 55, 72, 73, 108, 147, 166, 183–190, 385–401

T

Technocratic, 67, 93, 188, 190, 199, 200, 386, 401
Technology, 22, 24, 68, 70, 71, 76, 99, 108, 123, 138, 200–202, 358, 387, 407
Teleological, 51, 58, 63–65, 68, 69, 75, 133, 222
Teleological explanation, 51, 62, 64
Theoretical dynamics, 133
Theoretical ecology, 109, 131, 221–223, 282

Theoretical pluralism, 122
Theory building, 5, 22, 23, 89, 122, 126, 128, 130, 132, 164, 177, 248
Theory, ecological, 6, 7, 13, 16, 31, 47, 64, 68, 71, 73, 76, 87–91, 99, 119, 120, 126, 129, 130, 134, 279, 317, 379
Theory, ecosystem, 38, 72–74, 76, 102, 135, 183, 185–190, 246
Theory, systems, 5, 49, 72, 73, 108, 147, 166, 183–190, 242, 385–401
Thermodynamics, 38, 51, 88, 91, 108, 110, 185, 186, 188, 223, 378
Totality, viii, 48, 339, 361, 362
Triadic structure, 122
Trophic-dynamic, 102, 107, 186, 281
Trophic function, 282
Trophic level, 186, 216

U

Unit, ecological, xi, 11, 13–16, 37, 58, 66, 69, 104, 187, 189, 209, 211, 267, 268, 361, 409
Unit of nature, 38

V

Value, 38, 51, 63, 64, 68, 69, 74, 76, 147, 195, 217, 252, 291, 300, 330, 353, 405–407, 409, 410
Vegetation ecology, 88, 268
Vegetation formation, 266
Vegetation type, xi, 317–319
Visualiz(s)ation/visual, 151, 224, 251

W

Welfare economics, 379, 405, 407–408